Hypersonic Airbreathing Propulsion

Hypersonic Airbreathing Propulsion

William H. Heiser
United States Air Force Academy

David T. Pratt
University of Washington

With
Daniel H. Daley and Unmeel B. Mehta

EDUCATION SERIES
J. S. Przemieniecki
Series Editor-in-Chief
Air Force Institute of Technology
Wright-Patterson Air Force Base, Ohio

Published by
American Institute of Aeronautics and Astronautics, Inc.,
370 L'Enfant Promenade, SW, Washington, DC 20024-2518

American Institute of Aeronautics and Astronautics, Inc., Washington, DC

Library of Congress Cataloging-in-Publication Data

Heiser, William H.
   Hypersonic airbreathing propulsion / William H. Heiser, David T.
Pratt with Daniel H. Daley and Unmeel B. Mehta.
     p. cm.—(AIAA education series)
   Includes bibliographical references and index.
   1. Jet propulsion. 2. Airplanes—Ramjet engines—Design and
construction. 3. Airplanes—Scramjet engines—Design and
construction. 4. Hypersonic planes—Design and construction.
5. Aerothermodynamics. I. Pratt, David T. II. Title.
III. Series.
TL709.H38 1994   629.134′3535—dc20   93-25635
ISBN 1-56347-035-7

Fourth Printing

Texts Published in the AIAA Education Series

Re-Entry Vehicle Dynamics
  Frank J. Regan, 1984
Aerothermodynamics of Gas Turbine and Rocket Propulsion
  Gordon C. Oates, 1984
Aerothermodynamics of Aircraft Engine Components
  Gordon C. Oates, Editor, 1985
Fundamentals of Aircraft Combat Survivability Analysis and Design
  Robert E. Ball, 1985
Intake Aerodynamics
  J. Seddon and E. L. Goldsmith, 1985
Composite Materials for Aircraft Structures
  Brian C. Hoskins and Alan A. Baker, Editors, 1986
Gasdynamics: Theory and Applications
  George Emanuel, 1986
Aircraft Engine Design
  Jack D. Mattingly, William H. Heiser, and Daniel H. Daley, 1987
An Introduction to the Mathematics and Methods of Astrodynamics
  Richard H. Battin, 1987
Radar Electronic Warfare
  August Golden Jr., 1988
Advanced Classical Thermodynamics
  George Emanuel, 1988
Aerothermodynamics of Gas Turbine and Rocket Propulsion,
  Revised and Enlarged
  Gordon C. Oates, 1988
Re-Entry Aerodynamics
  Wilbur L. Hankey, 1988
Mechanical Reliability: Theory, Models and Applications
  B. S. Dhillon, 1988
Aircraft Landing Gear Design: Principles and Practices
  Norman S. Currey, 1988
Gust Loads on Aircraft: Concepts and Applications
  Frederic M. Hoblit, 1988
Aircraft Design: A Conceptual Approach
  Daniel P. Raymer, 1989
Boundary Layers
  A. D. Young, 1989
Aircraft Propulsion Systems Technology and Design
  Gordon C. Oates, Editor, 1989
Basic Helicopter Aerodynamics
  J. Seddon, 1990
Introduction to Mathematical Methods in Defense Analyses
  J. S. Przemieniecki, 1990

Space Vehicle Design
  Michael D. Griffin and James R. French, 1991
Inlets for Supersonic Missiles
  John J. Mahoney, 1991
Defense Analyses Software
  J. S. Przemieniecki, 1991
Critical Technologies for National Defense
  Air Force Institute of Technology, 1991
Orbital Mechanics
  Vladimir A. Chobotov, 1991
Nonlinear Analysis of Shell Structures
  Anthony N. Palazotto and Scott T. Dennis, 1992
Optimization of Observation and Control Processes
  Veniamin V. Malyshev, Mihkail N. Krasilshikov, and Valeri I. Karlov,
  1992
Aircraft Design: A Conceptual Approach
  Second Edition
  Daniel P. Raymer, 1992
Rotary Wing Structural Dynamics and Aeroelasticity
  Richard L. Bielawa, 1992
Spacecraft Mission Design
  Charles D. Brown, 1992
Introduction to Dynamics and Control of Flexible Structures
  John L. Junkins and Youdan Kim, 1993
Dynamics of Atmospheric Re-Entry
  Frank J. Regan and Satya M. Anandakrishnan, 1993
Acquisition of Defense Systems
  J. S. Przemieniecki, Editor, 1993
Practical Intake Aerodynamic Design
  E. L. Goldsmith and J. Seddon, Editors, 1993
Hypersonic Airbreathing Propulsion
  William H. Heiser and David T. Pratt, 1994
Hypersonic Aerothermodynamics
  John J. Bertin, 1994

Published by
American Institute of Aeronautics and Astronautics, Inc., Washington, DC

FOREWORD

This book and its companion volume, *Hypersonic Aerothermodynamics* by John J. Bertin, resulted from a series of discussions among faculty members of the Department of Aeronautics at the United States Air Force Academy in 1987. At that time, hypersonic, piloted flight was back in the public eye due to then President Reagan's announcement of a new program to develop and demonstrate the technology required to operate an airplane-like vehicle which could take off from a normal runway, and use airbreathing engines to climb and accelerate to sufficient altitude and speed to enter Earth orbit. Upon completion of its orbital mission, the vehicle would re-enter the atmosphere and operate as an airplane during descent and landing on a runway. This idea led to the National Aero-Space Plane (NASP) program. The single-stage-to-orbit concept envisioned in the NASP has, and will continue to require, the best efforts and intellectual talents the nation has available to make it a reality.

The advent of the NASP program was not the only factor that led to these volumes. The last significant hypersonic, manned vehicle program was the Space Shuttle which underwent engineering development in the 1960's. By the late 1980's, much of the talent involved in that program had long since been applied to other areas. The need for a modern treatment of hypersonic aerothermodynamics and airbreathing propulsion analysis and design principles for the academic, industrial and government communities was clear. As a result, the Air Force's Wright Laboratory and the NASP Joint Program Office, both located at Wright-Patterson Air Force Base, Ohio, entered into a cooperative effort with the Department of Aeronautics at the Academy to fund and provide technical and editorial oversight and guidance as these books were developed.

We sincerely hope that these volumes will serve as up-to-date sources of information and insight for the many students, engineers, and program managers involved in the exciting study and application of hypersonic flight in the years ahead.

G. KEITH RICHEY
Chief Scientist
Wright Laboratory

ROBERT R. BARTHELEMY
Director
NASP Joint Program Office

EDWARD T. CURRAN
Director
Aeropropulsion and Power
  Directorate
Wright Laboratory

MICHAEL L. SMITH
Professor and Head
Department of Aeronautics
United States Air Force Academy

PREFACE

A renaissance of interest and activity in hypersonic flight is happening because the remarkable performance improvements promised by airbreathing propulsion have been brought within our reach by the steady advance of the underlying technologies. *Hypersonic Airbreathing Propulsion* is intended to provide a broad and basic introduction to the elements needed to work in that field as it develops and grows. It is, to the best of our knowledge, the first fundamental, comprehensive, integrated treatment of the subject.

Hypersonic airbreathing propulsion can seem mysterious and forbidding at first because it deals with a regime of flight that lies far beyond ordinary experience and intuition. Indeed, many unfamiliar phenomena, such as chemical dissociation and supersonic mixing and combustion, are involved. Nevertheless, there are special rewards ahead because hypersonic airbreathing propulsion is subject to the usual laws of nature, and the desired results can be obtained once those laws are correctly applied to the situation at hand. The results are often even more appealing and easily understood than their subsonic or supersonic counterparts. The journey through this textbook will therefore only reaffirm and extend your understanding and appreciation of the basics.

We believe strongly that upper level engineering textbooks should empower the reader or student to actually do things they could not do before. To that end, you will find this textbook to be almost entirely self-contained. For one thing, almost every example analytical result presented here can be reproduced using the tools provided. For another, homework problems that develop and stretch understanding are furnished for almost every individual subject. Finally, an extensive array of PC-based, user-friendly, computer programs is provided in order to facilitate repetitious and/or complex calculations. These codes are, in fact, so general that they have application far beyond the bounds of this textbook.

A special issue of hypersonic airbreathing propulsion is that most experimental data are classified for either military or proprietary reasons and are therefore scarce. We have tried to compensate for this by squeezing the most out of the open literature and tailoring our approach so that future revelations are easily incorporated.

Hypersonic Airbreathing Propulsion was written primarily for use in both undergraduate and graduate courses in airbreathing propulsion. Readers familiar with *The Aerothermodynamics of Gas Turbine and Rocket Propulsion* by Gordon C. Oates will find the flow of our development quite similar to his. This was not our initial concept but an evolutionary outcome that was probably inevitable because the technical content and intended audiences have much in common, and because of the undisputed success of his approach. This has, in fact, been a voyage of discovery for us and, as a result, even the most experienced workers will find original and useful material throughout.

In addition to the customary material of Chap. 2 concerning fundamental concepts and laws, and Chaps. 4–7 concerning the behavior of ramjet and scramjet components, cycles, systems, and variants, our prior experience with related fields as well as our desire for comprehensiveness led us to include significant coverage of four other areas, as follows. Chapter 1 contains general background, including brief and pointed summaries of the history of hypersonic airbreathing flight and the current situation. Chapter 3 contains a complete development of the analysis of hypersonic aerospace vehicles, with special emphasis on the contribution of the airbreathing propulsion system to achieving success. Chapter 8 contains information and methods for the estimation of the performance of many types of hypersonic airbreathing propulsion systems. Chapter 9 contains a wide range of topics of special interest to the design and development of ramjet and scramjet engines.

Students taking courses based on this material should have completed at least compressible fluid mechanics and thermodynamics. It is preferable that they have also taken boundary layer theory, equilibrium chemistry, heat transfer, and mechanics and strength of materials. It would be ideal if they have also studied combustion reactions and aircraft or jet propulsion.

There is enough material here to occupy a full academic year. However, the topics are sufficiently independent that a suitable selection can be made for an academic quarter or semester. The treatment, in fact, invites the use of specific topics for individual study, as well as for the insertion of subjects of special interest to the instructor.

ACKNOWLEDGMENTS

Hypersonic Airbreathing Propulsion continues a firmly established and fruitful AIAA Education Series tradition. Like its eminent ancestors, *The Aerothermodynamics of Gas Turbine and Rocket Propulsion* by Gordon C. Oates and *Aircraft Engine Design* by Jack D. Mattingly, William H. Heiser, and Daniel H. Daley, it was sponsored by the U.S. Air Force and written and field-tested on cadets at the U.S. Air Force Academy.

No work of this magnitude is completed without large amounts and many kinds of help. The most important are presented next.

Major funding and important encouragement for this four year project were generously provided by the National Aero-Space Plane Joint Program Office (NASP JPO) and the USAF Wright Laboratory, through the auspices of Director Dr. Robert R. (Bart) Barthelemy and Chief Scientist Dr. G. Keith Richey, respectively. Additional vital support and hospitality were furnished by the Department of Aeronautics of the U.S. Air Force Academy, arranged primarily by Col. Michael L. Smith, Department Head, and Lt. Col. Thomas R. Yechout, Deputy for Plans and Programs, and by the University of Tennessee Space Institute.

The textbook benefited greatly from periodic meetings with a formal evaluation committee chaired by Dr. Richey. The propulsion reviewers were Dr. Edward T. (Tom) Curran, Director of the Aero Propulsion and Power Directorate of the Wright Laboratory, and Mr. Edward S. Gravlin, of the NASP JPO. We are indebted to them not only for their conscientious committee work, but for their sincere personal interest and active participation in the project from beginning to end.

Brig Gen (Ret) Daniel H. Daley, former Head of the Department of Aeronautics of the USAF Academy and incomparable authority on thermodynamics, propulsion, teaching, writing, and, now, basic physical constants and systems of units, patiently read every word, offered innumerable constructive suggestions, and improved everything he touched.

Dr. Unmeel B. Mehta, former Leader of the Computational Fluid Dynamics (CFD) Technical Support Team for the NASP JPO and Research Scientist at the NASA Ames Research Center, provided the sections on CFD that are so critical to correctly portraying the flavor and power of modern hypersonic airbreathing propulsion technology.

There are many other people whose advice, counsel, and other concrete contributions improved the final product. They include, in alphabetical order within organizations, Mr. Griffin Y. Anderson, Dr. J. Philip Drummond, Dr. Wayne D. Erickson and Dr. Scott D. Holland of the NASA Langley Research Center, Mr. Richard L. Balent, Mr. John L. Leingang, Mr. Donald J. Stava and Dr. Frank D. Stull of the USAF Wright Laboratory, Mr. Steven L. Barson, Dr. Pankaj Goel, and Dr. Herbert Lander of Rocketdyne, Dr. Frederick S. Billig and Dr. David M. Van Wie of the Johns Hopkins University, Mr. Howard L. Bowman of the USN Weapons Center, Dr. Robert E. Breidenthal and Mr. Kaveh Ghorbanian of the University of Washington, Ms. Cheryl Gumm and Mr. James O. Young of the USAF Flight Test Center, Mr. Thomas M. Graziano and Lt. Col. Robert V. Pieri of the USAF Academy, Dr. Peter Jacobs of the University of Queensland, Dr. Dietrich E. Koelle and Dr. Heribert Kuczera of MBB-Deutsche Aerospace, Mr. David L. Kors of Aerojet, Dr. Peter H. Kutschenreuter, Jr., of General Electric, Dr. Marion L. Laster of the USAF Arnold Engineering Development Center, Mr. Henry J. Lopez of Allied Signal Aerospace, Dr. Bonnie J. McBride of the NASA Lewis Research Center, Mr. Takashige Mori of Mitsubishi Heavy Industries, Dr. William L. Oberkampf of the Sandia National Laboratories, Dr. David Riggins of the University of Missouri-Rolla, Lt. Col. Gilbert M. Souchet of the French Air Force, and Dr. Sheng-Tao Yu of Sverdrup Technology. We also include in this category the many students at the USAF Academy and the University of Washington who were innocent victims in the cause of improving our understanding of the subject and thus its presentation.

We thoroughly enjoyed and appreciated collaborating with Dr. John Bertin of the Sandia National Laboratories, the author of the companion AIAA Education Series textbook *Hypersonic Aerothermodynamics.* His knowledge of and insight into his field are unsurpassed, and his work encouraged us to do our best as well.

There are several people associated with the production of this textbook who can only be described as indispensable because they determine the quality of the final presentation. They include Dr. John S. Przemieniecki, Senior Dean of the Air Force Institute of Technology and Editor-in-Chief of the AIAA Education Series, Ms. Jeanie K. Duvall and Mr. Timothy Valdez of the University of Texas at Austin, and Ms. Jeanne A. Godette, Ms. Christine Kalmin, and Mr. John A. Newbauer of AIAA headquarters.

A particularly rewarding feature of creating a textbook is the opportunity to share ideas with able, helpful friends, new and old. Your interest and energy created a supportive and stimulating climate that brought sustaining joy to the enterprise. We sincerely appreciate your personal investment in our work.

Finally, and most importantly, we thank our spouses, Leilani Heiser and Marilyn Pratt, for their unwavering support and encouragement before, during, and after this demanding effort. We will make it up to you somehow, someday.

TABLE
OF
CONTENTS

? Air   Burner   Equilibrium   Gas Tables   Performance   Trajectory   Quit

Equilibrium properties of atmospheric air

Hypersonic Airbreathing Propulsion (HAP) Computer Programs

HAP(Air)
   Equilibrium thermophysical properties of 21% O 79% N air.
   Normal and oblique shock waves, stagnation conditions, and isentropic compression and expansion.

HAP(Burner)

Aerothermodynamic design and analysis of burner-isolator system for calorically perfect gases.

HAP(Equilibrium)

Equilibrium properties of combustion products for various hydrocarbon fuels at assigned conditions.

Equilibrium and frozen composition isentropic expansion.

HAP(Gas Tables)

Equivalent to traditional compressible flow appendices for calorically perfect gases.

Isentropic flow.

Constant area, frictional flow (Fanno flow).

Constant area, frictionless heating (Rayleigh flow).

Normal shock waves.

Oblique shock waves.

Multiple oblique shock waves.

Conical shock waves.

Prandtl-Meyer flow.

HAP(Performance)

Stream thrust analysis of overall and component performance, and station properties of hypersonic airbreathing engines.

HAP(Trajectory)

Absolute values and dimensionless ratios of physical properties of the standard atmosphere for specified conditions, including burner entry state for calorically perfect gases.

1
GENERAL
BACKGROUND

1.1 INTRODUCTION

The subject of this book is the hypersonic airbreathing engine, a device that employs the enveloping atmosphere to propel aerospace vehicles at sustained speeds greatly in excess of the local speed of sound. This remarkable capability will complete the work on aviation begun by the Wright brothers in 1903 by making possible flight at virtually any speed and altitude, including the astounding prospect of escaping the sensible atmosphere of the Earth and coasting into a nearby permanent orbit. Perhaps more importantly for society, this will complete the shrinking of the planet that began with the *jet age*, that electrifying period of time in which the aircraft jet engine brought humankind together to an extent previously unimagined.

The purpose of this chapter is to provide a setting for the technical material on hypersonic airbreathing engines, two types of which are ramjets and scramjets, that constitutes the remainder of the book. This chapter consists of roughly equal parts of historical background and a narrative description of the most important features of the hypersonic airbreathing engines themselves. For readers new to the field, this information is meant to furnish the framework necessary to encourage and support further study. For readers experienced in the field, this information should indicate the level and tone of the approach, and may even provide some new insights.

There are several references that are so fascinating and valuable that they are essential reading for anyone truly involved in this field. They are, in fact, so pertinent and rich that they are recommended to anyone who wishes to be fully informed about the history and technology of hypersonic airbreathing flight. We have found them very useful, and they have helped to shape this book.

The first of these is *Ramjets*, an AIAA Selected Reprint Series volume edited by Gordon L. Dugger of the Applied Physics Laboratory of the Johns Hopkins University,[1.1] an organization that has played a leading role throughout the development of ramjets and scramjets. This document reproduces 14 of the most important papers that had appeared by 1969, and contains a complementary bibliography of another 107 references arranged into six topical areas. It should also be noted that many of the selected reprints contain extensive bibliographies of their own.

A recent major and very impressive work is a two volume (1391 page) set on the history of hypersonic flight authored and edited by Richard P. Hallion and published by the Aeronautical Systems Division of the Wright-Patterson Air Force Base in 1987.[1.2,1.3] Volume I is entitled *The Hypersonic Revolution: From Max Valier to Project Prime, 1924–1967*, and Volume II is entitled *The Hypersonic Revolution: From Scramjet to the National Aero-Space Plane, 1964–1986*. The special feature of these books is that they contain case studies written by top program managers describing the administration, planning, proposing, designing, and developing of several of the most important hypersonic systems of those times. The reader should therefore expect to discover the organizational, financial, political, and personal, as well as the technological history of the field in these pages.

There is also a substantial amount of interesting background information related to hypersonic airbreathing propulsion to be found in various periodicals, two of which will be cited here. First, many of the original or pioneering and now classical papers authored by the leading figures of hypersonics were published in the *Journal of the Aerospace Sciences* (earlier, *Aeronautical Sciences*). Perusing these volumes can be a rewarding and even an eye-opening experience. A thought-provoking yet typical example is the Twenty-First Wright Brothers Lecture, delivered by H. Julian Allen and entitled *Hypersonic Flight and the Re-entry Problem*.[1.4] Second, many practical or applications articles appear in *Aerospace America*, the monthly magazine of the AIAA membership. In particular, approximately annually there appears an issue at least partially devoted to a review of all aspects of airbreathing propulsion, including ramjets and scramjets. A recent article representative of this category is *Ramjets experience renewed interest worldwide*, written by T. D. Myers and Gordon Jensen.[1.5]

1.2   HISTORICAL OVERVIEW

It is not our purpose to compose a rigorous, comprehensive history of everything related to ramjets and scramjets. Instead, we have attempted to define and reach the limited goals that are really appropriate to this book, as described below.

To begin with, it is important to confirm that hypersonic airbreathing propulsion was not born yesterday, but has been on the minds of humankind for a long time. Furthermore, it serves many useful purposes to describe some of the milestone accomplishments that have taken place along the way, the main of which are to establish a mental picture of the flavor and pace of the field, to witness some of the lessons being learned, and, above all, to capture some of the excitement that makes aerospace so alluring. Finally, since we are dealing with something that has not yet happened but surely will, namely piloted hypersonic airbreathing flight, it is important to demonstrate that this aspiration has a compelling attraction.

1.2.1 The Beginnings

According to William H. Avery,[1.1] René Lorin of France in 1913 was the first person to recognize the possibility of using ram pressure in a propulsive device, although he concluded correctly that the engine performance would be inadequate because he was concerned only with flight at subsonic speeds where ram pressure is low.

An inherent weakness of ramjets is their inability to generate any thrust at all while standing still (or static) because they rely on the pressure increase associated with the slowing down of the oncoming air. An interesting aside, then, is that the turbojet, which uses an integral compressor in order to produce the necessary pressure increase, is a solution found by Guillaume to the static thrust and low speed performance problems, but not until 1921. The ramjet therefore has seniority, even though the turbojet is far more abundant today owing to its inherent adaptability and practicality.

Albert Fono of Hungary was issued a German patent in 1928 on a propulsion device that contained all the elements of a modern ramjet, and was specifically intended for supersonic flight.[1.1] Figure 1.1, which was reproduced from the original patent,[1.6] clearly shows the convergent-divergent inlet and nozzle identified with supersonic flow, as well as a low speed combustion chamber. There is no evidence, however, that any were built.

René Leduc of France was issued a patent in 1935 on a piloted aircraft propelled by a ramjet of his own design.[1.1] Figure 1.2, which was reproduced from the original patent,[1.7] shows his ingenious approach to integrating engine and airframe, as well as the unusual accommodations for the pilot. Although work on this project was suspended while France was occupied during World War II, it resumed afterward and resulted in an experimental aircraft similar in concept to that of the patent and designated the Leduc 010. A landmark event for ramjet propulsion took place on April 21, 1949, when the Leduc

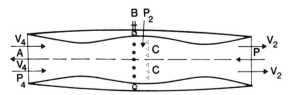

Fig. 1.1 The Fono ramjet, as reproduced from the original German patent.[1.6] Courtesy of the Johns Hopkins University Applied Physics Laboratory.

010 was released from its parent Languedoc aircraft and made its first powered flight. Figures 1.3 and 1.4 contain historic photographs of the Leduc 010, which was able to reach a Mach number of 0.84 at 26 kft (7.9 km) while still climbing.[1.8]

Leduc went on to develop and flight-test improved versions of his concept, including the Leduc 016 and the Leduc 021, shown in Fig. 1.5, which reached a Mach number of about 0.9.[1.8]

1.2.2 The Nord-Aviation Griffon II Turboramjet Aircraft

According to General Noël Daum,[1.1] the Nord-Aviation company of France embarked in 1953 on a project inspired by the ramjet work of Leduc and aimed at the realization of a practical airplane that could fly up to Mach numbers in excess of 2. The resulting airplane, known as the Griffon II, is shown in flight in Fig. 1.6.

Recognizing that the available turbojets performed well at subsonic flight speeds while ramjets performed well at supersonic speeds, the two were married into the first "combined cycle" engine in order to obtain the best of both worlds. Figure 1.7 shows a cutaway of the Griffon II aircraft. From this diagram it can be seen that the ramjet was wrapped around a SNECMA Atar 101 E3 dry turbojet, and that they shared both the inlet and nozzle. By controlling the fuel flows to the two engines, the fraction of the total thrust generated by the ramjet varied from 0 under static conditions to over 80 percent at a flight Mach number of 2.

The performance of the Griffon II was ample testimony to the success of the turboramjet engine. For example, it flew at a Mach number of 2.1 at an altitude of 61 kft (18.6 km) while still able to

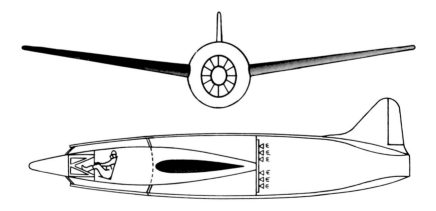

Fig. 1.2 The Leduc ramjet powered piloted aircraft, as reproduced from the original French patent.[1.7] Courtesy of the Johns Hopkins University Applied Physics Laboratory.

Fig. 1.3 The Leduc 010 mounted on a Languedoc transport aircraft prior to being carried aloft for a test flight. Courtesy of the Musée de l'Air et de l'Espace, Le Bourget, France.

Fig. 1.4 The Leduc 010 in flight. Courtesy of the Musée de l'Air et de l'Espace, Le Bourget, France.

Fig. 1.5 The Leduc 021 in flight. Courtesy of the Musée de l'Air et de l'Espace, Le Bourget, France.

accelerate and climb. Moreover, the Griffon II established a world speed record for the 100 km closed circuit of 1020 mph (1640 km/h) on February 24, 1959, demonstrating that the turboramjet provided high thrust in the turn as well as the straightaway.

1.2.3 Ramjet Powered Missiles

The ramjet has always been competitive with the rocket for missile propulsion in the atmosphere because it is equally simple in construction and has greater range for the same propellant weight. These attributes are particularly attractive for military applications because simplicity and low initial cost are essential features of devices that must function on demand and never return. It should be no surprise, then, that many types of ramjet powered missiles for many purposes have been built over the years, even though each one has had to solve the problem of reaching the "takeover velocity" at which the ramjet had sufficient thrust to continue the mission.

The beginnings of serious work on ramjets in the United States can be found in this arena. By 1946, for the purpose of providing improved defense for ships at sea, missile developments requiring ramjet power plants were being carried out at the Boeing Airplane Company, Grumman Aviation Corporation, the Johns Hopkins University, Marquardt Aviation Corporation, and the Massachusetts Institute of Technology. The Marquardt Aviation Corporation, in fact,

Fig. 1.6 The Griffon II turboramjet aircraft in flight. Courtesy of the Musée de l'Air et de l'Espace, Le Bourget, France.

mounted 20 in (51 cm) diameter engines on the wingtips of a Lockheed F-80 fighter and accomplished the first pure ramjet propulsion of a piloted airplane in 1946,[1.1] Fig. 1.8.

During the 1951–1960 time period, the U.S. Air Force sponsored the Lockheed Aircraft Company X-7 and X-7A supersonic reusable pilotless flight research vehicle program. These vehicles were carried aloft by and dropped from modified bomber aircraft, boosted to ramjet takeover velocity by a solid rocket, and propelled into sustained flight by Marquardt Aviation Corporation or Wright Aeronautical Corporation ramjets ranging in diameter from 20–36 in (51–91 cm).

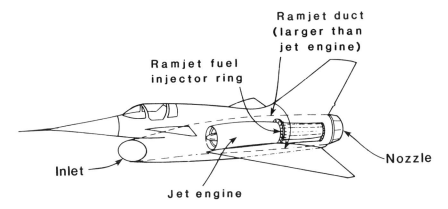

Fig. 1.7 Cutaway of the Griffon II turboramjet aircraft. Courtesy of the Musée de l'Air et de l'Espace, Le Bourget, France.

Fig. 1.8 The Lockheed F-80 with Marquardt ramjets attached to the wingtips. Courtesy of the U. S. Air Force.

From the airbreathing propulsion standpoint, this program was significant because it provided much useful ramjet performance information for Mach numbers up to 4.31, or 2881 mph (4636 km/h).[1.2] The Lockheed X-7 is shown in Fig. 1.9.

Of the many production military missiles that have employed ramjet propulsion, the most impressive may very well be the Bomarc, a joint venture of the Boeing Airplane Company and the Michigan Aeronautical Research Center, after which it was named. The Bomarc A, shown in Fig. 1.10, was deployed in 1955 for ground-to-air defense of the continental United States against enemy bombers. Before launch it weighed about 16,000 lbf (71,000 N) and stood about 43.6 ft (13.3 m) high. After a solid propellent rocket booster accelerated it to takeover velocity, two podded Marquardt liquid fuel ramjets having about 14,000 lbf (62,000 N) of thrust each could propel it about 435 mi (700 km) at an altitude of about 69 kft (21 km) and a Mach number of about 3.0.[1.9] In 1972, when the intercontinental ballistic missile had become the primary threat to the United States, the Bomarc B was retired from service.

Another imposing military missile powered by a ramjet was the Talos, shown in Fig. 1.11, developed as a surface-to-air naval fleet defense weapon by the Applied Physics Laboratory of the Johns Hopkins University. It was manufactured by the Bendix Corpora-

Fig. 1.9 The Lockheed X-7 supersonic research vehicle. Courtesy of the U. S. Air Force Flight Test Center.

tion Missile Systems Division and deployed in 1959. Before launch it weighed about 7000 lbf (31,000 N) and stood about 31.3 ft (9.5 m) high. After a solid propellant rocket booster accelerated it to takeover velocity, a McDonnell Aircraft Corporation liquid fuel ramjet having about 20,000 lbf (89,000 N) of thrust could propel it about 75 mi (120 km) at an altitude of about 87 kft (27 km) and a Mach number of about 2.5.[1.9] Talos missiles launched from the nuclear cruiser USS Long Beach were credited with the destruction at more than 65 mi (105 km) range of two North Vietnamese MIGs in 1968. The Talos was replaced by the Aegis system around 1980.

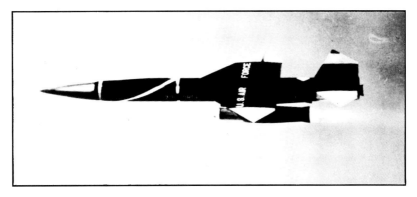

Fig. 1.10 The Bomarc A ramjet powered air defense missile. Courtesy of the U. S. Air Force.

As of 1990, six military ramjet missiles were operational.[1.5] Two of these are so-called first-generation missiles, namely Britain's Bloodhound and China's variant of the Bloodhound, the HY-3/C101, both used for surface-to-air combat. Their propulsion system is similar in concept to the Bomarc and their speed is supersonic, but their range is only about 50 mi (80 km).[1.10]

Another two are second-generation missiles, namely Britain's Sea Dart and the former Soviet Union's SA-4 Ganef, both featuring internal liquid fuel ramjets and tandem solid propellant booster rockets, and both also used for surface-to-air combat. The latter has a range in excess of about 44 mi (70 km) at an altitude of about 79 kft (24 km) and a Mach number of about 2.5.[1.10]

The remaining two are third-generation missiles, namely the former Soviet Union's SA-6 Gainful and France's ASMP (Air-Sol-Moyenne-Portée). They feature an integral solid rocket booster contained within the space that becomes the ramjet combustion chamber after the boost phase. The SA-6 is a surface-to-air weapon with a range of about 37 mi (60 km) at an altitude of about 59 kft (18 km) and a Mach number of about 2.8.[1.10] The ASMP, shown in Fig. 1.12, is the highest performance missile in the world for its size and weight, and the first air-to-surface ramjet missile to be deployed.[1.9] It has a range of about 155 mi (250 km) and a Mach number of about 3.0.[1.10]

Fig. 1.11 The Talos ramjet powered surface-to-air missile. Courtesy of the U. S. Air Force.

1.2.4 The NASA Hypersonic Ramjet Experiment and the X-15 Rocket Research Airplane

Ramjets are the leading choice for airbreathing propulsion applications where the flight Mach number is roughly in the range 3–6. Their distinctive feature is that combustion of fuel with air takes place after the flow has been slowed internally to subsonic speeds. When the flight Mach number exceeds about 6, it is no longer profitable to decelerate the flow to that extent, and combustion must take place at locally supersonic conditions. Engines that operate in this way are known as supersonic combustion ramjets, or scramjets for short.

From the mid-1950's to the early 1960's a great deal of progress had been made toward a developmental scramjet engine by means of analysis and component testing.[1.3] As a result, a surprisingly wide variety of experimental scramjet engines were built and ground-tested in "direct-connect" and/or "freejet" facilities during the mid- to late 1960's in order to determine such characteristics as their packaged (or installed) performance, internal performance, component behavior, operating limits, heat transfer, and durability. Some of the organizations participating in this work in the United States were the General Applied Science Laboratory, the General Electric Company,

Fig. 1.12 The ASMP air-to-surface missile being carried on the centerline pylon of a Mirage 2000. Courtesy of the French Air Force.

the Johns Hopkins University, the Marquardt Aviation Corporation, the NASA Langley Research Center, and the United Aircraft Research Laboratories. These programs incorporated many different approaches and concepts, and examined the hopes and fears of scramjet propulsion.[1.11]

The flagship of the developmental scramjet engines was the Hypersonic Ramjet Experiment or Hypersonic Research Engine (HRE, in either case), a project funded by the National Aeronautics and Space Administration (NASA) and carried out largely by the Garrett Corporation from May 1964 to April 1975. Interestingly, the real impetus for this work came from a very unlikely source, the North American X-15 rocket research airplane program. Quoting from Hallion[1.2]:

> NASA's major X-15 follow-on project involved a Langley-developed Hypersonic Ramjet Experiment. NASA Flight Research Center advanced planners had long wanted to extend the X-15's speed capabilities, perhaps even to Mach 8, by adding extra fuel in jettisonable drop tanks and some sort of thermal protection system. Langley researchers had developed a design configuration for a proposed hypersonic ramjet engine. The two groups now came together to advocate modifying one of the X-15's as a Mach 8 research craft that could be tested with a ramjet fueled by liquid hydrogen. The proposal became more attractive when the landing accident to the second X-15 in November 1962 forced the rebuilding of the aircraft. The opportunity to make the modifications was too good to pass up. In March 1963 the Air Force and NASA authorized North American to rebuild the airplane with a longer fuselage. Changes were to be made in the propellant system; two large drop tanks and a small tank for liquid hydrogen within the plane were to be added; the drop tanks could be recovered via parachute and refurbished, as with the Space Shuttle's solid-fuel boosters nearly two decades later. Forty weeks and $9 million later, North American delivered the modified plane, designated the X-15A-2, in February 1964.

Thus, the HRE program was born. The overall goal, in brief, was to test a complete, regeneratively cooled, flightweight scramjet on the X-15A-2 rocket research airplane. The X-15A-2 flew several times with a dummy ramjet attached to its stub ventral fin, including the flight of October 3, 1967, during which the airplane reached its maximum Mach number of 6.72, or 4520 mph (7273 km/h).[1.2] The X-15A-2 is shown in this configuration in Plate 1 (at end of book). Unfortunately, the opportunity for flight-testing an operational ramjet was lost, first when the cost of repairing the damage to the X-15A-2 that occurred during its record-making flight was found to be excessive, and finally when the entire X-15 program was terminated in 1968.[1.2]

Even though earthbound, the HRE program did produce some significant accomplishments. A complete flightweight, regeneratively cooled Structural Assembly Model (SAM) scramjet was built and tested to conditions simulating Mach 7 flight in the NASA Langley High Temperature Structures Tunnel. A water-cooled, "boiler-plate" Aerothermodynamic Integration Model (AIM) was built and tested to conditions simulating flight at Mach numbers of 5, 6, and 7 in the NASA Lewis Hypersonic Tunnel Facility.[1.3, 1.11] A photograph of the AIM is reproduced in Fig. 1.13.

In hindsight, the expectations of the experimental scramjet engine programs of the 1960's and 1970's proved to be overly optimistic, and largely illuminated critical unknowns. Such technical issues as the difficulty of attaining efficient mixing and combustion, the importance of including external engine drag as well as internal thrust in engine performance accounting, the design complications imposed by the extraordinarily wide operating range, and the need for realistic ground-test simulation facilities combined with the financial pressures of the times to greatly diminish support for scramjet engine development.[1.3]

Fig. 1.13 The Hypersonic Research Engine Aerothermodynamic Integration Model installed in the NASA Lewis Research Center's Plumbrook Hypersonic Tunnel Facility. Courtesy of the NASA Lewis Research Center.

1.2.5 Russian Subscale Model Flight Testing

Recently, the Russian Central Institute of Aviation Motors (CIAM) designed and launched a subscale model of a combined or "dual-mode" ramjet and scramjet airbreathing engine mounted on the nosetip of a rocket for the purpose of captive testing. The axisymmetric device had an inlet diameter of 8.9 in (22.6 cm) and a length of 47.2 in (120 cm). Ramjet or subsonic combustion operation started at a flight Mach number of about 3.5, and transition to scramjet or supersonic combustion operation took place at a flight Mach number of about 5.0. Operation at a maximum flight Mach number of about 5.5 at an altitude of about 85 kft (26 km) was reached before the available hydrogen fuel was consumed.[1.12]

This event is noteworthy for at least three reasons. First, the 15 s period of sustained scramjet operation in the atmosphere is the longest on public record. Second, it demonstrated ramjet to scramjet transition under realistic conditions. Third, it employed unpiloted flight to provide the harsh environmental conditions that are difficult or impossible to reproduce on the ground. Thus, this technique is a legitimate candidate for the low risk, low cost, early testing and/or demonstration of hypersonic airbreathing propulsion concepts.

1.2.6 Looking Back

H. Julian Allen echoes the sentiments of many aircraft designers in his opening sentence of the Twenty-First Wright Brothers Lecture,[1.4] "Progress in transportation has been brought about more by revolutionary than by evolutionary changes in methods of propulsion." This was dramatically demonstrated in reverse when the natural progression of airbreathing propulsion from turbojets to ramjets to scramjets was abandoned in the early 1970's. Witness, for example, Fig. 1.14, which presents the entire history of the Absolute World Air Speed Record for airplanes as determined according to the standardized rules of the International Aeronautics Federation.[1.13] This information demonstrates both the decisive influence of propulsion on speed and the termination of progress at the limit of turbojet capability. The question is not whether or how the limits of flight will be extended, but when.

1.3 LOOKING AHEAD

Visionaries have been attracted to airbreathing ramjet and scramjet engines through the decades because of the high speed sustained atmospheric flight they promise. This was certainly the case during the 1960's and 1970's when interest in their development flourished, and then waned.

In recent times we have witnessed a dramatic rebirth of activity

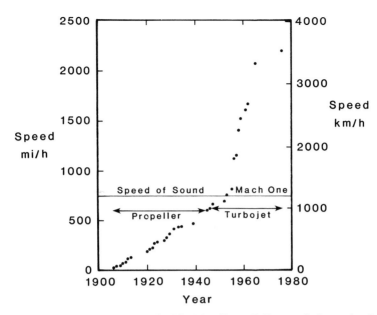

Fig. 1.14   The Absolute World Air Speed Record for airplanes from the Santos-Dumont Type 14-bis in 1906 to the Lockheed SR-71A in 1976 as determined by the International Aeronautics Federation.[1.13] In years where successive records were set, only the last is plotted.

involving ramjets and scramjets, but toward different ends. Many organizations in many countries are now energetically pursuing these engines because they offer to reduce the cost and increase the dependability of transporting payloads to Earth orbits. One good way to understand the basis of these potential benefits is to examine the weight breakdowns of current aircraft and space transportation systems.

Table 1.1 summarizes the present situation. The glaring fact revealed by these numbers is that oxygen, which is not carried for the airbreathing propulsion of aircraft, represents most of the takeoff weight of rockets. The Saturn V, for example, prior to lifting off for the moon, weighed over 6 million lbf, of which over 4 million lbf was liquid oxygen and only 250,000 lbf was payload. One view of this situation is that bringing oxygen on a trip through the atmosphere is like bringing a canteen of water to a fish.

The large fraction of oxygen required to propel the vehicle (and itself) obviously reduces the fraction available for payload, which means that the total weight of the system at takeoff is a large multiple of the payload to be delivered. In the case of Table 1.1, that multiple for rockets is 25, which explains why the vehicles found on the launching pad are so heavy. This, in turn, increases the size of

Table 1.1 Typical takeoff weight fraction breakdowns of current aircraft and multi-stage rocket transportation systems.

Takeoff Weight Fraction	Aircraft	Rocket
Payload	15%	4%
Empty	55%	7%
Fuel	30%	24%
Oxygen	0%	65%

everything associated with rocket propulsion systems, including manufacturing, assembly, transportation, and launching facilities. More importantly, although less obvious at first, the large oxygen fraction substantially reduces the fraction available for empty weight, which includes structure, propulsion, tankage, power, controls, and instrumentation. It is this, more than anything else, that makes airplanes different from rockets, because reduced empty weight translates directly into reduced ruggedness and flexibility. This, in turn, increases the attention that must be paid to every detail of every launch, while reducing the margin for error. It also prevents the inclusion of such attractive features as cost saving recovery systems and life saving escape systems. A simple but far-reaching observation is that the principal characteristic we associate with "airplane-like" operations is the dependability in the face of widely varying usage and prevailing conditions that is the direct result of using the available empty weight to provide ruggedness, flexibility, and, finally, productivity.

Taken together, these factors have kept the rocket launch rate low, which has led to customized design, manufacture, and operations; slow progress down the learning curve; and undependable schedules. The bottom line is that delivery costs to low Earth orbit are in the range of $3000–$10,000 per lbm of payload, a substantial barrier to the full realization of the benefits of operating in or traveling through space.[1.14]

One proposition, then, is to develop an aerospace plane or transatmospheric vehicle that relies primarily upon airbreathing propulsion and literally flies into near Earth orbit. The weight saved by leaving the oxygen behind would be used, as in ordinary aircraft, to make the vehicle smaller while increasing the empty weight fraction and therefore the ruggedness, flexibility, and productivity. Cryogenic hydrogen would be substituted for liquid hydrocarbon fuels in order to further reduce takeoff weight due to the higher combustion energy per pound, and to provide vitally needed cooling by virtue of its very low storage temperature. Wings and wheels would be among the added features, which means that ordinary airfields may be used instead of unique facilities, the engines need not be large enough to lift and accelerate the vehicle directly against the force of gravity, and the entire vehicle would return home and be reusable. The smaller vehicle would be more easily serviced or repaired, and the cargo compartment would be compatible with a variety of payload combinations, assuring relatively quick reaction and/or turnaround.

This proposition is being turned into reality today. Several nations are pursuing experimental aerospace plane programs based on these broad outlines. Some interesting variants exist because of their separate experience bases and preferences, but they share the goal of increasing accessibility to space by improving dependability while reducing orbital payload delivery costs by a factor of 10. They also share the opportunity to develop the technologies that enable other types of hypersonic flight, such as commercial transportation from point to point on the Earth. Two of the leading examples of these efforts are described next.

1.3.1 The German Sänger Space Transportation System

The German Hypersonics Technology Program was initiated in 1988 and is aimed at providing autonomous space launch capability from the European mainland while reducing orbital payload delivery costs

Fig. 1.15 Contemporary version of the two-stage Sänger Space Transportation System, shown with the Horus upper stage in place.[1.15]

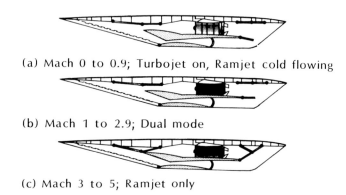

(a) Mach 0 to 0.9; Turbojet on, Ramjet cold flowing

(b) Mach 1 to 2.9; Dual mode

(c) Mach 3 to 5; Ramjet only

Fig. 1.16 Contemporary version of the EHTV turboramjet propulsion system shown in three modes of operation, featuring common air inlet and separate nozzle assemblies and external expansion.[1.15]

by an order of magnitude. The reference configuration is called the Sänger Space Transportation System, in honor of the famous rocket and hypersonic flight pioneer Eugen Sänger, who first designed vehicles intended to fly to and from space.[1.15, 1.16] As shown in Fig. 1.15, it is a reusable, *two-stage*, blended wing/body vehicle that employs horizontal takeoff and landing on conventional runways. The Sänger system emphasizes cost, safety, reliability, and flexibility, and uses advanced state-of-the-art technologies.

The piloted first or lower stage has also been known as the European Hypersonic Transport Vehicle (EHTV) because it has wide commonality with a potential hypersonic passenger aircraft. For example, the EHTV maximum Mach number is 6.8, the total takeoff mass is 807,000 lbm (366,000 kg), the payload is 247,000 lbm (112,000 kg), and the length and wing span are 277 ft (84.5 m) and 136 ft (41.4 m), respectively. The EHTV is propelled up to the stage separation Mach number of 6.8 by a hydrogen-fueled turboramjet propulsion system, such as that shown in Fig. 1.16.

There are two identical second or upper stages, similar in scale to the Space Shuttle Orbiter and powered by hydrogen-oxygen rockets. Horus (Hypersonic Orbital Reusable Upper Stage) is a piloted vehicle intended for Space Station support, and has a nominal payload of 6600 lbm (3000 kg) to a 280 mi (450 km) Earth orbit. Cargus is an unpiloted, reusable vehicle intended for the delivery and retrieval of inert cargo, and has a nominal payload of 18,700 lbm (8500 kg) to a 124 mi (200 km) low Earth orbit.

For subsequent use, the distributions of the masses of the three reference Sänger vehicles, in absolute and percentage forms, are compiled in Table 1.2. The vastly differing payload and empty mass values for the Horus and Cargus result from the differing demands

Table 1.2 Contemporary values for the mass distributions of the three reference Sänger vehicles.[1.16]

	EHTV Mass klbm (Mg)	Percent
Payload	247	30.6
(Stage II)	(112)	
Empty	344	42.6
	(156)	
Fuel	216	26.8
(LH_2)	(98)	
Total	807	100.0
	(366)	

	Horus Mass klbm (Mg)	Percent
Payload	6.6	2.7
(450 km orbit)	(3.0)	
Empty	61.9	25.1
	(28.1)	
Fuel & Oxidizer	178.4	72.2
($LH_2 + LO_2$)	(80.9)	
Total	247	100.0
	(112)	

	Cargus Mass klbm (Mg)	Percent
Payload	18.7	7.6
(200 km orbit)	(8.5)	
Empty	54.9	22.2
	(24.9)	
Fuel & Oxidizer	173.3	70.2
($LH_2 + LO_2$)	(78.6)	
Total	247	100.0
	(112)	

of piloted and reusable versus unpiloted and expendable missions. Note also that the Sänger design philosophy has resulted in an overall payload percent of only about 0.8 percent (3.0 Mg/366 Mg) for the Horus and about 2.3 percent (8.5 Mg/366 Mg) for the Cargus. Nevertheless, the robustness (i.e., reusability, reliability, and flexibility) of the system is expected to make it cost effective in the long run.

A space transportation system based on the Sänger concept could be operational during the first quarter of the 21st century, most likely under the auspices of the European Space Agency (ESA). An interesting sidelight is that an experimental hypersonic flight demonstrator that could fly in the 20th century is considered to be an essential verification element of the German Hypersonics Technology Program because of the enormous difficulty of providing realistic flight conditions in ground tests and the immense importance of demonstrating the performance of the engine as integrated into the vehicle.

1.3.2 The United States National Aero-Space Plane

The most vigorous, focused, flourishing, and publicized current hypersonic airbreathing aerospace vehicle effort is the National Aero-Space Plane (NASP) program of the United States. Institutionally, this militarily classified program is a joint program of the Department of Defense and the National Aeronautics and Space Administration (NASA) that is managed by the NASP Joint Program Office (JPO) and executed by a consortium of aerospace vehicle and propulsion companies. Commencing in 1986, it brought a two decade hiatus of activity in hypersonic flight to an end, attracted several thousand skilled workers to the field, and has invested about $400 million per year of combined government and industry funding in research and development.

The goal of the NASP program is to develop and demonstrate the feasibility of a piloted, horizontal takeoff and landing aircraft that will utilize conventional airfields, accelerate to hypersonic speeds, achieve orbit in a *single stage,* deliver useful payloads to space, return to Earth with propulsive capability, and have the operability, flexibility, supportability, and economic potential of airplanes.[1.17] To this end, the NASP program has managed and supported a spectrum of activity from basic research through a proposed experimental vehicle flight test program. The latter vehicle is currently referred to as the X-30, in keeping with the earlier designations of the family to which it belongs. A major raison d'etre for the X-30 is, of course, to serve as a testbed for the realistic testing of the airbreathing engines. An artist's conception of an X-30 vehicle in flight (prior to scorching off of the paint) is found in Plate 2 (at end of book).

The NASP development program has concentrated on five key areas of technology: airbreathing engines; high specific strength, high temperature materials; vehicle aerodynamics; airframe/propulsion integration; and subsystems. Additionally, considerable attention has been paid to the enabling capabilities of computational fluid dynamics, computational structural mechanics, and ground and flight testing infrastructure. An example of the type of progress being made can be seen in Fig. 1.17, which shows a two-dimensional inlet designed for a Mach 5 hypersonic cruise aircraft mounted in the NASA Lewis Research Center's 10×10 ft supersonic wind tunnel for freejet flowfield, performance, and stability testing.

Firm numbers are not readily available for the proportions of the X-30, but public presentations and open testimony suggest that a version capable of reaching orbit in a single stage would be somewhat larger than a DC-9, which has a total takeoff mass of about 120,000 lbm (55,000 kg) and a length of about 120 ft (37 m). The empty mass percent, fuel mass percent, and payload percent at takeoff are estimated to be in the range of 20–30 percent, 70–80 percent, and 2–4 percent, respectively.

Depending upon the future aggressiveness of the NASP program

Fig. 1.17 A Mach 5 hypersonic cruise aircraft two-dimensional inlet ready for freejet testing in the NASA Lewis Research Center's 10×10 ft supersonic wind tunnel in 1989. Courtesy of the NASA Lewis Research Center.

and the size of the first step to be taken, an X-30 could fly early in
the 21st century. That first step could be anywhere between a scale
model scramjet carried on or launched from an aircraft or missile and
the X-30 itself, and includes the possibility of a vehicle dedicated to
exploring the new unknowns of hypersonic airbreathing propulsion.
If the X-30 is successful, an operational space transportation vehicle
could be ready in the first quarter of the 21st century.

Whatever happens, however, the NASP program has revitalized
hypersonics in general and airbreathing propulsion in particular in
the United States. This textbook is a major beneficiary of the National Aero-Space Plane program. Quite simply, it would not have
been possible to produce at this time without the motivation and
stimulation of the NASP airbreathing propulsion community.

1.4 TECHNICAL OVERVIEW

The purpose of this section is to describe the general appearance
and behavior of ramjet and scramjet engines as well as the elements
and functions of their component parts. This material is intended to
provide a clear mental image of the physical hardware involved, but
will also lead to a summary of the special challenges of designing and
building engines for hypersonic flight.

1.4.1 Ramjet Engines

The essential features of ramjet two-dimensional or planar geometry
engines are diagrammed in Fig. 1.18. Ramjets are the engines of
choice for flight in the Mach number range 3–6, and are predominantly used for supersonic flight. It is convenient to describe them
by following the airflow from the undisturbed freestream at the far
left until it leaves the influence of the engine at the far right. Note
that neither ramjets nor scramjets need be axially symmetric about
a centerline because they contain no rotating machinery. In fact, it
is often convenient to make the outside surface of the vehicle serve
as the inside surface or boundary of the engine.

The first step in any conventional thermal power cycle is compression, which the ramjet accomplishes by slowing or decelerating the
oncoming airflow. The flow is usually compressed in several steps,
including passing through one or more oblique shock waves generated
by the forebody of the vehicle or of the diffuser, deceleration of the
supersonic flow in a convergent duct, transforming the supersonic
flow into subsonic flow through a normal shock wave system, and
further decelerating the subsonic flow in a divergent duct. Fuel is
then injected into the subsonic flow in the burner, where it vaporizes
(if initially liquid), mixes, and burns. The hot, high pressure flow
then accelerates back to a supersonic exit speed in the convergent-

divergent nozzle and finally exhausts into the atmosphere. A reaction force or thrust is generated by the flow passing through the ramjet because the high temperature exhaust flow has more velocity and momentum leaving than it did entering. This reaction force is known as the internal or uninstalled thrust of the engine.

It is important to recognize that there are also forces on the external surface of the engine, usually known as the cowl. These forces are almost always opposite to the internal thrust, and are referred to as the external drag or installation penalty. Designers must give internal thrust and external drag equal weight because the measure of engine performance that really matters is the difference between them, the quantity known as the net thrust or installed thrust.

1.4.2 Scramjet Engines

The velocity of the oncoming air, as seen from the frame of reference of the vehicle or engine, also represents relative kinetic energy. When the airflow is decelerated by the scramjet, the relative velocity and kinetic energy both decrease, and conservation of energy requires that any missing kinetic energy will reappear as internal energy, with the result that the pressure, temperature, and density of the flow entering the burner are considerably higher than in the freestream. When the flight Mach number exceeds about 6, this effect becomes so pronounced that it is no longer advantageous to decelerate the flow to subsonic speeds. Depending upon flight conditions and details of the diffuser operation, the adverse consequences can include pressures too high for practical burner structural design, excessive performance losses due to the normal shock wave system, excessive wall heat transfer rates, and combustion conditions that lose a large fraction of the available chemical energy to dissociation.

A logical way to solve this problem is to only partially compress and decelerate the oncoming flow, avoiding in particular the normal shock wave system, with the result that the flow entering the burner

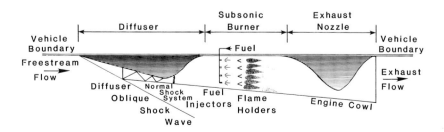

Fig. 1.18 Schematic diagram of a two-dimensional or planar geometry ramjet engine.

is supersonic. The resulting engine is known as a supersonic combustion ramjet, or scramjet for short. The essential features of two-dimensional or planar geometry scramjet engines are diagrammed in Fig. 1.19. Although scramjets have much in common with ramjets, the ensuing discussion will focus on the differences in the name of clarity.

Even though the diffuser is responsible for some of the desired compression and deceleration, much of it is invariably accomplished by oblique shock waves emanating from the vehicle forebody located upstream of the engine. This allows the engine to take advantage of the inevitable compression of the freestream by the vehicle and reduces the burden upon the diffuser. Moreover, since the diffuser exit flow is supersonic, the geometry is entirely convergent. Fuel is injected into the supersonic flow just downstream of the diffuser, and the emphasis is upon achieving rapid and thorough mixing (especially when all the entering oxygen is to be consumed) because the time available for the combustion process is short. The heat loads are highest in the burner, primarily because of the combination of the high energy of the oncoming flow and the local high gas density due to compression, rather than to the ongoing combustion. The exhaust nozzle need only be divergent because the accelerating flow is supersonic throughout, and some of the acceleration can take place outside the confining duct by using the afterbody of the vehicle as a free expansion surface.

1.4.3 Integration of Engine and Vehicle

An early lesson of high speed flight was that proper aerodynamic integration of the ramjet or scramjet with the remainder of the vehicle is crucial to success[1.2, 1.3] It was found, for example, that making hypersonic engines axisymmetric and attaching them to the vehicle by means of pylons or struts (see Fig. 1.9 and Plate 1) can produce

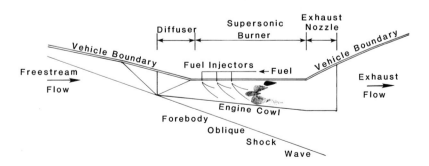

Fig. 1.19 Schematic diagram of a two-dimensional or planar geometry scramjet engine.

enough external drag on the pylon and the cowl to virtually cancel the internal thrust, and creates internal passages so narrow that the flow is dominated by wall effects and difficult to manage. Furthermore, this configuration cannot easily capitalize on the vehicle surfaces for compression and expanson.

The recent engineering design work on aerospace planes has shown that success is also dependent upon careful integration of engine and airframe structures, materials, cooling, controls, and subsystems. Workers in the field half-jokingly refer to the result as an "engineframe," and their point is well taken. We are a long way from the world of turbojets that can be easily transferred between several different types of aircraft and treated like a thrust producing plug-in unit.

Indeed, the characteristic look of contemporary hypersonic aerospacecraft springs from the need for good aerodynamic integration. As the vehicles fly to the very high altitudes necessary to avoid the excessive pressures that would result from hypersonic speeds, they also fly where the air is very rare, the density being only one-hundredth or one-thousandth or less of the sea level value. But airbreathing engines require airflow in order to generate the thrust that lifts and accelerates the vehicle, and they cannot be allowed to suffocate.

One way to capture the required airflow is to use the entire forebody underneath the vehicle as a compression surface, shaping it carefully for high efficiency, and recognizing that, since the air has no warning of the presence of the vehicle until it encounters the first oblique shock wave, a streamtube of air enormously larger than the physical opening of the engine inlet can be directed into the engine, as depicted in Fig. 1.20. Once inside the engine, the air that has been compressed and burned would then require a nozzle exit area even larger than the original freestream capture area in order to make good on the available thrust, but a conventional nozzle of this geometry would be too cumbersome to carry along. Instead, the entire afterbody underneath the vehicle is used as a free expansion surface,

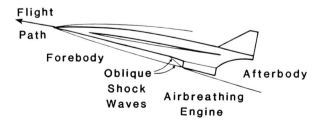

Fig. 1.20 Schematic diagrams of aerospace plane flowfield and configuration.

as depicted in Fig. 1.20. This obviously means that the entire underside of the vehicle must be carefully designed to accommodate engine performance under all flight conditions. The bottom of the aerospace plane may therefore be regarded as the engine, and the top as the airplane, a far cry from the early days of placing jet engines on pods to completely isolate them from the airplane aerodynamics, and probably the limiting case of airbreathing engine territorial conquest.

1.4.4 Hypersonic Airbreathing Propulsion Challenges

Based on the foregoing introductory discussions, it is now both possible and worthwhile to compile a list of the special problems that will be repeatedly encountered in hypersonic airbreathing propulsion. More or less in order of importance, they are:

* Operating efficiently and reliably over an extraordinarily large range of flight conditions, including Mach numbers from 0 to 25 (orbital speed) and altitudes from sea level to the top of the atmosphere.

This is compounded by the fact that practical aerospace plane configurations are likely to allow only one "hole" for the engine, and that ramjets and scramjets produce no thrust while standing still or when the atmosphere is too thin, and further complicated because the airframe and engine are totally integrated, not only with regard to generating net thrust, but also inevitably from the standpoint of the ordinary "aerodynamic" effects such as lift and drag forces and stability and control moments.

* Accomplishing stable, efficient mixing and combustion in a supersonic flow within a burner of reasonable size.

* Providing the structural integrity necessary for a reusable system despite the extremely hostile environmental conditions.

* Developing the analytical tools that enable confident control over the engine design and reliable prediction of the actual behavior.

* Proving that the aerospace plane and engine are ready for routine operations by means of analysis, ground testing, and flight testing of experimental vehicles.

REFERENCES

1.1 Dugger, G. L., *Ramjets,* AIAA Selected Reprint Series, Vol. VI, New York, June 1969.

1.2 Hallion, R. P., *The Hypersonic Revolution,* Vol. I: From Max Valier to Project Prime, 1924–1967, Aeronautical Systems Division, Wright-Patterson AFB, OH, 1987.

1.3 Hallion, R. P., *The Hypersonic Revolution,* Vol. II: From Scramjet to the National Aero-Space Plane, 1964–1986, Aeronautical Systems Division, Wright-Patterson AFB, OH, 1987.

1.4 Allen, H. J., "Hypersonic Flight and the Re-entry Problem," The Twenty-First Wright Brothers Lecture, *Journal of the Aeronautical Sciences*, Vol. 25, No. 4, Apr. 1958.

1.5 Myers, T. D., and Jensen, G., "Ramjets experience renewed interest worldwide," *Aerospace America*, Washington, DC, July 1990.

1.6 German Patent No. 554,906, Dr.-Ing. Albert Fono, May 1928.

1.7 French Patent No. 439,805, René Leduc, Dec. 1935.

1.8 Leduc, R., "Early Work and Latest Realizations With Ramjet Engines," *Jet Propulsion*, Vol. 27, No. 1, Jan. 1957.

1.9 Pretty, R. T., and Archer, D. H. R., *Jane's Weapon Systems, 1969–70*, Jane's Publishing Company, London, England, 1969.

1.10 Blake, B., *Jane's Weapon Systems, 1987–1988*, Jane's Publishing Company, London, England, 1987.

1.11 Waltrup, P. J., Anderson, G. Y., and Stull, F. D., "Supersonic Combustion Ramjet (Scramjet) Engine Development in the United States," The 3rd International Symposium on Air Breathing Engines, Munich, Germany, Mar. 1976.

1.12 Kandebo, S. W., "Franco-Russian Tests May Spur New Projects," *Aviation Week & Space Technology*, McGraw-Hill, New York, Dec. 14/21, 1992.

1.13 Berliner, D., *Victory Over The Wind*, Van Nostrand Reinhold, New York, 1983.

1.14 U. S. Congress, Office of Technology Assessment, *Access to Space: The Future of U.S. Space Transportation Systems*, OTA-ISC-415, U.S. Government Printing Office, Washington, DC, Apr. 1990.

1.15 Koelle, D. E., and Kuczera, H., "Sänger Space Transportation System Status Report 1988," 39th Congress of the International Astronautical Federation, Paper IAF-88-192, Bangalore, India, Oct. 1988.

1.16 Kuczera, H., Krammer, P., and Sacher, P., "The German Hypersonics Programme – Status Report 1991," AIAA Third International Aero-Space Plane Conference, Paper 91-5001, Dec. 1991.

1.17 Barthelemy, R. R., "The National Aero-Space Plane Program," AIAA First National Aero-Space Plane Conference, Paper 89-5001, July 1989.

2
TECHNICAL
BACKGROUND

2.1 INTRODUCTION

The purpose of this chapter is to facilitate and support the remainder of the work of this book. This goal will be accomplished in a variety of ways, such as characterizing the fundamental features of hypersonic flight, presenting realistic approximations to nature, and developing the relevant equations of compressible fluid mechanics and thermodynamics in their most useful forms.

This chapter should be taken seriously because its results determine the course of everything that follows. The concepts and physical models presented here are, in fact, what make possible a direct, coherent, and clear treatment of hypersonic airbreathing propulsion. Readers are therefore urged to reflect upon the possible implications of each model or simplifying assumption. More importantly, readers are strongly urged to return to this chapter whenever necessary in order to solidify their grasp of the subject matter.

Note should also be taken of six complementary textbooks that enlarge upon the material to come. They not only provide additional detail and insight, but also offer slightly different points of view, which often help to make things fall into place.

Hypersonic Aerothermodynamics, by John J. Bertin,[2.1] and *Hypersonic Airbreathing Propulsion* are companion volumes of the AIAA Education Series. The former is especially valuable because it emphasizes the practical design of hypersonic aerospace vehicles, and provides an abundance of data relating to their external flowfields. The two textbooks are as consistent as possible in terms of terminology, symbols, units, reference quantities, and physical models.

Hypersonic and High Temperature Gas Dynamics, by John D. Anderson, Jr.,[2.2] is especially useful for understanding the basics of hypersonic aerodynamics, including a comprehensive treatment of classical derivations, numerical techniques, and fundamental data.

The Aerothermodynamics of Gas Turbine and Rocket Propulsion, by Gordon C. Oates,[2.3] also appears in the AIAA Education Series and provides an excellent introduction to airbreathing propulsion. Although it focuses upon subsonic and supersonic, rather than hypersonic, flight, it has an approach and spirit very similar to this textbook, and can therefore provide worthwhile support for the reader.

And, because it continues to be mentioned so frequently by workers in this field, *The Dynamics and Thermodynamics of Compressible Fluid Flow*, by Ascher H. Shapiro,[2.4] must be included. The main reason for this is the careful and complete development of one-dimensional compressible flow behavior, a key element in the understanding of airbreathing propulsion.

Finally, for those interested in reviewing or learning basic principles, *Introduction to Flight*, by John D. Anderson, Jr.,[2.5] is recommended for aeronautics and *Engineering Thermodynamics*, by William C. Reynolds and Henry C. Perkins,[2.6] is recommended for thermodynamics.

This textbook is accompanied by a floppy disk containing a wide variety of menu-driven computer programs useful for the aerothermodynamic analysis of compressible flows. They are designated by the acronym HAP (for Hypersonic Airbreathing Propulsion) and a complete listing of the program titles can be found at the end of the Table of Contents. These computer programs performed the bulk of the calculations presented herein, and, being very general and flexible, have wide application elsewhere. We urge you to use the accommodating capabilities of HAP whenever possible and believe that you will find the experience rewarding and enjoyable.

2.2   UNITS

The selection of a system of units for a textbook on hypersonic propulsion is not a simple matter. Although the trend is toward the *universal* use of Système d'Unités International (SI) units, the situation is changing very slowly in the United States, and the British engineering (BE) system of units continues to prevail in the open literature. Fortunately, *both* these systems take the proportionality factor c to equal 1 in Newton's law of motion, so that

$$F = c \cdot ma = ma$$

which simplifies mathematical relations involving this equation. For these reasons, the BE system was chosen as primary, and the SI system is also prominently featured. For example, while all numerical results are first given in BE system units, they are also parenthetically displayed in their corresponding SI form. Furthermore, the propulsion community is especially fond of formulating equations and results in terms of dimensionless quantities, such as ratios and efficiencies, which are independent of the system of units. Finally, convenient tables of fundamental physical quantities and unit conversion factors are provided in App. A.

The fundamental units of the SI system are the kilogram (kg) of mass, the meter (m) of length, and the second (s) of time. The unit of force is derived from the law of motion as given above, and has the dimensions $kg \cdot m/s^2$. This force unit group is known as the newton (N), so that

1 N (force unit)

$$= 1 \text{ kg (mass unit)} \times 1 \text{ m/s}^2 \text{ (acceleration unit)}$$

The fundamental units of the BE system are the pound (lbf) of force, the foot (ft) of length, and the second of time. The unit of mass is derived from the law of motion as given above, and has the dimensions $lbf \cdot s^2/ft$. Although this mass unit group is sometimes referred to as the slug, no alias or pseudonym will be used in this textbook, so that

$1 \text{ lbf} \cdot s^2/ft$ (mass unit)

$$= 1 \text{ lbf (force unit)} \div 1 \text{ ft/s}^2 \text{ (acceleration unit)}$$

Since the pound mass (lbm) is very commonly used for mass measurements in airbreathing propulsion (largely because related physical and thermochemical properties are customarily tabulated in that form), it is important to establish the unit conversion factor between the mass units of lbm and $lbf \cdot s^2/ft$ when using the BE system. Noting that 1 lbf is defined as the force of gravity (weight) acting on 1 lbm at standard sea level, it follows that 1 lbf accelerates a 1 lbm at 32.17 ft/s^2, and therefore that

$$1 \text{ lbf} \cdot s^2/ft \text{ (mass unit)} = 32.17 \text{ lbm (mass unit)}$$

Thus, the $1 \text{ lbf} \cdot s^2/ft$ mass units appearing in any forthcoming BE calculations will *automatically be replaced* by 32.17 lbm in order to convert the results to lbm.

When dealing with airbreathing propulsion quantities expressed in the lbm unit, it is particularly important to keep in mind the fact that 1 lbm at (or near) the surface of the Earth *weighs* 1 lbf. The thrust specific fuel consumption, when reported as pounds mass of fuel per hour per pound force of thrust, may therefore be regarded as pounds weight of fuel per hour per pound force of thrust, and traditionally appears with the units of 1/hour (1/h). Similarly, air mass flow rate specific thrust, when reported as pounds force of thrust per pound mass of air per second, can be interpreted as pounds force of thrust per pound weight of air per second, with units of seconds.

In the case of rocket propulsion, propellant weight flow rate specific impulse is given as pounds force of thrust per pound weight of propellants per second, and also has units of seconds.

This is also a suitable place to mention *computational accuracy*.

All example calculations to be presented have been carried out to at least four significant figures. Nevertheless, in order to emphasize the fact that the results are based on engineering tools, and are intended simply to produce reasonably accurate answers, *they are reported to only three significant figures.* An exception to this rule occasionally occurs when computing ratios that are very close to 1. In these cases it is necessary to report the results to more significant figures in order to obtain comparable accuracy.

Rounding off can cause apparent discrepancies to occur, such as those in the first two columns of the last two rows of Table 9.1. We felt the better part of valor was to present the results as we found them.

2.3 ATMOSPHERIC FLIGHT

Figure 2.1 depicts the system of forces that act upon an aerospace vehicle in flight in what is known as the "natural axis system." It highlights the central assumption that the instantaneous installed engine thrust (of magnitude T) is in the same direction as the instantaneous vehicle velocity (of magnitude V). This is a close approximation to reality, even in the extreme case of rockets, that serves to greatly simplify the analyses that follow. Moreover, it makes the results much more understandable and intuitive, with little sacrifice of accuracy.

The remaining forces shown in Fig. 2.1 have traditional definitions and interpretations.[2.5] Thus, the vehicle lift L is perpendicular to the instantaneous velocity, and the vehicle drag D is parallel to the instantaneous velocity. The weight W or gravitational force mg is, of course, directed at the center of the Earth. The angle between the vehicle velocity and the horizontal θ is also called the flight path angle.

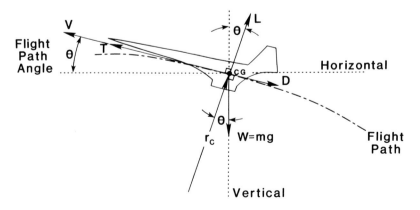

Fig. 2.1 Forces acting on an aerospace vehicle in flight.

Referring to Fig. 2.1 and applying Newton's second law along an axis parallel to the instantaneous velocity, we find that

$$m\frac{dV}{dt} = T - D - mg\sin\theta \qquad (2\text{-}1)$$

which, from the propulsion standpoint, can be interpreted to mean that the engine thrust produces vehicle acceleration while overcoming the vehicle drag and supporting part of the vehicle weight.

Applying Newton's second law along an axis perpendicular to the instantaneous velocity, we find that

$$\frac{mV^2}{r_c} = mg\cos\theta - L \qquad (2\text{-}2)$$

The left-hand side term, which is the normal acceleration force due to the curvature r_c of the flight path, is often called the *centrifugal force*. The centrifugal force is obviously important to orbiting satellites, for which it exactly balances the local force of gravity. Conversely, there are many situations in which the radius of curvature of the flight path is sufficiently large and the flight path angle θ sufficiently small that Eq. (2-2) reduces to

$$L = mg = W \qquad (2\text{-}3)$$

and simplifies analysis considerably. An example of this type of flight would be horizontal cruise (i.e., constant direction and velocity), for which Eq. (2-3) is exactly true.

2.4 THE ATMOSPHERE

Since the vehicle lift and drag and the installed thrust of the airbreathing engine all result from interactions with the surrounding atmosphere, it is important to understand the nature of this environment from the outset. One basic source of quantitative information is the *U.S. Standard Atmosphere, 1976*,[2.7] a document that is based on experimental data and periodically updated by experts representing a group of U.S. government agencies. The most relevant information from this publication can be found in App. B, and is used in the computer program HAP(Trajectory). You will find there the "standard day" distribution of pressure p, temperature T, density ρ, and speed of sound a as a function of geometrical height h, both BE (kft) and SI (km) units. As promised earlier, the information is presented entirely in the convenient form of ratios, all quantities being referenced to their standard sea level values.

2.4.1 Static Temperature of the Atmosphere

The variation of the ratio of static temperature T to the sea level value T_{SL} with geometric altitude, taken from App. B, is shown in Fig. 2.2. Like many of its counterparts, it consists of a series of straight line segments, either vertical (isothermal) or inclined (gradient). Of special interest for high speed flight is the stratosphere, which consists of those exactly or nearly isothermal segments lying between approximately 36 kft (11 km) and 169 kft (52 km). Most aircraft today fly below the stratosphere in the troposphere, or at the very bottom of the stratosphere. You will soon see why hypersonic vehicles prefer higher altitudes.

2.4.2 Absolute Viscosity of Air

Hypersonic flight involves a number of important viscous effects that require accurate prediction of the wall skin friction and boundary layer thickness, shape, transition, and separation. It is therefore important to have some means available to evaluate the absolute viscosity μ of the working medium, air. For such purposes we will use the relationship most frequently found in the literature, namely Sutherland's law,[2.1] which is

$$\mu = 2.27 \times 10^{-8} \left(\frac{T^{\frac{3}{2}}}{T + 199} \right) \text{ lbf·s/ft}^2 \qquad (2\text{-}4\text{a})$$

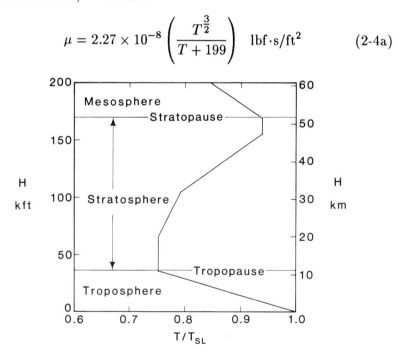

Fig. 2.2   Variation of standard day ratio of static temperature T to the sea level value T_{SL} with geometric altitude, Ref. 2.7 and App. B.

where T is in °R, or

$$\mu = 1.46 \times 10^{-6} \left(\frac{T^{\frac{3}{2}}}{T + 111} \right) \ \text{N} \cdot \text{s}/\text{m}^2 \qquad (2\text{-}4\text{b})$$

where T is in K.

For the static pressure range ordinarily encountered in atmospheric flight (0.01–100 atm), these equations are valid up to temperatures of at least 5400 °R (3000 K).[2.1] The reasons that the absolute viscosity is a function only of temperature under these conditions are that the air behaves as a perfect gas, in the sense that intermolecular forces are negligible, and that viscosity itself is a momentum transport phenomenon caused by the random molecular motion we associate with thermal energy or temperature.[2.2]

2.4.3 Thermal Conductivity of Air

As flight speed increases, heat transfer to the vehicle becomes a significant issue effect. Re-entry vehicles, for instance, represent an extreme situation where ablative shields are necessary to ensure survival during the passage through the atmosphere. Since wall heat transfer depends on the thermal conductivity k of the air, a means for its evaluation is also needed. In this case, a relationship similar to Sutherland's law is again found to be useful,[2.1] namely

$$k = 2.39 \times 10^{-7} \left(\frac{T^{\frac{3}{2}}}{T + 202} \right) \ \text{BTU}/(\text{s} \cdot \text{ft} \cdot °\text{R}) \qquad (2\text{-}5\text{a})$$

where T is in °R, or

$$k = 1.99 \times 10^{-3} \left(\frac{T^{\frac{3}{2}}}{T + 112} \right) \ \text{J}/(\text{s} \cdot \text{m} \cdot \text{K}) \qquad (2\text{-}5\text{b})$$

where T is in K.

For the range of static pressures ordinarily encountered in atmospheric flight (0.01–100 atm), these equations are valid up to at least 3600 °R (2000 K).[2.1] The reasons that the thermal conductivity is only a function of temperature under these conditions are the same as for absolute viscosity.

2.4.4 Static Pressure of the Atmosphere

The variation of the ratio of the static pressure p to the sea level value p_{SL} with geometric altitude, taken from App. B, is shown in

Fig. 2.3. Unlike static temperature, which does not change dramatically within the atmosphere, static pressure decreases more or less exponentially with increasing geometric altitude.

It is easy to see why this must be true. Combining the buoyancy equation for the stationary atmosphere,

$$dp = -\rho g dH \qquad (2\text{-}6)$$

and the perfect gas law gives

$$\frac{dp}{p} = -\frac{g}{RT} dH \qquad (2\text{-}7)$$

which can be integrated to yield

$$\frac{p_2}{p_1} = e^{-\frac{g}{R} \int_{H_1}^{H_2} \frac{dH}{T}} \qquad (2\text{-}8)$$

where the fact that g and the perfect gas constant R are very close to constant for these geometric altitudes has been used. Equation (2-8) shows that the relationship is exactly exponential for the isothermal regions. It also shows that the relationship is nearly exponential

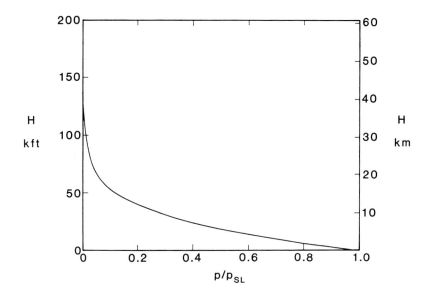

Fig. 2.3 Variation of the ratio of the standard day static pressure p to the sea level value p_{SL} with geometric altitude, Ref. 2.7 and App. B.

in the gradient regions, where the static temperature varies only gradually with geometric altitude.

Equation (2-8) can also be used to estimate the increase in geometric altitude that will reduce the static pressure by a factor of 2. Substituting representative values of g (32.2 ft/s^2 or 9.81 m/s^2), R (53.3 ft·lbf/lbm·°R or 287 N·m/kg· K), and T (400 °R or 222 K) reveals that the necessary increase in geometric altitude is about 15 kft (4.5 km). This result can be confirmed by inspection of Fig. 2.3 and/or App. B.

The material in the next section will demonstrate why the rapid decrease of static pressure with increasing geometric altitude is fortuitous for hypersonic flight.

2.4.5 Dynamic Pressure and Hypersonic Vehicle Flight Trajectories

The dynamic pressure q_0 of the atmosphere, as seen from the moving vehicle, is defined by the expression

$$q_0 = \frac{\rho_0 V_0^2}{2} \tag{2-9}$$

Please note carefully that, *hereinafter*, the subscript o denotes the undisturbed freestream flow conditions far ahead of the vehicle as seen from the *reference frame of the vehicle*. Thus, for example, p_0, T_0, and ρ_0 are the static properties of the atmosphere at the flight altitude, while V_0 is the *magnitude* of the velocity of the atmosphere relative to the vehicle (and, incidentally, vice versa). Also, for the moment, assume that supersonic flight includes $1 < M_0 < 5$, and hypersonic flight includes $M_0 > 5$.

Since

$$V_0 = a_0 M_0 \tag{2-10}$$

and a_0, the local speed of sound, is given by[2.4]

$$a_0 = \sqrt{\gamma_0 R_0 T_0} \tag{2-11}$$

then

$$q_0 = \frac{\rho_0 a_0^2 M_0^2}{2} = \frac{p_0}{R_0 T_0} \cdot \gamma_0 R_0 T_0 \cdot \frac{M_0^2}{2} = \frac{\gamma_0 p_0 M_0^2}{2} \tag{2-12}$$

where the perfect gas law has been used. Eqs. (2-9) and (2-12) allow the dynamic pressure to be calculated directly for any combination of vehicle altitude and velocity. Since Mach number is a more illuminating indicator of the governing aerodynamics being experienced by the vehicle, Eq. (2-12) is the preferred form here.

Aeronautical engineers are quite accustomed to the use of dynamic pressure as a scaling factor for the pressures and forces exerted on conventional aircraft.[2.5] For example, the lift and drag of airplanes are almost always given by the expressions

$$L = q_0 C_L S \qquad (2\text{-}13)$$

and

$$D = q_0 C_D S \qquad (2\text{-}14)$$

where S is a suitable reference area (such as the wing area), and C_L and C_D are the coefficients of lift and drag, respectively. Those familiar with the design of high performance aircraft will also recognize that they have structural limits that are globally specified in terms of dynamic pressure, the current upper limit being about 2000 lbf/ft^2 (95,000 N/m^2).

Remarkably enough, dynamic pressure is also a useful scaling factor for the pressures and forces experienced by hypersonic vehicles. The most popular method for calculating the behavior of inviscid hypersonic flows along surfaces is the modified Newtonian model, which states simply that

$$p - p_0 = q_0 C_{pt} \sin^2 \theta_b \qquad (2\text{-}15)$$

where p is the local surface static pressure, θ_b is the angle of inclination between the direction of the undisturbed freestream flow and the local surface of the vehicle, and C_{pt} is a stagnation point pressure coefficient that depends very weakly upon M_0 (e.g., for air C_{pt} is 1.73 at M_0 equal to 3, 1.80 at 5, and 1.84 at ∞).[2.1,2,2] Since the local pressures are proportional to q_0 according to Eq. (2-15), then so must be such integrals over the entire body surface as lift and drag.

It stands to reason, then, that if q_0 is too large, the structural forces and the drag on the vehicle can be excessive. On the other hand, if q_0 is too small, the wing area required for sustained flight may become unreasonably large. This explains why hypersonic vehicles end up being designed to operate within a fairly narrow range of q_0, approximately 500–2000 lbf/ft^2 (20,000–90,000 N/m^2). For reasons that will become gradually clearer, the tendency is to fly these vehicles along the trajectory of their highest allowable q_0. These observations about hypersonic atmospheric flight will prove very useful in the analyses that follow.

The next logical step, therefore, is to see what flight trajectories of constant q_0 look like. This is the purpose of Fig. 2.4, which contains standard day geometric altitude versus Mach number trajectories for the expected range of design values of q_0. They can be generated in a straightforward way by selecting desired combinations of dynamic pressure and geometric altitude, using the latter to obtain static pressure from App. B, and calculating the flight Mach number from Eq. (2-12) in the form

$$M_0 = \sqrt{\frac{2q_0}{\gamma_0 p_0}} \qquad (2\text{-}16)$$

The computer program HAP(Trajectory) is based upon this method, and may also be used to generate Fig. 2.4.

Several features of Fig. 2.4 are worth emphasizing. First, if it were not for the exponential decrease of static pressure with geometric altitude, truly enormous flight altitudes would be required for hypersonic vehicles. Second, for the same reason, the extent of the flight corridor for the expected design range of q_0 is relatively small. This agrees with the ballpark estimate of Sec. 2.4.4, which, taken together with Eq. (2-12) and the fact that T is essentially constant for these altitudes, shows that the constant q_0 trajectories of Fig. 2.4 should be about 15 kft (4.5 km) apart at the same M_0. Third, regardless of q_0, comparison with Fig. 2.2 reveals that virtually all hypersonic flying will take place in the stratosphere. Finally, since the speed of sound depends on static temperature and therefore on altitude, lines of constant Mach number are not quite lines of constant velocity. For example, when M_0 is 25, the values of V_0 corresponding to q_0 of 500, 1000, and 2000 lbf/ft^2 are 26,800, 27,000, and 26,300 ft/s, respectively.

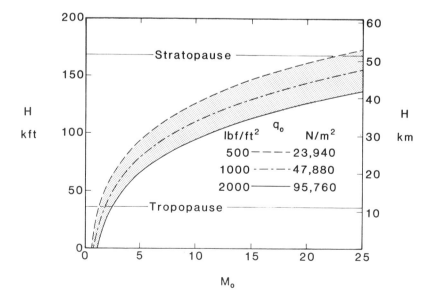

Fig. 2.4 Standard day geometric altitude versus flight Mach number trajectories for constant dynamic pressure.

2.4.6 Freestream Mass Flow per Unit Area

Airbreathing engines generate thrust in direct proportion to the rate at which they are able to capture and process the surrounding atmosphere. In order to produce enough thrust to reach and sustain high speed flight, it follows that hypersonic airbreathing engines must have the ability to collect large amounts of air. The mass flow rate of air entering the engine is most conveniently given by the expression

$$\dot{m}_0 = \rho_0 V_0 A_0 \qquad (2\text{-}17)$$

where $\rho_0 V_0$ is the mass flow per unit area of the undisturbed freestream far ahead of the vehicle, and the capture area A_0 is the geometric area of the freestream tube that eventually enters the engine inlet. The capture area can be larger or smaller than the physical opening of the engine inlet, and is determined by the entire flowfield upstream of the engine face.

The freestream mass flow per unit area for any flight altitude and Mach number can be calculated from the equation

$$\rho_0 V_0 = \rho_0 a_0 M_0 \qquad (2\text{-}18)$$

using the standard day information contained in App. B, or by using HAP(Trajectory). Better yet, contours of constant freestream mass flow per unit area can easily be generated by using Eq. (2-18) in the form

$$M_0 = \frac{\text{constant}}{\rho_{\text{SL}} a_{\text{SL}}} \cdot \frac{\rho_{\text{SL}}}{\rho_0} \cdot \frac{a_{\text{SL}}}{a_0} \qquad (2\text{-}19)$$

because the right-hand side is a function only of altitude. The results of this analysis are displayed in Fig. 2.5, where the range of constant q_0 trajectories of Fig. 2.4 is repeated for reference in the background. It is immediately obvious that flying faster at constant q_0 dramatically reduces the available freestream mass flow per unit area, simply because it means flying high where the density of the atmosphere is low.

This effect can be seen even more distinctly if Eq. (2-9) is employed in the form

$$\rho_0 V_0 = \frac{2q_0}{V_0} = \frac{2q_0}{a_0 M_0} \qquad (2\text{-}20)$$

Since a_0 hardly changes over these flight altitudes, Eq. (2-20) reveals that the freestream mass flow per unit area primarily varies inversely with Mach number along any trajectory of constant dynamic pressure. This useful rule of thumb can be confirmed by further inspection of Fig. 2.5. Such a trend unfortunately runs counter to the needs of the airbreathing engine, which must maintain mass flow and thrust as flight speed increases. This calls for resourcefulness on the part of the engine designers, and certainly explains the appearance

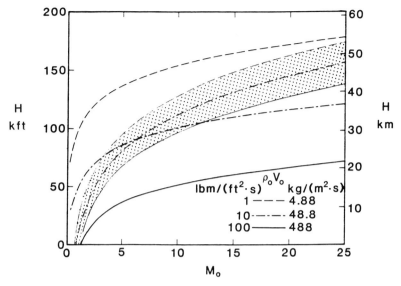

Fig. 2.5   Standard day geometric altitude versus flight Mach number contours for constant freestream mass flow per unit area. The range of constant dynamic pressure trajectories of Fig. 2.4 also appears.

of long forebody compression surfaces on hypersonic vehicles, their most important function being to make the freestream capture area A_0 much larger than the physical opening of the engine inlet.

Equation (2-20) also reveals that the freestream mass flow per unit area is proportional to dynamic pressure at any given Mach number. Therefore, as noted in Sec. 2.4.5, one flies at as high a dynamic pressure as other considerations will allow in order to maintain engine thrust.

2.4.7 Freestream Reynolds Number

The first indication of the type of viscous effects to be experienced by an aerospace vehicle in flight can be obtained by evaluating the freestream Reynolds number from the expression

$$Re_{\mathrm{L,0}} = \frac{\rho_0 V_0 L}{\mu_0} = \frac{\rho_0 a_0 M_0 L}{\mu_0} \tag{2-21}$$

where L is the length scale of the vehicle. In the discussions and calculations that follow, L will be taken to be 100 ft (30.5 m), a value representative of modern hypersonic vehicles. However, you may encounter situations where a considerably different value of L is appropriate. Once the length scale has been agreed upon, Eqs.

(2-21) and (2-4), and the standard day information of App. B can be combined or HAP(Trajectory) used to calculate the freestream Reynolds number for any flight altitude and Mach number. Again, as in the case of freestream mass flow per unit area, contours of constant freestream Reynolds number can easily be generated by using Eq. (2-21) in the form

$$M_0 = \text{constant} \cdot \frac{\mu_{\text{SL}}}{\rho_{\text{SL}} a_{\text{SL}} L} \cdot \frac{\mu_0}{\mu_{\text{SL}}} \cdot \frac{\rho_{\text{SL}}}{\rho_0} \cdot \frac{a_{\text{SL}}}{a_0} \qquad (2\text{-}22)$$

because the right-hand side is a function only of altitude.

The results of this analysis are displayed in Fig. 2.6, where the constant q_0 trajectories of Fig. 2.4 are repeated in the background for reference. Since the absolute viscosity of air does not change appreciably over the altitudes of interest to airbreathing propulsion, Eq. (2-21) reveals that contours of constant freestream Reynolds number and constant freestream mass flow per unit area should be almost identical in shape. This is confirmed by comparing Figs. 2.5 and 2.6 together.

One of the first and most important questions to arise in the design process is whether the vehicle boundary layers are likely to be laminar, transitional, or turbulent. Laminar boundary layers are of-

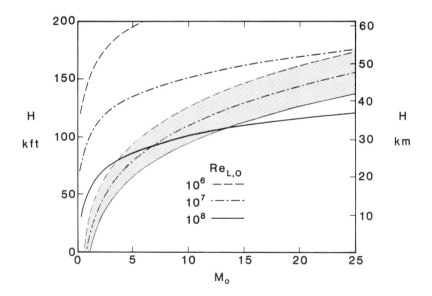

Fig. 2.6 Standard day geometric altitude versus flight Mach number contours for constant freestream Reynolds number, with length scale L equal to 100 ft (30.5 m). The range of constant dynamic pressure trajectories of Fig. 2.4 also appears.

ten preferred over turbulent because they have considerably less skin friction (i.e., shear drag) and wall heat transfer, and can be computed with greater certainty. However, they are much more prone to separate when encountering adverse pressure gradients or shock waves, resulting in considerably increased local values of pressure (or form) drag and wall heat transfer, and they can have a much larger displacement thickness, which reduces the flow rate of air that can be captured by a downstream engine inlet.

The traditional method for estimating whether transition from laminar to turbulent boundary layer flow will take place is to compare the expected freestream Reynolds numbers with critical (or transition) Reynolds numbers based upon the most relevant experience. Selecting such values is no easy task, because the transition process is one of the most complex found in nature, and especially so for hypersonic flows. One reason for this is that the transition process, in which small disturbances imposed upon an otherwise laminar flow grow into mature turbulence, can take place over sizable distances, and certainly does not take place at a point. Another reason is that the transition process depends at least upon the local Reynolds number, Mach number, streamwise pressure gradient, freestream turbulence content, three-dimensional crossflows, gas chemistry, and upon roughness, heat and mass transfer, and curvature of the wall.[2.1, 2.2, 2.8]

Nevertheless, the available data generally agree that natural transition to turbulent flow will take place on an aerospace vehicle having $Re_{L,o} > 10^7$, provided that the freestream Mach number is less than about 10. Boundary layer stability theory predicts that the critical Reynolds number will increase with freestream Mach number, and this is found to be true in practice. In fact, once the freestream Mach number exceeds about 20, natural transition to turbulent flow may not take place until $Re_{L,o} > 10^8$.[2.2, 2.8]

Returning to Fig. 2.6 with this information in mind, it can be seen that turbulent flow will dominate at lower freestream Mach numbers and laminar flow will dominate at higher Mach numbers, regardless of the constant dynamic pressure trajectory under consideration. The designer must therefore be prepared to deal with the entire spectrum of boundary layer flows, from laminar to turbulent, including the possibility of extended regions of boundary layer transitional flow.

2.4.8 Freestream Knudsen Number

Rarefied gas or free molecular effects become important when the altitude is high enough and the density low enough that the mean free path between air molecules becomes comparable to the length scale of the vehicle. The freestream Knudsen number, defined as *the mean free path between the freestream air molecules divided by the length scale of the vehicle*, is customarily used to signal the onset of the different regimes of free molecular flows. The first regime to be encountered is characterized by velocity slip (the gas velocity at the wall is no longer zero) and temperature slip (the gas temperature at the wall is no longer the surface temperature), and occurs when the Knudsen number reaches approximately 0.03. For Knudsen numbers in excess of about 0.2, the continuum Navier-Stokes flow equations no longer apply.[2.1,2.2]

Fortunately, even at 200 kft (61 km), the highest anticipated altitude of powered flight using airbreathing engines (see Fig. 2.4), the mean free path of standard day air is less than 10^{-3} ft,[2.7] so that the Knudsen number will not exceed 10^{-5} for hypersonic airbreathing flight. For this reason, rarefied gas effects are not likely to occur and will not be considered further in this textbook.

2.4.9 Representative Atmospheric Properties

Now and then in the course of the development presented here, it will be *convenient* (but not *necessary*) to use *representative* values of several atmospheric properties rather than, for example, the detailed descriptions found in App. B. This will only be done when the approximation is justifiable and profitable, and it will always be made clear when this option is being exercised. Representative values will be designated by the additional subscript R (e.g., $T_{O,R}$). Please note that the mathematical analyses will not depend upon this approximation, and they can therefore accept any description of the atmosphere. You are, in fact, encouraged to explore the consequences of this approach at your leisure.

Judging from the information presented in Sec. 2.4.1, the primary candidate for this treatment is the static temperature of the atmosphere, which does not vary significantly over the expected corridor of hypersonic flight. The representative value of static temperature is chosen to be:

$$T_{O,R} = 400\,^\circ R \quad (222\ K)$$

which corresponds to the qualitative stratospheric average. We have already seen that several atmospheric properties depend only upon temperature, and these can now be assigned representative values of their own, including:

$$\mu_{O,R} = 3.03 \times 10^{-7}\ \text{lbf} \cdot \text{s/ft}^2 \tag{2-4a}$$

$$(1.45 \times 10^{-5}\ \text{N} \cdot \text{s/m}^2) \tag{2-4b}$$

$$k_{O,R} = 3.18 \times 10^{-6}\ \text{BTU}/(\text{s} \cdot \text{ft} \cdot \,^\circ R) \tag{2-5a}$$

$$\left[1.98 \times 10^{-2}\ \text{J}/(\text{s} \cdot \text{m} \cdot \text{K})\right] \tag{2-5b}$$

$$a_{O,R} = 980\ \text{ft/s} \quad (299\ \text{m/s}) \tag{2-11}$$

The last of these, namely the representative atmospheric speed of sound, yields the greatest immediate benefit because it leads to an unambiguous relationship between flight Mach number and flight velocity, at a small cost in accuracy. This relationship is summarized in Table 2.1, where those measures of flight velocity most frequently found in the hypersonic literature have been included.

2.4.10 Caveat Designer

Let the designer beware. Although many worthwhile purposes are served by establishing and employing such generalized models of the atmosphere as published in Ref. 2.7, it is obvious that the atmosphere defies any simple description. The properties of the atmosphere not only vary with altitude, but also with longitude and latitude, and from year to year, season to season, day to night, and moment to moment. Reference 2.7 makes it clear that these variations can be significant, and they have been dramatized by measurements taken on Space Shuttle flights that reveal large density excursions over small altitude changes that are known as *potholes in the sky*.[2.1]

It may be useful someday for you to know that the military specification MIL-STD-210A defines expected "cold" and "hot" day deviations from standard conditions for aircraft design. The differences average about ±50 °R (30 K), but are much greater near sea level and the lower stratosphere.

The lesson here is that vehicles intended for hypersonic flight must be designed to either tolerate any foreseeable variations in atmospheric properties, or operate in ways that compensate when sufficiently severe atmospheric conditions are encountered.

Table 2.1 **Relationship between flight Mach number and flight velocity for the representative atmospheric speed of sound of 980 ft/s.**

Flight Mach Number M_0	kft/s	Flight Velocity, V_0 km/s	mi/s
1	0.9800	0.2987	0.1856
1.020	1	0.3048	0.1894
3.348	3.281	1	0.6214
5.388	5.280	1.609	1

2.5   EQUILIBRIUM AIR CHEMISTRY

When air flowing at high speed is decelerated or brought to rest and some or all of its kinetic energy is converted to enthalpy, the air molecules receive a considerable amount of internal energy. For hypersonic speeds, this internal energy is sufficient to excite the normally dormant vibration and dissociation states of the air molecules, thereby changing the chemical composition of the air and absorbing some of the internal energy that would otherwise appear as random molecular motion or thermal energy. Even though the newly excited states consume an appreciable fraction of the internal energy, the resulting equilibrium temperatures can still be 5400 °R (3000 K) or higher.

A special feature of hypersonic flows therefore is that air chemistry effects must often be incorporated in order for any analysis or calculation to be realistic. The behavior of high temperature air in chemical and thermodynamic equilibrium can be accounted for in a number of ways, including the use of tables[2.9] or graphs (usually Mollier charts),[2.10] and direct solution of the equations of statistical thermodynamics, the latter being the approach of this textbook.

The determination of the equilibrium state of air will be accomplished by means of the computer program HAP(Air) that resides on the floppy disk accompanying this textbook. The solution is achieved by minimizing the Gibbs function for the ensemble of particles, subject to the conservation of atom populations. The numerical method is iterative and robust, and the underlying thermodynamic data were generated by the NASA-Lewis Research Center. For further information you are referred to Sec. 6.3.

2.5.1 Perfect Equilibrium and Perfect Gas Assumptions

Almost all calculations of this type are based upon the assumptions of perfect chemical and thermodynamic equilibrium and the behavior of all constituents as perfect gases. Before we proceed further, it is worth pausing to review the justification of these models of the behavior of air under typical hypersonic flight conditions.

Perfect equilibrium for air would result if the reaction rate (or collision rate between molecules) were infinite. Even though this can never be exactly true, it will be approached in practice if the time taken for the particles to travel past the vehicle is much larger than the time they characteristically require to return to equilibrium when disturbed. For vehicles with a length scale of 100 ft (30.5 m) and flight speeds of 10,000 ft/s (3048 m/s), the residence time is only 10^{-2} s. Nevertheless, even for the low density air of the upper stratosphere the equilibrium time for air is less than 10^{-2} s,[2.7] so that the flow is never far from local equilibrium.

Air will behave as a *perfect gas*, in the sense that intermolecular forces are negligible, if the molecules are sufficiently far apart. One method for estimating the departure of the behavior of a gas from perfect behavior is based upon the principle of corresponding states.[2.6] It has been found experimentally that the compressibility factor

$$Z = \frac{p}{\rho RT}$$

is very much the same function of reduced pressure (the ratio of static pressure to the critical pressure) and the reduced temperature (the ratio of static temperature to the critical temperature) for many gases, including air. The generalized compressibility chart, a plot of Z versus reduced pressure for a series of reduced temperatures, has been very useful in predicting the properties of many substances. Looking at a typical chart reveals that, as long as the reduced pressure is less than about 2.0 and the reduced temperature is greater than about 1.8, the compressibility factor Z will be within a few percent of 1.0.[2.6] Since the critical pressure for air is 37.3 atm and the highest expected static pressure is 100 atm, the reduced pressure will never exceed 2.68. Since the critical temperature for air is 238 °R (132 K), and the lowest temperature associated with the standard atmosphere is 390 °R (217 K), the reduced temperature will never be less than 1.64. And, because the nature of the compression process is that pressure and temperature increase together, the worst combination of high reduced pressure and low reduced temperature never occurs. We may safely conclude, therefore that the compressibility factor remains very close to unity and the perfect gas assumption for air is reasonable.

The expression *real gas* is reserved for situations in which changes in equilibrium chemical composition due to dissociation become important.

Before closing this topic, it is interesting to note that perfect equilibrium (which requires that the molecules be not too far apart) and perfect gas behavior (which requires that the molecules be not too close together) are somewhat in conflict. Our analyses are the beneficiary of the fortunate fact that the air we deal with happens to be at neither of these extremes.

2.5.2 Equilibrium Behavior of Air

The material contained in this section is entirely based upon computations performed using HAP(Air). The chemical composition of the *representative* air used in these computations was 79 percent N_2 and 21 percent O_2 by moles (or number of molecules). You may choose to repeat the process for other representations of air using HAP(Equilibrium), by including such trace constituents as water or argon, but the general conclusions to be reached will not change.

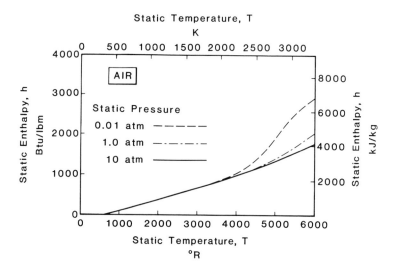

Fig. 2.7   Equilibrium static enthalpy of air as a function of static temperature and static pressure, as computed by HAP(Air).

2.5.2.1 Equilibrium static enthalpy of air.   Figure 2.7 presents the equilibrium relationship between static enthalpy and static temperature for air for several different static pressures, where the static enthalpy is defined as

$$h = e + \frac{p}{\rho} \qquad (2\text{-}23)$$

e being the internal energy. The temperatures and pressures shown there bracket any forseeable values to be found in practice. The fact that enthalpy increases more and more rapidly as temperature increases at a given pressure conforms with our expectations that real gas effects or new modes of molecular internal energy storage would be excited at higher temperatures. This effect becomes more pronounced as the static pressure is lowered because the reduced molecular collision frequency allows more dissociation to exist.

Although the enthalpy is an extremely useful property for aerothermodynamic analysis, it does not provide penetrating insight into the experiences of the individual air molecules. For that we must turn to other measures.

2.5.2.2 Equilibrium chemical constituents of air.   Figure 2.8 presents the equilibrium behavior of several key constituents of representative air as a function of static temperature at a static pressure of 0.01 atm. The lowest expected pressure was chosen for study in order to highlight the chemical changes that take place. Referring to Fig. 2.7, it is obvious that higher pressures merely delay their onset.

Returning to Fig. 2.8, it can be seen that the most important chemical reaction in the range of our interest is the dissociation of molecular oxygen (O_2) into atomic oxygen (O), which begins at approximately 3600 °R (2000 K), and is essentially complete by 6300 °R (3500 K). The rate of change of dissociation is greatest in the vicinity of 5000 °R (about 2800 K), which coincides with the steepest slope on both Figs. 2.7 and 2.8. The dissociation of oxygen evidently absorbs a considerable amount of internal molecular energy.

The dissociation of molecular nitrogen (N_2) into atomic nitrogen (N), which begins at about 6300 °R (3500 K), is insignificant over our range of interest and can be disregarded. Figure 2.8 shows that small mole fractions of nitric oxide (NO) can be formed under these conditions, the peak being 0.024 in the vicinity of 5000 °R (2800 K). Although this development has a relatively small impact on the thermochemistry of air, it can have a remarkably large impact on engine design because NO contributes to the formation of smog near the ground as well as to the destruction of the ozone layer. Nitric oxide has therefore been identified as a leading atmospheric pollutant, and the amount allowed as engine exhaust emission is tightly controlled by public opinion and law.

As noted above, similar results would occur at higher static pressures, but they would take place at higher temperatures.

2.5.2.3 Equilibrium specific heat at constant pressure of air. Oftentimes it is more convenient to do the aerothermodynamic analysis of engine

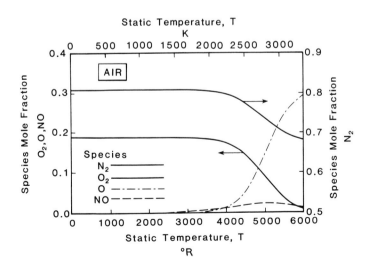

Fig. 2.8 **Equilibrium mole fraction composition for several constituents of air as a function of static temperature for a static pressure of 0.01 atm, as computed by HAP(Air).**

performance using the equilibrium specific heat at constant pressure

$$C_p = \left(\frac{\partial h}{\partial T}\right)_p \qquad (2\text{-}24)$$

rather than the static enthalpy. The specific heat at constant pressure for air can be found by taking the local slope of the constant pressure lines on Fig. 2.7 or directly from HAP(Air), and the results of this procedure are given in Fig. 2.9.

Figure 2.9 reveals that the C_p of our representative air exhibits three distinct types of behavior over the range of temperatures and pressures of interest.

For temperatures between 390 °R (217 K) and about 720 °R (400 K), C_p is neither a function of temperature nor of pressure, remaining very near its standard atmospheric value of 0.240 BTU/(lbm·°R) [1005 J/(kg·K)]. A classical thermodynamicist would say that air behaves as a calorically perfect gas within these temperature limits. And students of the kinetic theory of gases would recognize that this is the temperature range where the diatomic molecules of air have five degrees of freedom fully excited, namely three degrees of translation and two degrees of rotation.

For temperatures between about 720 °R (400 K) and about 3000 °R (1700 K), C_p depends upon temperature only, rising to a value of about 0.300 BTU/(lbm·°R) [1256 J/(kg·K)]. A classical

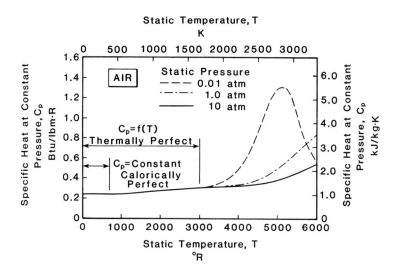

Fig. 2.9 Equilibrium specific heat at constant pressure of air as a function of static temperature and static pressure, as computed by HAP(Air).

thermodynamicist would say that air behaves as a thermally perfect gas within these temperature limits. From the standpoint of the kinetic theory of gases, this is the temperature range over which two addition degrees of freedom of vibration of the diatomic molecules become excited.

For temperatures above about 3000 °R (1700 K), C_p depends strongly upon both temperature and pressure because chemical reactions have become important. In fact, the temperature at which the thermally perfect model for air breaks down coincides with the beginning of the formation of NO, as seen on Fig. 2.7. The astounding values of C_p that occur at higher temperatures are clearly due to the dissociation of O_2. Once the real gas realm of chemical reaction and dissociation has been reached, no simple model for C_p will suffice, and the equilibrium properties must be obtained by calculation.

2.5.2.4 Equilibrium ratio of specific heats of air.

The equilibrium ratio of specific heats of the working medium is sometimes directly useful in aerothermodynamic analysis, and always a good indicator of how the flow will act. In a manner similar to that outlined above, HAP(Air) may be used to compute the equilibrium internal energy e as a function of static temperature and volume (or density), and the equilibrium specific heat at constant volume (or density) determined from the expression

$$C_v = \left(\frac{\partial e}{\partial T} \right)_v \tag{2-25}$$

C_p and C_v information may then be combined to allow the ratio of specific heats

$$\gamma = \frac{C_p}{C_v} \tag{2-26}$$

to be found.

These quantities may also be directly computed by HAP(Air). The results of these computations are presented in Fig. 2.10, where the general features echo those already discussed with regard to Figs. 2.7 through 2.9. Most importantly, however, we can see that air behaves in accordance with the kinetic theory of gases in the calorically and thermally perfect regions for which it is predicted that

$$\gamma = \frac{n + 2}{n} \tag{2-27}$$

where n is the number of degrees of freedom.[2.4] Thus, for the calorically perfect regime where n is 5, Eq. (2-27) yields a γ of $7/5 = 1.40$, which is almost exactly correct. Also, for the thermally perfect regime, where n verges toward 7, Eq. (2-27) yields a γ of $9/7 = 1.286$, which is the value it has reached when the formation of NO begins.

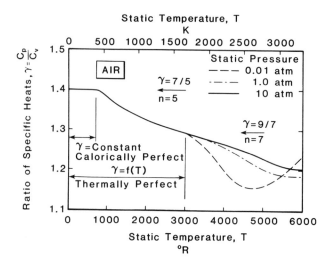

Fig. 2.10 Equilibrium ratio of specific heats of air as a function of static temperature and static pressure, as computed by HAP(Air).

As we have seen before, the realm of chemical reaction and dissociation is extraordinarily complicated, and cannot be treated with a simple model. Nevertheless, our intuitions should be reassured by the fact that the new states of excitation appear to be many new degrees of freedom as far as γ is concerned.

2.5.2.5 Closure. However commonplace and prosaic air may seem, it can be a very complex substance under the right circumstances, some of which can occur in hypersonic flight. We have focused in this section upon its equilibrium properties for a number of mutually supportive reasons. To begin with, it is important to understand the general behavior of air because it is the fluid upon which the airbreathing engines will depend. Furthermore, it is important to become familiar with and develop confidence in the computational tools that will be used in the remainder of this textbook. Finally, we will be faced with even more complex chemistry when the combustion reactions of fuels with air are considered, and the simpler case of air alone is good preparation for that time.

2.6 THE GOVERNING AEROTHERMODYNAMIC EQUATIONS

Real progress in the understanding and control of hypersonic flows and hypersonic airbreathing engines eventually rests upon the ability to analyze or compute their behavior. The selection of a closed set of governing equations requires unusual discretion because the range of their application is quite broad. For students, they must be reasonably transparent and be well suited to the solution of certain *classical* problems that breed a solid intuitive grasp of the subject. For workers in the field, they must embody the dominant physical phenomena and therefore lead to realistic results. For a textbook, they must be manageable and directly related to fundamental derivations found elsewhere.

Several generations of aerothermodynamicists have found that these purposes are best served by applying the conceptual model known as *one-dimensional flow,* in which the principal assumption is that the fluid properties are constant across the flow and, therefore, depend upon a single spatial coordinate (the axial dimension).[2.3, 2.4] Two advantages of this approach for our purposes should be immediately obvious. First, the analysis is greatly simplified because the fluid properties vary only in the streamwise direction. Second, airbreathing engine flows are particularly susceptible to this approach because they are usually confined within definite boundaries, either gaseous or solid.

Nevertheless, there are always variations of flow properties in directions transverse to the axial. In hypersonic airbreathing engines they would be generated at least by: (1) viscous forces and heat transfer occurring at the wall that result in profiles (transverse variations) that develop or evolve as the flow proceeds; (2) transverse pressure gradients that accompany streamline curvature; (3) the unavoidable presence of reflecting oblique expansion waves and shock waves making their way downstream; and (4) the mixing and/or combustion of separate streams of fluid introduced through finite jets or orifices. It is therefore of paramount importance that the method or process for averaging the properties across the flow be chosen carefully so that meaningful results can be obtained. Several authors have written comprehensively and lucidly about the consequences of different types of averaging or weighing of flow properties (e.g., area average, mass average, or continuity average) on the results,[2.3, 2.11] and you will find their conclusions to be interesting and sometimes surprising.

Conversely, there are conditions for which the one-dimensional approach can be used with great confidence. One of these would be where the bulk of the transverse variations occur within a vanishingly small *boundary layer* region, the remainder of the flow having very nearly uniform conditions. Another would be in *fully developed flow,* where the shapes of the property profiles vary only slowly with axial position, and a careful determination of proper weighing functions need be done only once. Finally, and most importantly, there is the *stream thrust average,* which assumes that the fluid properties are perfectly uniform at each axial station, and corresponds to imagining the actual flow to mix completely at constant axial area with no further external intervention (such as friction or heat transfer).

Although the stream thrust is obviously conserved in this thought model, a thermodynamicist would recognize that the mixed state has already undergone all possible dissipation and therefore represents a lower limit of availability for the flow. The stream thrust average is therefore often utilized, as it will be in this textbook, because it is analytically simple and thermodynamically conservative.

Although the one-dimensional approach can never be perfectly correct, the alternatives are both hopelessly complex and completely unwieldy for reaching a basic understanding built upon fundamental principles. We should be thankful that one-dimensional analysis exists and is readily available for our purposes.

2.6.1 Assumptions

In addition to presuming that the flow can be treated with acceptable accuracy by the one-dimensional approach described above, the following assumptions are used in the analyses that follow:

a. The flow is *steady* in the sense that any unsteady terms that could appear in the governing equations are negligible compared to those necessary to describe the steady flow. This is an extremely good approximation for hypersonic flows because the steady flow inertial terms are enormous. It is exactly true for airbreathing engine performance evaluation, which is always done at steady state. It is not true, however, for situations that involve high speed oscillations or rapid changes in engine operation such as inlet starting or unstarting, although a one-dimensional approach including unsteady effects can be useful even in such complex circumstances.

b. The *thermochemistry* of the flow is known, in the sense that the atomic composition of the mixture is set and that all intensive properties of the fluid can be determined once any two intensive properties have been found. This determination can be almost trivial, as in the case of calorically perfect gases, fairly difficult, as in the case of equilibrium or frozen composition flows, or very demanding, when reaction rate chemistry must be taken into account.

c. The effects of gravitational, acceleration, electrical, and magnetic *fields* on the motion or energy of the fluid are negligible.

The derivations of the equations presented below are frequently found in foundational textbooks,[2.3, 2.4, 2.6] and will not be repeated here. Nonetheless, it is hoped that the form in which they are displayed is sufficiently unmistakable that they can be accepted at face value.

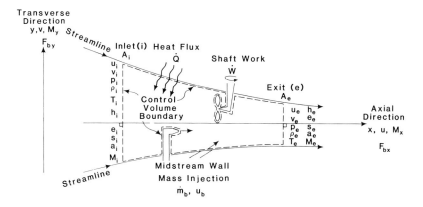

Fig. 2.11 Geometry and nomenclature for finite control volume one-dimensional flow analysis.

2.6.2 General Finite Control Volume Analysis

Referring to Fig. 2.11, where the geometry has been chosen to be suitable for the *finite control volume* one-dimensional analysis of air-breathing engines, it can be seen that allowance has been made for the introduction of several flows of known properties through the walls or midstream injectors. Since each entering and leaving flow will be treated as one-dimensional, summation over the individual streams is more appropriate to the analysis than integration.

2.6.2.1 Conservation of mass (continuity). The equation of conservation of mass for the control volume of Fig. 2.11 is

$$\rho_i u_i A_i + \sum \dot{m}_b = \rho_e u_e A_e \qquad (2\text{-}28)$$

where it should be emphasized that the subscript i denotes the inlet flow, e the exit flow, and b a flow entering through any other control volume boundary (including walls and injectors).

2.6.2.2 Conservation of momentum (Newton's principle). Applying New-ton's law to the control volume of Fig. 2.11 yields two equations, one for the axial or x direction, and one for the transverse or y direction. The former is

$$p_i A_i + (\rho_i u_i A_i)u_i + \sum \dot{m}_b u_b + F_{bx} = p_e A_e + (\rho_e u_e A_e)u_e \qquad (2\text{-}29)$$

where F_{bx} is the total axial force (pressure and viscous) acting on the control volume boundary other than the inlet and exit, and u_b is the axial component of velocity of a flow entering through any control volume boundary other than the inlet. The latter is

$$(\rho_i u_i A_i)v_i + \sum \dot{m}_b v_b + F_{by} = (\rho_e u_e A_e)v_e \qquad (2\text{-}30)$$

where F_{by} is the total transverse force (pressure and viscous) acting on the control volume boundary other than the inlet and exit, and v_b is the transverse component of velocity of a flow entering through any control volume boundary other than the inlet. Please note that the selection of the transverse direction is arbitrary, but it is presumed here that an orientation of interest has been deliberately chosen. In most cases that choice will be quite obvious. It should also be obvious that this analysis assumes that all velocities into the page (the other transverse direction) are inconsequential, but if this is not so, it can be corrected with little effort.

2.6.2.3 Conservation of energy (first law). Applying the first law of thermodynamics to the control volume of Fig. 2.11 yields

$$\rho_i u_i A_i \left(h_i + \frac{u_i^2 + v_i^2}{2} \right) + \sum \dot{m}_b \left(h_b + \frac{u_b^2 + v_b^2}{2} \right) + \dot{W} + \dot{Q}$$

$$= \rho_e u_e A_e \left(h_e + \frac{u_e^2 + v_e^2}{2} \right) \qquad (2\text{-}31)$$

where $\dot{W}$ is the total rate at which the shaft does work on the control volume fluid, and $\dot{Q}$ is the total rate at which heat is added to the control volume fluid from the surroundings. The former becomes important in cases where machinery, such as compressors or turbines, are present within the control volume, and the latter is important when the flow is hot and the walls must be cooled by removing heat in order to survive. Please note that it will be necessary to base the enthalpy on a standard reference state (ordinarily 77 °F/25 °C, 1 atm) in order to insure consistency whenever chemical changes or reactions can occur.

2.6.2.4 Equations of state. Assumption b of Sec. 2.6.1 translates to a series of relationships between the intensive properties of the fluids involved for the given atomic composition of the mixture. The usual version of this series is:

$$\rho = \rho(p, T) \qquad (2\text{-}32)$$

$$h = h(p, T) \qquad (2\text{-}33)$$

$$e = e(p, T) \qquad (2\text{-}34)$$

As already noted, the intensive properties h and e are based upon standard reference states for each species when chemical changes are expected.

The chemical composition of the fluid is an important by-product of these evaluations, even though it does not appear explicitly in the conservation equations in our formulation.

2.6.2.5 Entropy and the Gibbs equation.   Although the static entropy of the fluid s does not appear explicitly in these equations, and is therefore not involved in their solution, it is frequently evaluated because it serves as an effective indicator of the state of the flow. Experience has also shown that constant entropy or isentropic flow is an important (and often *best case*) baseline from which useful qualitative and/or quantitative comparisons can be made.

The primary means for determining or specifying the entropy change for a pure substance is the *Gibbs equation*

$$T ds = dh - \frac{dp}{\rho} \qquad (2\text{-}35)$$

2.6.2.6 Mach number.   The speed of sound or acoustic velocity is the speed at which a planar pressure disturbance small enough to be isentropic progresses through a gas. For an equilibrium chemically reacting mixture it can be shown that the speed of sound is given by

$$a^2 = \left(\frac{\partial p}{\partial \rho} \right)_s = \frac{\gamma p}{\rho} \left\{ \frac{1 + \frac{1}{p} \left(\frac{\partial e}{\partial \frac{1}{\rho}} \right)_T}{1 - \rho \left(\frac{\partial h}{\partial p} \right)_T} \right\} \qquad (2\text{-}36)$$

where the ratio of specific heats and the partial derivatives are evaluated at chemical equilibrium.[2.2]

Since the derivatives in this equation can also be evaluated from the known equations of state, it is therefore possible to evaluate the Mach number from

$$M^2 = \frac{V^2}{a^2} = \frac{u^2 + v^2}{a^2} \qquad (2\text{-}37)$$

2.6.2.7 General solution.   Simply by counting unknowns and available equations it can be seen that we have now assembled the desired closed set of governing equations. Referring again to Fig. 2.11, the typical problem to be solved begins with all information given except the properties of the flow at the exit. Thus, there are normally 10 unknowns, which can be solved for by the 10 equations (2-28) through (2-37).

2.6.2.8 Calorically perfect gas.   There are many times when it is appropriate and/or instructive to employ the calorically perfect gas version of the equations of state rather than the exact set given in Sec. 2.6.2.4. This approach is quite accurate, for example, when working with air in subsonic, transonic, or supersonic flows for which the static temperatures and pressures remain within the calorically perfect range (see Figs. 2.9 and 2.10).

The governing aerothermodynamic equations are *dramatically* simplified when the perfect gas constant and the specific heats may be treated as fixed, so much so that closed form analytical expressions can be found for even very complex flow situations. Thus, it is appealing to use this model even when the gas is not exactly calorically perfect, because the benefit of revealing the physics or general character of the flow being examined greatly exceeds the cost of slightly decreased accuracy. We will extend the use of this model into the thermally perfect regime by using *average* perfect gas constant and specific heat values *valid over the range of flow conditions under consideration,* known hereinafter as *mean* values. The behavior of gases based upon *mean* perfect gas constants and specific heats will be referred to as *calorically perfect* for the remainder of this textbook.

Returning to Figs. 2.9 and 2.10, it can be seen that it should be possible to select suitable *mean* values of the specific heats for analyzing the point-to-point behavior of air for temperatures to about 3000 °R (1700 K) and beyond. The perfect gas constant of air depends only upon its molecular weight, which does not vary until its chemical composition changes, and very little afterward.

The equations of state of a calorically and thermally perfect gas equivalent to Eqs. (2-32) through (2-34) are[2.2, 2.6]:

$$p = \rho RT \tag{2-38}$$

$$dh = C_p dT \tag{2-39}$$

$$de = C_v dT \tag{2-40}$$

where h and e are no longer based on a standard reference state. The Gibbs equation, Eq. (2-35) becomes

$$ds = C_p \frac{dT}{T} - R \frac{dp}{p} \tag{2-41}$$

Equation (2-36) reduces to

$$a^2 = \gamma RT \tag{2-42}$$

[see Eq. (2-11)], and Eq. (2-37) therefore becomes

$$M^2 = \frac{V^2}{\gamma RT} = \frac{u^2 + v^2}{\gamma RT} \qquad (2\text{-}43)$$

Please note that Eqs. (2-23), (2-26), and (2-38) through (2-40) can be combined to prove that

$$C_p - C_v = R \qquad (2\text{-}44)$$

and that

$$\frac{C_p}{R} = \frac{\gamma}{\gamma - 1} \qquad (2\text{-}45)$$

To summarize, when carrying out a calorically perfect gas analysis, the ten unknowns are obtained by the solution of the ten equations (2-28) through (2-31) plus (2-38) through (2-43). Please bear in mind that these are the *mean* perfect gas constant and specific heat values valid over the range of interest.

2.6.2.9 Examples of finite control volume analysis. The five Example Cases that follow are based upon the steady, one-dimensional flow of a calorically perfect gas as shown in Fig. 2.11. They are meant to show the breadth and power of one-dimensional analysis, while demonstrating several key features of hypersonic flows that play prominent roles in ramjets and scramjets and familiarizing the reader with applying the set of governing equations.

The reader should be aware that the Example Cases are also found in the subroutines of HAP(Gas Tables).

Example Case 2.1: Total Enthalpy, Total Temperature, and Total Pressure

Consider a one-dimensional flow without energy interactions with the surroundings, i.e.,

$$\dot{W} = \dot{Q} = \dot{m}_b = 0$$

Under these conditions, the energy equation, Eq. (2-31), reduces to

$$h_i + \frac{V_i^2}{2} = h_e + \frac{V_e^2}{2} = h_t \qquad (2\text{-}46)$$

which means that the *total enthalpy* h_t is a fixed property of constant energy flows, even if the gas is *not* calorically perfect.

For the specific case of an aerospace vehicle traversing the atmosphere, this result may be written in the form

$$h_t = h_0 + \frac{V_0^2}{2} \tag{2-47}$$

Since the hypersonic flight velocity can be very large, while the static enthalpy of the atmosphere remains essentially fixed, it is clear that the total enthalpy can be formidable. The total enthalpy is also equal to the enthalpy that would exist locally if the flow were brought completely to rest without energy interactions, and is therefore also known as the *stagnation enthalpy*.

Over the range for which suitably accurate *mean* values of the perfect gas constant and specific heats can be defined, Eqs. (2-39) and (2-46) can be combined to yield

$$C_p T_i + \frac{V_i^2}{2} = C_p T_e + \frac{V_e^2}{2} = C_p T_t \tag{2-48}$$

where T_t is the *total temperature*. This expression may be rearranged with the help of Eqs. (2-43) and (2-45) to show that

$$T_t = T_i \left(1 + \frac{\gamma - 1}{2} M_i^2\right) = T_e \left(1 + \frac{\gamma - 1}{2} M_e^2\right) \tag{2-49}$$

Thus, the total temperature

$$T_t \doteq T \left(1 + \frac{\gamma - 1}{2} M^2\right) \tag{2-50}$$

is a fixed property of constant energy flows of calorically perfect gases.

It is tempting to equate the total temperature with the static temperature that would exist if the flow were brought completely to rest without energy interactions, or the *stagnation temperature*. Unfortunately, the stagnation enthalpy of hypersonic flows is so large that acceptably accurate mean values of the perfect gas constant and specific heats cannot be specified a priori, and the stagnation temperature must be determined from the exact set of equations of state. *Nevertheless*, the total temperature is an extremely convenient device for the *point-to-point* analysis even of hypersonic flows, over ranges for which mean values of the gas constants can be specified, provided that Eq. (2-50) is understood to be its fundamental definition.

It is interesting and productive to portray Eq. (2-48) in terms of the *dimensionless static enthalpy* $C_p T / C_p T_{ti}$ and the *dimensionless kinetic energy* $V^2 / 2 C_p T_{ti}$, or

$$\frac{C_p T}{C_p T_{ti}} + \frac{V^2}{2 C_p T_{ti}} = 1 \tag{2-51}$$

as shown in Fig. 2.12. The state of the flow *must* be somewhere on this straight line. For the *special* case of constant area flow with friction but without energy interaction, this line is known in the classical literature as a *Fanno line*. However, the same relation applies to the *much more general* situations to be encountered here, so the line will *not* be referred to by that name. The breaks at either end of the line are meant to remind us that it may not be possible to find *mean* values of the gas constants valid over the entire range.

For the specific case of an aerospace vehicle traversing the atmosphere, Eq. (2-50) may be written in the form

$$\frac{T_{t0}}{T_0} = 1 + \frac{\gamma - 1}{2} M_0^2 \tag{2-52}$$

where γ is the mean value for air for the range of interest. The value of this formulation is that it reveals that the ratio of the total temperature T_{t0} to the local static temperature T_0 increases with the square of the flight Mach number, which is a rather strong dependence.

Equations (2-41), (2-45), and (2-49) may now be combined and manipulated to show that

$$\frac{s_e - s_i}{R} = \ell n \left\{ \frac{p_i}{p_e} \cdot \left(\frac{T_e}{T_i} \right)^{\frac{\gamma}{\gamma-1}} \right\} = \ell n \left\{ \frac{p_i}{p_e} \left(\frac{1 + \frac{\gamma-1}{2} M_i^2}{1 + \frac{\gamma-1}{2} M_e^2} \right)^{\frac{\gamma}{\gamma-1}} \right\} \tag{2-53}$$

Thus, the grouping known as the *total pressure*

$$p_t \doteq p \left(1 + \frac{\gamma - 1}{2} M^2 \right)^{\frac{\gamma}{\gamma-1}} \tag{2-54}$$

is a fixed property of constant energy *isentropic* flows of calorically perfect gases. Since the entropy can only increase for this type of flow, Eq. (2-53) shows that the total pressure can only decrease as the flow proceeds. Because γ for air is 1.40 or less, the exponent of this expression is at least 3.50, which means that the dependence of the total pressure on flight Mach number is *remarkably* strong for a natural process.

It is tempting to equate the total pressure with the static pressure that would exist if the flow were brought completely to rest isentropically and without energy interactions, or the *stagnation pressure*. Unfortunately, the stagnation enthalpy of hypersonic flows is so large that acceptably accurate mean values of the perfect gas constant and specific heats cannot be specified, and the stagnation pressure must be determined from the exact set of equations of state. *Nevertheless,*

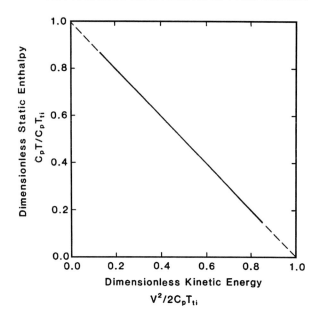

Fig. 2.12 Dimensionless energy diagram for the constant energy flow of a calorically perfect gas, showing the relationship between dimensionless static enthalpy and dimensionless kinetic energy, Eq. (2-51).

the total pressure is an extremely convenient device for the point-to-point analysis even of hypersonic flows, over ranges for which mean values of the gas constants can be specified, provided that Eq. (2-54) is understood to be its fundamental definition.

Note should also be taken of the fact that, according to Eqs. (2-53) and (2-54),

$$\pi = \frac{p_{te}}{p_{ti}} = e^{-(s_e - s_i)/R} \leq 1 \tag{2-55}$$

which means that the *total pressure ratio* π depends exponentially upon the entropy increase. In other words, the total pressure ratio should be expected to decrease very rapidly as entropy, the natural indicator of the effects of dissipation or *flow losses*, increases.

Example Case 2.2: Impulse Function and Stream Thrust Function

Consider a one-dimensional flow for which no mass is injected, i.e.,

$$\dot{m}_b = v = 0 \qquad u = V$$

and all properties at the inlet and exit planes are known. In this situation Eq. (2-28) becomes

$$\dot{m} = \rho_i u_i A_i = \rho_e u_e A_e \qquad (2\text{-}56)$$

and Eq. (2-29) becomes

$$F_{bx} = (p_e A_e + \dot{m} u_e) - (p_i A_i + \dot{m} u_i) \qquad (2\text{-}57)$$

so that the total force acting on the flow within the control volume can be directly determined.

In the regime where the fluid behaves as a calorically perfect gas, the axial momentum equation may be rearranged into several useful forms. For example, with the help of Eqs. (2-38), (2-43), and (2-56), Eq. (2-57) becomes

$$F_{bx} = p_e A_e \left(1 + \gamma M_e^2\right) - p_i A_i \left(1 + \gamma M_i^2\right) = I_e - I_i \qquad (2\text{-}58)$$

where

$$M = \frac{u}{a} \qquad (2\text{-}59)$$

and

$$I = pA \left(1 + \gamma M^2\right) \qquad (2\text{-}60)$$

The latter expression is known in compressible fluid mechanics as the axial *impulse function*. The axial force imposed upon the flow between two axial stations is often most conveniently calculated as the difference between their respective impulse functions. The impulse function formulation emphasizes the fact that the ratio of the contribution of momentum flux to that of pressure forces is in the ratio γM^2. Thus, in high axial Mach number flows, the greatest part of the axial impulse function is due to momentum flux.

Next, dividing Eq. (2-57) through by Eq. (2-56), and using Eq. (2-38), we find that

$$\frac{F_{bx}}{\dot{m}} = u_e \left(1 + \frac{RT_e}{u_e^2}\right) - u_i \left(1 + \frac{RT_i}{u_i^2}\right) = Sa_e - Sa_i \qquad (2\text{-}61)$$

where

$$Sa = \frac{I}{\dot{m}} = u \left(1 + \frac{RT}{u^2}\right) \qquad (2\text{-}62)$$

The latter expression is known in compressible fluid mechanics as the *stream thrust function* (Sa originally standing for mass flow rate specific thrust of air). The stream thrust function is the parameter that most easily leads to the determination of mass flow rate specific thrust, a quantity often used in performance evaluations in order to eliminate dependence on absolute size.

Finally, dividing Eq. (2-62) by $\sqrt{C_p T_{ti}}$ and rearranging, we find that the *dimensionless stream thrust function* for *axial flow* (i.e., $u = V$)

$$\Phi = \frac{Sa}{\sqrt{C_p T_{ti}}} = \sqrt{\frac{1}{2}\left(\frac{2C_p T_{ti}}{V^2}\right)\left\{2\left(\frac{V^2}{2C_p T_{ti}}\right) + \frac{\gamma-1}{\gamma}\left(\frac{C_p T}{C_p T_{ti}}\right)\right\}}$$

(2-63)

Note that Φ depends upon the same dimensionless quantities that are found in Eq. (2-51). Equation (2-63) is portrayed in Fig. 2.13 for typical values of Φ and γ, where it can be seen that the curve is an inclined parabola passing through the origin. For a given Φ, the state of the flow *must* be somewhere on this curved line. For the *special* case of constant area flow with heat interaction but without friction this line is known in the classical literature as a *Rayleigh line*. However, the same relation applies to the *much more general* situations to be encountered here, so the line will *not* be referred to by that name. The breaks at the right end of the line are meant to remind us that it may not be possible to find *mean* values of the gas

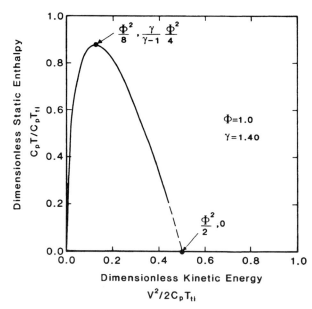

Fig. 2.13 **Dimensionless energy diagram for the constant stream thrust flow of a calorically perfect gas, showing the relationship between dimensionless static enthalpy and dimensionless kinetic energy, Eq. (2-63).**

constants valid over the entire range, although one may judiciously select the range for which they are suitably accurate.

Example Case 2.3: Normal Shock Waves

Normal shock waves are discontinuities in one-dimensional, constant throughflow area, axial flows that are subject to the three constraints of constant mass flow, constant energy, and constant stream thrust. The upstream and downstream or *jump* conditions for normal shock waves can therefore be obtained for calorically perfect gases by the simultaneous solution of Eqs. (2-51) and (2-63) for constant values of $C_p T_{ti}$ and Φ. A typical *graphical* solution is portrayed in Fig. 2.14, which combines Figs. 2.12 and 2.13.

Straightforward mathematical manipulations of Eqs. (2-51) and (2-63) yield the more familiar closed form, *algebraic* solutions for the jump conditions of normal shock waves. For example, the two dimensionless kinetic energy solutions are

$$\frac{V^2}{2C_p T_{ti}} = \left\{ \frac{\gamma\Phi}{(\gamma+1)\sqrt{2}} \pm \sqrt{\frac{1}{2}\left(\frac{\gamma\Phi}{\gamma+1}\right)^2 - \left(\frac{\gamma-1}{\gamma+1}\right)} \right\}^2 \qquad (2\text{-}64)$$

and the respective dimensionless static enthalpy solutions are

$$\frac{C_p T}{C_p T_{ti}} = 1 - \frac{V^2}{2C_p T_{ti}} \qquad (2\text{-}65)$$

In order to prove that these two solutions correspond to supersonic and subsonic flow, it is first necessary to observe that straight lines emanating from the origin of Fig. 2.14 are lines of constant Mach number, larger slopes belonging to lower Mach numbers. Next, it can be seen from the geometry of this diagram that there is only one value of Φ for which the constant stream thrust function line is tangent to the constant energy line *at point c* and for which there is one solution (rather than two or zero). This condition arises when the discriminant of Eq. (2-64) is zero, and consequently

$$\Phi_c = \sqrt{2\left(1 - \frac{1}{\gamma^2}\right)} \qquad (2\text{-}66)$$

($\Phi_c = 0.990$ for $\gamma = 1.40$), and

$$\frac{V_c^2}{2C_p T_{ti}} = \frac{\gamma-1}{\gamma+1} \qquad (2\text{-}67)$$

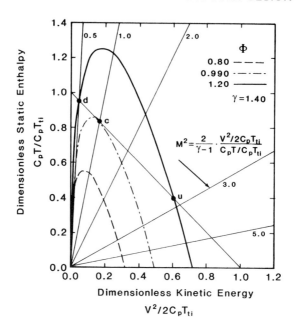

Fig. 2.14 Graphical solution for the jump conditions of normal shock waves based upon Eqs. (2-51) and (2-63). The straight lines emanating from the origin are lines of constant Mach number, increasing from left to right as indicated by their labels.

$(V^2/2C_pT_{ti} = 0.167$ for $\gamma = 1.40)$. Combining Eqs. (2-50) and (2-67) reveals that at point c

$$M_c = 1 \qquad (2\text{-}68)$$

Finally, returning to the geometry of Fig. 2.14, it follows that all values of Φ that have two solutions must have one that is subsonic and one that is supersonic. The normal shock therefore takes the flow from the upstream, supersonic condition (point u) to the downstream, subsonic condition (point d).

Before moving on, it is worth noting in anticipation of coming topics that the graphical presentation of Fig. 2.14 has the special virtue that constant values of such primal physical quantities as static enthalpy, static temperature, kinetic energy, velocity, Mach number, and functions of Mach number only all appear as straight lines.

Example Case 2.4: Constant Area Heating and Thermal Choking

Consider a frictionless, one-dimensional flow of constant area for which there is only a total heat interaction $\dot{Q}$ with the surroundings, i.e.,

$$F_{bx} = \dot{W} = \dot{m}_b = v = 0 \qquad A_i = A_e \qquad \dot{m}_i = \dot{m}_e = \dot{m} \qquad u = V$$

Under these circumstances, which correspond to classical Rayleigh flow, Eq. (2-28) becomes

$$\rho_i u_i = \rho_e u_e \qquad (2\text{-}69)$$

and Eq. (2-29) becomes

$$p_i + \rho_i u_i^2 = p_e + \rho_e u_e^2 \qquad (2\text{-}70)$$

which means that the stream thrust function is constant. Taking the working fluid to be calorically perfect, we must therefore reach the conclusion that

$$\Phi = \frac{Sa}{\sqrt{C_p T_{ti}}} = \Phi_i = \text{constant} \qquad (2\text{-}71)$$

where Φ_i (the *inlet* dimensionless stream thrust function) as given by Eq. (2-63) and $C_p T_{ti}$ (the *inlet* total enthalpy) may be evaluated from the known inlet conditions.

Equation (2-31) becomes

$$C_p T_i + \frac{V_i^2}{2} + \frac{\dot{Q}}{\dot{m}} = C_p T_{ti} \left(1 + \frac{\dot{Q}}{\dot{m} C_p T_{ti}} \right) = C_p T_e + \frac{V_e^2}{2} = C_p T_{te} \quad (2\text{-}72)$$

or

$$\frac{T_{te}}{T_{ti}} = \frac{C_p T_e}{C_p T_{ti}} + \frac{V_e^2}{2 C_p T_{ti}} = 1 + \frac{\dot{Q}}{\dot{m} C_p T_{ti}} \doteq \tau_e \geq 1 \qquad (2\text{-}73)$$

where τ_e is the *total temperature ratio* and the primary indicator of *dimensionless heating*. Equations (2-71) and (2-73) may be combined to find the *algebraic* solution for the two possible exit conditions, namely

$$\frac{V_e^2}{2 C_p T_{ti}} = \left\{ \frac{\gamma \Phi_i}{(\gamma + 1)\sqrt{2}} \pm \sqrt{ \frac{1}{2} \left(\frac{\gamma \Phi_i}{\gamma + 1} \right)^2 - \tau_e \left(\frac{\gamma - 1}{\gamma + 1} \right) } \right\}^2 \qquad (2\text{-}74)$$

and

$$\frac{C_p T_e}{C_p T_{ti}} = \tau_e - \frac{V_e^2}{2 C_p T_{ti}} \qquad (2\text{-}75)$$

Equations (2-71) and (2-73) can also be solved *graphically*, as shown in Fig. 2.15, where the positive root of Eq. (2-74) belongs to supersonic inlet conditions and the negative root to subsonic inlet conditions. Since τ_e may be treated as an independent variable, Fig.

2.15 reveals that heating the flow drives it toward an exit Mach number of 1 regardless of whether it was initially subsonic or supersonic. Thus, in the typical example of Fig. 2.15, a τ_e of 1.20 reduces the supersonic branch Mach number from an inlet value of 2.74 to an exit value of 1.89, and increases the subsonic branch Mach number from 0.493 to 0.598.

The remarks about the tangent point solution made in the previous Example Case also apply here, in particular that point c is the sonic condition, at which the discriminant of Eq. (2-74) must be zero, or

$$\tau_c = \frac{\Phi_i^2}{2\left(1 - \dfrac{1}{\gamma^2}\right)} \tag{2-76}$$

($\tau_c = 1.47$ for $\Phi_i = 1.20$ and $\gamma = 1.40$), and

$$\frac{V_c^2}{2C_pT_{ti}} = \frac{1}{2}\left(\frac{\gamma\Phi_i}{\gamma+1}\right)^2 \tag{2-77}$$

($V_c^2/2C_pT_{ti} = 0.245$ for $\Phi_i = 1.20$ and $\gamma = 1.40$).

Most importantly, Fig. 2.15 clearly demonstrates that for any inlet condition there is a maximum amount of energy that can be added before the exit Mach number reaches 1. There are *no* solutions to the governing equations for $\tau_e > \tau_c$. Once that point has been reached, it is physically impossible to add any more energy without forcing the upstream conditions to change, and the flow is said to be *thermally choked*. The conditions necessary for thermal choking can and do occur in practice, especially when the inlet Mach number is not far from 1, and should not be considered merely as an intellectual curiosity.

Finally, it is also possible to show that the entropy always increases as the flow is heated. This is best done by substituting the above results into the Gibbs equation and rearranging in order to show that[2.3, 2.4]

$$\frac{s_e - s_i}{C_p} = \ell n\left\{\left(\frac{M_e}{M_i}\right)^2\left(\frac{1 + \gamma M_i^2}{1 + \gamma M_e^2}\right)^{\frac{\gamma+1}{\gamma}}\right\} \tag{2-78}$$

Since the grouping

$$\frac{M^2}{(1 + \gamma M^2)^{\frac{\gamma+1}{\gamma}}}$$

has a maximum when M is 1, and since heating always drives M toward 1, it follows that heating is always accompanied by increased entropy.

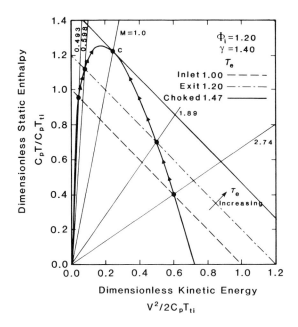

Fig. 2.15 Graphical solution for the frictionless, constant area heating based upon Eqs. (2-71) and (2-73). The straight lines emanating from the origin are lines of constant Mach number, increasing from left to right as indicated by their labels.

Example Case 2.5: Ideal Exit Velocity and Mass Flow Parameter

Consider the one-dimensional, isentropic flow of a calorically perfect gas for which there are no interactions with the surroundings and the transverse velocity is negligible, i.e.,

$$\dot{W} = \dot{Q} = \dot{m}_b = v = 0 \qquad u = V$$

According to Example Case 2.1, under these conditions there is no change of total temperature or total pressure from inlet to exit. From the former we find that

$$V_e^2 = 2C_p T_{ti}\left(1 - \frac{T_e}{T_{ti}}\right) \tag{2-79}$$

and from the latter

$$\frac{T_e}{T_{ti}} = \left(\frac{p_e}{p_{ti}}\right)^{\frac{\gamma-1}{\gamma}} \tag{2-80}$$

so that the *ideal exit velocity* can be evaluated for any given value of exit pressure p_e from

$$V_e = \sqrt{2C_p T_{ti} \left\{ 1 - \left(\frac{p_e}{p_{ti}} \right)^{\frac{\gamma-1}{\gamma}} \right\}} \qquad (2\text{-}81)$$

This expression will be very familiar to anyone who has dealt with rocket nozzle exhaust conditions, where the total conditions are those of the combustion chamber. A standard measure of performance is the ideal *vacuum* exhaust velocity, or that which would occur at altitudes for which $(p_e/p_{ti})^{(\gamma-1)/\gamma}$ is much less than 1. Equation (2-81) shows that the ideal vacuum exhaust velocity is equal to $\sqrt{2C_p T_{ti}}$.

The exit mass flow per unit area is given by the expression

$$\frac{\dot{m}_e}{A_e} = \rho_e V_e = \frac{p_e V_e}{R T_e} = M_e \left(\frac{p_e}{p_{ti}} \right) \sqrt{\frac{T_{ti}}{T_e}} \cdot p_{ti} \sqrt{\frac{\gamma}{R T_{ti}}} \qquad (2\text{-}82)$$

where Eqs. (2-38) and (2-42) have been used. This equation can, with the help of Eqs. (2-50) and (2-54), be rearranged into the grouping known as the *mass flow parameter*

$$MFP_e = \frac{\dot{m}_e \sqrt{T_{ti}}}{p_{ti} A_e} = \sqrt{\frac{\gamma}{R}} \cdot M_e \left(1 + \frac{\gamma-1}{2} M_e^2 \right)^{-\frac{\gamma+1}{2(\gamma-1)}} \qquad (2\text{-}83)$$

which greatly facilitates the application of the conservation of mass equation to a wide variety of situations. The mass flow parameter depends only upon Mach number and gas properties and, as Fig. 2.16 shows, has a maximum at a Mach number of 1. This result is a staple of the compressible flow literature, and means that the highest possible mass flow per unit area occurs when the flow velocity equals the local speed of sound, where the flow is said to be *choked*.

Equation (2-83) can be applied at any axial station by using the *local* values of total temperature and total pressure. It is therefore frequently used either to determine the throughflow area required for a given Mach number directly, or to determine the Mach number existing at a given throughflow area by iteration or by using computerized root-finding methods.

2.6.3 General Differential Control Volume Analysis

Referring to Fig. 2.17, the same observations can be made for *differential control volume* analysis as for the finite control volume analysis of Sec. 2.6.2. In fact, the entire analysis follows the same lines, with some minor exceptions, the main of which is that it is assumed that transverse velocities are negligible at the inlet and exit of each differential control volume element (i.e., $u = V$). This approximation is justified by the fact that the differential analysis is usually applied to flows that are completely confined by essentially axial ducts. You will also notice that the multiple mass injection has been replaced

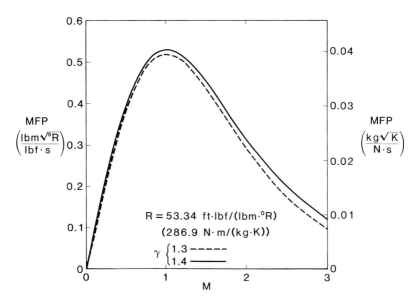

Fig. 2.16 Mass flow parameter as a function of Mach number for a calorically perfect gas with the molecular weight of air and two typical values of the ratio of specific heats.

by a single injection flow that represents the portion attributable to the differential control volume under consideration. Finally, the designations $\bar{d}\dot{W}$ and $\bar{d}\dot{Q}$ are used to remind the reader that W and Q are not properties of the surroundings, and hence the infinitesimal increments cannot be integrated to give the changes in W and Q.

The conservation equations that follow have been arranged with the traditionally dependent flow property terms on the left-hand side and the traditionally independent source terms on the right-hand side.

2.6.3.1 Conservation of mass (continuity). The equation of conservation of mass for the differential control volume of Fig. 2.17 is

$$\frac{d\rho}{\rho} + \frac{dV}{V} = \frac{d\dot{m}_b}{\dot{m}} - \frac{dA}{A} \tag{2-84}$$

2.6.3.2 Conservation of momentum (Newton's principle). Applying Newton's law and conservation of mass to the differential control volume of Fig. 2.17 yields this equation for the axial direction

$$\frac{dp}{p} + \frac{\rho V^2}{p} \cdot \frac{dV}{V} = \frac{(u_b - V)\, d\dot{m}_b + dF_{bx}}{pA} \tag{2-85}$$

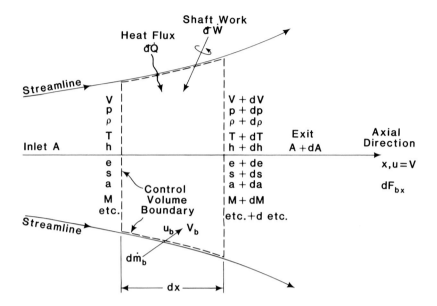

Fig. 2.17 Geometry and nomenclature for differential control volume one-dimensional flow analysis.

where dF_{bx} signifies only the contribution of the wall shear stress and the drag of protuberances to the differential axial force. The contribution of the normal (pressure) force on the wall to the differential axial force has already been included in Eq. (2-85).

2.6.3.3 Conservation of energy (first law). Applying the first law of thermodynamics and conservation of mass to the differential control volume of Fig. 2.17 yields

$$\frac{dh}{h} + \frac{V^2}{h} \cdot \frac{dV}{V} = \frac{\left(h_b - h + \dfrac{V_b^2 - V^2}{2}\right) d\dot{m}_b + đ\dot{W} + đ\dot{Q}}{\dot{m}h} \qquad (2-86)$$

2.6.3.4 General solution. Recognizing that the equations of state (Sec. 2.6.2.4) and the Gibbs equation (Sec. 2.6.2.5) also apply to the differential control volume analysis, we can see that a closed set of governing equations has been gathered. Returning to Fig. 2.17, the typical problem to be solved begins with all information given except the properties of the flow at the exit. Thus, there are normally nine unknowns, which can be solved for by the nine equations (2-84) through (2-86) and (2-32) through (2-37).

Once the changes across a differential control volume can be solved for, it follows that they can be integrated in the axial direction in order to find the changes across a finite control volume. One may therefore very well contemplate the relationship between the finite and differential control volume approaches, and wonder what is to be gained by the additional effort that must accompany the latter. The answer to this implied question is that the differential control volume analysis requires *less input information* and provides a *finer grain of output information*. In particular, the differential axial momentum equation already contains the effects of the pressure forces normal to the control volume boundary, and the integrated results expose the process that must take place in order to get from the beginning to the desired end, oftentimes placing precise demands on the internal details of designs.

2.6.3.5 Calorically perfect gas. When the gas can be treated as calorically perfect so that Eqs. (2-38) through (2-43) apply, the set of governing equations is transformed into a particularly functional and comprehensible set of statements about the behavior of the flow. The four central differential equations are assembled below in their most transparent arrangement.

$$\frac{d\rho}{\rho} + \frac{dV}{V} = \frac{d\dot{m}_b}{\dot{m}} - \frac{dA}{A} \qquad\qquad \text{Mass} \qquad (2\text{-}87)$$

$$\frac{dp}{p} + \gamma M^2 \frac{dV}{V} = \frac{(u_b - V)\, d\dot{m}_b + dF_{bx}}{pA} \qquad \text{Momentum} \quad (2\text{-}88)$$

$$\frac{dT}{T} + (\gamma - 1)M^2 \frac{dV}{V} = \frac{dT_t}{T} \qquad\qquad \text{Energy} \qquad (2\text{-}89)$$

$$= \frac{\left(h_b - h + \dfrac{V_b^2 - V^2}{2}\right) d\dot{m}_b + đ\dot{W} + đ\dot{Q}}{\dot{m}C_p T}$$

$$\frac{dp}{p} - \frac{d\rho}{\rho} - \frac{dT}{T} = 0 \qquad\qquad \text{Perfect gas} \quad (2\text{-}90)$$

In most situations, and especially the usual case for which the right-hand side forcing functions are known, Eqs. (2-87) through (2-90) constitute a closed set that can be directly solved for the differential changes of and the new values of p, ρ, T, and V. Once this has been done, Eqs. (2-39) through (2-43) allow the new values of h, e, s, a, and M to be found. The results may then be used either to follow the flow via integration as it proceeds from inlet to exit, or to determine the sensitivity of the local behavior to imposed conditions.

Several auxiliary relationships have frequently been found to be helpful in the analysis and understanding of these flows and they are compiled below for convenient application.

$$\frac{dM}{M} = \frac{dV}{V} - \frac{1}{2}\frac{dT}{T} \qquad \text{Mach number} \qquad (2\text{-}91)$$

$$\frac{dT_t}{T_t} = \frac{dT}{T} + \frac{(\gamma - 1)M\,dM}{\left(1 + \dfrac{\gamma - 1}{2}M^2\right)} \qquad \text{Total temperature} \qquad (2\text{-}92)$$

Please note that Eq. (2-92) is Eq. (2-89) stated in different terms.

$$\frac{dp_t}{p_t} = \frac{dp}{p} + \frac{\gamma M\,dM}{\left(1 + \dfrac{\gamma - 1}{2}M^2\right)} \qquad \text{Total pressure} \qquad (2\text{-}93)$$

Before moving on to their solution, we should point out that the form of the governing differential equations for calorically perfect gases has been strongly shaped by the experiences and traditions of compressible fluid mechanics. For one thing, the logarithmic differential form (i.e., dp/p, dV/V, etc.) immediately reveals the *relative* rate of change of the property in question. For another, the equation set is quite amenable to numerical integration and even yields closed form algebraic solutions for some very important types of flow. Finally, stating the coefficients of the logarithmic differentials in terms of the local Mach number clearly displays the various personalities of subsonic, transonic, and supersonic flows, and exposes mathematical singularities that control the fundamental nature of the flow.

2.6.3.6 Examples of differential control volume analysis. The four Example Cases that follow are based upon the steady, one-dimensional flow of a calorically perfect gas as shown in Fig. 2.17. They illustrate some very important phenomena of compressible flows, illuminate the special value of differential control volume analysis, and, in several instances, validate the results of the finite control volume analysis.

The reader should be aware that the Example Cases are also found in the subroutines of HAP(Gas Tables).

Example Case 2.6: Total Enthalpy, Total Temperature, and Total Pressure

Consider a one-dimensional flow without energy interactions with the surroundings, i.e.,

$$d\dot{W} = d\dot{Q} - d\dot{m}_b = 0$$

Under these conditions, the energy equation, Eq. (2-86), reduces to

$$d\left(h + \frac{V^2}{2}\right) = 0 \qquad (2\text{-}94)$$

or

$$h + \frac{V^2}{2} = \text{constant} = h_t \qquad (2\text{-}95)$$

where, as in Example Case 2.1, h_t is the total or stagnation enthalpy that occurs when the flow is brought to rest.

In the regime where the fluid behaves as a calorically perfect gas, Eq. (2-89) immediately reveals that

$$dT_t = 0 \qquad (2\text{-}96)$$

or

$$T_t = \text{constant} \qquad (2\text{-}97)$$

Equation (2-41) may now be integrated to show that

$$\frac{p_{te}}{p_{ti}} = e^{-(s_e - s_i)/R} \qquad (2\text{-}98)$$

where the total pressure has again been defined as

$$p_t = p\left(1 + \frac{\gamma - 1}{2}M^2\right)^{\frac{\gamma}{\gamma - 1}} \qquad (2\text{-}99)$$

The flow will in general not be isentropic, however, because the differential control volume boundary friction and drag force dF_{bx} can generate entropy. In this case, fortunately, the differential control volume approach allows the entropy increase to be evaluated. Combining Eqs. (2-41), (2-88), and (2-97) yields

$$\frac{ds}{R} = -\frac{dF_{bx}}{pA} \qquad (2\text{-}100)$$

so that

$$\frac{s_e - s_i}{R} = -\int_i^e \frac{dF_{bx}}{pA} \qquad (2\text{-}101)$$

which can, in principle, be integrated from inlet to any desired exit location and used to determine the total pressure from Eq. (2-98). Please note that dF_{bx} caused by friction or drag is less than zero.

Example Case 2.7: Constant Area Heating and Thermal Choking

Consider a frictionless flow of constant area for which there is only a heat interaction with the surroundings, i.e.,

$$d\dot{W} = dF_{bx} = d\dot{m}_b = dA = 0$$

This classical Example Case of Rayleigh flow illustrates the general method of solution of Eqs. (2-87) through (2-90). The first three are solved for differential velocity change and become, respectively,

$$\frac{d\rho}{\rho} = -\frac{dV}{V} \qquad (2\text{-}102)$$

$$\frac{dp}{p} = -\gamma M^2 \frac{dV}{V} \qquad (2\text{-}103)$$

$$\frac{dT}{T} = \frac{dT_t}{T} - (\gamma - 1)M^2 \frac{dV}{V} \qquad \left(\frac{dT_t}{T_t} = \frac{d\dot{Q}}{\dot{m}C_pT_t} \right) \quad (2\text{-}104)$$

Substituting these into the fourth gives

$$\frac{dV}{V} = \left(\frac{1 + \dfrac{\gamma - 1}{2}M^2}{1 - M^2} \right) \frac{dT_t}{T_t} \qquad (2\text{-}105)$$

This, in turn, is substituted into Eq. (2-91) in order to obtain an equation that can be directly integrated to find the relationship between Mach number and heat (or total temperature) added, namely

$$\frac{dM}{M} = \frac{(1 + \gamma M^2)\left(1 + \dfrac{\gamma - 1}{2}M^2\right)}{2(1 - M^2)} \cdot \frac{dT_t}{T_t} \qquad (2\text{-}106)$$

The result of a finite amount of heating would, of course, be the same as that of Example Case 2.4, but the differential equation contains much useful information in its raw form. For example, it shows that heating always drives the Mach number toward 1 and thermal choking, but never through 1 because of the singularity in the denominator. Moreover, the numerator shows that the slope (the rate at which the Mach number approaches 1) is much larger for supersonic conditions than for subsonic. The direct connection between this and Example Case 2.4 is established by combining

$$\frac{d(C_pT)}{C_pT_t} = \left(\frac{1 - \gamma M^2}{1 - M^2}\right)\frac{dT_t}{T_t}$$

derived from Eqs. (2-104) and (2-105) with

$$\frac{d\left(\frac{V^2}{2}\right)}{C_pT_t} = \left\{\frac{(\gamma - 1)M^2}{1 - M^2}\right\}\frac{dT_t}{T_t}$$

derived from Eq. (2-105) in order to obtain

$$\frac{d(C_pT)}{C_pT_t} = \left\{\frac{1 - \gamma M^2}{(\gamma - 1)M^2}\right\}\frac{d(\frac{V^2}{2})}{C_pT_t}$$

Substituting numerical values from Fig. 2.15 will confirm that this is indeed the slope at any point along the Rayleigh line.

A remarkably simple result is now obtained from Eqs. (2-93), (2-103), (2-105), and (2-106), namely

$$\frac{dp_t}{p_t} = -\frac{\gamma M^2}{2}\cdot\frac{dT_t}{T_t} \tag{2-107}$$

which means that the total pressure *always* decreases due to constant area heating, all the more so as Mach number increases. This decrease of total pressure is commonly known as the *Rayleigh heating loss,* and is an inevitable fact of thermodynamic life. In order to limit the Rayleigh heating loss in the combustors of conventional aircraft gas turbine engines, the entry Mach number is reduced by increasing the throughflow area as much as possible. The result is a typical combustor entry Mach number of less than 0.05 which, when taken together with a typical total temperature doubling, causes less than a 0.2 percent reduction in total pressure. This luxury is not available to scramjets.

Example Case 2.8: Constant Pressure Heating

Consider a frictionless flow for which there is only a heat interaction with the surroundings and the throughflow area is deliberately varied in such a way that static pressure is held constant, i.e.,

$$đW = dF_{bx} = dṁ_b = dp = 0$$

This is an especially interesting case because of the contrast it offers with the constant area heating of Example Case 2.7 and because it could be more advantageous in practice. From the analytical stand-point, it is special because two of the principal flow quantities (p and V) will be found to remain constant and because the most important results are obtained in closed form.

Under these conditions, Eq. (2-88) becomes

$$dV = 0 \qquad (2\text{-}108)$$

This carries the important implication that a line of constant veloc-ity (or kinetic energy) *also represents a line of frictionless constant pressure heating* on a diagram such as Fig. 2.15. Furthermore, Eq. (2-89) becomes

$$dT = dT_t \qquad (2\text{-}109)$$

and Eq. (2-91) becomes

$$\frac{dT}{T} = -2\frac{dM}{M} \qquad (2\text{-}110)$$

When this is substituted into Eq. (2-92), a differential equation link-ing total temperature to Mach number is found, namely

$$\frac{dM}{M} = -\frac{1}{2}\left(1 + \frac{\gamma - 1}{2}M^2\right)\frac{dT_t}{T_t} \qquad (2\text{-}111)$$

which can be integrated to yield

$$\frac{M_e}{M_i} = \frac{1}{\sqrt{\tau_e\left(1 + \dfrac{\gamma - 1}{2}M_i^2\right) - \dfrac{\gamma - 1}{2}M_i^2}} \qquad (2\text{-}112)$$

where $\tau_e = T_{te}/T_{ti}$. Figure 2.18 contains the outcome of calculations based upon this relationship. As you can see, the Mach number decreases steadily with heating or energy addition (total temperature increase) and passes continuously through the sonic condition *without difficulty.* The slope of the curves is only a function of local Mach number and increases with local Mach number.

Since

$$\pi = \frac{p_{te}}{p_{ti}} = \left(\frac{1 + \dfrac{\gamma - 1}{2}M_e^2}{1 + \dfrac{\gamma - 1}{2}M_i^2}\right)^{\frac{\gamma}{\gamma - 1}} \qquad (2\text{-}113)$$

then incorporation of Eq. (2-112) leads to

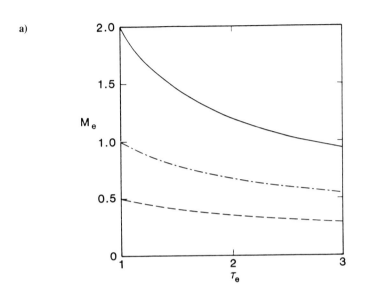

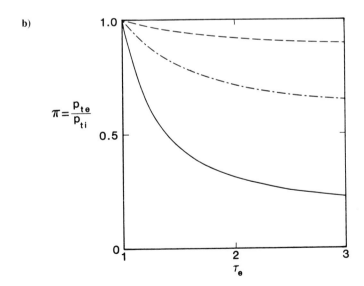

Fig. 2.18 Local Mach number, total pressure ratio, and area ratio as functions of total temperature ratio and inlet Mach number for constant pressure heating.

c)

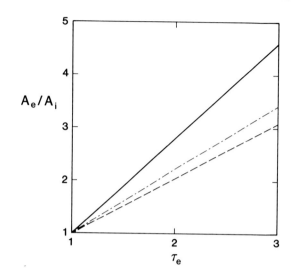

Fig. 2.18 (continued) **Local Mach number, total pressure ratio, and area ratio as functions of total temperature ratio and inlet Mach number for constant pressure heating.**

$$\pi = \frac{1}{\left\{1 + \frac{\gamma - 1}{2} M_i^2 \left(1 - \frac{1}{\tau_e}\right)\right\}^{\frac{\gamma}{\gamma - 1}}} \qquad (2\text{-}114)$$

Figure 2.18 also contains calculated results based upon this equation. The total pressure again decreases with increased heating (or total temperature), and more rapidly as the inlet Mach number is increased. This is another manifestation of the Rayleigh heating loss, and Eqs. (2-93) and (2-111) lead to

$$\frac{dp_t}{p_t} = -\frac{\gamma M^2}{2} \cdot \frac{dT_t}{T_t} \qquad (2\text{-}115)$$

which is the same as for constant area (or any other type of) heating. It should also be noted that Eq. (2-113) shows that the total pressure can never fall below

$$p_{te} = \frac{p_{ti}}{\left(1 + \frac{\gamma - 1}{2} M_i^2\right)^{\frac{\gamma}{\gamma - 1}}} = p_i \qquad (2\text{-}116)$$

which corresponds to an exit Mach number of zero.

Finally, the variation of area necessary to maintain constant pressure can be found by combining Eqs. (2-87) and (2-90) into

$$\frac{dA}{A} = \frac{dT}{T} \tag{2-117}$$

or

$$\frac{A_e}{A_i} = \frac{T_e}{T_i} \tag{2-118}$$

Since V is constant, then

$$\frac{A_e}{A_i} = \frac{\gamma R T_e}{V_e^2} \cdot \frac{V_i^2}{\gamma R T_i} = \left(\frac{M_i}{M_e}\right)^2 \tag{2-119}$$

or, using Eq. (2-112),

$$\frac{A_e}{A_i} = \tau_e \left(1 + \frac{\gamma-1}{2}M_i^2\right) - \frac{\gamma-1}{2}M_i^2 \tag{2-120}$$

which shows that the throughflow area increases with total temperature and inlet Mach number. The linear dependence of throughflow area on total temperature is shown in Fig. 2.18, where it can also be seen that rather large increases of throughflow area may be required.

Example Case 2.9: Choking with Friction and Area Variation

Consider a flow that has no energy interactions with the surroundings, i.e.,

$$đ\dot{W} = đ\dot{Q} = d\dot{m}_b = 0$$

Under these conditions, Eqs. (2-87) through (2-91) become, respectively,

$$\frac{d\rho}{\rho} = -\frac{dV}{V} - \frac{dA}{A} \tag{2-121}$$

$$\frac{dp}{p} = -\gamma M^2 \frac{dV}{V} + \frac{dF_{bx}}{pA} \tag{2-122}$$

$$\frac{dT}{T} = -(\gamma-1)M^2 \frac{dV}{V} \tag{2-123}$$

$$\frac{dV}{V} = \frac{\dfrac{dF_{bx}}{pA} + \dfrac{dA}{A}}{M^2 - 1} \tag{2-124}$$

$$\frac{dM}{M} = \left(\frac{1 + \dfrac{\gamma-1}{2}M^2}{M^2 - 1}\right)\left(\frac{dF_{bx}}{pA} + \frac{dA}{A}\right) \tag{2-125}$$

The final result shows that continuous passage through the sonic point when friction is present does not occur at constant area, but, since $dF_{bx} < 0$, requires that

$$\frac{dA}{A} = -\frac{dF_{bx}}{pA} > 0 \qquad \text{at } M = 1 \qquad (2\text{-}126)$$

Moreover, when the throughflow area is constant, it shows that frictional forces always drive Mach number toward 1, a standard result for classical Fanno line behavior.

2.6.4 The H-K Diagram

The great utility of graphically displaying flow processes in terms of dimensionless static enthalpy versus dimensionless kinetic energy has been amply demonstrated by the elementary example cases already considered. This method of presentation will prove even more valuable in explaining and illustrating the more complex internal flow behavior of ramjet and scramjet engines. In order to ease communications, this diagram will hereinafter simply be called *the H-K diagram* (H for dimensionless static enthalpy and K for dimensionless kinetic energy).

Please note that the H-K diagram, for all its other virtues, is *not a state diagram* because only one axis is an intensive thermodynamic property. In other words, there is no necessary relationship between a point on the H-K diagram and the other intensive thermodynamic properties of the fluid, such as static pressure or static entropy. Nevertheless, the H-K diagram will provide more than enough information to reveal the things we really need to know about the flow. Also, under some frequently encountered conditions, such as one-dimensional flow with known $\dot{m}$, A, and T_{ti}, the H-K diagram *is a* state diagram.

An especially useful generalization for moving about the H-K diagram, based upon the assemblage of preceding example cases, is that heating, friction, and area decrease *all* act separately and together to drive the Mach number toward 1. They may therefore in an intuitive sense be said to *block, constrict, obstruct, restrict, or occlude* the flow. Similarly, cooling, reverse friction (i.e., any streamwise force), and area increase could be said to *unblock, enlarge, relieve, open, or free* the flow. You will find these mental images helpful in what follows.

2.6.5 Aerothermodynamics of Scramjets and Ramjets

We have now developed the background necessary to easily visualize and comprehend the general operation of scramjet and ramjet engines. For simplicity, it has been assumed that the air behaves as a calorically perfect gas with $\gamma = 1.40$ and that the mass flow remains constant. The discussion will center on Figs. 2.19 and 2.20, which show the succession of states of the air on the H-K diagram using T_{t0} as the reference total temperature. The dimensionless stream thrust function at a given Mach number is not some arbitrary value but is *specified* by Eq. (2-63) as rearranged into the Mach number form

$$\Phi = \sqrt{\frac{\tau_e(\gamma - 1)M^2}{1 + \frac{\gamma - 1}{2}M^2}} \left(1 + \frac{1}{\gamma M^2}\right) \qquad (2\text{-}127)$$

Please note that a general characteristic of Figs. 2.19 and 2.20 is that decelerating (i.e., reducing the velocity or kinetic energy of) a supersonic flow at constant total temperature (i.e., constant energy) always reduces the value of the dimensionless stream thrust function Φ, which is equivalent to there being a net axial force directed against the flow. Similarly, decelerating a subsonic flow or accelerating a supersonic flow always increases Φ, while accelerating a subsonic flow always decreases Φ. These are consequences of the previously proven fact that the point of tangency occurs at $M = 1.0$. You should be able to reconcile these stream thrust observations with the customary versions of how inlets and nozzles work.

Figure 2.19 shows the H-K diagram for the air being processed by a *scramjet* with $\tau_e = 1.40$ that is powering a vehicle at a freestream Mach number of 10.0, where $\Phi_0 = 1.390$. The air is first decelerated and *compressed* from the freestream condition (point 0) to the burner entry condition (point 1) by means of a combination of isentropic compression and oblique shock waves. The purposes of this compression are to provide a large enough static temperature ratio T_1/T_0 for satisfactory thermodynamic cycle efficiency (usually in the range of 6–8 and 6.50 for this example) and to produce high enough values of p_1 and T_1 to support complete and stable combustion in the burner. Even when these criteria have been met in hypersonic flight, the burner entry Mach number remains supersonic, as Fig. 2.19 indicates.

The air is then *heated* in a combustion process that releases the chemical energy of the fuel. The heating is represented in this type of analysis by an increasing total temperature, in this case by a factor of 1.40. The precise path of this process depends upon the philosophy of the burner design, and two of many possible different types are depicted in Fig. 2.19. The first, joining point 1 to point 2, is frictionless, constant area heating, which *is* a Rayleigh line of constant Φ. The second, joining point 1 to point 3, is frictionless, constant pressure heating, which was found in Example Case 2.8 to be a line of constant velocity. There is clearly no danger of reaching point c and thermal choking for the constant area combustor in this scenario.

The heated air is then accelerated and *expanded* from a burner exit condition such as point 2 or 3 to the freestream static pressure at point 4. Because there are total pressure losses in the scramjet, the Mach number at point 4 can never be quite as large as the freestream Mach number [see Eq. (2-54)], but it is large enough that the kinetic energy and velocity at point 4 exceed that of point 0, which means that the scramjet produces net thrust. As a corollary, the total pressure losses and therefore the precise location of point 4 also depend upon the type of burner design.

Figure 2.20 shows the H-K diagram for the air being processed by a *ramjet* with $\tau_e = 1.40$ that is powering a vehicle at a freestream Mach number of 3.0, where $\Phi_0 = 1.224$. The air must be first decelerated and *compressed* from the freestream condition (point 0) almost to the stagnation condition in order to meet the burner entry criteria at point 1 as described for the scramjet. In fact, Eq. (2-52) reveals that the largest possible T_1/T_0 is

$$\frac{T_1}{T_0} = \frac{T_{t0}}{T_0} = 1 + \frac{\gamma - 1}{2}M_0^2 \qquad (2\text{-}128)$$

which is 2.80 for $M_0 = 3.0$. Experience shows that the compression is best accomplished by a series of processes, including a combination of isentropic compression and oblique shock waves from point 0

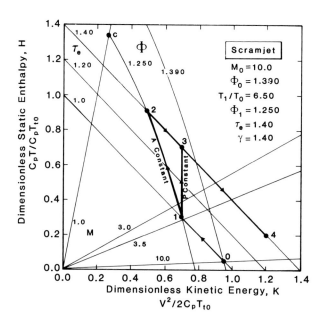

Fig. 2.19 The *H-K* diagram for a scramjet.

to point u, a normal shock wave from point u to point d, and subsonic diffusion from point d to point 1. The normal shock wave is ordinarily produced by a convergent-divergent diffuser that is capable of providing a range of shock Mach numbers depending upon the position of the normal shock wave in the divergent portion of the duct.

The combustion *heating* of the air is again portrayed as being either frictionless, constant area (point 1 to point 2) or frictionless, constant pressure (point 1 to point 3). Since the Mach number is very small in either case, points 2 and 3 lie close together. Please note that a constant area process starting from point d would have reached point c and thermal choking long before the desired amount of energy could have been added. Of course, further deceleration can always bring the burner entry Mach number far enough down and Φ far enough up that the desired energy can be added without thermal choking, just as it is in turbojet combustors.

The heated air is then accelerated and *expanded* from a burner exit condition such as point 2 or 3 to the freestream static pressure at point 4. The passage of the constant total temperature line through Mach 1 in Fig. 2.20 shows that a convergent-divergent nozzle having a choked throat is necessary for this process (see Example Case 2.5). The throughflow area (or blockage) of the throat ultimately determines the strength of the normal shock wave (i.e., the value of Φ at which the normal shock takes place) and therefore also the overall

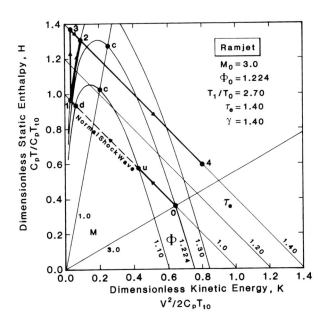

Fig. 2.20 The *H-K* diagram for a ramjet.

total pressure loss of the flow through the ramjet. Finally, the kinetic energy and velocity at point 4 exceed those of point 0, which means that the ramjet produces net thrust.

2.7 COMPUTATIONAL FLUID DYNAMICS

To instill a sound intuitive understanding of hypersonic airbreathing propulsion systems, the conceptual model of one-dimensional flow is used throughout this textbook. This model by itself does not provide the detail necessary for designing and analyzing such systems. A realistic model of three-dimensional flow is necessary. The governing aerothermodynamics equations for this flow are also derived by the control volume analysis leading to integral equations, which can be transformed into partial differential equations. These equations have been presented in the companion textbook, *Hypersonic Aerothermodynamics*.[2.1] They are enormously complicated, because they are highly nonlinear with coupled physical and chemical phenomena and because they represent phenomena with large variations in spatial and temporal scales. Because of these complexities, computational fluid dynamics (CFD) is the only available technology for solving these equations. How the fluid dynamics are computed starting from these equations is presented in textbooks such as *Numerical Computation of Internal and External Flows*[2.12] and *Computational Fluid Mechanics and Heat Transfer*.[2.13] This possibility of simulating fluid dynamics makes CFD one of the most powerful weapons for understanding phenomena and for designing and analyzing realistic propulsion systems. The objective of this section is to explain why CFD is such a powerful tool, to present caveats about this tool, and to assert that the credibility of a design is no better than the credibility of the tool used for developing that design.

As the phrase indicates, *computational fluid dynamics* encompasses two disciplines, computation and fluid dynamics. Together, these disciplines are used to numerically simulate the real fluid dynamics through modeling. The CFD technology is critical in efforts to lower the development costs and to improve and extend the performance and effectiveness of flight vehicles, of airbreathing engines, of ocean vehicles, of parachutes, and of manufacturing processes involving fluid flows. This technology is necessary for operating wind-tunnel test facilities more efficiently and for enhancing flight-test operations and flight safety during flight vehicle development programs. Furthermore, this technology can revolutionize the design and analysis processes, when fluid dynamics is coupled with other disciplines such as electromagnetics, optics, structures, and flight dynamics.

In the 1970's the Space Shuttle was built largely with off-the-shelf technology, utilizing the classic approach of flight vehicle development. Two avenues were used in the design process: compu-

tation and measurement. The computation was based on simplifying assumptions about the physics, the governing equations, and
the shapes. The design and analysis tools were developed based on
measured data. Computational fluid dynamics was not one of these
tools. Only after the design was frozen was CFD used to analyze the
Shuttle flowfields. That the classic approach did not do a good job
of predicting the aerodynamic characteristics became evident during
the Space Shuttle flight program.[2.1, 2.14] For example, the Orbiter
experienced nose-up pitching moments during entry at hyperveloci-
ties that exceeded the established limits.[2.15] Because the Shuttle was
designed using tolerances (based on measurement scatters) and variations (from model to full-scale) in aerodynamic coefficients, it was
possible to tolerate very large discrepancies in preflight aerodynamic
data.[2.16]

A hypersonic airbreathing propulsion system extends from the
nose of the flight vehicle to the tail through the engine. When
nonequilibrium real-gas processes arise, not only the fluid properties but also full-scale size are necessary for complete simulations.
The usefulness of ground-based facilities in determining the fluid
dynamics at flight conditions is limited, because of the following
reasons.[2.17-2.19] First, there are fundamental difficulties in creating complete simulations in ground-based facilities. (These limitations are discussed in Sec. 9.8.) Second, details of the upper atmosphere such as local composition, temperature, and turbulence are
not known sufficiently to properly establish the relationship between
flight conditions and ground-based data. Moreover, flight data are
not available for developing tools based on simplified assumptions
about the physics, the governing equations, and the shapes. The design of hypersonic propulsion systems, therefore, has to depart from
the classic approach by making much more extensive use of CFD as
a design and analysis tool.

2.7.1 The Role of CFD in Design and Analysis

Computational fluid dynamics is used in design and analysis under
the following conditions. The design specifications of a fluid dynamics system are essentially determined by the performance quantities
and, to some extent, by the global flowfields. The performance estimates are required for predicting the operational performance of the
system, making design evaluations, determining design sensitivities
and optimization, and establishing a design data base. Increments
in performance estimates are required for design tradeoff studies.
The global flowfields are useful for understanding the fluid dynamics
and also for making tradeoff studies. Often, results based on complex CFD methods are used to develop or calibrate simple methods.
Furthermore, computations enhance the credibility and usefulness of

ground-based and flight tests conducted for design and analysis, and they reduce costs for conducting these tests.

Prefabrication (that is, before a test model is fabricated) computations can provide a sanity check of the proposed test program. These computations can help determine appropriate instruments, the required measurement precision, and their proper placement. They can help determine the type, quality, and quantity of test data necessary for code certification. The qualitative and possibly quantitative design of model support and other devices and their interference effects can be obtained with CFD. Pretest computations can make the experimental test matrix more relevant and less of a "fishing expedition." Furthermore, computed results can help to fill gaps in the test data base. This augmented data base is useful, for example, when quantities are to be integrated over some domain in which sufficient test data are not available.

If CFD-design technology is not available for a class of fluid dynamics systems, this technology is developed. The development of the first of this class of systems and of this technology are undertaken simultaneously during the design phase and during the flight-test program[2.20] (Fig. 2.21). During the design phase, the computational fluid dynamicist collaborates with the designer and with the exper-

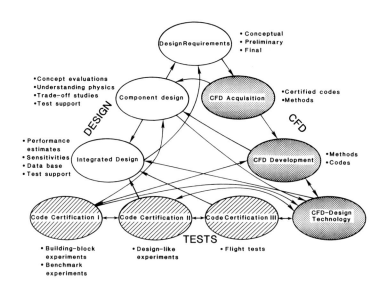

Fig. 2.21 The CFD-design technology development triad, consisting of CFD, design, and tests. Under each of these disciplines, three related functions are conducted.

imental fluid dynamicist who is conducting the ground-based tests. During the flight-test program, the computational fluid dynamicist further develops the CFD-design technology by again collaborating with the experimental fluid dynamicist who is conducting the flight tests. The development of this technology is an iterative and a staged process.

On the one hand, during this development process the numerics and physics models used for simulating the reality are tested for their validity. First, the numerical modeling of governing equations is checked by conducting grid refinement studies and by making sure that the physical laws are enforced during computations. These laws are the conservation laws and the second law of thermodynamics. Then, the modeling of physics is validated by comparing computed results with test data. Initially, the physics observed in unit (building-block and benchmark) experiments is simulated. Subsequently, design-like experiments are considered. Ultimately, the physics associated with a fluid dynamics system in flight is compared with that simulated. During each of these validation-of-physics modeling processes, the numerical modeling is also verified. The validation of a physics model is only accomplished when detailed surface and flowfield comparisons with test data verify the model's ability to accurately simulate the critical physics of the flow over a range of specified parameters. The acceptable level of accuracy of simulations resulting from chosen numerics and physics models, and their criticality, depend on the requirement set by the utility of these simulations. As a progression is made from unit experiments to flight experiments, these models are put through processes of validation for different types of flow problems. The objective is to have a model that is applicable to various types of flow problems associated with a class of fluid dynamics systems for which the CFD-design technology is being developed.

On the other hand, a CFD code is developed for a type of flow problem. This code undergoes code certification, which is defined as "the process of evaluating a computer code in terms of its logic, numerics, fluid dynamics, and results, to ensure compliance with specific requirements."[2.20] These requirements are dictated by the use for which the code is developed. For example, the requirements for conducting research and those for designing propulsion systems are not identical.

The primary limitation of ground-based testing is that the complete verification of CFD thrust estimates computed and the complete simulation of flight physics are not feasible. The primary limitation of flight testing is the lack of suitable test instruments and feasibility of gathering adequate data. On the other hand, the primary limitation of CFD is that it is dependent on appropriate modeling of the physics of the boundary layer, mixing, and combustion.

2.7.2 Caveats About CFD

By definition, simulation is not reality. A numerical simulation is acceptable if it reproduces the reality to the level required for a specific utility of the simulation. The departure of the simulation from the reality is an error in the simulation. An *estimate* of this error is the uncertainty in this simulation. Both the computational (numerics) and fluid dynamics aspects of CFD contain uncertainties that affect the computed results. Moreover, an additional uncertainty is introduced by the human element. These uncertainties are discussed at some length in Ref. 2.21, and they are briefly mentioned below (Fig. 2.22).

2.7.2.1 Computational uncertainties. Once the modeling equations (which include the initial and boundary conditions) are determined, numerical algorithms are developed to solve them and computer codes are constructed. There are two sources of uncertainties of computation: equivalence and numerical accuracy. Note that a computer program may contain coding errors, which are not uncertainties but mistakes.

The computational model needs to describe the "reality" contained in the theoretical (mathematical or empirical) model. A departure from equivalence of the two realities introduces uncertainties. For example, numerical algorithms for multidimensional problems often use solution procedures based on one-dimensional waves, without taking into account multidimensional wave propagation. Under certain conditions, these algorithms may produce a contact discontinuity instead of a shock wave, and vice versa.

There are two main sources of uncertainty related to numerical accuracy. First, an algorithm consists primarily of an approximation of the mathematical model owing to discretization. In the limit of

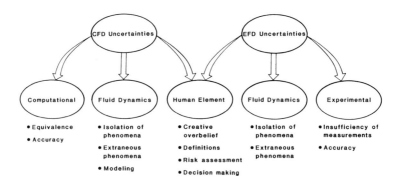

Fig. 2.22 Computational fluid dynamics (CFD) and experimental fluid dynamics (EFD) have many common sources of uncertainties.

the spatial and temporal grid (used to locate computational points in space and time) sizes approaching zero, a consistent discretization would not have any discretization errors. In practice, this limit cannot be taken. For instance, algorithms for combustor flows may modify the combustion phenomena owing to numerical dissipation (diffusion). Second, a solution procedure used in an algorithm may contain an approximation. For example, the solution accuracy may be dependent on the convergence criteria used in any iterative procedure.

2.7.2.2 Fluid dynamics uncertainties. There are three sources of uncertainties related to fluid dynamics: (1) isolation of phenomena, (2) extraneous phenomena, and (3) modeling of phenomena.

The first uncertainty is caused by isolation of fluid dynamics phenomena, which can be either deliberate or unavoidable. In order to understand certain phenomena, it is customary to set up a unit problem demonstrating these phenomena, assuming that there is either absolutely no influence or perfectly known influence on these phenomena by other natural phenomena. Sometimes lack of knowledge leads to isolation of phenomena. On the other hand, unavoidable isolation of phenomena takes place when it is not possible to address all relevant phenomena simultaneously. In either case, an approximation or an uncertainty is introduced. The boundary-layer transition from a laminar to turbulent flow is considered to be dependent on Mach number, Reynolds number, and the wall temperature, without considering the effects of chemical kinetics. At Mach numbers greater than 8, this isolation of phenomena may have an effect on the location of transition and on the length of the transition region. Another example is that of utilizing a smaller number of chemically reacting species than those known to occur in reality.

The second uncertainty is caused by the insertion of extraneous phenomena. When the reality of interest either cannot be simulated or is difficult to simulate, sometimes an alteration other than a simplification (isolation) of this reality is made so that this modified reality can be simulated. This introduction of extraneous phenomena may perturb the manifestation of existing phenomena. An example of extraneous phenomena uncertainty is the simulation of ground-based combustor flow with chemical reactions in addition to those expected under flight conditions.

The third uncertainty is caused by improper modeling of the phenomena under consideration. A model describes reality in mathematical or empirical terms or both. The uncertainty is related to the validity of the model. There are various sources of uncertainty of modeling: the basic flow equations, the transition model, the transition length model, the relaminarization model, the turbulence model (momentum and heat fluxes), the relationship between viscous stress and strain rate, the relationship between the first and second coefficient of viscosity, the chemical reaction rates, the vibrational and radiation excitation rates, the surface chemical reaction rates (surface catalysis), the gas and transport properties, the upstream flow conditions, and the flow dimensionality. The modeling uncertainty also includes the uncertainty of the range of validity of the model.

Note that models are generally based on test data. Uncertainties are also inherent in experimental fluid dynamics. Both test measurements and test fluid dynamics contain uncertainties. In the case of measurements, ground-based or flight, there are fluid dynamics system and measurement-sensor interaction uncertainties, and measurand and derived data uncertainties. Moreover, the insufficiency of data also introduces uncertainties. The fluid dynamics uncertainties arise when testing is done under conditions other than the operating conditions of the fluid dynamics system. These uncertainties arise owing to isolation of phenomena and extraneous phenomena. For example, the ground-based facilities may manifest phenomena other than or in addition to those likely to occur in flight.

2.7.2.3 Human element uncertainties.   There are four types of human element uncertainties: phenomenon of creative overbelief,[2.22] uncertainties about definitions, uncertainties about risk assessment, and uncertainties in decision making. The first two types can be eliminated with systematic questioning, whereas the latter two types of uncertainties are difficult to eliminate. The latter two principally arise when CFD is used in the design process.

Usually a person develops an emotional attachment to his or her creation and tends to visualize this creation as a reality, though based on insufficient evidence. The competitive market generally encourages overselling and fostering of creative overbelief. A success-oriented approach to design does not acknowledge inherent risks, for budgetary as well as political reasons. Uncertainties about definitions are caused by ambiguity of meaning and interpretation. An example of the former is the attitude that measurements are the reality; an example of the latter is the false assignment of significance to what has been measured.

Consider the following definition of CFD code calibration: "The comparison of CFD code results with experimental data for realistic geometries that are similar to the ones of design interest"; this comparison is "made in order to provide a measure of the code's capability to predict specific parameters that are of importance to the design objectives without necessarily verifying that all the features of the flow are correctly modeled."[2.23] A mere comparison of computed results and experimental data leading to a measure is not sufficient to justify declaring the code to be a calibrated code. In this definition, the phrase "to provide a measure of the code's capability to predict" is ambiguous. The condition of acceptability of the measure needs to be spelled out. For example, if the measure, that is, the difference between the computed results and the test data, is comparable in magnitude to the measured parameter, then this measure and hence the code generating this parameter are unacceptable.

Frequently, a statement to the effect that "an excellent agreement is obtained between computations and test data" is made. This statement is not fully satisfactory. "Excellent" needs to be quantified or defined for engineering applications. Questions such as the following must be addressed in order to eliminate uncertainties caused by different interpretations: What is the level of credibility of computed results? What are the limitations of these results? Under what conditions are such results acceptable?

Uncertainties about risk assessment arise from disagreements over what constitutes a risk and what is considered to be an acceptable risk. For example, what are acceptable risks owing to computational uncertainties, fluid dynamic uncertainties, and measurement uncertainties within the flight envelope of a hypersonic propulsion system? Uncertainties in decision making arise because of insufficient information. For example, how does one determine some of the fluid dynamics uncertainties without flight-test data?

2.7.3 Credibility of Design

Since the advent of the computer, the development of a reliable computer code that can perform its intended function has been a challenge. Reliability is essential because of the criticality of performance, operability, and safety of a fluid dynamics system. Just as reliability, or the lack of it, is a characteristic of this code, credibility, or lack of it, is a characteristic of the output of this code. The credibility of the output determines the credibility of the system. In other words, the credibility of the design of this system is no better than the credibility of the tools used for the design. A design challenge is to determine credible CFD results. Briefly, a discussion of this challenge is presented below, along with an example, and a suggestion for addressing it may be found in Ref. 2.24.

The question of the utmost importance is what is the level of credibility of the computed results, or what are the quantified uncertainties associated with those results for designing a fluid dynamics system that will meet specific operational goals. These uncertainties determine the margin to be built into the design; they are essential for establishing the success risks and safety risks associated with fluid dynamics systems and are critical for developing a risk reduction plan so that their magnitudes are reduced to an acceptable level. The computational and fluid dynamics models need to be developed and validated with test data for possible wide applications. Models cannot be validated with measurements nor can the uncertainties in computed results be determined with them, unless measurement uncertainties are known. On the other hand, codes are developed for specific types of applications and, as such, must be certified; that is, codes must be put through a process that assures or informs with certainty the potential user that the codes generate results with a known level of uncertainty.

The process of establishing credibility of computed results also determines reliability and limits of applicability of the code generating these results. The level of credibility required is decided by the degree to which system specifications are sensitive to performance quantities and, in turn, the degree to which performance quantities are sensitive to CFD uncertainties.[2.20]

When strengths, weaknesses, range of applicability, and limitations of a technology are known, it becomes a useful design tool. A designer is effective only if he knows these characteristics of his design tools. When the tool is certified, it becomes a credible design tool. However, the use of certified codes does not necessarily establish the credibility of the design. Given the same CFD code and the same flow conditions, two designers may compute two different sets of results if they are not trained in the proper use of the code. The designers need to be "programmed," as it were, to properly use CFD if they are to produce good designs.

An example of the development of a credible CFD design tool is provided by Oberkampf, Walker, and Aeschliman.[2.25-2.26] These investigators established the level of credibility of computed results obtained from a code called SPRINT for a class of hypersonic vehicle. The shape of this vehicle consists of 10 percent spherically blunted cone with a slice on the windward side (Fig. 2.23). The slice is parallel to the axis and begins at 0.7 of the length of the vehicle, measured from the nose. The vehicle can have three different flaps attached to the aft portion of the slice, providing deflection angles 10, 20, and 30 deg. For the same freestream condition, a grid refinement study was undertaken to establish the level of numerical accuracy of the computed performance quantities (forces and moment), and a measurement-uncertainty analysis was carried out to establish the credibility of measured performance quantities that were used for verifying the modeled physics and numerics in the code. In surveying the literature, one rarely finds such an example of the development of a CFD design tool, wherein both computational and measurement uncertainties are addressed.

The computational grid was doubled twice in each direction of the three dimensions while holding the other two dimensions fixed. These grid refinement studies were done for two angles of attack (α), the end points of the intended range of applicability of the code,

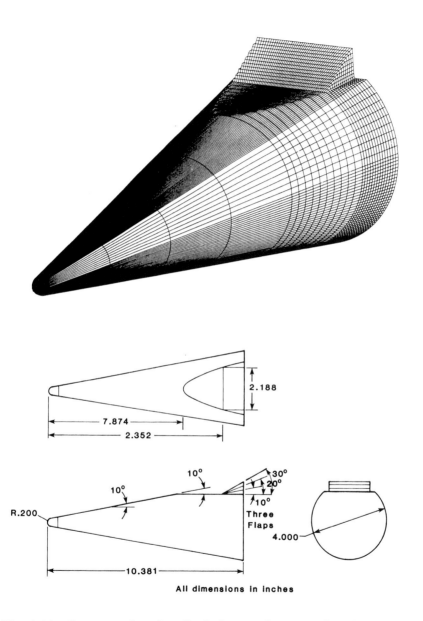

Fig. 2.23   Computational and wind-tunnel test model. A computa-
tional-grid system is shown on the surface of the model.

$\alpha = 0$ and $\alpha = 16$ deg. Based on the solutions presented in Ref. 2.26 for the intermediate and the largest number of grid points, the Richardson extrapolation[2.27] is used to obtain an estimated exact solution, as the number of grid points in all three dimensions simultaneously approached infinity. For example, the estimated exact value of the axial force coefficient C_a on the forecone (the cone forward of the sliced portion) at $\alpha = 16$ deg. is 7.906×10^{-2}. Utilizing this value, the uncertainties, that is, the estimated errors in the computed values, are plotted in Fig. 2.24. A requirement for code certification was that forces and moments should have uncertainties less than 1 percent from the estimated exact values. This requirement dictates how to use the SPRINT code. For instance, this requirement led to a grid size of $385 \times 49 \times 49$ on the forecone.

The instrumentation uncertainty and test section flowfield nonuniformity uncertainty associated with measured performance parameters were estimated. The former uncertainty considered the precision error, neglecting the bias errors, by comparing measured performance parameters for the same physical location of the test model in the test section. The latter uncertainty, which is not known in most of the existing hypersonic ground-based facilities[2.18] and which is hardly analyzed in the literature, was obtained by placing the model at two different axial locations in the test section and at four different roll angles from 0 to 270 deg. For example, the total measurement uncertainty, σ, in C_a is 0.535×10^{-3}. Of this amount, the uncertainty due to precision error is 63 percent, and that caused by flow nonuniformity is 37 percent. Therefore, the measured C_a is the best estimate of the true value with 95 percent probability that this value is believed to lie within $\pm 2\sigma$ of the measured value.

Figure 2.25 shows measured and computed values of C_a for the

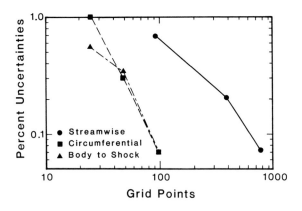

Fig. 2.24 **Percent uncertainties in axial force coefficients as a function of the number of computational grid points.**

slice-only configuration and one with a flap deflected by 10 deg. For the former configuration, the computed force is within $\pm 2\sigma$ of the measured value, except near zero angle of attack. To explain this exception, Ref. 2.25 questions whether the true test conditions were duplicated in computations. For the latter configuration, the agreement between the computed and measured C_a is not within $\pm 2\sigma$. This disagreement is a consequence of having a physics model in the SPRINT code that did not allow for reversed flow in the axial direction ahead of the flap. Note that this reversed flow was observed in tests. To obtain solutions, this deficiency in the physics model was compensated for by changing the modeled numerics. Specifically, the numerical damping and stabilizing parameters were changed. Both physics and numerics models affected these solutions.

Therefore, the SPRINT code is certified to compute forces and moment, with less than 1 percent uncertainty or within measurement uncertainties reported in Ref. 2.25, provided the following conditions

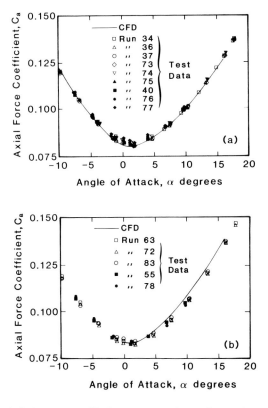

Fig. 2.25   Axial force coefficients as a function of angle of attack for (a) slice-only configuration, and those for (b) the configuration of Fig. 2.23 with 10-deg. flap deflection.

are met: (1) the vehicle considered must be a member of the class of vehicles represented in Fig. 2.23; (2) this vehicle should be exposed to freestream conditions close to those considered during the certification process; and (3) the numerics (including the grid size and the numerical damping and stabilizing parameters) are kept identical to those chosen during this process to compute these performance parameters with less than 1 percent computational uncertainty. To increase the range of applicability of this code, this process has to be repeated for other classes of vehicle shapes and at other freestream conditions.

2.7.4 Closure

Computational fluid dynamics is a formidable tool in the hands of appropriately qualified designers. It may be used in the design process to achieve one of the following objectives: understanding the physics, making tradeoff studies, determining design sensitivities and optimization, making design evaluations, developing the design data base, supporting ground-based and flight tests, and expanding safely the flight envelope. With CFD the designers can often address problems for which no design experience, test database, or test techniques exist. Moreover, CFD has the potential for representing and computing the physics associated with hypersonic, free-flight conditions more accurately than it can be represented and measured in ground-based test facilities. Of course, computational fluid dynamics as a design tool has some limitations, just as other tools have. Only by thoroughly knowing these limitations can the designers properly use CFD as a tool. The CFD-design technology for hypersonic propulsion systems is vigorously being developed. The significance of this technology in designing such systems is illustrated in Secs. 5.8, 6.6, and 7.6, along with a few examples. Moreover, these examples are presented with a view to the further development of the intuitive understanding of these systems.

2.8   DEFINING HYPERSONIC FLOW

A precise definition of hypersonic flow conditions is elusive because the term merely connotes very high Mach numbers and is inherently qualitative. In fact, the term appears to have been first used in 1946 by the renowned aerodynamicist H. S. Tsien, who coined it to replace the less explicit expressions being used at that time, such as *superaerodynamic* or *hypervelocity*.[2.28] The authors of textbooks that deal even partially with hypersonic flows invariably devote some effort to providing a quantitative description of the Mach number boundary between supersonic and hypersonic flows.[2.1, 2.2, 2.4]

For propulsion engineers, defining and locating a useful boundary is a relatively straightforward matter. There are, in fact, two entirely separate approaches that lead to the same conclusion. First, the hypersonic propulsion viewpoint can be captured by means of the H-K diagram of Fig. 2.26, which was obtained by adding the local Mach numbers and traditional designations for the regimes of compressible fluid mechanics to Fig. 2.12.

Referring to Fig. 2.26 and observing that

$$\frac{\text{Kinetic energy}}{\text{Static enthalpy}} = \frac{K}{H} = \frac{V^2}{2C_pT_{t0}} \cdot \frac{C_pT_{t0}}{C_pT} = \frac{\gamma - 1}{2}M^2 \qquad (2\text{-}129)$$

(an interesting relationship in itself, and the origin of the term that appears in legions of ensuing compressible flow equations) we see that in subsonic aerodynamics most of the total temperature is invested in static enthalpy, and changes of Mach number are largely due to changes of velocity. In transonic aerodynamics there are substantial fractions of both static enthalpy and kinetic energy, and the Mach number changes rapidly because both the speed of sound and the velocity are changing. In the hypersonic region the bulk of the total temperature is invested in kinetic energy, and the Mach number

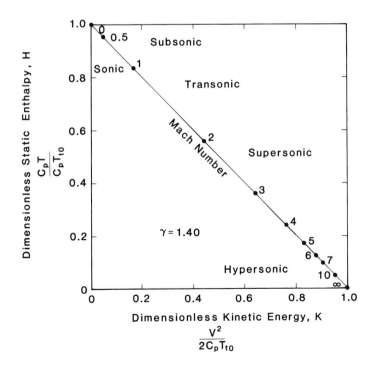

Fig. 2.26 H-K diagram for the constant energy flow of a calorically perfect gas.

changes largely because the static temperature and speed of sound are changing.

This leads to one internal flow or propulsion-based definition of the hypersonic regime, namely that this is where most of the total temperature exists as kinetic energy, and changes in Mach number have little effect on the kinetic energy or velocity of the flow. Figure 2.26 suggests that the boundary for this regime is a Mach number in the range of 5–7, for which the kinetic energy fraction is 0.83–0.91. Small wonder that airbreathing engines that operate under hypersonic conditions are known in the trade as *kinetic energy machines.*

Second, the hypersonic propulsion viewpoint can be captured by estimating the lowest freestream Mach number that would cause real gas effects to occur (at stagnation). Thus, rewriting Eq. (2-52) as

$$M_0 = \sqrt{\frac{2}{\gamma - 1}\left(\frac{T_{t0}}{T_0} - 1\right)} \qquad (2\text{-}130)$$

and substituting representative values of $\gamma = 1.40$, $T_0 = 400$ °R (222 K), and $T_{t0} = 3000$ °R (1700 K), we find that $M_0 \cong 6$.

Interestingly, these approaches lead to similar quantitative results as the various aerodynamics-based definitions. Thus, we are safe in concluding that hypersonic flows have Mach numbers in excess of 5–7.

REFERENCES

[2.1] Bertin, J. J., *Hypersonic Aerothermodynamics,* AIAA Education Series, Washington, DC, 1994.

[2.2] Anderson, J. D., Jr., *Hypersonic and High Temperature Gas Dynamics,* McGraw-Hill, New York, 1989.

[2.3] Oates, G. C., *The Aerothermodynamics of Gas Turbine and Rocket Propulsion,* Revised and Enlarged Edition, AIAA Education Series, Washington, DC, 1988.

[2.4] Shapiro, A. H., *The Dynamics and Thermodynamics of Compressible Fluid Flow,* Ronald Press, New York, 1953.

[2.5] Anderson, J. D., Jr., *Introduction to Flight,* McGraw-Hill, New York, 1989.

[2.6] Reynolds, W. C., and Perkins, H. C., *Engineering Thermodynamics,* McGraw-Hill, New York, 1977.

[2.7] Anonymous, *U.S. Standard Atmosphere, 1976,* U.S. Government Printing Office, Washington, DC, Oct. 1976.

[2.8] Stetson, K. F., "On Predicting Hypersonic Boundary Layer Transition," Flight Dynamics Lab, Air Force Wright Aeronautical Labs, AFWAL-TM-84-160-FIMG, Wright-Patterson AFB, OH, Mar. 1987.

2.9 Hilsenrath, J., and Klein, M., "Tables of Thermodynamic Properties of Air in Chemical Equilibrium Including Second Virial Corrections from 1500 to 15,000 K," AEDC-TR-65-68, Arnold Engineering Development Center, TN, 1965.

2.10 Moeckel, W. E., and Weston, K. C., "Composition and Thermodynamic Properties of Air in Chemical Equilibrium," NACA TN-4265, Aug. 1958.

2.11 Kutschenreuter, P. H., Jr., and Balent, R. L., "Hypersonic Inlet Performance from Direct Force Measurements," *Journal of Spacecraft and Rockets,* Vol. 2, No. 2, Mar. 1965.

2.12 Hirsch, C., *Numerical Computation of Internal and External Flows,* Vol. 1: Fundamentals of Numerical Discretization, 1988, and Vol. 2: Computational Methods for Inviscid and Viscous Flow Models, 1990, John Wiley, Chichester (England) and New York.

2.13 Anderson, D. A., Tannehill, J. C., and Pletcher, D., *Computational Fluid Mechanics and Heat Transfer,* Hemisphere Publication, Washington/McGraw-Hill, New York, 1984.

2.14 Hamilton, J. T., Wallace R. O., and Dill, C. C., "Launch Vehicle Aerodynamic Data Base Development Comparison with Flight Data," *Shuttle Performance: Lessons Learned,* NASA CP-2283, Part 1, 1983.

2.15 Griffith, B. J., Maus, J. R., and Best, J. T., "Explanation of the Hypersonic Longitudinal Stability Problem – Lessons Learned," *Shuttle Performance: Lessons Learned,* NASA CP-2283, Part 1, 1983.

2.16 Silveria, M. A., "The Beginning of a New Aerodynamic Research Program," *Shuttle Performance: Lessons Learned,* NASA CP-2283, Part 2, 1983.

2.17 *Review of Aeronautical Wind Tunnel Facilities,* Committee on Assessment of National Aeronautical Wind Tunnel Facilities, Aeronautical and Space Engineering Board, National Research Council, National Academy Press, Washington, DC, 1988.

2.18 *Hypersonic Technology for Military Application,* Committee on Hypersonic Technology for Military Application, Air Force Studies Board, National Research Council, National Academy Press, Washington, DC, 1989.

2.19 *Requirements for Hypersonic Test Facilities,* Report of the Ad Hoc Committee, United States Air Force Scientific Advisory Board, Department of the Air Force, Washington, DC, May 1989.

2.20 Mehta, U. B., "Computational Requirements for Hypersonic Flight Performance Estimates," *Journal of Spacecraft and Rockets,* Vol. 27, No. 2, Mar.–Apr. 1990.

2.21 Mehta, U. B., "Some Aspects of Uncertainty in Computational Fluid Dynamics Results," Transactions of ASME, *Journal of Fluids Engineering,* Vol. 113, Dec. 1991.

[2.22] Morkovin, M. V., "A Harangue to Young Authors by a Weary Reviewer," Guest Editorial, Transactions of ASME, *Journal of Fluids Engineering*, Vol. 96, June 1974.

[2.23] Bradley, R. G., "CFD Validation Philosophy," *Symposium on Validation of Computational Fluid Dynamics*, Paper No. 1, AGARD CP-437, 1988.

[2.24] Mehta, U. B., "The Aerospace Plane Design Challenge: Credible Computations," *Journal of Aircraft*, Vol. 30, No. 4, July–Aug. 1993.

[2.25] Oberkampf, W. L., and Aeschliman, D. P., "Joint Computational and Experimental Aerodynamics Research on a Hypersonic Vehicle, Part 1: Experimental Results," *AIAA Journal*, Vol. 30, No. 8, Aug. 1992.

[2.26] Walker, M. M., and Oberkampf, W. L., "Joint Computational and Experimental Aerodynamics Research on a Hypersonic Vehicle, Part 2: Computational Results," *AIAA Journal*, Vol. 30, No. 8, Aug. 1992.

[2.27] Richardson, L. F., and Gaunt, J. A., "The Deferred Approach to the Limit," *Transactions of the Royal Society of London, Series A: Mathematical and Physical Sciences*, Vol. 226, pp. 229–361, 1927.

[2.28] Tsien, H. S., "Similarity Laws of Hypersonic Flows," *Journal of Mathematics and Physics*, Vol. 25, No. 3, 1946.

PROBLEMS

2.1   An aerospace vehicle is flying as illustrated in Fig. 2.1. If we treat the vehicle as a moving mass, energy considerations lead to the conclusion that

$$(T - D)V \quad = \quad W \cdot \frac{dH}{dt} \quad + \quad \frac{W}{g} \cdot \frac{d}{dt}\left(\frac{V^2}{2}\right) \quad = \quad W \cdot \frac{dz_e}{dt}$$

$$\begin{matrix} \textit{rate of} \\ \textit{mechanical} \\ \textit{energy input} \end{matrix} \qquad \begin{matrix} \textit{storage} \\ \textit{rate of} \\ \textit{potential} \\ \textit{energy} \end{matrix} \qquad \begin{matrix} \textit{storage rate} \\ \textit{of kinetic} \\ \textit{energy} \end{matrix}$$

where

$$z_e = \text{Energy height} = H + \frac{V^2}{2g}$$

Show that this same result can be obtained from Eq. (2-1), which was based on force considerations. Reconcile the two approaches for the case of a constant speed climb, during which altitude changes but velocity does not.

2.2   An aerospace vehicle is flying as illustrated in Fig. 2.1. Suppose it is climbing much more slowly than it is moving forward, so that r_c is approximately the radius of the Earth, r_E. Equation (2-2) reveals that the centrifugal force associated with the flight path curvature reduces the need for lift, which, in turn, usually has the beneficial effect of reducing the drag of the lifting surfaces. Use Eq. (2-2) to show that

$$\frac{L}{W} = 1 - \frac{V_0^2}{g r_E}$$

For the representative atmosphere of Sec. 2.4.9, calculate and plot L/W as a function of flight Mach number for $0 < M_0 < 25$.

2.3   According to Ref. 2.7, the perfect gas constants used in establishing the properties of the standard atmosphere were not changed over the first 80 km (262 kft) of geometric altitude, where mixing dominates. Therefore, if the information presented in App. B is internally consistent, the functions

$$\frac{p}{p_{SL}} \cdot \frac{\rho_{SL}}{\rho} \cdot \frac{T_{SL}}{T}$$

and

$$\frac{a}{a_{SL}} \cdot \sqrt{\frac{T_{SL}}{T}}$$

should be identical to 1 at any geometric altitude.
Is the information of App. B internally consistent?

2.4   Starting from Eq. (2-8) for the behavior of atmospheric static pressure and employing the reference temperature of 400 °R (222 K) show that

$$\frac{p}{p_{SL}} \cong e^{-kH} \qquad k = 0.047/\text{kft} \ (0.15/\text{km})$$

Plot this relationship versus the standard atmosphere of App. B and comment on its validity.
Now substitute this result into Eq. (2-16) to show that

$$M_0 = \sqrt{\frac{2q_0}{\gamma_0 p_{SL} e^{-kH}}}$$

Use this equation to superimpose contours of constant q_0 on Fig. 2.4 and comment on its validity.

2.5 One of the most commonly used hypersonic boundary layer transition prediction methods (even though it does not include many important influences and is therefore only a rule of thumb) is that[2.8]

$$\frac{\sqrt{Re_{L,O}}}{M_0} = \text{constant} \sim 300$$

for freestream Mach numbers in excess of about 6.

Develop an expression for this boundary similar to Eq. (2-22) and overlay the calculated result on Fig. 2.4. If this transition model is correct, are the conclusions reached in Sec. 2.4.7 still valid?

2.6 One of the most popular means for calculating the local heat transfer of compressible flows is the reference temperature method, which extends incompressible formulas by correcting transport properties for compressibility effects (an approximate engineering approach based on intuitive insight).[2.1, 2.2] For fully turbulent hypersonic flow over a flat plate with the wall temperature cooled to nearly the freestream static temperature, the reference temperature method would predict that the local wall heat flux

$$\dot{q}_w \propto \frac{\rho^* V_0^3}{(Re^*)^{0.2}} \propto \rho^* V_0^3 \left(\frac{\mu^*}{\rho^* V_0}\right)^{0.2} \propto V_0^{2.8} (\mu^*)^{0.2} \left(\frac{p_0}{RT^*}\right)^{0.8}$$

where the starred transport properties are evaluated at

$$T^* = T_0(1 + 0.032 M_0^2)$$

Develop an expression for constant $\dot{q}_w$ similar to Eq. (2-22) and overlay some contours of constant $\dot{q}_w$ on Fig. 2.4. HAP(Trajectory) will prove useful to this exercise.

If this heat transfer model is correct, what is the likelihood that the local aerodynamic heating will become a dominant factor at very high Mach numbers?

2.7 Employ the computer program HAP(Air) to generate the equilibrium constituents of representative air as a function of static temperature for static pressures of 1 and 100 atm and plot them as shown in Fig. 2.8.

(a) Does this information confirm the assertion that the effect of higher static pressures is to delay the onset of dissociation and chemical reactions to higher temperatures?

(b) Are these results in agreement with the information found in Figs. 2.7, 2.9, and 2.10?

2.8 Suppose that the mole fraction composition of air is:

$$N_2 = 78\% \qquad O_2 = 20\% \qquad A = 1\% \qquad H_2O = 1\%$$

and the static pressure is 0.01 atm. Employ the computer program Hap(Equilibrium) to generate information equivalent to Figs. 2.7 through 2.10 for this mixture.

Do the new constituents have a significant impact upon the equilibrium constituents or thermodynamic properties of air?

2.9   Employ the computer program HAP(Air) to determine the equilibrium speed of sound for representative air at a pressure of 1 atm and over the temperature range of Fig. 2.7. Compare this with the calorically perfect, nonreacting gas value calculated from Eq. (2-42) and the corresponding values of γ and R taken from HAP(Air). What can you say generally or specifically about these results?

2.10   Nail down the concept of converting kinetic energy to enthalpy by calculating the difference between static and total enthalpy for a vehicle flying at $0 < M_0 < 25$. Use Eq. (2-47) and the representative atmosphere of Sec. 2.4.9, and find the answer in both BE and SI units.

2.11   Nail down the concept of isentropic compression by using Eq. (2-54) to calculate the total pressure for a vehicle flying along any constant q_0 trajectory at $0 < M_0 < 25$. Assume $\gamma_0 = 1.40$.

2.12   For the isentropic, one-dimensional flow of a calorically perfect gas in an axial duct, use the concept of mass flow parameter (MFP) to show that the "static pressure MFP" is given by the expression

$$\frac{\dot{m}\sqrt{T_t}}{pA} = \sqrt{\frac{\gamma}{R}} \cdot M \cdot \sqrt{1 + \frac{\gamma - 1}{2} M^2}$$

Use this result to construct a chart of "static pressure MFP" as a function of Mach number for $\gamma = 1.40$ and $R = 53.3$ ft·lbf/(lbm·°R) [287 N·m/(kg· K)]. This chart is particularly helpful to experimentalists, who often find it most convenient to determine local Mach number from local measurements of wall static pressure. Note that the MFP axis is not dimensionless.

2.13   Compare the results obtained from the finite control volume and differential control volume analyses with those obtained from HAP(Gas Tables) for the following Example Cases:

(a) Isentropic flow

(b) Normal shock waves

(c) Constant area heating (Rayleigh flow)

(d) Ideal exit velocity

(e) Mass flow parameter

(f) Constant pressure heating

2.14 Consider the frictionless, one-dimensional flow of a calorically perfect gas for which there is only a heat interaction with the surroundings and the throughflow area is varied in such a way that static temperature remains constant, i.e.,

$$dF_{bx} = d\dot{m}_b = đ\dot{W} = dT = 0$$

(a) Find an expression for the area ratio A_e/A_i in terms of γ, $(T_{te} - T_{ti})/T_i$, and M_i.

(b) Find an expression for the total pressure ratio p_{te}/p_{ti} in terms of the same variables.

2.15 Consider the frictionless, one-dimensional flow of a calorically perfect gas for which there is only a heat interaction with the surroundings and the throughflow area is varied in such a way that Mach number remains constant, i.e.,

$$dF_{bx} = d\dot{m}_b = đ\dot{W} = dM = 0$$

(a) Find an expression for the area ratio A_e/A_i in terms of γ, M, and $T_{te}/T_{ti} = \tau_e$.

(b) Find an expression for the total pressure ratio p_{te}/p_{ti} in terms of the same variables.

2.16 Consider the one-dimensional frictional flow of a calorically perfect gas for which there are no energy interactions with the surroundings and the throughflow area is varied in such a way that the velocity remains constant, i.e.,

$$d\dot{m}_b = đ\dot{W} = đ\dot{Q} = dV = 0$$

Assume that the duct is of circular cross section and that the skin friction shear stress is given by

$$\tau_b = C_f \cdot \frac{\rho V^2}{2} \quad \text{or} \quad dF_{bx} = -\frac{C_f \rho V^2 \pi D}{2} \cdot dx$$

where the skin friction coefficient C_f is constant.

(a) Show that the duct diameter is given by

$$D = D_i + \gamma M^2 C_f x$$

(b) Find an expression for the total pressure ratio p_{te}/p_{ti} in terms of γ, M, C_f, and x/D_i.

2.17   Construct the H-K diagram for parameters of your own choice for a:

(a) Scramjet (see Fig. 2.19)

(b) Ramjet (see Fig. 2.20)

Take notes about what lessons you learned along the way.

2.18   Increase your awareness of the H-K diagram by drawing *both* branches of *choked* flow for:

(a) Isentropic flow (convergent-divergent nozzle)

(b) Rayleigh flow (Where does $dH/dK = 0$?)

(c) Fanno flow

2.19   Use HAP(Gas Tables) to show a sequence of oblique shock waves on the H-K diagram for intervals of 1, 2, and 5 deg. for:

(a) $M_1 = 3$

(b) $M_1 = 5$

(c) $M_1 = 10$

2.20   Starting from the total pressure MFP [Eq. (2-83)], show that the quantity

$$\frac{p_t A}{\dot{m}}\sqrt{\frac{\gamma}{RT_{t0}}} = \left(1 + \frac{K}{H}\right)^{\frac{\gamma}{\gamma-1}}\sqrt{\frac{\gamma-1}{2}\cdot\frac{H}{K}\cdot H}$$

and draw lines of constant values on the H-K diagram.

(a) For flows of constant M, how does $p_t A$ vary with τ?

(b) Prove that the lines are tangent to lines of constant τ at $M = 1$

(c) Describe the physical meaning of these results.

2.21   Starting from the static pressure MFP (Problem 2.12), show that the quantity

$$\frac{pA}{\dot{m}}\sqrt{\frac{\gamma}{RT_{t0}}} = H\sqrt{\frac{\gamma-1}{2K}}$$

and draw lines of constant values on the H-K diagram.

(a) For flows of constsnt K, how does pA vary with τ?

(b) For flows of constant M, how does pA vary with τ?

(c) Describe the physical meaning of these results.

2.22   For the H-K diagram of Fig. 2.26, show that:

(a) The minimum rate of change of Mach number with K occurs when K is 0.5.

(b) The minimum rate of change of Mach number with K occurs when the Mach number is

$$\sqrt{\frac{2}{\gamma - 1}}$$

(c) Explain the physics of this phenomenon.

3
HYPERSONIC AEROSPACE
SYSTEM PERFORMANCE

3.1 INTRODUCTION

A simple fact of life is that new types of airbreathing engines
are not developed for their own sake. The costs to society, in terms
of capital, time, risk, and, most importantly, the energy of some of
its most highly trained and skilled engineers, can only be justified
if some greater purpose is served. In the case of aerospace systems,
support for an engine development is usually furnished only if there
is reasonable promise that it will allow either an existing mission to
be carried out more effectively (e.g., the evolution of very high bypass
turbofan engines for low fuel consumption of transport aircraft) or
an entirely new mission to be accomplished (e.g., the development of
afterburning turbojet engines for supersonic flight).

More precisely, then, interest in any proposed type of airbreathing
engine depends upon what kind of aerospace system (i.e., vehicle plus
engine plus fuel plus payload plus mission) it makes possible. Hypersonic airbreathing engines to power either sustained atmospheric
cruise vehicles or aerospace planes flying to orbit must face the same
test. The principal goal of this chapter, therefore, is to provide the
basis for a broad and balanced approach to the evaluation of the
overall performance of hypersonic aerospace systems.

The methodology to be employed should be called "top-down,"
because it is structured to first yield a quantitative assessment of the
performance of the total aerospace system, and then down through
successive layers to reveal the assets and liabilities of the major components and their underlying technologies. This approach has much
in common with the management and tracking of aerospace system
development programs, and can indeed be used for that purpose.
Nevertheless, it is exploited here because it excels at revealing what
is really going on, particularly with regard to the performance required from the airbreathing engine.

3.2 AIRBREATHING ENGINE PERFORMANCE MEASURES

The members of the airbreathing propulsion community have
long found it convenient to describe their engines in terms of a handful of global performance parameters, and there is no need for them
to change their habits because these traditional figures of merit retain their usefulness even in hypersonic flight. They are described

below in terms of the physical quantities connected with the generalized "control volume" or "black box" airbreathing engine depicted in Fig. 3.1. Please note that this idealized model conforms to the one-dimensional description of Sec. 2.6, and presumes that the exhaust flow is perfectly expanded to the surrounding atmospheric pressure, the condition that the designer attempts to attain because it maximizes the thrust produced.[3.1] The entry and exhaust velocities are taken to be parallel to the uninstalled thrust.

The *uninstalled thrust* F is defined as the total thrust exerted by the engine, assuming ideal external flow. For the airbreathing engine of Fig. 3.1, it is equal and opposite to the difference in momentum fluxes between the entering and leaving flows.[3.1]

3.2.1 Specific Thrust

The air mass flow rate specific thrust, commonly known simply as the *specific thrust,* is defined as

$$\frac{\text{Uninstalled thrust}}{\text{Entry air mass flow rate}} = \frac{F}{\dot{m}_0} \qquad (3\text{-}1)$$

This relationship emphasizes the fact that the total uninstalled or internal thrust of an airbreathing engine is proportional to the total entry mass flow rate of air, all other things being equal. Thus, if both the uninstalled thrust required for a given task and the specific thrust of the type of engine to be employed are known, the mass flow rate of air that must enter the engine can be found directly from Eq. (3-1).

3.2.2 Specific Fuel Consumption

The thrust specific fuel mass flow rate, commonly known as the thrust specific fuel consumption or simply as the *specific fuel consumption,* is defined as

$$\frac{\text{Fuel mass flow rate}}{\text{Uninstalled thrust}} = S = \frac{\dot{m}_f}{F} \qquad (3\text{-}2)$$

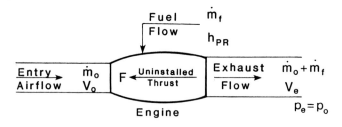

Fuel Flow $\dot{m}_f$

h_{PR}

Entry Airflow $\dot{m}_0$ V_0 $F \xleftarrow{}$ Uninstalled Thrust Exhaust Flow $\dot{m}_0 + \dot{m}_f$ V_e

$p_e = p_0$

Engine

Fig. 3.1 Schematic diagram of an idealized airbreathing engine.

This relationship emphasizes the fact that the fuel mass flow rate of an airbreathing engine is proportional to the total uninstalled or internal engine thrust, all other things being equal. Thus, if both the uninstalled thrust required for a given task and the specific fuel consumption of the type of engine to be employed are known, the fuel mass flow rate that must be provided by the vehicle can be found directly from Eq. (3-2).

3.2.3 Specific Impulse

The fuel weight flow rate specific thrust, commonly known simply as the *specific impulse,* is defined as

$$\frac{\text{Uninstalled thrust}}{\text{Fuel weight flow rate}} = I_{sp} = \frac{F}{g_0 \dot{m}_f} \tag{3-3}$$

where g_0 is the standard sea level value of the acceleration of gravity. Aside from the conversion of fuel mass flow rate to sea level fuel weight flow rate, specific impulse is merely the inverse of specific fuel consumption. However, both are used in practice to bring out different qualities of airbreathing engines. Specific fuel consumption is preferred when the issue is fuel economy, and specific impulse is preferred when the issue is adequate thrust.

Before closing this topic, it is worth emphasizing two characteristics that specific thrust, specific fuel consumption, and specific impulse share that have made them so helpful through the years. First, they deal with those properties of the engine that are most important to the customer, namely thrust, fuel consumption, and airflow. Second, they are all ratios of total engine properties, and are therefore independent of the size of the engine. This allows engines of the same type but different sizes to be compared fairly, and also leads to the concept of scalable or "rubber" engines that are sized to fit the application under consideration.

3.2.4 Fuel/Air Ratio

Although it is not, strictly speaking, a performance measure, the ratio of fuel mass flow rate to air mass flow rate, or fuel/air ratio for short, frequently appears in equations that link the performance measures to one another, and is important in its own right as a general indicator of the combustion conditions in the burner. The fuel/air ratio is defined as

$$\frac{\text{Fuel mass flow rate}}{\text{Entry air mass flow rate}} = f = \frac{\dot{m}_f}{\dot{m}_0} \tag{3-4}$$

3.2.4.1 Stoichiometric fuel/air ratio. The ideal upper limit for the fuel/air ratio is the value that corresponds to complete mutual combustion of all the oxygen present in the air with all the reactants available in the fuel, and is known as the stoichiometric fuel/air ratio. The underlying notion is that anything less would not take full advantage of the available oxygen, and anything more would waste the fuel that could not be burned.

Everything we need to know about the stoichiometric fuel/air ratio can be found from the basic principles of chemical reactions. Since almost all the fuels under consideration for hypersonic flight are hydrocarbons, the general chemical equation for their complete combustion with the representative air of Sec. 2.5.2 is

$$C_xH_y + \left(x + \frac{y}{4}\right)\left(O_2 + \frac{79}{21}N_2\right)$$
$$\rightarrow xCO_2 + \frac{y}{2}H_2O + \frac{79}{21}\left(x + \frac{y}{4}\right)N_2 \qquad (3\text{-}5)$$

where it has been assumed that the combustion is ideal in the sense that the products include only carbon dioxide and water. Equation (3-5) ensures that all carbon, hydrogen, and oxygen atoms are consumed in this stoichiometric chemical reaction.

Based upon the left-hand side of Eq. (3-5), the general expression for the stoichiometric fuel/air ratio is

$$f_{st} = \frac{36x + 3y}{103(4x + y)} \qquad (3\text{-}6)$$

where the atomic weights of the participating elements have been incorporated (H=1, C=12, N=14, and O=16).

A typical ramjet fuel is octane, C_8H_{18}, for which $x = 8$ and $y = 18$, and Eq. (3-6) shows that $f_{st} = 0.0664$. A typical scramjet fuel is hydrogen, for which $x = 0$ and $y = 2$, and Eq. (3-6) shows that $f_{st} = 0.0291$. One obvious conclusion is that the fuel/air ratios likely to be encountered in practice are substantially less than 1.

3.2.5 Airbreathing Engine Overall Efficiency

One can imagine that the function of the airbreathing engine, when viewed as a thermodynamic cycle, is to convert the chemical energy stored in the fuel into mechanical energy for the aerospace system. This leads to another performance measure called overall efficiency that, although less intensely pursued in the literature, is particularly revealing for hypersonic flight.

The rate at which the engine makes mechanical energy available to the aerospace system is known as the *thrust power,* and is given by the expression

$$\text{Thrust power} = FV_0 \tag{3-7}$$

where it has been assumed, in keeping with Secs. 2.3 and 3.2, that the uninstalled thrust, like the installed thrust, is parallel to the direction of flight.

Placing an unambiguous, fair value on the rate at which the chemical reactions make energy available to the engine cycle requires extra thought. The standard practice in the propulsion and power industry is to replace the actual combustion process with a fictitious one in which the pressure is constant and there are no heat or work interactions.[3.2] The energy made available by the chemical reactions is then defined as the heat that must be removed from the final combustion products in order to return them to the same temperature as the initial reactants (at the same pressure). When expressed on a per pound or kilogram mass of fuel basis, this quantity is known as the *heat of reaction* or *heating value* of the fuel (h_{PR}). Finally, since the heating value varies relatively little over the normal ranges of initial pressures and temperatures of the reactants, it is common practice to evaluate it at the standard reference state (77 °F/25 °C, 1 atm). The heats of reaction of values of some typical gaseous fuels reacting with air are found in Table 3.1.

This approach is both equitable and sensible, and will accordingly be used for hypersonic airbreathing engine analysis. The rate at which the chemical reactions make energy available to the engine cycle therefore becomes

$$\text{Chemical energy rate} = \dot{m}_f h_{PR} \tag{3-8}$$

Based on the foregoing, the *overall efficiency* of the airbreathing engine cycle is defined as

Table 3.1   Heats of reaction or heating values for typical hydrocarbon fuels reacting with air at the standard reference state, all reactants and products being gaseous, Ref. 3.2.

Fuel	Heats of Reaction, h_{PR}	
	BTU/lbm Fuel	kJ/kg Fuel
Hydrogen, H_2	51,571	119,954
Methane, CH_4	21,502	50,010
Ethane, C_2H_6	20,416	47,484
Hexane, C_6H_{14}	19,391	45,100
Octane, C_8H_{18}	19,256	44,786

$$\text{Overall efficiency} = \eta_o = \frac{\text{Thrust power}}{\text{Chemical energy rate}} = \frac{FV_0}{\dot{m}_f h_{PR}} \qquad (3\text{-}9)$$

It should be emphasized that the overall efficiency is a direct indicator of how well the engine uses the energy originally deposited in the fuel tanks or, conversely, how much fuel must be put on board in order to provide the energy needed for a given mission. Moreover, commonsense application of the second law of thermodynamics leads to the conclusion that overall efficiency cannot exceed 1; otherwise the chemical energy of the fuel could be restored and the surplus of mechanical energy used to create perpetual motion.

3.2.5.1 Thermal efficiency and propulsive efficiency. It can be very enlightening to further break down the airbreathing engine overall efficiency into its "grass roots" constituents as follows:

$$\eta_o = \underbrace{\frac{\text{Engine mechanical power}}{\text{Chemical energy rate}}}_{\substack{\text{Thermal efficiency} \\ \eta_{th}}} \cdot \underbrace{\frac{\text{Thrust power}}{\text{Engine mechanical power}}}_{\substack{\text{Propulsive efficiency} \\ \eta_p}}$$

This word *equation* reveals that our purpose is to follow the energy along its "food chain" from chemical (in the fuel tank) to mechanical (generated by the engine) to the aerospace system (thrust power). Provided that the exhaust flow is perfectly expanded to atmospheric pressure, as shown in Fig. 3.1, the mechanical power generated by the engine manifests itself only as a change in kinetic energy of the flow, so that

$$\begin{aligned} \text{Engine mechanical power} &= (\dot{m}_0 + \dot{m}_f)\frac{V_e^2}{2} - \dot{m}_0\frac{V_0^2}{2} \\ &= \dot{m}_0\left\{(1+f)\frac{V_e^2}{2} - \frac{V_0^2}{2}\right\} \end{aligned}$$

and therefore that

$$\eta_o = \eta_{th} \cdot \eta_p = \underbrace{\frac{(1+f)\frac{V_e^2}{2} - \frac{V_0^2}{2}}{f h_{PR}}}_{\eta_{th}} \cdot \underbrace{\frac{FV_0}{\dot{m}_0\left\{(1+f)\frac{V_e^2}{2} - \frac{V_0^2}{2}\right\}}}_{\eta_p} \qquad (3\text{-}10)$$

where Eq. (3-4) for f was incorporated.

Thermodynamic analysis of engine cycles teaches us that *thermal efficiency* primarily increases with cycle pressure (and temperature) ratio, although it cannot exceed 1. This is the driving force behind high pressure ratio aircraft engine cycles. Furthermore, since the uninstalled thrust is merely the change in momentum flux from entry to exhaust, then

$$F = (\dot{m}_0 + \dot{m}_f)V_e - \dot{m}_0 V_0 = \dot{m}_0 \left\{ (1 + f)V_e - V_0 \right\} \qquad (3\text{-}11)$$

so that the *propulsive efficiency* portion of Eq. (3-10) becomes

$$\eta_p = \frac{2\left\{ (1 + f)\dfrac{V_e}{V_0} - 1 \right\}}{\left\{ (1 + f)\left(\dfrac{V_e}{V_0}\right)^2 - 1 \right\}} \qquad (3\text{-}12)$$

which shows that propulsive efficiency increases as the ratio of the exhaust velocity to the freestream velocity decreases. This is the driving force behind high bypass ratio aircraft engine cycles, which spread the available engine mechanical power across more air in order to reduce this velocity ratio. The propulsive efficiency can, in fact, slightly exceed 1 when the fuel/air ratio is sufficiently large because of the momentum of the added fuel mass flow. However, we found in Sec. 3.2.4.1 that the fuel/air ratio is small compared to 1 even for stoichiometric combustion. Therefore, a reasonable approximation and more revealing formula for propulsive efficiency is obtained by neglecting the fuel/air ratio when compared to 1 in Eq. (3-12), with the result that

$$\eta_p = \frac{2}{\dfrac{V_e}{V_0} + 1} \qquad (3\text{-}13)$$

This is the most transparent and frequently encountered form and, since exhaust velocity must exceed inlet velocity in order to obtain positive thrust, this approximation for propulsive efficiency never exceeds 1.

In closing, overall efficiency can be seen to be a powerful and intuitively appealing measure of airbreathing engine performance. It is anchored in the first law of thermodynamics, and is limited by the second law to values between 0 and 1. The other airbreathing engine performance measures offer no such comforts. As an added benefit, the effects of the energy food chain of thermal and propulsive efficiencies will be plainly seen in even the most elaborate engine performance analyses that follow. From now on, therefore, overall efficiency and its constituent efficiencies will be in our vanguard.

3.2.6 Performance Measure Interrelationships

Equations (3-1) through (3-4), and (3-9) may be rearranged to yield a complete set of exact interrelationships between the airbreathing

Table 3.2 Interrelationships between the various airbreathing engine performance measures. For example, in order to express the specific thrust $F/\dot{m}_0$ in terms of the specific fuel consumption S and the fuel/air ratio f, read across the $F/\dot{m}_0$ row to the S column to find that $F/\dot{m}_0 = f/S$.

$\dfrac{F}{\dot{m}_0} =$	$\dfrac{F}{\dot{m}_0}$	$\dfrac{f}{S}$	$g_o f I_{sp}$	$\dfrac{f h_{PR}}{V_0} \cdot \eta_o$
$S =$	$\dfrac{f}{F/\dot{m}_0}$	S	$\dfrac{1}{g_o I_{sp}}$	$\dfrac{V_0}{h_{PR}\eta_o}$
$I_{sp} =$	$\dfrac{1}{g_o f} \cdot \dfrac{F}{\dot{m}_0}$	$\dfrac{1}{g_o S}$	I_{sp}	$\dfrac{h_{PR}}{g_o V_0} \cdot \eta_o$
$\eta_o =$	$\dfrac{V_0}{f h_{PR}} \cdot \dfrac{F}{\dot{m}_0}$	$\dfrac{V_0}{h_{PR}S}$	$\dfrac{g_o V_0}{h_{PR}} \cdot I_{sp}$	η_o

engine performance measures. The results of these manipulations are presented in Table 3.2.

3.2.7 Airbreathing Engine Performance Measure Examples

It is generally true that specifying the flight speed, the fuel heating value, and any two performance measures allows all the others to be calculated. The two examples that follow will demonstrate extreme cases of this assertion, as well as providing a first look at the levels and trends of typical ramjet and scramjet engine performance.

Example 3.1

A ramjet is being flown at a velocity of 2000 ft/s (610 m/s) and is burning a hydrocarbon fuel with a heating value of 19,000 BTU/lbm (44,200 kJ/kg). The uninstalled specific thrust $F/\dot{m}_0$ is 75 lbf·s/lbm (736 N·s/kg) and the specific fuel consumption S is 2.2 lbm/(lbf·h) [62.3 g/(kN·s)]. We wish to determine all the remaining engine performance measures.

$$f = \frac{F}{\dot{m}_0} \cdot S = \frac{75 \times 2.2}{3600} = 0.0458 \qquad \text{Table 3.2}$$

$$\eta_o = \frac{V_0}{h_{PR}S} = \frac{2000 \times 3600}{19,000 \times 778.2 \times 2.2} = 0.221 \qquad \text{Table 3.2}$$

$$I_{sp} = \frac{1}{g_0 S} = \frac{3600 \times 32.17}{32.17 \times 2.2} = 1640 \text{ s} \qquad \text{Table 3.2}$$

$$\frac{V_e}{V_0} = \left(\frac{1}{1+f}\right)\left(\frac{F}{\dot{m}_0 V_0} + 1\right) = \frac{1}{1.0458}\left(\frac{75 \times 32.17}{2000} + 1\right) = 2.11$$
$$\tag{3-11}$$

$$\eta_p = \frac{2\left\{(1+f)\dfrac{V_e}{V_0} - 1\right\}}{\left\{(1+f)\left(\dfrac{V_e}{V_0}\right)^2 - 1\right\}} = 0.660 \qquad\qquad (3\text{-}12)$$

$$\eta_p \cong \frac{2}{\dfrac{V_e}{V_0} + 1} = 0.643 \qquad\qquad (3\text{-}13)$$

$$\eta_{th} = \frac{\eta_o}{\eta_p} = 0.335 \qquad\qquad (3\text{-}10)$$

Example 3.2

A scramjet is flying at a velocity of 8000 ft/s (2440 m/s) and is burning hydrogen with a heating value of 51,600 BTU/lbm (120,000 kJ/kg). The fuel/air ratio is 0.0291 (stoichiometric) and the overall efficiency is 0.40. We wish to determine all the remaining engine performance measures.

$$\frac{F}{\dot{m}_0} = \frac{f h_{PR}}{V_0} \cdot \eta_o = \frac{0.0291 \times 51,600 \times 778.2 \times 0.40}{8000}$$
$$= 58.4 \text{ lbf} \cdot \text{s/lbm} \quad (573 \text{ N} \cdot \text{s/kg}) \qquad \text{Table 3.2}$$

$$S = \frac{f}{F/\dot{m}_0} = \frac{0.0291 \times 3600}{58.4} = 1.79 \text{ lbm/(lbf} \cdot \text{h)} \quad [50.8 \text{ g/(kN} \cdot \text{s)}]$$
$$\text{Table 3.2}$$

$$I_{sp} = \frac{1}{g_0 S} = \frac{3600 \times 32.17}{32.17 \times 1.79} = 2010 \text{ s} \qquad \text{Table 3.2}$$

$$\frac{V_e}{V_0} = \left(\frac{1}{1+f}\right)\left(\frac{F}{\dot{m}_0 V_0} + 1\right) = \frac{1}{1.0291}\left(\frac{58.4 \times 32.17}{8000} + 1\right) = 1.20$$
$$\tag{3-11}$$

$$\eta_p = \frac{2\left\{(1+f)\dfrac{V_e}{V_0} - 1\right\}}{\left\{(1+f)\left(\dfrac{V_e}{V_0}\right)^2 - 1\right\}} = 0.975 \tag{3-12}$$

$$\eta_p \cong \frac{2}{\dfrac{V_e}{V_0} + 1} = 0.909 \tag{3-13}$$

$$\eta_{th} = \frac{\eta_o}{\eta_p} = 0.410 \tag{3-10}$$

3.3 ROCKET PERFORMANCE MEASURES

Hypersonic aerospace systems often employ a combination of air-breathing and rocket propulsion in order to take advantage of their relative strengths or compensate for their inherent weaknesses. Rockets must be considered, at a minimum, in situations where airbreathing engines cannot provide any thrust at all, such as when ramjets are standing still or scramjets are flown beyond the sensible atmosphere.

Rocket performance measures must be fundamentally different from those of airbreathing engines. For one thing, there is no entry airflow, so that specific thrust and fuel/air ratio are meaningless quantities. For another, rocket engine behavior is not easily susceptible to the "closed cycle" reasoning of thermodynamics, so that the prior overall, thermal, and propulsive efficiency discussions are not applicable.

Briefly, then, the only performance measures that are directly relevant to rockets are *specific fuel* (plus oxidizer) *consumption* and *specific impulse*. And, because, rockets are usually thought of in terms of the thrust they supply, the overwhelming choice of practitioners is specific impulse.

There are interrelationships that provide insight into specific impulse and specific fuel consumption. Referring again to Fig. 3.1, and recognizing that for rockets the entry air flow is 0, we see that the fuel (plus oxidizer) mass flow rate (denoted by $\dot{m}_f$) equals the exhaust mass flow rate, and the uninstalled thrust is merely the exhaust momentum flux, it follows that

$$I_{sp} = \frac{F}{g_0 \dot{m}_f} = \frac{\dot{m}_f V_e}{g_0 \dot{m}_f} = \frac{V_e}{g_0} = \frac{1}{g_0 S} \tag{3-14}$$

so that there is a one-to-one correspondence between specific impulse, specific fuel consumption, and exhaust velocity.

As an example, consider a typical hydrogen-oxygen rocket engine with a specific impulse of 460 s. Equation (3-14) immediately reveals that the exit velocity must be a remarkably large 14,800 ft/s (4510 m/s). Equation (3-14) also shows that the corresponding specific fuel consumption is 7.83 lbm/(lbf·h) [222 g/(kN·s)], which is quite high compared to the airbreathing engine performance measure example results of the previous section.

3.4 AEROSPACE SYSTEM PERFORMANCE MEASURES

The total initial mass of an aerospace system is an excellent indicator of the cost of accomplishing a given mission. On one hand, the acquisition cost certainly increases with the amount of material and the number and size of the parts involved. On the other, the operational, maintenance and repair costs increase with the size and complexity of the total system, particularly when special launching and landing facilities or fuels are required. Total initial mass, therefore, is widely used as the first basis for the comparison of competing aerospace systems, in spite of the fact that it is an imperfect measure and that any final decision requires far more sophisticated and elaborate economic analyses.

Since the gravitational force or weight of a fixed mass decreases as the geometric altitude increases and the force of gravity decreases, mass is preferred to weight for clarity in aerospace analyses. This removes an important source of ambiguity that can occur when one aerospace system is compared to another.

The initial mass is the total mass of the aerospace system just before it begins its mission. The initial mass m_i is also simply the sum of the payload mass m_p, the empty mass m_e, and the fuel mass m_f, or

$$m_i = m_p + m_e + m_f \qquad (3\text{-}15)$$

It should be obvious that what is included in each of these mass categories must be precisely defined and rigorously applied in order to achieve consistent and understandable results. Generally speaking, the most difficult questions revolve around the definition of payload mass. Agreement must be made in advance, for example, whether or not payload mass will include the crew and their personal equipment, or easily removed devices that are at least partially dedicated to the support of the specific payload under consideration. Once the payload mass category is adequately specified, the empty mass category will necessarily include anything but payload and fuel.

Nevertheless, it is not essential to have an absolutely pure definition of each mass category. Provided that everything that adds to the initial mass is accounted for, and that all parties apply the working definitions properly, correct and practical results will be obtained.

With the mass category definitions in hand, it is usually possible to determine the payload mass directly from the mission or system specification. Thus, it will be assumed henceforth that the payload mass m_p is known, and that the goal is to find the empty mass m_e and fuel mass m_f needed to perform the mission, and therefore the initial mass m_i.

3.4.1 Fuel Mass Fraction

The discussions of Sec. 3.2 underscore the fact that the function of the airbreathing engine is to convert the chemical energy stored in the fuel into mechanical energy for the aerospace system. The relationships derived there make it possible to determine the amount of fuel that will be consumed for any given type of flight with any given engine cycle. Since the analysis depends strongly upon whether or not the kinetic energy and/or potential energy of the aerospace system is changing, two different cases will be considered separately. The first case is cruise, for which they are essentially constant. The second case is transatmospheric flight (or boosting), for which they both increase.

Please note that the installed thrust of the airbreathing engine T is less than the uninstalled thrust F by an amount equal to the *installation or external drag* D_e. This bookkeeping system is widely employed because the external drag depends upon the "throttle setting" of the engine, and is therefore largely determined by the design and operation of the engine. Experience has shown that it is convenient to represent the installation drag as a fraction of the uninstalled engine thrust because the latter is a fixed reference quantity and because the resulting algebra is straightforward. Thus,

$$D_e = \phi_e F \qquad (3\text{-}16)$$

where ϕ_e is the engine *installation drag* or *loss coefficient*, and

$$T = F - D_e = F(1 - \phi_e) \qquad (3\text{-}17)$$

3.4.1.1 Hypersonic cruise aircraft: airbreathing propulsion.
An aircraft is said to be cruising when it is flying in a straight line at constant speed and altitude. Neither the velocity nor the geometric altitude of the aircraft are changing with time. When Eq. (2-1) is applied to this case it is evident that the installed thrust of the engine instantaneously equals the drag of the aircraft, and this result may be combined with Eqs. (3-9) and (3-17) and rearranged to yield

$$D = T = F(1 - \phi_e) = \frac{\eta_o h_{PR}(1 - \phi_e)}{V_0} \cdot \frac{dm_f}{dt} \qquad (3\text{-}18)$$

where the interchangeability of V and V_0 has been used. Similarly, when Eq. (2-2) is applied to this case, omitting the negligible effects of centrifugal forces, it is evident that the lift of the aircraft instantaneously equals the weight mg_o of the aircraft, and this result may be combined with Eqs. (2-13), (2-14), and (3-18) to yield

$$mg_o \cdot \frac{C_D}{C_L} = \frac{\eta_o h_{PR}(1 - \phi_e)}{\dfrac{dR}{dt}} \cdot \frac{dm_f}{dt} \tag{3-19}$$

where dR is the incremental range or distance flown during the time interval dt. Since the rate at which the mass of the aircraft decreases is the negative of the rate at which fuel is consumed, Eq. (3-19) can be rearranged into the differential equation form

$$\frac{dm}{m} = -\left\{ \frac{g_o}{\eta_o h_{PR}(1 - \phi_e)\dfrac{C_L}{C_D}} \right\} \cdot dR \doteq -B \cdot dR \tag{3-20}$$

Equation (3-20), which amounts to equating the thrust power to the power dissipated by the drag of the aircraft plus engine, is obviously a statement about the use of the fuel chemical energy. Although Eq. (3-20) could be numerically integrated over any finite range of flight, it is more convenient and instructive to consider the frequently encountered situation in which B is either constant or a suitable average over the range can be chosen, and Eq. (3-20) is integrated in closed form to solve for the fraction of the initial mass remaining at the end of the flight

$$\frac{m_{\text{final}}}{m_i} = exp\left\{ -\frac{g_o R}{\eta_o h_{PR}(1 - \phi_e)\dfrac{C_L}{C_D}} \right\} = e^{-BR} \tag{3-21}$$

This is a version of the famous *Breguet range formula*, the exponent being the range factor B times the range or distance flown R.[3.3]

   From the airbreathing propulsion standpoint, the quantity of greatest interest is the fraction of the initial aircraft mass that must be fuel in order to accomplish this flight, known as the *fuel mass fraction*

$$\Pi_f = \frac{m_i - m_{\text{final}}}{m_i} = 1 - \frac{m_{\text{final}}}{m_i} = 1 - e^{-BR} \tag{3-22}$$

This result certainly displays the correct trends, fuel mass fraction decreasing as engine overall efficiency, fuel heating value, and ratio of lift coefficient to drag coefficient increase, and increasing as the

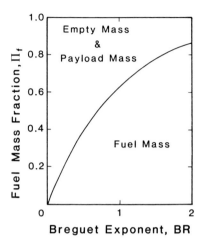

Fig. 3.2   Fuel mass fraction as a function of the Breguet exponent *BR* for hypersonic cruise aircraft.

engine external drag coefficient and range increase. The general nature of Eq. (3-22) is illustrated in Fig. 3.2, where it can be seen that BR must be less than 1 in order to have much left for payload or structure.

Examples

Consider an aircraft system being designed to cruise at hypersonic speed for a distance of 5000 mi (8047 km). The airbreathing engine has an overall efficiency of 0.50 and the $(1 - \phi_e)C_L/C_D$ is 5.0 under these flight conditions. The fuel mass fraction is to be calculated for two different types of fuel.

3.3 Hydrogen fuel, $h_{PR}= 51,600$ BTU/lbm (120,000 kJ/kg)

$$BR = \frac{32.17 \times 5000 \times 5280}{0.50 \times 51,600 \times 778.2 \times 32.17 \times 5.0} = 0.263 \quad (3\text{-}20)$$

$$\Pi_f = 1 - e^{-0.263} = 0.231 \quad (3\text{-}22)$$

3.4 Hydrocarbon fuel, $h_{PR}= 19,000$ BTU/lbm (44,200 kJ/kg)

$$BR = \frac{32.17 \times 5000 \times 5280}{0.50 \times 19,000 \times 778.2 \times 32.17 \times 5.0} = 0.714 \quad (3\text{-}20)$$

$$\Pi_f = 1 - e^{-0.714} = 0.510 \quad (3\text{-}22)$$

Equation (3-22) makes possible the rapid exploration of a number of hypersonic cruise aircraft options. For example, when solved for airbreathing engine overall efficiency, it becomes

$$\eta_o = \frac{g_0 R}{h_{PR}(1 - \phi_e)\dfrac{C_L}{C_D}\ell n\left(\dfrac{1}{1 - \Pi_f}\right)} \qquad (3\text{-}23)$$

This relationship has been plotted in Fig. 3.3 for a typical hydrocarbon fuel vehicle. This information allows the minimum airbreathing engine overall efficiency needed to cruise a given distance with a given fuel mass fraction to be found, and reveals that the required overall efficiency is very sensitive to the available fuel mass fraction. The overall efficiency can, of course, be converted into any of the other airbreathing engine performance measures by means of Table 3.2. However, additional information about the flight speed and/or the fuel/air ratio will have to be supplied.

3.4.1.2 Hypersonic cruise aircraft: rocket propulsion.   The fuel (plus oxidizer) mass fraction for hypersonic cruise aircraft powered by rockets can be obtained by the same procedure as in the preceding section, except that Eq. (3-14) is used in place of Eq. (3-9) at the beginning.

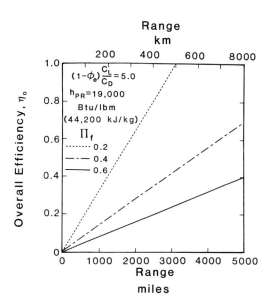

Fig. 3.3   Required airbreathing engine overall efficiency as a function of hypersonic aircraft cruise range and fuel mass fraction, for a typical set of hydrocarbon fuel system parameters.

When this is done, Eq. (3-22) still applies, but

$$B = \left\{ \frac{1}{I_{sp}V(1 - \phi_e)\dfrac{C_L}{C_D}} \right\} \tag{3-24}$$

Figure 3.2 therefore also pertains to the case of rocket propulsion, although the cruise speed must be specified in order to evaluate B.

Similarly, for the case of rocket propulsion, the specific impulse equivalent of Eq. (3-23) is

$$I_{sp} = \frac{R}{V(1 - \phi_e)\dfrac{C_L}{C_D}\ell n \left(\dfrac{1}{1 - \Pi_f}\right)} \tag{3-25}$$

This relationship has been plotted in Fig. 3.4 for a typical rocket vehicle. Since realistic values of specific impulse are certainly less than 500 s (and are usually 300–400 s), this example makes it immediately clear that rocket powered cruise vehicles for ranges in excess of several hundred miles are not an attractive option. Nevertheless, the same information illustrates that rocket propulsion is practical for short range, high speed flight, and has indeed found many applications there.

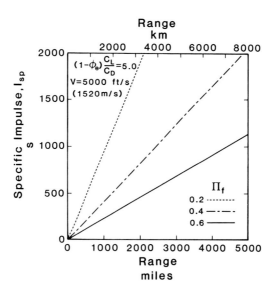

Fig. 3.4   Required rocket engine specific impulse as a function of hypersonic aircraft cruise range and fuel (plus oxidizer) mass fraction, for a typical set of system parameters.

3.4.1.3 Transatmospheric vehicles: airbreathing propulsion. Transatmospheric vehicles generally accelerate from a static ground base through the atmosphere in order to reach a stationary Earth orbit. From the propulsion standpoint, the outstanding feature of transatmospheric flight is the enormous amount of kinetic plus potential energy that must be imparted to each unit of terminal mass. This situation stands in stark contrast to that of hypersonic cruise, and therefore has a substantially different analysis.

The starting point is, however, the same, namely Eq. (2-1), which, when multiplied by the instantaneous flight velocity, becomes

$$mV\frac{dV}{dt} = m\frac{d}{dt}\left(\frac{V^2}{2}\right) = (T - D)V - mg\frac{dr}{dt} \qquad (3\text{-}26)$$

where r is the radial distance from the center of the Earth, and $dr/dt = V \sin\theta$. Equation (3-26) can be combined with Eqs. (3-9) and (3-17) and rearranged to yield

$$\frac{dm}{m} = -\left\{\frac{d\left(\dfrac{V^2}{2}\right) + gdr}{\eta_o h_{PR}\left(1 - \dfrac{D + D_e}{F}\right)}\right\} \qquad (3\text{-}27)$$

where the interchangeability of V and V_0 has been used. This equation plays the same role for transatmospheric flight that Eq. (3-20) does for hypersonic cruise, but it cannot be applied to the latter case because the numerator and denominator are both 0 under cruise conditions. Nevertheless, like Eq. (3-20), it is obviously a statement about the use of the fuel chemical energy, and amounts to equating the thrust power to the rate at which the kinetic plus potential energy of the system is increasing plus the power dissipated by the drag of the aircraft plus engine. In fact, the group of terms in the denominator

$$\eta_o\left(1 - \frac{D + D_e}{F}\right)$$

is often referred to by workers in the field as the *effective overall efficiency* because it combines the internal performance of the engine together with the external performance of the installation and airframe. This expression for effective overall efficiency emphasizes the fact that, all other things being equal, it is advantageous to make the ratio of thrust to drag as large as possible in order to save fuel.

Equation (3-27), like Eq. (3-20), can be numerically integrated over any finite flight path, but it is again more convenient and instructive to consider the situation in which the denominator is either constant or a suitable average over the flight can be chosen, and Eq. (3-27) is integrated in closed form to solve for the fraction of the initial mass remaining at the end of the mission

$$\frac{m_{\text{final}}}{m_i} = exp\left\{-\frac{\left(\dfrac{V_{\text{final}}^2}{2} - \dfrac{V_i^2}{2}\right) + \int\limits_{i}^{\text{final}} g\, dr}{\eta_o h_{PR}\left(1 - \dfrac{D + D_e}{F}\right)}\right\} \qquad (3\text{-}28)$$

It is an interesting and fortunate result that this expression has the same "Breguet-like" nature as Eq. (3-21), despite the contrast in the types of flight involved. The corresponding fuel mass fraction is

$$\Pi_f = 1 - exp\left\{-\frac{\left(\dfrac{V_{\text{final}}^2}{2} - \dfrac{V_i^2}{2}\right) + \int\limits_{i}^{\text{final}} g\, dr}{\eta_o h_{PR}\left(1 - \dfrac{D + D_e}{F}\right)}\right\} \qquad (3\text{-}29)$$

This result also displays the correct trends, in that the fuel mass fraction required to accomplish the mission increases as the total kinetic plus potential energy difference increases, and as the effective overall efficiency and fuel heating value decrease.

Before the design consequences of Eq. (3-29) can be quantitatively examined, it will be transformed to reflect the kinetic and potential energy increases that accompany typical transatmospheric missions. This is easily done for the case of aerospace planes that travel in a single stage from an airbase on Earth to stationary orbit, as follows.

Figure 3.5 depicts a mass m in a stationary or circular orbit of radius r about the Earth. The *orbital energy* of the mass is defined here as the sum of its orbital kinetic and potential energies relative to an inertial reference frame at the surface, or

$$\text{Orbital energy} = \frac{mV^2}{2} + \int_{r_o}^{r} mg\, dr$$

Dividing both sides of this equation by the mass m, it can be seen that the *mass specific orbital energy* oe is given by the expression

$$oe = \frac{\text{Orbital energy}}{m} = \frac{V^2}{2} + \int_{r_o}^{r} g\, dr \qquad (3\text{-}30)$$

Aside from the negligible contribution of the initial velocity squared, the right-hand side of Eq. (3-30) is identical to the numerator of the exponent found in Eqs. (3-28) and (3-29), so that for transatmospheric travel we can write

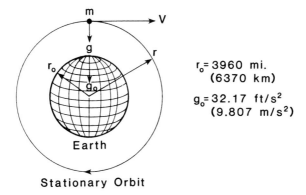

Fig. 3.5   Schematic representation of an object in a stationary orbit about the Earth.

$$\Pi_f = 1 - exp\left\{ -\frac{oe}{\eta_o h_{PR}\left(1 - \frac{D + D_e}{F}\right)} \right\} \qquad (3\text{-}31)$$

Since the gravitational field of the Earth follows the inverse square law, then,

$$\int_{r_o}^{r} g\,dr = \int_{r_o}^{r} g_o \left(\frac{r_o}{r}\right)^2 dr = g_o r_o \left(1 - \frac{r_o}{r}\right) \qquad (3\text{-}32)$$

When Eq. (2-2) is applied to stationary orbits above the sensible atmosphere where lift is negligible, it can be used to show that

$$\frac{V^2}{2} = \frac{gr}{2} = \frac{g_o r}{2}\left(\frac{r_o}{r}\right)^2 = \frac{g_o r_o}{2}\left(\frac{r_o}{r}\right) \qquad (3\text{-}33)$$

Equations (3-30), (3-32), and (3-33) can now be combined to give the desired, elementary relationship for the specific orbital energy

$$\frac{oe}{g_o r_o} = \underbrace{\frac{1}{2}\frac{r_o}{r}}_{\text{Kinetic}} + \underbrace{\left(1 - \frac{r_o}{r}\right)}_{\text{Potential}} = 1 - \frac{1}{2}\frac{r_o}{r} \qquad (3\text{-}34)$$

This result applies, of course, to any planetary body, and it is quite interesting to see it portrayed as in Fig. 3.6, which shows the specific orbital energy as well as the contributions from specific kinetic energy and specific potential energy as a function of the orbital radius ratio r/r_o. For objects in low Earth orbit, about 100 miles

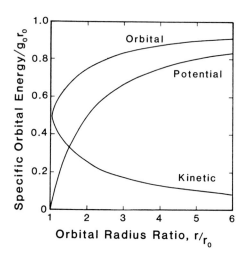

Fig. 3.6 Specific orbital energy, specific kinetic energy, and specific potential energy for stationary planetary orbits as a function of orbital radius ratio.

(161 km) above the surface of the Earth, r/r_o is only about 1.025 and the specific orbital energy overwhelmingly consists of specific kinetic energy. In fact, the ratio of kinetic to potential is about 20 for this case. Hence, the engines on the aerospace plane are largely there to generate kinetic energy, and are accordingly referred to in the field as "accelerators." For objects in geostationary Earth orbit, about 22,500 miles (36,200 km) from the center of the Earth, r/r_o is about 5.68, and the specific orbital energy is mostly specific potential energy. The ratio of kinetic to potential is less than 0.11 for this case.

Equation (3-34) also allows the required magnitude of the specific orbital energy to be calculated for a given r/r_o from an easily obtained physical quantity, namely the product of the radius of the planet and the acceleration due to gravity at the surface of the planet. In the case of the Earth, using the data shown in Fig. 3.5, this product is 673×10^6 (ft/s)^2 [62.5×10^6 (m/s)^2]. In the case of low Earth orbits, where the orbital radius ratio is very nearly 1, Eq. (3-33) reveals that the necessary orbital velocity is 25,900 ft/s (7900 m/s). While such great velocities have become quite commonplace during the space age, they must still be regarded as extraordinarily large for airplanes driven by airbreathing engines. Using the representative speed of sound of Sec. 2.4.9 of 980 ft/s (299 m/s), the low Earth orbital velocity corresponds to a Mach number of over 26, which explains the choice of Mach number range found throughout this textbook.

Finally, Eqs. (3-31) and (3-34) are combined to produce the desired formulation for the fuel mass fraction, namely

$$\Pi_f = 1 - exp\left\{-\frac{g_o r_o \left(1 - \frac{1}{2}\frac{r_o}{r}\right)}{\eta_o h_{PR}\left(1 - \frac{D + D_e}{F}\right)}\right\} \qquad (3\text{-}35)$$

which will be used to evaluate the performance of transatmospheric vehicles.

Examples

Consider an aerospace plane being designed to fly as a single stage to a low Earth orbit having $r/r_o = 1.03$. An appropriate average for the effective overall efficiency for the trip is 0.35. The fuel mass fraction is to be calculated for two different types of fuel.

3.5 Hydrogen fuel, $h_{PR}=51{,}600$ BTU/lbm (120,000 kJ/kg)

$$\left[\frac{g_o r_o \left(1 - \frac{1}{2}\frac{r_o}{r}\right)}{\eta_o h_{PR}\left(1 - \frac{D + D_e}{F}\right)}\right] = \frac{673 \times 10^6 \left(1 - \frac{1}{2 \times 1.03}\right)}{0.35 \times 51{,}600 \times 778.2 \times 32.17}$$

$$= 0.766$$

$$\Pi_f = 1 - e^{-0.766} = 0.535 \qquad (3\text{-}35)$$

3.6 Hydrocarbon fuel, $h_{PR}=19{,}000$ BTU/lbm (44,200 kJ/kg)

$$\left[\frac{g_o r_o \left(1 - \frac{1}{2}\frac{r_o}{r}\right)}{\eta_o h_{PR}\left(1 - \frac{D + D_e}{F}\right)}\right] = \frac{673 \times 10^6 \left(1 - \frac{1}{2 \times 1.03}\right)}{0.35 \times 19{,}000 \times 778.2 \times 32.17}$$

$$= 2.08$$

$$\Pi_f = 1 - e^{-2.08} = 0.875 \qquad (3\text{-}35)$$

Equation (3-35) makes possible the rapid exploration of a number of transatmospheric vehicle options. For example, when solved for airbreathing engine overall efficiency, it becomes

$$\eta_o = \frac{g_o r_o \left(1 - \frac{1}{2}\frac{r_o}{r}\right)}{h_{PR}\left(1 - \frac{D + D_e}{F}\right) \ell n\left(\frac{1}{1 - \Pi_f}\right)} \qquad (3\text{-}36)$$

This relationship has been plotted in Fig. 3.7 for a typical situation. This information allows the minimum airbreathing engine overall efficiency needed to reach a given orbital radius from the Earth to be

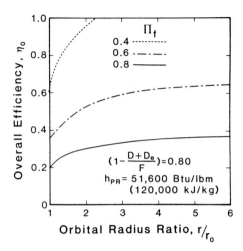

Fig. 3.7   Required airbreathing engine overall efficiency as a function of orbital radius ratio from the Earth and fuel mass fraction, for a typical set of hydrogen fuel system parameters.

found, and reveals that the required overall efficiency is extremely sensitive to the available fuel mass fraction. The overall efficiency can, of course, be converted into any of the other airbreathing engine performance measures by means of Table 3.2. However, additional information about appropriate values for the flight speed and/or the fuel/air ratio will have to be supplied.

3.4.1.4 Transatmospheric vehicles: rocket propulsion.   The fuel (plus oxidizer) mass fraction for transatmospheric vehicles powered by rockets can be obtained by the same procedure as in the preceding section, except that Eq. (3-14) is substituted for Eq. (3-9) at the beginning. When this is done, Eq. (3-27) becomes

$$\frac{dm}{m} = -\left\{ \frac{d\left(\dfrac{V^2}{2}\right) + g\,dr}{g_0 I_{sp} V \left(1 - \dfrac{D + D_e}{F}\right)} \right\} \tag{3-37}$$

If we restrict our attention to low Earth orbits, for which r/r_0 is very close to 1 and the potential energy is very much less than the kinetic energy, the equivalent of Eq. (3-35) is

$$\Pi_f = 1 - exp\left\{ -\frac{\sqrt{g_0 r_0}}{g_0 I_{sp} \left(1 - \dfrac{D + D_e}{F}\right)} \right\} \tag{3-38}$$

and the equivalent of Eq. (3-36) is

$$I_{sp} = \frac{\sqrt{g_0 r_0}}{g_0 \left(1 - \dfrac{D + D_e}{F}\right) \ln \left(\dfrac{1}{1 - \Pi_f}\right)} \tag{3-39}$$

This relationship has been plotted in Fig. 3.8 for typical low Earth orbit rocket vehicle parameters. Since realistic values of specific impulse are certainly less than 500 s (and are usually 300–400 s), this example shows that we should expect present-day fuel (plus oxidizer) mass fractions of the order of 0.9. This confirms the information contained in Table 1.1, and also shows that the specific impulse must be increased to at least 1000 s in order to make an appreciable fraction of the initial mass available for empty mass and payload mass.

3.4.2 Empty Mass Fraction

The empty mass of an aerospace system consists of such things as the vehicle structure and control surfaces, avionics and guidance equipment, wheels and landing gears, propulsion engines, fuel system (including tanks, pumps, valves, piping, and safety-related items), instrumentation and control apparatus, housekeeping and environmental control machines, thermal management devices, crew quarters and necessities (including food, water, and oxygen), and escape and

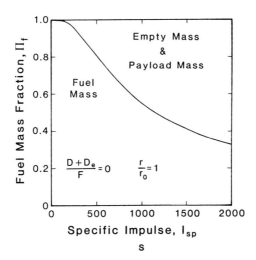

Fig. 3.8 Required rocket engine fuel (plus oxidizer) mass fraction as a function of specific impulse, for a typical low Earth orbit mission.

rescue equipment. In fact, as noted earlier, the empty mass includes everything but the payload and fuel.

Faced with this formidable array of matter to be accounted for, it must be obvious that the empty mass cannot be evaluated as crisply as the payload or fuel. Nevertheless, experience has shown that reasonable estimates, or probable upper and lower bounds, can be put in advance upon the ratio of empty mass to initial mass, or the *empty mass fraction*

$$\Pi_e = \frac{m_e}{m_i} \tag{3-40}$$

In the actual real world of aerospace, the design process begins with best estimates (or even guesses) for the empty mass fraction, in order to determine whether or not the proposed system is even worth pursuing. The realistically achievable empty mass fraction is only discovered much later, after many iterative design loops have been carried out, and the desired detail and credibility have been attained. The target empty mass fraction is therefore one of the last properties of the aerospace system to be determined, and even then is subject to considerable variability because it is based upon technological projections that may not happen on schedule (or ever).

Strong encouragement that an intelligent initial estimate for the empty mass fraction of an aerospace system can be made may be found in Fig. 3.9, which contains a sequence of data points for production, high performance fighter aircraft.[3.4,3.5] Despite the fact that the fighter aircraft displayed in Fig. 3.9 have a wide variety of missions and that they represent several different generations of design done by entirely different companies, their empty mass fractions all easily fall within 0.10 of the historical correlation also shown there.

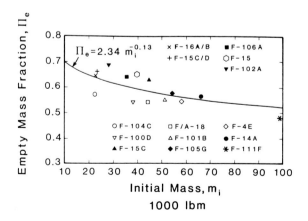

Fig. 3.9 **Empty mass fraction as a function of initial mass for production, high performance fighter aircraft, Ref. 3.4.**

Similar correlations with comparable accuracy are available for other types of aircraft, such as cargo, passenger, or turboprop. One may enjoyably speculate about the reasons behind these agreeable results, but it would be difficult to argue that they do not exist or that they are not part of the arsenal of successful preliminary designers.

It will therefore hereinafter be assumed that a reasonably accurate, experience based estimate of the empty mass fraction can be made for *any* aerospace system under consideration. Of course, every attempt should and will be made to improve the initial estimate in order to increase confidence in the results and to reduce the number of iterations needed to converge to a solution.

3.4.3 Initial Mass Ratio

Because the empty mass and fuel mass naturally expressed themselves as fractions of the initial mass, Eq. (3-15) can now be cast into the extremely compact, transparent, and productive form

$$\frac{m_i}{m_p} = \Gamma = \frac{1}{1 - \Pi_e - \Pi_f} \tag{3-41}$$

The left-hand side of Eq. (3-41) is one of the most direct and important measures of aerospace system performance, namely the ratio of the initial mass to the payload mass, also known as the *initial mass ratio*. A goal of design is to make the initial mass ratio as small as possible. Therefore, the initial mass ratio may be used in its raw form to rank-order aerospace systems intended to accomplish the same mission. Furthermore, since the payload mass is presumed known, the initial mass ratio may be used to calculate the initial takeoff or launch mass of a system. Finally, the inverse of the initial mass ratio will be recognized as the *payload mass fraction*, a figure of merit frequently employed in commercial aviation.

The right-hand side of Eq. (3-41) shows that the initial mass ratio increases as either the empty mass fraction or the fuel mass fraction increases. When the sum of the empty mass fraction and the fuel mass fraction reaches 1, there is no room left for any payload, and the initial mass ratio is infinite. Even though this limit is never approached in practice, this example drives home the fact that Eq. (3-41) is extremely nonlinear in the sense that a very small increase in either the empty mass fraction or the fuel mass fraction can cause a very large increase in the initial mass ratio. Fortunately, the reverse is also true. It will be well to bear in mind, then, that the leading indicator is not by how much empty mass fraction plus fuel mass fraction changes, but by how much its difference from 1 changes.

Equation (3-41) is possibly the most fascinating and potent arrow in the entire quiver of design because it integrates the contributions of all the participants together while maintaining their separate identities. In order to see this clearly, imagine that Eq. (3-41) is being used to calculate the initial mass as an indicator of system cost. To begin with, the initial mass is directly proportional to the payload mass, which is a requirement either explicitly or implicitly stated by the customer. The initial mass may therefore be made larger (or smaller) by increasing (or decreasing) the demands of the mission. Continuing, the empty mass fraction is largely the province of structures and materials specialists representing either the entire aerospace system or its assorted components. Finally, the fuel mass fraction has been found to be the result of the combined efforts of propulsion and vehicle performance specialists, as well as the severity of the mission. A corollary is that a successful system is usually the result of balancing requirements and capabilities, and insuring that the technical goals are understood and met at every level.

Equation (3-41) may be used in its indirect form

$$\Pi_e = 1 - \Pi_f - \frac{1}{\Gamma} \tag{3-42}$$

in order to determine the empty mass fraction required by an allowable initial mass ratio and a given fuel mass fraction.

Occasionally, other measures of aerospace system performance are of interest, and they can generally be derived from the material above. For example, in some situations the fuel is regarded as "free," and the desired performance measure is the ratio of initial "hardware" mass (empty mass plus payload mass) to payload mass. Using our definitions and equations, it follows that this performance measure can be expressed as

$$\frac{m_e + m_p}{m_p} = \frac{m_i - m_f}{m_p} = \frac{m_i}{m_p}\left(1 - \frac{m_f}{m_i}\right) = \Gamma(1 - \Pi_f)$$

$$= \frac{m_e}{m_p} + 1 = \frac{m_e}{m_i} \cdot \frac{m_i}{m_p} + 1 = \Pi_e\Gamma + 1 \tag{3-43}$$

and evaluated from already known quantities.

Examples

3.7 Consider the hydrogen fuel, airbreathing, hypersonic cruise aircraft of Example 3.3. Assuming that an empty mass fraction of 0.50 is possible for this application, calculate the resulting initial mass ratio.

$$\Gamma = \frac{1}{1 - \Pi_e - \Pi_f} = \frac{1}{1 - 0.50 - 0.231} = 3.72 \tag{3-41}$$

3.8 Consider the hydrocarbon fuel, airbreathing, hypersonic cruise aircraft of Example 3.4. Assuming that the initial mass ratio cannot exceed 8.0 for this application, calculate the allowable empty mass fraction.

$$\Pi_e = 1 - \Pi_f - \frac{1}{\Gamma} = 1 - 0.510 - \frac{1}{8.0} = 0.365 \qquad (3\text{-}42)$$

3.9 Consider the hydrogen fuel, airbreathing, transatmospheric vehicle of Example 3.5. Assuming that an empty mass fraction of 0.40 is possible for this application, calculate the resulting initial mass ratio.

$$\Gamma = \frac{1}{1 - \Pi_e - \Pi_f} = \frac{1}{1 - 0.40 - 0.535} = 15.4 \qquad (3\text{-}41)$$

3.10 Consider the hydrocarbon fuel, airbreathing, transatmospheric vehicle of Example 3.6. Assuming that the initial mass ratio cannot exceed 20 for this application, calculate the allowable empty mass fraction.

$$\Pi_e = 1 - \Pi_f - \frac{1}{\Gamma} = 1 - 0.875 - \frac{1}{20} = 0.0750 \qquad (3\text{-}42)$$

3.11 Consider the rocket powered, transatmospheric vehicle of Fig. 3.8 having a specific impulse of 400 s. Assuming that the initial mass ratio cannot exceed 20 for this application, calculate the allowable empty mass fraction.

$$\Pi_e = 1 - \Pi_f - \frac{1}{\Gamma} = 1 - 0.867 - \frac{1}{20} = 0.0830 \qquad (3\text{-}42)$$

3.4.4 Required Airbreathing Engine Overall Efficiency

The availability of Eq. (3-41) makes it possible to rapidly explore the feasibility of a great variety of hypersonic aerospace system options in *terms of the airbreathing engine overall efficiency*. This, in turn, permits the direct evaluation of either the airbreathing engine overall efficiency required to accomplish a mission, or the impact of efficiency gains or losses on other design parameters.

3.4.4.1 Hypersonic cruise aircraft: airbreathing propulsion. Combining Eqs. (3-23) and (3-41), we find that

$$\eta_o = \frac{g_0 R}{h_{PR}(1 - \phi_e)\dfrac{C_L}{C_D}\ell n\left(\dfrac{1}{\Pi_e + \frac{1}{\Gamma}}\right)} \qquad (3\text{-}44)$$

for hypersonic cruise aircraft propelled by airbreathing engines.

Examples of the use of this relationship are found in the parametric studies presented in Figs. 3.10 and 3.11, which are comparable to Fig. 3.3 for a range of 5000 mi (8045 km). The most compelling conclusions arising from this information are that the required airbreathing engine overall efficiency can be greatly reduced by increasing either the heating value of the fuel (switching from hydrocarbon to hydrogen) or the amount of fuel (decreasing the empty mass fraction). Moreover, bearing in mind that the initial mass ratio will not be allowed to exceed 10 (a payload mass fraction of 0.10) for profitable commercial operation, and that the lowest probable empty mass fraction is about 0.35 (the KC-135), it appears that an airbreathing engine overall efficiency in excess of 0.45 will be required if a hydrocarbon fuel is used. In contrast, using hydrogen fuel would require an overall efficiency in the vicinity of only 0.20 for an empty mass fraction of 0.45 (the Concorde).

3.4.4.2 Transatmospheric vehicles: airbreathing propulsion. Combining Eqs. (3-36) and (3-41), we find that

$$\eta_o = \frac{g_0 r_0 \left(1 - \frac{1}{2}\frac{r_0}{r}\right)}{h_{PR}\left(1 - \frac{D + D_e}{F}\right)\ln\left(\frac{1}{\Pi_e + \frac{1}{\Gamma}}\right)} \tag{3-45}$$

for transatmospheric vehicles propelled by airbreathing engines.

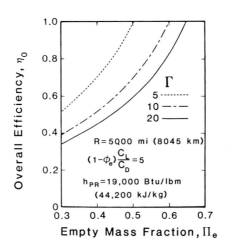

Fig. 3.10 **Required hydrocarbon fuel airbreathing engine overall efficiency as a function of empty mass fraction and initial mass ratio for a typical hypersonic cruise aircraft mission.**

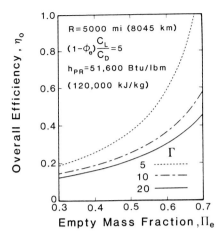

Fig. 3.11 Required hydrogen fuel airbreathing engine overall efficiency as a function of empty mass fraction and initial mass ratio for a typical hypersonic cruise aircraft mission.

An example of the use of this relationship is found in the parametric study presented in Fig. 3.12, which is comparable to Fig. 3.7 for a low Earth orbital radius ratio of 1.03. Once again, the required airbreathing engine overall efficiency is quite sensitive to empty mass fraction. Since an initial mass ratio of about 10 will be required in order to achieve sufficient productivity, the required overall efficiency will be in the vicinity of 0.45 if the empty mass fraction is 0.45. If the empty mass fraction can be reduced to 0.30, the required overall efficiency need only be about 0.30.

3.4.5 Multiple-Stage Vehicles

Until this point, we have dealt exclusively with single-stage vehicles because they are given first consideration in any design study. This is due to the fact that their operation and support are decidedly less complicated than that of multiple-stage vehicles. Nevertheless, staging can be beneficial because unneeded mass is discarded at opportune times during the flight, which reduces total energy or fuel mass required, and leaves more room for empty mass and/or payload mass. Therefore, as the rocket community has amply demonstrated, *staging* is a pragmatic method for increasing the initial mass ratio or payload mass fraction, and one that is *essential* in marginal situations. Referring to Chap. 1, we can see that the leading candidates for the aerospace plane include both single-stage to orbit (SSTO) and two-stage to orbit (TSTO) vehicles.

It is therefore important that we also have tools available for analyzing the performance of multiple-stage vehicles. Fortunately,

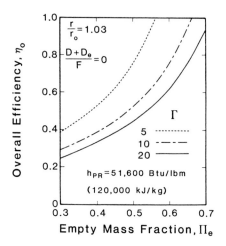

Fig. 3.12 Required hydrogen fuel airbreathing engine overall efficiency as a function of empty mass fraction and initial mass ratio for a typical near Earth orbit transatmospheric mission.

the study of single stages accomplished most of the necessary work, and very little remains to be done. The methodology for extending the single-stage results to multiple stages will be demonstrated now for the case of two stages, and the same principles can be applied to an arbitrary number of stages.

For a general two-stage vehicle, we can write

$$m_{i1} = m_{p1} + m_{e1} + m_{f1} \tag{3-46}$$

and

$$m_{i2} = m_{p2} + m_{e2} + m_{f2} \tag{3-47}$$

where the subscripts 1 and 2 refer to the first and second stages, respectively. Comparing the individual terms of these equations to their single-stage equivalents, it may be seen that

$$m_p = m_{p2} \qquad m_{p1} = m_{i2} \qquad m_i = m_{i1} \tag{3-48}$$

where the terms without numerical subscripts refer to the entire vehicle and therefore have the same meaning as for a single-stage vehicle. Consequently, it follows directly that

$$\Gamma = \frac{m_i}{m_p} = \frac{m_{i1}}{m_{p2}} = \frac{m_{i1}}{m_{p1}} \cdot \frac{m_{p1}}{m_{p2}} = \frac{m_{i1}}{m_{p1}} \cdot \frac{m_{i2}}{m_{p2}} = \Gamma_1 \Gamma_2 \tag{3-49}$$

where

$$\Gamma_1 = \frac{m_{i1}}{m_{p1}} = \frac{1}{1 - \Pi_{e1} - \Pi_{f1}} \tag{3-50}$$

and

$$\Gamma_2 = \frac{m_{i2}}{m_{p2}} = \frac{1}{1 - \Pi_{e2} - \Pi_{f2}} \qquad (3\text{-}51)$$

so that the two-stage vehicle initial mass ratio is simply the product of the initial mass ratios of the separate stages.

These general relationships will now be applied to the relatively straightforward but extremely important case of transatmospheric flight to low Earth orbit. Be mindful, however, that they are very general, and may be applied in arbitrarily complex situations.

Example 3.12: Two Airbreathing Engine Stages

Consider a two-stage vehicle being designed to travel to a low Earth orbit that employs airbreathing propulsion for both stages. Combining Eqs. (3-31), (3-50), and (3-51), we find that

$$\Gamma = \Gamma_1\Gamma_2 = \frac{1}{\left\{e^{-\alpha\Lambda_1} - \Pi_{e1}\right\}\left\{e^{-(1-\alpha)\Lambda_2} - \Pi_{e2}\right\}}$$

where

$$\Lambda_1 = \frac{oe}{\left\{\eta_o h_{PR}\left(1 - \dfrac{D + D_e}{F}\right)\right\}_1}$$

and

$$\Lambda_2 = \frac{oe}{\left\{\eta_o h_{PR}\left(1 - \dfrac{D + D_e}{F}\right)\right\}_2}$$

and α is the *energy split,* or the fraction of the total orbital energy oe provided by the first stage, and $1 - \alpha$ is the fraction of the orbital energy provided by the second stage. Note that Eq. (3-27) leads to the conclusion that any of the required energy not furnished by one stage must be furnished by the other, so that the two stages taken together must provide the same total orbital energy.

Suppose, further, that the two stages have the same system performance parameters, or

$$\Lambda_1 = \Lambda_2 = \Lambda$$

and that they have identical empty mass fractions, or

$$\Pi_{e1} = \Pi_{e2} = \Pi_e$$

Under these conditions, the desired expression for the initial mass ratio becomes

$$\Gamma = \frac{1}{\left\{e^{-\alpha\Lambda} - \Pi_e\right\}\left\{e^{-(1-\alpha)\Lambda} - \Pi_e\right\}}$$

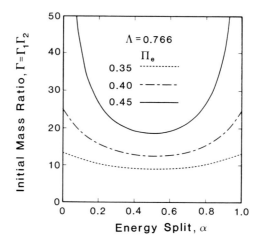

Fig. 3.13 Initial mass ratio of a typical two airbreathing engine stage, hydrogen fuel, low Earth orbit vehicle as a function of energy split and empty mass fraction.

The behavior of this expression is illustrated in Fig. 3.13, where the dependence of the initial mass ratio on energy split and empty mass fraction is depicted. The numerical values conform to the hydrogen fuel transatmospheric vehicle of Example 3.5, and may be regarded as typical.

With regard to energy split, both the form of the governing equation and the shape of the lines agree that initial mass ratio is symmetrical about $\alpha = 0.5$ for a given empty mass fraction. More importantly, $\alpha = 0.5$ is also the point at which the initial mass ratio reaches its *minimum* value of

$$\Gamma = \frac{1}{\left(e^{-\frac{\Lambda}{2}} - \Pi_e \right)^2}$$

With regard to the empty mass fraction, several important conclusions emerge. First, the minimum initial mass ratio is very sensitive to the empty mass fraction, for the reasons stated at the beginning of Sec. 3.4.3. Second, the benefits of staging are proportionately much greater for the more marginal systems (i.e., larger initial mass ratios and empty mass fractions), providing support for the remarks made at the beginning of this section. Third, the sensitivity to energy split is proportionately much greater for the more marginal systems, and drives them toward an equal split.

We could have just as easily used this approach to understand the influence of orbital radius ratio, airbreathing engine overall efficiency, or fuel type on the initial mass ratio, with similar results, or, in the manner of Eq. (3-45), solved for the airbreathing engine overall efficiency required in order to achieve success. The essential point is that these analytical tools are insightful and flexible.

Example 3.13: Airbreathing Engine First Stage/Rocket Engine Second Stage

Consider a two-stage vehicle being designed to travel to a low Earth orbit that employs airbreathing propulsion for the first stage and rocket propulsion for the second stage. In view of the likelihood that a rocket engine may be inevitable for propulsion at the high speed, high altitude end of the flight trajectory and/or orbital insertion, this could provide a very attractive option for development. A case in point is the German Sänger Space Transportation system which, as detailed in Table 1.2, has this configuration.

The approach of Example 3.12 may be applied again to this situation, provided that we restrict our attention to orbits for which the orbital radius ratio r/r_0 is very close to 1 so that the potential energy may be ignored in the rocket analysis. Combining Eqs. (3-31) and (3-50), we find that

$$\Gamma_1 = \frac{1}{e^{-\alpha\Lambda_1} - \Pi_{e1}}$$

where

$$\Lambda_1 = \frac{g_0 r_0 \left(1 - \frac{1}{2}\frac{r_0}{r}\right)}{\left\{\eta_0 h_{PR}\left(1 - \frac{D + D_e}{F}\right)\right\}_1}$$

and α is the energy split, as above. Combining Eqs. (3-38) and (3-51), we find that

$$\Gamma_2 = \frac{1}{e^{-(1-\sqrt{\alpha})\Lambda_2} - \Pi_{e2}}$$

where

$$\Lambda_2 = \frac{\sqrt{g_0 r_0}}{g_0 \left\{I_{sp}\left(1 - \frac{D + D_e}{F}\right)\right\}_2}$$

Finally, combining these results with Eq. (3-49) leads to the desired expression for the initial mass ratio

$$\Gamma = \Gamma_1 \Gamma_2 = \frac{1}{\left\{e^{-\alpha\Lambda_1} - \Pi_{e1}\right\}\left\{e^{-(1-\sqrt{\alpha})\Lambda_2} - \Pi_{e2}\right\}}$$

The behavior of this expression is illustrated in Fig. 3.14, where the dependence of the initial mass ratio on stage performance parameters and empty weights is depicted. The numerical values of the airbreathing first stage conform to the hydrogen fuel transatmospheric vehicle of Example 3.5. The numerical values of the rocket propelled second stage conform to the example of Fig. 3.8 at a specific impulse of 400 s. For added realism, the empty weight fractions center about the projections for the Sänger Space Transportation System reported in Table 1.2, which are approximately 0.43 for the first stage and 0.23 for the second stage.

With regard to energy split, both the form of the governing equation and the shape of the lines agree that the initial mass ratio is not symmetrical about $\alpha = 0.5$, but is tilted slightly in favor of the first stage supplying the majority of the orbital energy. The minimum value of initial mass ratio, which does not have a simple mathematical expression in this instance, occurs at an energy split in the vicinity of 0.55.

With regard to the empty mass fractions, the foremost conclusion is that they exert considerable influence on the attainable initial mass ratio. In this example, the minimum initial mass ratio varies from only 9.9, for the optimistic (lower) estimates of empty mass fraction, to 14.2, for the pessimistic (higher) estimates of empty mass fraction. This sensitivity is an obvious source of both hope and fear for the designers of such vehicles.

3.5 RECAPITULATION

There could hardly be a technical journey more satisfying than the one in this chapter. The stated goal, namely to provide a broad

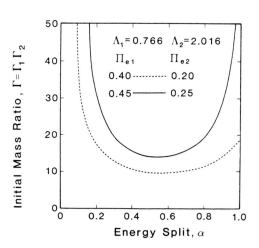

Fig. 3.14 Initial mass ratio of a typical two-stage airbreathing/rocket low Earth orbit vehicle as a function of energy split and empty mass fractions.

and balanced approach to the evaluation of the overall performance of hypersonic aerospace systems, has been conclusively reached.

The development relied upon only a handful of fundamental principles, all of which were easily recognized. The results represent nature faithfully, and may be applied to an extremely wide variety of hypersonic aerospace system circumstances. Their usefulness depends, in fact, largely upon our imagination.

The performance measures may be evaluated to any desired degree of accuracy, although the selected level should be consistent with the available database. The examples distributed through this chapter proved the flexibility of the methodology, as well as its ability to give insightful, reasonable answers. The same will be true of the problems at the end of the chapter.

Most importantly, however, this approach allows us to focus sharply upon the contribution of each of the major underlying technologies to hypersonic flight. In particular, the pivotal role played by the *overall efficiency and performance of the airbreathing engine* shines through. This is an important outcome in its own right, and even more so because it leads to the work of the remainder of this textbook, which is to understand the behavior of ramjets and scramjets and therefore predict and control their overall performance.

REFERENCES

[3.1] Oates, G. C., *The Aerothermodynamics of Gas Turbine and Rocket Propulsion,* Revised and Enlarged Edition, AIAA Education Series, Washington DC, 1988.

[3.2] Reynolds, W. C., and Perkins, H. C., *Engineering Thermodynamics,* McGraw-Hill, New York, 1977.

[3.3] Anderson, J. D. Jr., *Introduction to Flight,* McGraw-Hill, New York, 1989.

[3.4] Mattingly, J. D., Heiser, W. H., and Daley, D. H., *Aircraft Engine Design,* AIAA Education Series, New York, 1987.

[3.5] Raymer, D. P., *Aircraft Design: A Conceptual Approach,* AIAA Education Series, Washington, DC, 1989.

PROBLEMS

3.1   The expression for the stoichiometric fuel/air ratio f_{st} for the combustion of a general hydrocarbon fuel with representative air is sometimes more convenient to use when rewritten in terms of the fuel carbon/hydrogen ratio, x/y.

(a) Write Eq. (3-6) in terms of x/y and plot f_{st} as a function of x/y for all imaginable hydrocarbon fuels, from pure carbon to pure hydrogen.

(b) Superimpose on the above graph the heat of reaction or heating value h_{PR} of a wide variety of hydrocarbon fuels, including pure carbon (gaseous) and pure hydrogen, as a function of x/y.

(c) Although f_{st} and h_{PR} vary widely with x/y, their product $f_{st}h_{PR}$, which can be interpreted as the energy released per pound of *airflow*, is the critical quantity for propulsion. Use the data gathered above to superimpose $f_{st}h_{PR}$ as a function of x/y.

(d) Does this information reveal any trends? If so, why?

3.2   The goal of airbreathing engine designers is to achieve low specific fuel consumption and high specific thrust at the same time. For better or for worse, it is possible to demonstrate that the laws of nature make this a challenging task.

(a) Assuming that f is negligible in comparison to 1, combine Eqs. (3-2), (3-9), (3-10), (3-11), and (3-13) to show that specific fuel consumption must, in fact, *increase* with specific thrust in accordance with

$$S = \frac{1}{\eta_{th} h_{PR}} \left(\frac{F}{2\dot{m}_0} + V_0 \right)$$

(b) Strengthen your grasp on this fundamental concept by sketching a graph of the relationship between specific fuel consumption and specific thrust and answering the following questions. What is the influence of increasing thermal efficiency η_{th}, fuel heating value h_{PR}, or flight speed V_0 on this relationship? What can be done to an engine cycle to make specific thrust increase while η_{th}, h_{PR}, and V_0 remain constant?

3.3   Extend Table 3.2 to include a row and column for V_e/V_0 based on Eq. (3-11).

3.4   Complete Example 3.2 by calculating all the scramjet performance measures for flight speeds from 5000 ft/s (1524 m/s) to 25,000 ft/s (7620 m/s), and plotting the results.

(a) At what flight speed does a rocket with a specific impulse I_{sp} of 400 s become superior to the scramjet?

(b) How can the scramjet be effective at very high flight speeds where the velocity ratio V_e/V_0 is only slightly greater than 1?

(c) How can the exact propulsive efficiency exceed 1?

3.5   Reproduce Fig. 3.3 for a hypersonic cruise aircraft that uses hydrogen fuel instead of hydrocarbon fuel. Take the fuel heating value

h_{PR} to be 51,600 BTU/lbm (120,000 kJ/kg) and $(1 - \phi_e)C_L/C_D$ to be 4.0 (reflecting the increased drag of the increased volume associated with liquid hydrogen fuel). Does this information suggest the possibility of practical hypersonic travel?

3.6 Consistent bookkeeping is essential to the successful development of aerospace systems. In particular, all forces must be accounted for once, and only once, even though different accounting schemes are employed. Special discipline and meticulousness are therefore called for when different organizations are responsible for different parts of the aerospace system, as is the case when airbreathing engines are integrated into hypersonic aerospace systems.

The "effective overall efficiency" of Sec. 3.4.1.3 is a good (if trivial) example of a solid bookkeeping system in the sense that $D + D_e$ includes all the external drag of the vehicle plus engine, and any change in the portion of the external boundary arbitrarily assigned to one or the other can change either D or D_e, but not their sum.

A more complex situation can arise in highly integrated hypersonic aerospace systems because changing the arbitrary assignment of the external boundary can change the "uninstalled thrust." Satisfy yourself that our bookkeeping system is consistent by showing that the "effective overall efficiency" is invariant to the assignment of the external boundary.

3.7 Prove that the specific kinetic energy is one half the difference between the specific potential energy and the specific escape energy (i.e., $r/r_o \rightarrow \infty$) for any stationary planetary orbit.

3.8 Show that the specific kinetic energy associated with the rotational speed of the surface of the Earth is negligible compared to the lowest feasible specific orbital energy $g_o r_o/2$.

3.9 The overall efficiency η_o of scramjet engines is seldom constant over wide variations of flight speed V_0. Explore the behavior of the fuel mass fraction Π_f for the case of an airbreathing transatmospheric vehicle flying to a near Earth orbit, where $r/r_o = 1.0$ and potential energy effects are negligible, and the engine overall efficiency is given by the expression

$$\eta_o = \eta_R \left\{ 1 - \left(\frac{V_0}{V_R} \right)^2 \right\}$$

where η_R and V_R are reference constants.

(a) Write expressions for specific thrust $F/\dot{m}_0$, specific fuel consumption S, and specific impulse I_{sp} in terms of f and V_0/V_R, and sketch their behavior on a graph.

(b) Integrate Eq. (3-27), assuming that the remaining terms in the denominator of the exponent are constant and that the initial velocity is negligible, to show that

$$\Pi_f = 1 - \left\{1 - \frac{g_o r_o}{V_R^2}\right\}^K$$

where

$$K = \frac{V_R^2}{2\eta_R h_{PR}\left(1 - \frac{D + D_e}{F}\right)}$$

(c) Assuming that a typical value for K is 0.4, plot Π_f as a function of $g_o r_o/V_R^2$. What conclusions do you draw from this information?

3.10 There are other portrayals of performance that you may find revealing and/or convenient. For example:

(a) Calculate and replot Figs. 3.3 and 3.4 as fuel mass fraction Π_f versus range for several constant η_o or I_{sp} contours.

(b) Calculate and replot Fig. 3.7 as fuel mass fraction Π_f versus orbital radius ratio for several constant η_o contours.

3.11 Draw a universal plot of initial mass ratio Γ as a function of empty mass fraction Π_e for several constant Π_f contours based on Eq. (3-41). Collect data from any sources on this diagram and identify significant trends.

3.12 Using the definitions and equations of Sec. 3.4, show that the ratio of initial "hardware" mass (empty mass plus payload mass) to payload mass for a *two-stage* vehicle is given by the expression

$$\frac{m_e + m_p}{m_p} = \Pi_{e1}\Gamma_1\Gamma_2 + \Pi_{e2}\Gamma_2 + 1 = \Pi_{e1}\Gamma + \Pi_{e2}\Gamma_2 + 1$$

Reconstruct Figs. 3.13 and 3.14 in terms of this modified measure of performance and explain whether or not the conclusions previously reached on the basis of initial mass ratio are still valid.

3.13 Use the final result of Example 3.12 to show that the airbreathing engine overall efficiency required to accomplish a mission with a given oe, h_{PR}, $(D + D_e)/F$, Π_e, and Γ must exceed

$$\eta_o = \frac{oe}{2h_{PR}\left(1 - \frac{D + D_e}{F}\right)\ell n\left(\frac{1}{\Pi_e + \frac{1}{\sqrt{\Gamma}}}\right)}$$

Plot the required overall efficiency as a function of initial mass ratio for $r/r_o = 1.03$, $h_{PR} = 51{,}600$ BTU/lbm (120,000 kJ/kg), $(D + D_e)/F = 0.1$, and $\Pi_e = 0.30$, 0.40, and 0.50. What conclusions do you draw from this information?

3.14 Under many conditions, the performance of a single-stage vehicle is *superior* to a two-stage vehicle. Assuming that the performance parameters and empty mass fractions of all stages are alike, use the results of Example 3.12 to show that the ratio of the initial mass ratio for a single-stage vehicle to the *minimum* initial mass ratio for a two-stage vehicle is given by the expression

$$\frac{(e^{-\frac{\Lambda}{2}} - \Pi_e)^2}{e^{-\Lambda} - \Pi_e}$$

and therefore that a single-stage vehicle would be *preferred* when

$$\Lambda \leq 2\ell n \left(\frac{2}{1 + \Pi_e} \right)$$

or, with the help of Eq. (3-31), when

$$\Pi_f \leq 1 - \left(\frac{1 + \Pi_e}{2} \right)^2$$

Can you explain why this happens? How would you characterize in words the conditions for which a single-stage vehicle is superior? Which of the two-stage cases of Fig. 3.13 would be better off with a single-stage solution? You may find it helpful to plot the boundary of Π_f versus Π_e, along with the boundary of $\Pi_e + \Pi_f \leq 1$, and locate some familiar cases on the graph.

3.15 Repeat the development of Example 3.13 for a single-stage "hybrid" transatmospheric aerospace system, which combines airbreathing propulsion and rocket propulsion in the same vehicle. The airbreathing engine would operate alone for the first part of the flight, and the rocket for the second. Note that for this case, if we treat the vehicle as having two stages, it follows that

$$m_{i1} = m_i = m_p + m_e + m_{f1} + m_{f2} \qquad \text{"stage" 1}$$
$$m_{i2} = m_p + m_e + m_{f2} = m_i - m_{f1} \qquad \text{"stage" 2}$$

Show that the initial mass ratio of the "hybrid" aerospace system is given by the expression

$$\Gamma = \frac{1}{e^{-\{\alpha\Lambda_1 + (1 - \sqrt{\alpha})\Lambda_2\}} - \Pi_{e1}}$$

How does the "hybrid" aerospace system initial mass ratio compare with the two-stage results of Fig. 3.14? Can you explain why this happens?

3.16 Extend the work of Example 3.12 to prove that the minimum possible initial mass ratio for an airbreathing aerospace vehicle having n "equal" stages is given by the expression

$$\Gamma = \frac{1}{\left(e^{-\frac{\Lambda}{n}} - \Pi_e\right)^n}$$

(a) Try various combinations of Λ, Π_e, and n to convince yourself that the "best" solution is almost certain to have either one or two stages. Why doesn't the initial mass ratio keep going down with n?

(b) Show that, as n becomes very large, the initial mass ratio approaches

$$\Gamma = \frac{exp\left\{\Lambda(1 - \Pi_e)\right\}}{(1 - \Pi_e)^n}$$

so that the general tendency is that increasing the number of stages will only increase the initial mass ratio.

HYPERSONIC AIRBREATHING ENGINE PERFORMANCE ANALYSIS

4.1 INTRODUCTION

The next goal is now clearly defined, namely to develop the means to analyze the performance of hypersonic airbreathing engines based upon their individual characteristics and the prevailing flight conditions. The work of Chap. 3 demonstrated that it would be best to obtain performance in terms of the airbreathing engine overall efficiency η_o, although any equivalent performance measure of Table 3.2 could be used as an intermediary.

Fortunately, there are several diverse approaches that will enable us to reach our goal, each of which has its own special attractions and impediments. Taken together, however, they provide a complete portrait of how hypersonic airbreathing engines operate, and allow performance to be determined to any desired degree of accuracy, including rough initial estimates. Since this is a textbook, the ability of these approaches to expose the underlying physics and sharpen intuitive powers of reasoning will receive particular emphasis.

4.1.1 Engine Reference Station Designations

One of the most important requirements for the understanding of the behavior of hypersonic airbreathing engines is the precise but sensible definition of their geometrical configuration. This is an essential first step that has many important by-products, including a solid mental image of the whole and its component parts, a foundation for rigorous mathematical treatment, and the basis for unambiguous communication between associates.

Experience has shown that analysis and understanding are best facilitated by establishing *reference stations* at critical axial positions along the engine flowpath that are designated by both names and numbers.[4.1,4.2] The engine reference stations are located at the junction between the major identifiable components, and thus mark the end of one and the beginning of another. Each property of the flow will be represented by a single value at each engine reference station, so that this method automatically conforms to the *one-dimensional flow* approach of Sec. 2.6.

The engine reference station designation scheme employed hereinafter for airbreathing engine analyses is given in Fig. 4.1, where the processes taking place between the stations have been identified. Figures 1.20 and 1.21 contain some of the customary names for the components that are located between the stations. Although most of this information is or will soon become quite familiar, the omission of several station numbers deserves an early explanation. This is done for two reasons, the first of which is to make our terminology and results compatible with conventional airbreathing engine analysis,[4.1,4.2] and the second of which is to provide the flexibility for the eventual inclusion of such components to the engine as fans, compressors, turbines, mixers, injectors, ejectors, heat exchangers, and afterburners. Station 10 exists to account for the external expansion that is not (and need not) be considered in the case of axially symmetric airbreathing engines having fully enclosed exhaust nozzles.

Table 4.1, which relates the airbreathing engine reference station numbers to their names, is provided below in order to further clarify the reference station scheme.

4.2 THERMODYNAMIC CLOSED CYCLE ANALYSIS

The first approach to estimating the performance of hypersonic airbreathing engines is that of *heat engine* or *thermal engine* closed cycle analysis, often referred to as *Brayton cycle* analysis.[4.1,4.3] This method has endured because the departure point is the familiar terri-

Table 4.1 Airbreathing engine reference station locations.

Reference Station	Engine Location
0	Undisturbed or freestream conditions
	External compression begins
1	External compression ends
	Internal compression begins
	Inlet or diffuser entry
3	Inlet or diffuser exit
	Internal compression ends
	Burner or combustor entry
4	Burner or combustor exit
	Internal expansion begins
	Nozzle entry
9	Internal expansion ends
	Nozzle exit
	External expansion begins
10	External expansion ends

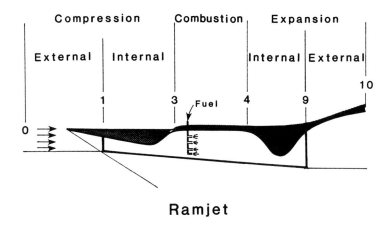

Ramjet

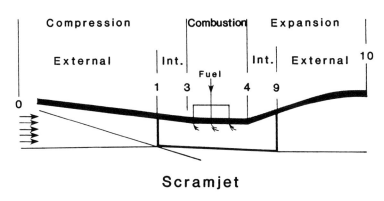

Scramjet

Fig. 4.1 Airbreathing engine reference station numbers and related terminology.

tory of classical thermodynamics, and because the results are always intuitively appealing and insightful. However, before the power and generality of thermodynamics can be unleashed on our behalf, the conditions under study must obey the cardinal rules. This is accomplished by the two steps described next.

First, it must be possible to treat the working medium (in this case, atmospheric air) as a *pure substance*. That is to say, specifying the values of any two independent intensive thermodynamic properties must fix the values of all other intensive thermodynamic properties.[4.3] This can be achieved in the present case by assuming that the air is in its *equilibrium* state at all times, and that the combustion process is replaced by a *heat addition* process that supplies energy equal to that released by combustion, but with no mass addition or change in the chemical constituents of the air. Even though it is true that neither of these assumptions is perfectly met in practice, the evidence of Sec. 2.5.1 regarding the expectation of only minor deviations from equilibrium, and of Sec. 3.2.4.1 regarding the expectation of small fuel/air ratios, suggests that the two underlying assumptions are a reasonable representation of nature in this situation. Please bear in mind that this in no way prevents the air from exhibiting the full range of complex equilibrium behavior already seen in Sec. 2.5.

Second, the working medium must experience a series of equilibrium processes that return it to its *original state*. This is accomplished in the airbreathing engine or Brayton cycle by means of the four processes indicated on the absolute static temperature–static specific entropy (alias *temperature-entropy* or *T-s*) diagram of Fig. 4.2, which are described in words below.

Point 0 to Point 3. Adiabatic compression from the freestream static temperature T_0 to the burner entry static temperature T_3. The irreversibilities or "losses" due to skin friction and shock waves cause the entropy to increase from the freestream value s_0 to the burner entry value s_3. When irreversibilities are absent there is no change of entropy and the adiabatic compression process will be called *isentropic* or *ideal*.

Point 3 to Point 4. Constant static pressure, frictionless heat addition from the burner entry static temperature T_3 to the burner exit static temperature T_4, but with no mass addition. The axial component of the control volume conservation of momentum equation, Eq. (2-29), shows that the velocity of the air is constant. The Gibbs equation, Eq. (2-35), allows the change of entropy to be determined directly by integration, whether or not the gas is calorically perfect.

The *combustion* or heat addition process could easily have been modeled by keeping some other flow property constant, such as throughflow area, Mach number, or static temperature, but constant pressure is preferable for at least four reasons. First, constant pressure is desirable from the aerodynamic standpoint because it avoids the possibility of boundary-layer separation, as well as the necessity of designing the structure primarily to withstand the peak pressure. Second, for scramjets this merely extends the traditional gas turbine and ramjet analyses where, fortuitously, the burner static pressure is also constant because the flow velocity is brought as close to 0 as possible [see Eq. (2-29) again]. Third, as we shall see in Chap. 6, constant area combustors often actually operate at constant pressure during the chemical energy release because of boundary-layer separation. Fourth, this adds gracefulness to the ensuing mathematical manipulations.

Point 4 to Point 10. Adiabatic expansion from the burner static pressure $p_3 = p_4$ to the freestream static pressure $p_{10} = p_0$, consis-

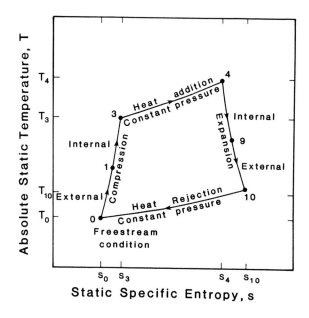

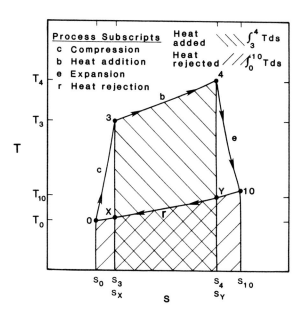

Fig. 4.2   Schematic versions of the airbreathing engine or Brayton cycle temperature-entropy or *T-s* diagram. Note the correspondence of cycle points with the engine reference stations of Fig. 4.1 and Table 4.1.

tent with the assumption of Sec. 3.2 that the exhaust flow is perfectly expanded to the surrounding atmospheric pressure. The irreversibilities or *losses* due to skin friction and shock waves cause the entropy to increase from the burner exit value s_4 to s_{10} at the end of the adiabatic expansion process. When irreversibilities are absent, there is no change of entropy and the adiabatic expansion process will be called *isentropic* or *ideal*.

Point 10 to Point 0. The thermodynamic cycle is closed via an imaginary constant static pressure, frictionless process in which sufficient heat is rejected from the exhaust air (possibly to the surrounding atmosphere) that it is returned to its original temperature-entropy state. Since this process is similar to that of the burner, we are led to the interesting conclusion that the velocity of the air is constant, or that the velocity of the air remaining at the end of the thermodynamic cycle is identical to that which produced thrust at the end of the adiabatic expansion process. From the energy viewpoint, the rejected heat is equivalent to (or the surrogate for) that portion of the heat added in the burner that is not converted into *cycle work*.

4.2.1 Thermodynamic Cycle Efficiency

When viewed as a heat or thermal engine, the *cycle work* of the engine, in the thermodynamic sense, appears as the difference in the kinetic energy between the incoming and outgoing flows. Note that the excess kinetic energy of the flow at Station 10 could, for example, be entirely removed by a device such as a turbine, in order to produce mechanical energy.

Referring again to the T-s diagram of Fig. 4.2, and assuming the heat addition and rejection interactions with the surroundings to be reversible, it is possible to directly evaluate several physical quantities of immediate interest. For example, the heat added per unit mass of air to the cycle from Point 3 to Point 4 is given by

$$\text{Heat added} = \int_3^4 T\,ds \qquad (4\text{-}1)$$

and the heat rejected per unit mass of air from the cycle from Point 10 to Point 0 is given by

$$\text{Heat rejected} = \int_0^{10} T\,ds \qquad (4\text{-}2)$$

Moreover, since there is no thermal energy exchange with the surroundings during the adiabatic compression and expansion processes, it follows that per unit mass of air

$$\text{Cycle work} = \text{Heat added} - \text{Heat rejected}$$

$$= \frac{V_{10}^2}{2} - \frac{V_0^2}{2} = \int_3^4 T\,ds - \int_0^{10} T\,ds \qquad (4\text{-}3)$$

and therefore that the *Thermodynamic cycle efficiency* is

$$\eta_{tc} = \frac{\text{Cycle work}}{\text{Heat added}}$$

$$= \frac{(V_{10}^2/2) - (V_0^2/2)}{\int_3^4 T\,ds} = 1 - \frac{\int_0^{10} T\,ds}{\int_3^4 T\,ds} \qquad (4\text{-}4)$$

Applying Eq. (4-4) to Fig. 4.2 confirms our expectation that the thermodynamic cycle efficiency generally improves as the ratio of the absolute temperature at which heat is added to the absolute temperature at which heat is rejected is increased, provided, of course, that the entropy does not increase excessively during the adiabatic compression and expansion processes.

It is sometimes helpful to eliminate entropy from Eqs. (4-3) and (4-4) in favor of enthalpy by means of the Gibbs equation [Eq. (2-35)], in which case they become, respectively,

$$\text{Cycle work} = \int_3^4 T\,ds - \int_0^{10} T\,ds = (h_4 - h_3) - (h_{10} - h_0) \qquad (4\text{-}5)$$

and

$$\eta_{tc} = 1 - \frac{h_{10} - h_0}{h_4 - h_3} \qquad (4\text{-}6)$$

It is also helpful to note that the heat added can be replaced in any of the above equations by the equivalent fuel chemical energy release of Eq. (3-8), so that per unit mass of air

$$\text{Heat added} = \frac{\eta_b \dot{m}_f h_{PR}}{\dot{m}_0} = \eta_b f h_{PR} = \int_3^4 T\,ds = h_4 - h_3 \qquad (4\text{-}7)$$

where the fuel heating value has been multiplied by the *combustion efficiency* η_b in order to account for the possibility that some of the chemical energy of the fuel is not released because of inadequate mixing or reaction time for complete combustion. Bear in mind, however, that the fuel contributes only heat to the cycle and no mass in this scenario.

These concepts are illustrated concretely in Fig. 4.3, which contains a realistic and typical airbreathing ramjet or scramjet dimensionless T-s diagram. In particular, the constant pressure heat addition and rejection processes were calculated for equilibrium air using HAP(Air). The effects of the increase of C_p with temperature are clearly visible during the constant pressure heat addition process where, in accordance with the Gibbs equation, the local slope of the curve is given by the expression

$$\frac{dT}{ds} = \frac{T}{C_p}$$

The exact thermodynamic cycle efficiency of Fig. 4.3 is directly calculated from Eq. (4-6) to be 0.549. This agrees with the result based on Eq. (4-4) and the eyeball estimation of the T-s area integrals that

$$\eta_{tc} = 1 - \frac{\displaystyle\int_0^{10} T\,ds}{\displaystyle\int_3^4 T\,ds} \cong 1 - \frac{3.10 \times 2.06}{11.15 \times 1.27} = 0.549$$

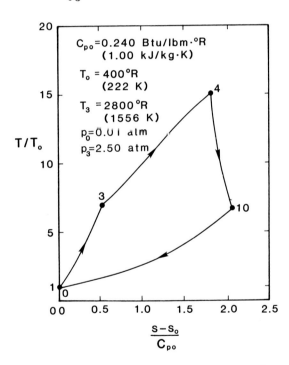

Fig. 4.3 A typical dimensionless temperature-entropy diagram for a ramjet or scramjet engine. This cycle corresponds for example, to the combustion of hydrogen and air, and the constant pressure heat addition and rejection processes were computed using HAP(Air). The following information about the enthalpy at each of the junction points will be useful in the calculations that follow:

$$h_0 = -33 \text{ BTU/lbm} \quad (-77 \text{ kJ/kg})$$
$$h_3 = 612 \text{ BTU/lbm} \quad (1423 \text{ kJ/kg})$$
$$h_4 = 1971 \text{ BTU/lbm} \quad (4584 \text{ kJ/kg})$$
$$h_{10} = 580 \text{ BTU/lbm} \quad (1349 \text{ kJ/kg})$$

Furthermore, Eq. (4-7) can be used along with the information of Fig. 4.3 to deduce that, if the fuel being used is hydrogen, then the *effective fuel/air ratio*

$$\eta_b f = \frac{\int_3^4 T ds}{h_{PR}} \cong \frac{11.15 \times 400 \times 1.27 \times 0.24}{51,600} = 0.0263$$

Several useful observations can be made regarding the quantitative results for the cycle of Fig. 4.3 and their implications for hypersonic flight. First, the thermodynamic cycle efficiency η_{tc} is remarkably high, suggesting that the engine overall efficiency η_o will also be high. Second, an average specific heat at constant pressure for the heat addition process can be estimated by means of the Gibbs equation to be

$$C_p \cong \frac{s_4 - s_3}{\ell n(T_4/T_3)} \cong \frac{1.27 \times 0.24}{\ell n(15.1/7.0)}$$

$$= 0.396 \text{ BTU/lbm} \cdot^\circ \text{R} \quad (1.66 \text{ kJ/kg} \cdot \text{K})$$

This value is consistent with the data of Fig. 2.9, and leads to the conclusion that a dramatically different (and incorrect) *T-s* diagram would have resulted if the behavior of the air had been calculated throughout the cycle using its freestream thermodynamic properties. Finally, and most importantly, comparison of this information with the chemical constituent behavior of Fig. 2.8 reveals that continually increasing the burner entry temperature will eventually result in dissociation of the air during the heat addition process, some portion of which will *not* be restored (or reassociated) during the rapid adiabatic expansion process that follows because the required equilibration time is large compared to the residence time of the gases. Any internal molecular energy invested in dissociation is therefore likely to be "lost" or unavailable for exhaust flow kinetic energy, thus reducing the thermodynamic cycle efficiency. The expansion process in this case is obviously not in thermodynamic equilibrium, and is therefore not susceptible to the present cycle analysis. It must also be emphasized that, as will be seen in Chap. 6, more complex but similar phenomena arise when the energy addition is due to actual combustion or chemical reactions.

Such are the pitfalls of classical thermodynamic closed cycle analysis.

4.2.2 Maximum Allowable Compression Temperature

The result of the foregoing is that the static temperature at the end of compression or the beginning of combustion cannot be increased indefinitely, but must be limited to a value that prevents excessive dissociation in the exhaust flow. As the burner entry and exit static temperatures increase, the "loss" of energy to unequilibrated dissociation at first gradually eats into the benefits usually expected from the higher thermodynamic cycle efficiencies and eventually overwhelms them. This is a serious limitation for real hypersonic airbreathing engines, one that must be avoided in practice.

The determination of the maximum allowable temperatures requires a combination of elaborate computations and experienced judgment because these temperatures depend on many interrelated variables, including flight altitude and Mach number, inlet losses, fuel type, fuel/air ratio, and burner and exhaust system geometry. It stands to reason, however, that because dissociation and reassociation are gas kinetic phenomena, the limit is best stated in terms of the static temperature at the end of compression or the beginning of combustion.

Fortunately, since air behaves as shown in Figs. 2.7 through 2.10 and the design fuel/air ratio is often at or near the stoichiometric value, the *maximum allowable compression temperature* T_3 is almost always found to be in the relatively narrow range of 2600–3000 °R (1440–1670 K), so that a rational *representative* estimate for T_3 is 2800 °R (1560 K), the value used in the example cycle of Fig. 4.3. Moreover, referring again to Figs. 2.7 through 2.10, we find that the entire adiabatic *compression* process will take place where the air behaves as a thermally perfect gas and dissociation effects are negligible. In fact, for the purposes of instruction, we will hereinafter treat the air during the adiabatic compression process as a calorically perfect gas having a constant ratio of specific heats of $\gamma_c = 1.36$.

4.2.3 Required Burner Entry Mach Number

Interestingly, the limit on compression temperature leads directly to restrictions on the burner entry Mach number M_3. In accordance with Example Case 2.1, the stagnation temperature of the inlet flow, which is a fixed property, is given by the expressions

$$T_t = T_0 \left(1 + \frac{\gamma_c - 1}{2} M_0^2\right) = T_3 \left(1 + \frac{\gamma_c - 1}{2} M_3^2\right) \qquad (4\text{-}8)$$

so that the burner entry Mach number must equal

$$M_3 = \sqrt{\frac{2}{\gamma_c - 1} \left\{ \frac{T_0}{T_3} \left(1 + \frac{\gamma_c - 1}{2} M_0^2\right) - 1 \right\}} \qquad (4\text{-}9)$$

for the given maximum allowable compression temperature T_3. The relationship of Eq. (4-9) is presented in Fig. 4.4 for two typical limiting values of T_3/T_0. Please note the following useful features of these results. First, when

$$M_0 < \sqrt{\frac{2}{\gamma_c - 1}\left(\frac{T_3}{T_0} - 1\right)} \qquad (4\text{-}10)$$

there is no solution because the allowable T_3 is higher than the stagnation temperature of the freestream flow. Second, when

$$M_0 > \sqrt{\frac{2}{\gamma_c - 1}\left\{\left(\frac{\gamma_c + 1}{2}\right)\frac{T_3}{T_0} - 1\right\}} \qquad (4\text{-}11)$$

the flow entering the burner must remain *supersonic* or the allowable T_3 will be exceeded. This makes the supersonic combustion ramjet, or scramjet, inevitable! Third, in the hypersonic limit of Sec. 2.8,

$$\frac{M_3}{M_0} = \sqrt{\frac{T_0}{T_3}} \qquad (4\text{-}12)$$

which leads to the frequently used rule of thumb that

$$\frac{M_3}{M_0} \cong \sqrt{\frac{400}{2800}} \cong 0.38$$

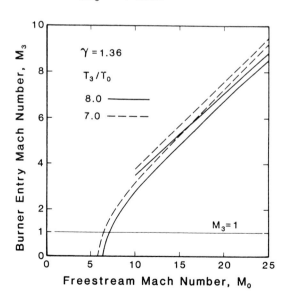

Fig. 4.4 **Burner entry Mach number as a function of flight Mach number and maximum allowable compression temperature. The straight lines are the hypersonic asymptotic limits as given by Eq. (4-12).**

Finally, note from Eq. (4-11) with $T_3/T_0 = 7$ that nature has arranged things so that supersonic combustion will also commence at $M_0 \cong 6.4$, well within the range designated as the boundary of hypersonic flight in Sec. 2.8. Thus, hypersonic flight and scramjet propulsion are virtually synonymous.

4.2.4 Airbreathing Engine Performance Measures

As the conclusion to this thermodynamic closed cycle analysis, general expressions will be derived for the airbreathing engine performance measures we seek. In keeping with the spirit of this approach, we will continue to focus upon the various airbreathing engine efficiencies, while remembering that the uninstalled thrust is given by Eq. (3-11) and that Table 3.2 is available for convenient conversions.

4.2.4.1 Thermal efficiency.
Recognizing that the fuel contributes no mass in this analysis, inspection of Eqs. (3-10), (4-4), and (4-7) reveals immediately that the airbreathing engine thermal efficiency

$$\eta_{th} = \frac{\dfrac{V_{10}^2}{2} - \dfrac{V_0^2}{2}}{fh_{PR}} = \frac{\dfrac{V_{10}^2}{V_0^2} - 1}{\dfrac{fh_{PR}}{V_0^2/2}} = \eta_b \eta_{tc} \qquad (4\text{-}13)$$

The reason for the difference between engine thermal efficiency η_{th} and thermodynamic cycle efficiency η_{tc} is that the latter is not penalized for the fuel chemical energy not released in the burner, whereas the former must account for all the chemical energy the fuel can make available.

Using the data of Fig. 4.3 as an example, and taking the combustion efficiency η_b to be 0.90, we find that the thermal efficiency is 0.494.

4.2.4.2 Propulsive efficiency.
Recognizing that the fuel contributes no mass in this analysis, it is not difficult to combine Eqs. (3-13) and (4-13) to prove that the airbreathing engine propulsive efficiency

$$\eta_p = \frac{2}{\dfrac{V_{10}}{V_0} + 1} = \frac{2}{\sqrt{\eta_{th} \cdot \dfrac{fh_{PR}}{V_0^2/2} + 1} + 1} \qquad (4\text{-}14)$$

Presuming, for the moment, that the T-s diagram is relatively unaffected by flight speed, inspection of Eq. (4-14) reveals that the propulsive efficiency continuously increases with flight speed and, as expected, approaches 1. Using the data of Fig. 4.3 as an example, we find that as the flight speed increases from 6000 to 10,000 ft/s (1829 to 3048 m/s), the velocity ratio V_{10}/V_0 decreases from 1.43 to 1.17 and the propulsive efficiency correspondingly increases from 0.824 to 0.921.

4.2.4.3 Overall efficiency.   Following Eq. (3-10), the airbreathing engine overall efficiency η_o is merely the product of Eqs. (4-13) and (4-14). An alternative but direct expression for η_o can also be obtained by combining Eqs. (3-9), (3-11), and (4-13), from which we find that

$$\eta_o = \eta_{th} \cdot \eta_p = \frac{2\left(\dfrac{V_{10}}{V_0} - 1\right)}{\dfrac{fh_{PR}}{V_0^2/2}} = \frac{2\left(\sqrt{\eta_{th} \cdot \dfrac{fh_{PR}}{V_0^2/2} + 1} - 1\right)}{\dfrac{fh_{PR}}{V_0^2/2}} \quad (4\text{-}15)$$

Using the data of Fig. 4.3 again as an example, either method of calculation shows that the engine overall efficiency increases from 0.407 to 0.455 as the flight speed increases from 6000 to 10,000 ft/s (1829 to 3048 m/s). The overall efficiency, of course, approaches the thermal efficiency of 0.494 as the flight speed becomes arbitrarily large.

When these results are compared with those of the requirement studies of Sec. 3.4, one cannot escape the conclusion that the scramjet does indeed offer the *promise* of competitive performance for likely hypersonic vehicle applications.

4.2.4.4 Specific impulse.   For completeness, and eventual comparison with rocket propulsion, the airbreathing engine specific impulse can be calculated from the Table 3.2 relationship

$$I_{sp} = \frac{h_{PR}}{g_0 V_0} \cdot \eta_o \quad (4\text{-}16)$$

For the data of Fig. 4.3, and assuming that the fuel being used is hydrogen, the specific impulse decreases from 2730 s to 1830 s as the flight speed increases from 6000 to 10,000 ft/s (1829 to 3048 m/s). Due to the inverse dependence of specific impulse upon flight speed, the specific impulse value changes much more rapidly than the overall efficiency, which makes the latter a more convenient performance measure for everyday purposes.

4.2.4.5 Some general conclusions.   Examination of Eqs. (4-13) through (4-15) leads to the conclusion that these three performance parameters are functions only of the thermal efficiency and the dimensionless parameter $fh_{PR}/(V_0^2/2)$, which is the ratio of the energy made available by the chemical reaction to the kinetic energy of the freestream

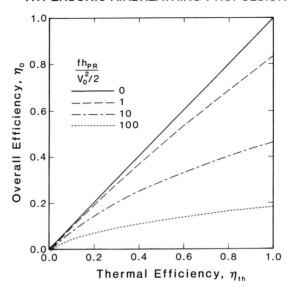

Fig. 4.5 Airbreathing engine overall efficiency as a function of thermal efficiency and the dimensionless parameter $fh_{PR}/(V_0^2/2)$.

air. Figure 4.5, which is based on those equations, was created in order to emphasize the generality of the interrelationships between the performance parameters.

An important and reaffirming conclusion brought forth by Fig. 4.5 is that the overall efficiency increases for a fixed thermal efficiency as the parameter $fh_{PR}/(V_0^2/2)$ decreases and, therefore, in accordance with Eq. (4-14), the propulsive efficiency increases.

Another intriguing result can be obtained by combining Eqs. (4-3), (4-7), and (4-15) in order to show that the overall efficiency can also be written as

$$\eta_o = \left(\frac{V_{10}}{V_0} - 1\right)\frac{V_0^2}{fh_{PR}} = \left\{\sqrt{\frac{2(\eta_b fh_{PR} - \int_0^{10} T\,ds)}{V_0^2} + 1} - 1\right\}\frac{V_0^2}{fh_{PR}}$$

$$(4\text{-}17)$$

This formulation can be used to confirm by inspection that the best engine overall efficiency for fixed heat addition and flight speed is achieved when the exhaust entropy s_{10} (or exhaust static temperature T_{10}) is minimized because that condition leads to the least heat rejection, and not necessarily when the burner entry temperature is maximized.

Finally, comparison of Eqs. (4-13), (4-15), and (4-16) leads to the conclusion that the maximum values of η_{tc}, η_{th}, η_o, and I_{sp} occur simultaneously for any given combination of η_b and $f h_{PR}/(V_0^2/2)$. Such are the joys of classical thermodynamic closed cycle analysis.

4.3 FIRST LAW ANALYSIS

The next level of understanding is reached by analyzing the behavior of the flow across each of the successive thermodynamic processes by means of simple but physically tenable models of the behavior of air. This course of action will be referred to as *first law analysis* because its main purpose is to provide a fair evaluation of the static enthalpy at each of the endpoints of the thermodynamic processes (or engine reference stations), and therefore, as seen in Sec. 4.2, the evaluation of any desired airbreathing engine performance measure.

Although the framework for first law analysis is identical to that of Sec. 4.2, the goal is to find closed form solutions for the performance of real ramjets and scramjets and use them to expose important trends and sensitivities, without recourse to elaborate thermochemical calculations. The inspiration for this method comes from an imaginative and timeless paper authored by C. H. Builder.[4.4]

4.3.1 Thermodynamic Process Assumptions

As we have seen above, the airbreathing engine performance measures can be determined for given values of the flight speed V_0, fuel/air ratio f, and fuel heating value h_{PR}, once the airbreathing engine thermal efficiency η_{th} is known. The most straightforward and convenient way to find the thermal efficiency from thermodynamic cycle quantities is by means of Eqs. (4-6) and (4-13), which depend upon the knowledge of the static enthalpy at each of the four junctions of the thermodynamic cycle and the combustion efficiency η_b. In order to determine the required static enthalpies, it will be necessary to circumnavigate the entire T-s diagram, starting at and returning to the known initial or freestream condition. And this, in turn, requires only two assumptions beyond those of Sec. 4.2, as described below.

First, it will be assumed that the equilibrium air behaves as a *calorically perfect gas* across each of the four individual thermodynamic processes of the Brayton cycle. Recognizing that the equilibrium properties of air can, in fact, vary significantly from point to point around the cycle, this assumption amounts to believing that reasonable or *representative* values can be assigned to the gas properties required to describe each separate thermodynamic process. This assumption is not as daring as it may first appear because the T-s diagram tends to maintain its shape once the maximum allowable compression temperature has been reached. The risk associated with this assumption is also lessened because the evaluation of the static enthalpies is, as you shall soon see, subject to some checks and balances. And, just as in previous cases, the benefits of increased clarity greatly outweigh the cost of decreased accuracy.

Second, it will be assumed that reasonable *empirical models* exist to describe the adiabatic compression, constant pressure combustion, and adiabatic expansion processes. To begin with, they will be described by *process efficiencies* that are taken to be *constant*. The process efficiencies for compression and expansion are extensions of the definitions ordinarily applied to gas turbine and ramjet components, wherein the actual or real change in static enthalpy is referenced to the ideal or isentropic change in static enthalpy that would accompany the same change in static pressure.[4.1] Referring to Fig. 4.2 for nomenclature, the *adiabatic compression process efficiency* definition is

$$\eta_c \doteq \frac{h_3 - h_x}{h_3 - h_0} \leq 1 \tag{4-18}$$

and the *adiabatic expansion process efficiency* definition is

$$\eta_e \doteq \frac{h_4 - h_{10}}{h_4 - h_Y} \leq 1 \tag{4-19}$$

The combustion process efficiency definition is as given in Eq. (4-7). This approach has a measure of risk because the process efficiencies really do depend upon flight speed, but more complex models can and will be easily substituted in Chap. 8. Meanwhile, this is an excellent starting point, and it will reveal several of the essential truths about the behavior of ramjets and scramjets.

4.3.2 Thermodynamic Process Analyses

For completeness, all flow properties of interest are solved for in the analyses of the four thermodynamic processes that follow. The order of solution is straightforward in the sense that, except at the very end, each quantity can be evaluated in turn and iteration is not required. Moreover, each step starts by stating the quantity to be found and the equation utilized. Reference to the nomenclature of Fig. 4.2 and the process descriptions at the start of Sec. 4.2 are encouraged.

Both the mathematical manipulations and the interpretation of results are greatly facilitated by using the *cycle static temperature ratio*

$$\psi = \frac{T_3}{T_0} \geq 1 \tag{4-20}$$

as the independent variable. This is a *natural* choice because, as we have seen, ψ is a principal determinant of thermodynamic cycle efficiency (Sec. 4.2) and can be directly used to impose the limit of maximum allowable compression temperature (Sec. 4.2.2).

Adiabatic Compression Process (Point 0 to Point 3).

1. T_X/T_0 [Definition of compression efficiency, Eq. (4-18)]

$$\eta_c = \frac{h_3 - h_X}{h_3 - h_0} = \frac{C_{pc}(T_3 - T_X)}{C_{pc}(T_3 - T_0)} = \frac{\psi - \dfrac{T_X}{T_0}}{\psi - 1}$$

Therefore,

$$\frac{T_X}{T_0} = \psi(1 - \eta_c) + \eta_c \geq 1 \tag{4-21}$$

2. $s_3 - s_0$ [The Gibbs equation, Eq. (2-35), Point 0 to Point X]

$$ds = \frac{dh}{T} = C_{pc}\frac{dT}{T}$$

Therefore,

$$s_3 - s_0 = s_X - s_0 = C_{pc}\ln\frac{T_X}{T_0} \geq 0 \tag{4-22}$$

3. p_3/p_0 [The Gibbs equation, Eq. (2-35), Point X to Point 3]

$$\frac{dh}{T} = C_{pc}\frac{dT}{T} = \frac{dp}{\rho T} = R_c\frac{dp}{p}$$

Therefore,

$$\frac{p_3}{p_0} = \left(\frac{T_3}{T_X}\right)^{(C_{pc}/R_c)} = \left(\psi\frac{T_0}{T_X}\right)^{(C_{pc}/R_c)} \geq 1 \tag{4-23}$$

Heat Addition Process (Point 3 to Point 4).

1. V_4 [Conservation of momentum, Eq. (2-29)]

$$V_4 = V_3 \tag{4-24}$$

2. T_4/T_3 [Conservation of energy, Eq. (2-31)]

$$h_4 - h_3 = C_{pb}(T_4 - T_3) = \eta_{bfh}\eta_{PR}$$

Therefore,

$$\frac{T_4}{T_3} = 1 + \frac{C_{p0}}{C_{pb}} \cdot \frac{\eta_b f h_{PR}}{\psi C_{p0} T_0} \geq 1 \tag{4-25}$$

3. $s_4 - s_3$ [The Gibbs equation, Eq. (2-35)]

$$ds = \frac{dh}{T} = C_{pb}\frac{dT}{T}$$

Therefore,

$$s_4 - s_3 = C_{pb} \ell n \frac{T_4}{T_3} \geq 1 \qquad (4\text{-}26)$$

Adiabatic Expansion Process (Point 4 to Point 10).

1. T_Y/T_4 [The Gibbs equation, Eq. (2-35), Point 4 to Point Y]

$$\frac{dh}{T} = C_{pe} \frac{dT}{T} = \frac{dp}{\rho T} = R_e \frac{dp}{p}$$

Therefore,

$$\frac{T_Y}{T_4} = \left(\frac{p_Y}{p_4}\right)^{(R_e/C_{pe})} = \left(\frac{p_0}{p_3}\right)^{(R_e/C_{pe})} \leq 1 \qquad (4\text{-}27)$$

2. T_{10}/T_4 [Definition of expansion efficiency, Eq. (4-19)]

$$\eta_e = \frac{h_4 - h_{10}}{h_4 - h_Y} = \frac{C_{pe}(T_4 - T_{10})}{C_{pe}(T_4 - T_Y)} = \frac{1 - (T_{10}/T_4)}{1 - (T_Y/T_4)}$$

Therefore,

$$\frac{T_{10}}{T_4} = 1 - \eta_e \left(1 - \frac{T_Y}{T_4}\right) \leq 1$$

or, combining this with Eqs. (4-21), (4-23), and (4-27),

$$\frac{T_{10}}{T_4} = 1 - \eta_e \left\{ 1 - \left[1 - \eta_c \left(1 - \frac{1}{\psi}\right)\right]^{\frac{C_{pc}}{R_c} \cdot \frac{R_e}{C_{pe}}} \right\} \leq 1 \qquad (4\text{-}28)$$

3. $s_{10} - s_4$ [The Gibbs equation, Eq. (2-35), Point Y to Point 10]

$$ds = \frac{dh}{T} = C_{pe} \frac{dT}{T}$$

Therefore,

$$s_{10} - s_4 = s_{10} - s_Y = C_{pe} \ell n \frac{T_{10}}{T_Y} = C_{pe} \ell n \left(\frac{T_{10}}{T_4} \cdot \frac{T_4}{T_Y}\right) \geq 0 \qquad (4\text{-}29)$$

Heat Rejection Process (Point 10 to Point 0).

1. $h_{10} - h_0$ [Specific heat at constant pressure, Eq. (2-39)]

$$dh = C_{pr} dT$$

Therefore,

$$h_{10} - h_0 = C_{pr}(T_{10} - T_0) = C_{pr}T_0 \left(\psi \cdot \frac{T_4}{T_3} \cdot \frac{T_{10}}{T_4} - 1 \right) \geq 0$$

$$(4\text{-}30)$$

2. $s_{10} - s_0$ [The Gibbs equation, Eq. (2-35)]

$$ds = \frac{dh}{T} = C_{pr}\frac{dT}{T}$$

Therefore,

$$s_{10} - s_0 = C_{pr} \ln \frac{T_{10}}{T_0} = C_{pr} \ln \left(\psi \cdot \frac{T_4}{T_3} \cdot \frac{T_{10}}{T_4} \right) \geq 0 \qquad (4\text{-}31)$$

Since the enthalpy differences required for Eq. (4-6) have already been determined, this final step is not, strictly speaking, necessary. Nevertheless, a worthwhile purpose can be served by rearranging Eq. (4-31) into the form

$$C_{pr} = \frac{s_{10} - s_0}{\ln \dfrac{T_{10}}{T_0}} = \frac{(s_{10} - s_4) + (s_4 - s_3) + (s_3 - s_0)}{\ln \left(\psi \cdot \dfrac{T_4}{T_3} \cdot \dfrac{T_{10}}{T_4} \right)} \qquad (4\text{-}32)$$

thus guaranteeing that the cycle is closed. This provides an independent check on the equilibrium specific heat at constant pressure of air C_{pr} during the heat rejection process, and one that is seldom capitalized upon. If the value of C_{pr} given by Eq. (4-32) is not in reasonable accord with that expected from, for example, the data of Fig. 2.9, then adjustments in the representative equilibrium thermodynamic properties from other parts of the cycle must be made.

Equation (4-32) was used to update C_{pr} and insure the integrity of the thermodynamic closed cycle analysis in *all* calculations that follow.

Thermodynamic Cycle Efficiency.

Equations (4-6), (4-7), (4-25), and (4-30) can be combined to yield an algebraic expression for the *thermodynamic cycle efficiency*, namely

$$\eta_{tc} = 1 - \frac{C_{pr}}{C_{p0}} \left\{ \left(\psi \cdot \frac{C_{p0}T_0}{\eta_b f h_{PR}} + \frac{C_{p0}}{C_{pb}} \right) \frac{T_{10}}{T_4} - \frac{C_{p0}T_0}{\eta_b f h_{PR}} \right\} \leq 1 \quad (4\text{-}33)$$

where T_{10}/T_4 is given by Eq. (4-28). Note in particular that the driving force of heat addition appears only in the dimensionless grouping $\eta_b f h_{PR}/C_{p0}T_0$. The remaining engine performance measures are, of course, obtained from Eqs. (4-13) through (4-16).

4.3.3 First Law Analysis Results

Figure 4.6 shows the T-s diagram for a first law analysis corresponding to the cycle of Fig. 4.3. Hope for this approach is strengthened by
the close comparison with the exact T-s diagram of Fig. 4.3, shown
as a dotted line. Since the area integrals that are used to evaluate
the heat addition and heat rejection are very similar, it should be
expected that their respective thermodynamic cycle efficiencies, and
therefore their engine performance measures, will also be very nearly
the same.

This expectation is realized when the thermodynamic quantities
accompanying Fig. 4.6 are substituted into Eq. (4-33). The resulting
thermodynamic cycle efficiency η_{tc} of 0.564 is in satisfactory agreement with its Sec. 4.2 counterpart of 0.549.

The remaining engine performance measures, calculated by
means of Eqs. (4-13) through (4-16), are compared below with the
thermodynamic cycle analysis results of Sec. 4.2.4.

	First Law Analysis		Thermodynamic Cycle Analysis	
V_0 (ft/s)	6000	10,000	6000	10,000
η_{th}	0.507	0.507	0.494	0.494
η_p	0.821	0.919	0.824	0.921
η_o	0.417	0.466	0.407	0.455

Although agreement between the results of the two approaches is
not perfect, the differences amount to less than a few percent, which
is the best that one can expect from analyses of this type, and less
than the errors due to factors that have yet to be included. This
exercise does demonstrate, nevertheless, that the first law analysis
can be used at least to determine the leading trends and sensitivities
of airbreathing engine performance, particularly when operating in
the proximity of the point at which the thermophysical properties of
air were chosen.

The first law analysis will therefore now be used to explore the
response of the airbreathing engine performance to the variation in
a number of potentially important parameters.

4.3.3.1 Influence of cycle static temperature ratio.

The data accompany-
ing Fig. 4.6 were employed in the first law analysis of Eqs. (4-20)
through (4-33), and then in Eqs. (4-13) and (4-15), in order to generate Fig. 4.7, which reveals the influence of the cycle static temperature ratio ψ on airbreathing engine overall efficiency η_o. The

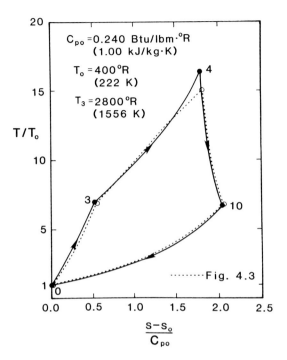

Fig. 4.6 **The first law analysis dimensionless temperature-entropy diagram corresponding to the typical airbreathing engine cycle of Fig. 4.3. The thermophysical properties of air used in the first law analysis are:**

$$\psi = 7.0$$

$$\frac{\eta_b \, \tilde{h}_{PR}}{C_{p0} T_0} = 14.1$$

$$\eta_c = \eta_b = \eta_e = 0.90$$

$$C_{p0} = 0.240 \text{ BTU/lbm} \cdot {}^{\circ}\text{R} \quad (1.00 \text{ kJ/kg} \cdot \text{K})$$

$$C_{pc} = 0.260 \text{ BTU/lbm} \cdot {}^{\circ}\text{R} \quad (1.09 \text{ kJ/kg} \cdot \text{K})$$

$$C_{pb} = 0.360 \text{ BTU/lbm} \cdot {}^{\circ}\text{R} \quad (1.51 \text{ kJ/kg} \cdot \text{K})$$

$$\frac{C_{pc}}{R_c} \cdot \frac{R_e}{C_{pe}} = 0.722$$

$$C_{pe} = 0.360 \text{ BTU/lbm} \cdot {}^{\circ}\text{R} \quad (1.51 \text{ kJ/kg} \cdot \text{K})$$

$$R_c = R_e$$

highest value of η_o for a flight speed of 10,000 ft/s (3048 m/s) is 0.467 and is located in the vicinity of $\psi = 8.0$. The highest value of η_o for a flight speed of 6000 ft/s (1829 m/s) is 0.417 and occurs at $\psi = 7.91$, the highest cycle static temperature ratio possible [see, for example, Eq. (4-8)]. Both of these curves are relatively flat over the range of ψ displayed, and clearly demonstrate that there is no benefit to be gained by increasing ψ indefinitely, even for equilibrium flow. Indeed, within the framework of this first law analysis, the results imply that $\psi < 8.0$ unless other considerations intervene.

Here again, as in Sec. 4.2.2, we have discovered that there is a rational upper limit to the desirable compression temperature or the cycle static temperature ratio ψ due, in this case, to compression and expansion losses. Fortunately, the numerical results are similar for the two cases, so that they merely reinforce each other. This time, however, we have capitalized upon the minimum entropy condition described in Sec. 4.2.4.5.

The presence of an optimum cycle static temperature ratio could have been suspected on the basis of the forms of Eqs. (4-28) and (4-33), which contain competing influences of ψ. Indeed, if, as a first approximation, the thermophysical properties of air are taken to be everywhere constant at their freestream values in this first law analysis, Eqs. (4-28) and (4-33) taken together reduce to the remarkably simple form

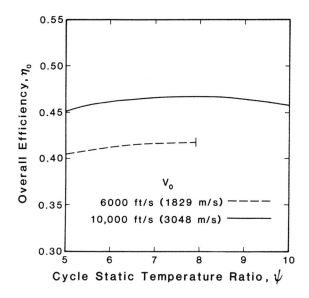

Fig. 4.7 Airbreathing engine overall efficiency as a function of cycle static temperature ratio and flight speed.

$$\eta_{tc} = \frac{C_{p0}T_0}{\eta_b f h_{PR}}(\psi - 1)\left\{ \eta_c \eta_e \left(1 + \frac{\eta_b f h_{PR}}{\psi C_{p0}T_0} \right) - 1 \right\} \qquad (4\text{-}34)$$

which can be analytically shown to have a maximum when

$$\psi = \sqrt{\frac{\eta_c \eta_e}{1 - \eta_c \eta_e} \cdot \frac{\eta_b f h_{PR}}{C_{p0}T_0}} \qquad (4\text{-}35)$$

For the present case, namely $\eta_c = \eta_b = \eta_e = 0.90$ and $\eta_b f h_{PR}/C_{p0}T_0 = 14.1$, Eqs. (4-34) and (4-35) yield an optimum static temperature ratio of 7.75 and a thermodynamic cycle efficiency of 0.615, for which the engine overall efficiency is 0.448 at a flight speed of 6000 ft/s (1829 m/s) and 0.505 at 10,000 ft/s (3048 m/s). These values are somewhat higher than the previous first law analysis results for the more realistic representation of air because they do not reflect the increase of C_{pb} with temperature that results in a lower T_4 and therefore a lower η_{tc}.

The Motivation for Turbojets.

The foregoing analysis can be correctly interpreted to mean that the engine cycle has a *preferred* amount of compression for optimum performance. The desired amount of compression can be obtained merely by decelerating the freestream flow, provided that

$$M_0 \geq \sqrt{\frac{2}{\gamma_c - 1}(\psi - 1)} \qquad (4\text{-}36)$$

In fact, the greater this inequality, the higher will be the Mach number entering the burner, in accordance with Eq. (4-9).

But what happens when even the stagnation temperature of the freestream is less than the desired burner inlet temperature, as in Eq. (4-10)? One solution is to provide a mechanical compression system for the entering stream that receives its power from a turbine driven by the exhaust stream, this being the configuration of the modern turbojet engine.

4.3.3.2 Influence of fuel heating value.   The data accompanying Fig. 4.6 were employed in the first law analysis of Eqs. (4-20) through (4-33), and then in Eqs. (4-13) and (4-15) in order to determine the influence of the heating value of the fuel on the airbreathing engine overall efficiency. The results tabulated below show the effect on overall efficiency of increasing or decreasing the fuel heating value (at constant combustion efficiency) by approximately 10 percent relative to the nominal case values of Fig. 4.6.

	Overall Efficiency, η_o	
V_0 (ft/s)	6000	10,000
$\dfrac{\eta_b f h_{PR}}{C_{p0} T_0} = 15.5$	0.416	0.469
Nominal case	0.417	0.466
$\dfrac{\eta_b f h_{PR}}{C_{p0} T_0} = 12.7$	0.416	0.461

The impact of fuel heating value on overall efficiency is evidently negligible, the small numerical differences being within the *noise level* of the first law analysis. This result is, however, unquestionably good news for the specific thrust and specific impulse, both of which will increase in proportion to the fuel heating value according to the Table 3.2 performance interrelationships.

4.3.3.3 Influence of thermodynamic process efficiencies. The data accompanying Fig. 4.6 were employed in the first law analysis of Eqs. (4-20) through (4-33), and then in Eqs. (4-13) and (4-15), in order to determine the influence of the thermodynamic process efficiencies η_c, η_b, and η_e on the airbreathing engine overall efficiency η_o. The results tabulated below show the response of the overall efficiency to a variation in each of the individual process efficiencies above and below their nominal case values of Fig. 4.6 by the amount 0.10. Only one change from the nominal case of Fig. 4.6 takes place at a time in the calculations presented here.

	Overall Efficiency, η_o	
V_0 (ft/s)	6000	10,000
Nominal case	0.417	0.466
$\eta_c = 1.00$	0.496	0.565
$\eta_c = 0.80$	0.340	0.375
$\eta_b = 1.00$	0.463	0.518
$\eta_b = 0.80$	0.370	0.415
$\eta_e = 1.00$	0.476	0.539
$\eta_e = 0.80$	0.355	0.392

The obvious conclusion is that all three thermodynamic process efficiencies carry heavy, if not equal, weights. In fact, in all cases, the percentage change in overall efficiency is greater than the percentage change in process efficiency. And, returning to the requirement studies of Sec. 3.4, this extent of improvement in any *one* process efficiency could move airbreathing engine performance from unacceptable to irresistible. The process efficiencies must therefore all be preserved or enhanced if the promise of satisfactory airbreathing engine performance is to be realized.

4.4   STREAM THRUST ANALYSIS

The thermodynamic cycle and first law analyses are unable to easily account for several phenomena that can have a significant influence on airbreathing engine performance, such as the mass, momentum, and kinetic energy fluxes contributed by the fuel, the geometry of the burner, and exhaust flows that are not matched to the ambient pressure. Such phenomena can usually be accounted for, while sacrificing some insight and generality and requiring substantially more initial information, by *one-dimensional flow* approaches that use the entire set of control volume conservation equations.

These methods lean heavily on momentum relationships, and often use the stream thrust function Sa (see Example Case 2.2) as a primary flow quantity. This technique will therefore be referred to as *stream thrust analysis,* and it will now be examined because of its intrinsic strengths as well as the interesting contrast it offers to the energy techniques of Secs. 4.2 and 4.3.

An important stopping point along the way to this development was a comprehensive report authored by E. T. Curran and R. R. Craig,[4.5] in which the stream thrust analysis is used to maximum advantage in order to understand and predict scramjet engine performance.

4.4.1 Uninstalled Airbreathing Engine Thrust

Stream thrust analysis is greatly simplified if the control volume under consideration is selected in such a way that the desired results arise effortlessly. The control volumes to be used in this analysis are shown in Fig. 4.8, where it should be noted that the *dividing streamlines* that constitute the boundary between internal and external flow coincide with the outside surface of the engine. The most important quantity to be evaluated is the *uninstalled thrust F*, which is defined as equal and opposite to the net axial force acting on the internal flow when the external flow is perfect (i.e., reversible). It can be shown that perfect external flow is equivalent to either the net axial force acting on the external flow or the integral of the axial projection of the gauge pressure (i.e., local static pressure minus freestream static pressure) plus frictional forces acting on the dividing streamlines being *zero*.[4.1] The thrust is transferred to the vehicle through the engine support (sometimes called the mount, pylon, sting, or thrust frame) by means of material stresses that also act on the control volume boundary.

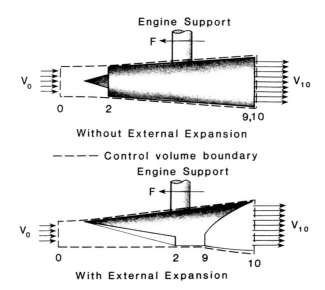

Fig. 4.8 Control volumes used in the evaluation of uninstalled airbreathing engine thrust.

This definition of uninstalled thrust is a common practice that leads to a convenient division of responsibilities between engine and vehicle designers. The bookkeeping system of the vehicle designers equates *installation drag* D_e to the net axial force acting on the external flow in the negative direction due to such irreversibilities as form and frictional drag. The *installed thrust* T is defined as the uninstalled thrust less the installation drag, and is therefore equal to the net axial force acting on the entire flow. Thus, this accounting system leaves the engine designers primarily responsible for the forces acting on the internal surfaces, although they certainly apply their talents to help minimize installation drag. You should bear in mind in this regard that installation drag can be generated along the free streamlines that constitute the control volume boundary as well as the external material surfaces. For example, in the typical supersonic or hypersonic flow situations depicted in Fig. 4.8, a commonsense conclusion is that, since A_{10} is usually larger than A_0, the static pressure along the external surfaces of the control volume will exceed the freestream static pressure, with the result that the gauge pressure will be positive and the integrated force on the external flow amounts to an installation drag.

Next, it is assumed that, because the flow is supersonic or hypersonic, the flow is undisturbed to the control volume inlet plane (Station 0). Furthermore, it is assumed that the flow properties at the control volume exit plane (Station 10) are represented by suitable one-dimensional averages, and that the average static pressure there is not necessarily equal to the freestream value. This situation may be easier to imagine for engines that have no external expansion, such as the upper portion of Fig. 4.8, and for which the exhaust flow is completely confined or surrounded by the internal surface of the nozzle, but the principle is the same whether or not external expansion is involved. You may find it helpful in this regard to envision the lower portion of Fig. 4.8 with a mirror image of the engine swung underneath so that the two taken together form a close approximation to an engine without external expansion. When the average static pressure at Station 10 exceeds the freestream static pressure, the exhaust flow is said to be *underexpanded,* and when the opposite is true, the exhaust flow is said to be *overexpanded.*

Finally, in order to simplify the equations and calculations that follow, it is assumed that the entire velocity at each engine station is aligned with the thrust or axial direction, and that the throughflow area is oriented perpendicular to that direction. These assumptions can be removed without great intellectual effort, but the additional mathematical and arithmetical complications are disproportionate to their benefits.

Returning to Fig. 4.8, it follows from the foregoing discussions and Eq. (2-29) that the net uninstalled engine thrust can be written as

$$F = \dot{m}_{10}V_{10} - \dot{m}_0 V_0 + (p_{10} - p_0)\,A_{10} \qquad (4\text{-}37\text{a})$$

$$= \{\dot{m}V + pA\}_{10} - \{\dot{m}V + pA\}_0 - p_0(A_{10} - A_0) \quad (4\text{-}37\text{b})$$

or, following Example Case 2.2, in terms of the uninstalled engine specific thrust as

$$\frac{F}{\dot{m}_0} = (1 + f)Sa_{10} - Sa_0 - \frac{R_0 T_0}{V_0}\left(\frac{A_{10}}{A_0} - 1\right) \qquad (4\text{-}38)$$

where the perfect gas law has been used in order to eliminate density.

This approach immediately displays two interesting features. First, the engine performance measure that naturally arises is the specific thrust, which can be converted into other measures of interest by means of Table 3.2. Second, the stream thrust function Sa and throughflow area A, being known at Station 0, need only be determined at Station 10 in order to reveal the engine specific thrust for given freestream conditions and f and A_0.

4.4.2 Component Analysis

In order to find the stream thrust function and throughflow area at Station 10, it is best to divide the engine up into several pieces and analyze them individually. Experience has shown that it is most advantageous to break the engine down into those parts or *components* that are associated with the three thermodynamic processes described in Sec. 4.2. There are several technical and historical reasons for this division, but the most important by far is that significantly different physical phenomena are at work in each and, consequently, the engine companies are organized into *component groups.*

The component groups have a wide variety of titles, and the boundaries between them can become blurred because they are defined differently from organization to organization. For example, the names of the parts associated with the compression process include, at least, compression surfaces, spikes, ramps, intakes, inlets, diffusers, cowls, and nacelles. In view of this situation, the three main components will be referred to in the stream thrust analysis by the thermodynamic processes they produce, namely compression, combustion, and expansion. This has the added benefit of retaining the station designations described in Sec. 4.1.

The development here will be similar to that of the first law analysis in the sense that the steps are followed in a straightforward sequence. The cycle static temperature ratio ψ will again be used as a principal independent parameter. In order to reduce the amount of thermophysical property data required for air, it will be assumed that the perfect gas constant R is the same at all stations, an approximation that can be justified because the molecular weight of air does not vary significantly from place to place. Please note that the perfect gas law was used repeatedly to eliminate density from the ensuing equations, and that the quantity p_{10}/p_0 is treated as an independent parameter.

While traversing the steps below, please bear in mind that the compression and expansion components receive the same treatment as that of Sec. 4.3.2, but the combustion component is subjected to the full-blown control volume conservation law analysis.

Compression Component (Station 0 to Station 3).

1. Sa_0 [Stream thrust function, Eq. (2-62)]

$$Sa_0 = V_0 \left(1 + \frac{RT_0}{V_0^2}\right) \tag{4-39}$$

2. T_3 [Eq. (4-20)]

$$T_3 = \psi T_0 \tag{4-40}$$

3. V_3 [Conservation of energy, Eq. (2-31)]

$$V_3 = \sqrt{V_0^2 - 2C_{pc}T_0(\psi - 1)} \tag{4-41}$$

4. Sa_3 [Stream thrust function, Eq. (2-62)]

$$Sa_3 = V_3 \left(1 + \frac{RT_3}{V_3^2}\right)$$ (4-42)

5. p_3/p_0 [Adiabatic compression process, Eqs. (4-21) and (4-23)]

$$\frac{p_3}{p_0} = \left\{\frac{\psi}{\psi(1 - \eta_c) + \eta_c}\right\}^{(C_{pc}/R)}$$ (4-43)

6. A_3/A_0 [Conservation of mass, Eq. (2-28)]

$$\frac{A_3}{A_0} = \psi \cdot \frac{p_0}{p_3} \cdot \frac{V_0}{V_3}$$ (4-44)

Combustion Component (Station 3 to Station 4).

Two separate solutions are given here, corresponding to either *constant pressure* or *constant area* combustion. The combustion energy release is modeled as heat addition *with* mass addition. Absolute enthalpies are employed to better represent flow properties over the range of interest. The following four quantities, for which it is assumed that reasonable values can be assigned, appear for the first time in these equations:

$\dfrac{V_{fx}}{V_3}$: ratio of fuel injection *axial* velocity to V_3

$\dfrac{V_f}{V_3}$: ratio of fuel injection *total* velocity to V_3

$C_f \cdot \dfrac{A_w}{A_3} = \dfrac{\text{Combustor drag}}{\dfrac{\rho_3 V_3^2}{2} \cdot A_3}$: burner effective drag coefficient

$C_{pb}(T - T^\circ) = h$: employs a reference temperature T° to estimate the *absolute* static enthalpy h (see Fig. 2.7)

h_f: absolute sensible enthalpy of fuel entering combustor (this quantity is usually much less than h_{PR} and is therefore neglected in the ensuing calculations)

Constant Pressure Combustion.

1. V_4 [Conservation of momentum, Eq. (2-29)]

$$V_4 = V_3 \left\{ \frac{1 + f \cdot \dfrac{V_{fx}}{V_3}}{1 + f} - \frac{C_f \cdot \dfrac{A_w}{A_3}}{2(1 + f)} \right\} \tag{4-45}$$

2. T_4 [Conservation of energy, Eq. (2-31)]

$$T_4 = \frac{T_3}{1 + f} \left\{ 1 + \frac{1}{C_{pb} T_3} \left[\eta_b f h_{PR} + f h_f + f C_{pb} T^\circ \right. \right.$$
$$\left. \left. + \left(1 + f \cdot \frac{V_f^2}{V_3^2} \right) \frac{V_3^2}{2} \right] \right\} - \frac{V_4^2}{2 C_{pb}} \tag{4-46}$$

3. A_4/A_3 [Conservation of mass, Eq. (2-28)]

$$\frac{A_4}{A_3} = (1 + f) \cdot \frac{T_4}{T_3} \cdot \frac{V_3}{V_4} \tag{4-47}$$

Constant Area Combustion.

1. V_4 [Conservation of momentum *and* energy, Eqs. (2-29) and (2-31)]

$$V_4 = \frac{-b \pm \sqrt{b^2 - 4ac}}{2a} \tag{4-48}$$

where:

$$a = 1 - \frac{R}{2 C_{pb}}$$

$$b = -\frac{V_3}{1 + f} \left\{ \left(1 + \frac{R T_3}{V_3^2} \right) + f \cdot \frac{V_{fx}}{V_3} - \frac{C_f}{2} \cdot \frac{A_w}{A_3} \right\}$$

$$c = \frac{R T_3}{1 + f} \left\{ 1 + \frac{1}{C_{pb} T_3} \left[\eta_b f h_{PR} + f h_f + f C_{pb} T^\circ \right. \right.$$
$$\left. \left. + \left(1 + f \cdot \frac{V_f^2}{V_3^2} \right) \frac{V_3^2}{2} \right] \right\}$$

2. T_4 [Conservation of momentum *and* energy, Eqs. (2-29) and (2-31)]

$$T_4 = \frac{c}{R} - \frac{V_4^2}{2C_{pb}} \qquad (4\text{-}49)$$

3. p_4/p_0 [Conservation of mass, Eq. (2-28)]

$$\frac{p_4}{p_0} = (1 + f) \cdot \frac{p_3}{p_0} \cdot \frac{T_4}{T_3} \cdot \frac{V_3}{V_4} \qquad (4\text{-}50)$$

Either Constant Pressure or Constant Area Combustion.

4. Sa_4 [Stream thrust function, Eq. (2-62)]

$$Sa_4 = V_4 \left(1 + \frac{RT_4}{V_4^2}\right) \qquad (4\text{-}51)$$

Expansion Component (Station 4 to Station 10).

1. T_{10} [Adiabatic expansion process, Eqs. (2-35) and (4-19)]

$$T_{10} = T_4 \left\{1 - \eta_e \left[1 - \left(\frac{p_{10}}{p_0} \cdot \frac{p_0}{p_4}\right)^{(R/C_{pe})}\right]\right\} \qquad (4\text{-}52)$$

2. V_{10} [Conservation of energy, Eq. (2-31)]

$$V_{10} = \sqrt{V_4^2 + 2C_{pe}(T_4 - T_{10})} \qquad (4\text{-}53)$$

3. Sa_{10} [Stream thrust function, Eq. (2-62)]

$$Sa_{10} = V_{10} \left(1 + \frac{RT_{10}}{V_{10}^2}\right) \qquad (4\text{-}54)$$

4. A_{10}/A_0 [Conservation of mass, Eq. (2-28)]

$$\frac{A_{10}}{A_0} = (1 + f) \cdot \frac{p_0}{p_{10}} \cdot \frac{T_{10}}{T_0} \cdot \frac{V_0}{V_{10}} \qquad (4\text{-}55)$$

Engine Performance Measures.

The results of calculations based on this method are substituted into Eq. (4-38) in order to find the specific thrust $F/\dot{m}_0$. The other performance measures are determined from the interrelationships of Table 3.2, and the thermal and propulsive efficiencies from Eq. (3-10). Although this equation set does not readily lend itself to purely analytical results, such as were found for the first law analysis, it was formulated for direct transcription into HAP(Performance). However, in order to emphasize the importance of constant *pressure* combustion, HAP(Performance) can only be used for that case. You will have to do the constant area case by hand, or write your own computer program.

4.4.3 Stream Thrust Analysis Results

Judging from the number and variety of independent parameters that appear in Eqs. (4-38) through (4-55), we are now in a position to explore a truly *enormous* number of possibilities, too many, in fact, for an introductory textbook. Rather than exhaust the available supply of combinations and permutations, we will instead use the stream thrust analysis primarily to shed light on the underlying behavior of the hypersonic airbreathing engine, as was done earlier in first law analysis.

Unless stated otherwise, the nominal case values used in the stream thrust analysis calculations are those compiled below. Wherever possible, they have the same values as in the previous first law analysis. Also, the combustion pressure will normally be taken to be constant, as before.

$$\psi = 7.0, \qquad\qquad V_0 = 10,000 \text{ ft/s } (3048 \text{ m/s})$$

$$\eta_c = \eta_b = \eta_e = 0.90$$

$$fh_{PR} = 1510 \text{ BTU/lbm } (3510 \text{ kJ/kg})$$

$$T_0 = 400 \text{ °R } (222 \text{ K}), \qquad f = 0.0$$

$$R = 1730 \text{ (ft/s)}^2/\text{°R } (289.3 \text{ (m/s)}^2/\text{ K})$$

$$C_{pc} = 0.260 \text{ BTU/lbm} \cdot \text{°R } (1.09 \text{ kJ/kg} \cdot \text{K})$$

$$C_{pb} = 0.360 \text{ BTU/lbm} \cdot \text{°R } (1.51 \text{ kJ/kg} \cdot \text{K})$$

$$C_{pe} = 0.360 \text{ BTU/lbm} \cdot \text{°R } (1.51 \text{ kJ/kg} \cdot \text{K})$$

$$\gamma_c = 1.362 \text{ , Eq. (2-45)}$$

$$\gamma_b = 1.238 \text{ , Eq. (2-45)}$$

$$\gamma_e = 1.238 \text{ , Eq. (2-45)}$$

$$h_f = 0.0 \qquad\qquad T^o = 400 \text{ °R } (222 \text{ K})$$

$$\frac{V_{fx}}{V_3} = 0.0 \qquad\qquad \frac{V_f}{V_3} = 0.0$$

$$C_f \cdot \frac{A_w}{A_3} = 0.0 \qquad\qquad \frac{p_{10}}{p_0} = 1.0$$

The computations that follow were done with HAP(Performance).

4.4.3.1 Influence of cycle static temperature ratio.

The influence of cycle static temperature ratio on overall airbreathing engine efficiency η_o is portrayed in Fig. 4.9, which should be compared with the first law information of Fig. 4.7.

The first conclusion to be reached is that the two methods predict essentially the same level of airbreathing engine performance, particularly in the neighborhood of the point at which the thermophysical property data were chosen, namely $\psi = 7.0$. In fact, at this point and a flight speed of 10,000 ft/s (3048 m/s), the stream thrust analysis $\eta_o = 0.470$ while the first law analysis $\eta_o = 0.466$. And at 6000 ft/s (1829 m/s), the stream thrust analysis $\eta_o = 0.419$ while the first law analysis $\eta_o = 0.417$.

The second conclusion to be reached is that the difference between the two methods is greatest when ψ exceeds about 8.0, which is fortunate because this has already been shown to be an undesirable regime of operation. One may speculate that the first law analysis, because of its method of closing the thermodynamic cycle, is more reliable in this range, but nothing should overshadow the fact that

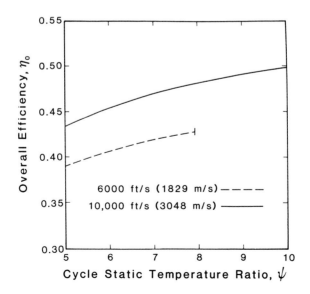

Fig. 4.9 Airbreathing engine overall efficiency as a function of cycle static temperature ratio and flight speed.

the two methods are in good agreement for the range of cycle static temperature ratios expected in practice.

4.4.3.2 Influence of fuel heating value.   The influence of fuel heating value on overall airbreathing efficiency as predicted by both stream thrust analysis and first law analysis is tabulated below. The results correspond to increasing or decreasing the fuel heating value (at constant combustion efficiency) by approximately 10 percent from the nominal case.

	Overall Efficiency, η_o			
V_0 ft/s	6000		10,000	
	First Law	Stream Thrust	First Law	Stream Thrust
$\dfrac{\eta_b f h_{PR}}{C_{p0} T_0} = 15.5$	0.416	0.414	0.469	0.468
Nominal case	0.417	0.419	0.466	0.470
$\dfrac{\eta_b f h_{PR}}{C_{p0} T_0} = 12.7$	0.416	0.425	0.461	0.472

The stream thrust results are clearly in close agreement with those obtained by means of first law analysis, the conclusion again being that fuel heating value alone has little impact on overall efficiency.

4.4.3.3 Influence of thermodynamic process efficiencies.   The influence of thermodynamic process efficiencies on overall airbreathing efficiency as predicted by both stream thrust analysis and first law analysis is tabulated below. The results correspond to increasing or decreasing the individual process efficiencies from their nominal values by the amount 0.10. Only one change from the nominal case takes place at a time in the calculations presented here.

The stream thrust results are clearly in close agreement with those obtained by means of first law analysis. If anything, the stream thrust analysis is more sensitive to η_c and η_e than first law analysis, which merely underlines the main conclusions of that study.

Taken together, the information found so far demonstrates that first law analysis and stream thrust analysis predict very similar performance for hypersonic airbreathing engines. This enhances the credibility of *both* approaches, and leads to the following exploration of the influence of several parameters only conveniently available through stream thrust analysis.

V_0 ft/s	Overall Efficiency, η_o			
	6000		10,000	
	First Law	Stream Thrust	First Law	Stream Thrust
Nominal case	0.417	0.419	0.466	0.470
$\eta_c = 1.00$	0.496	0.513	0.565	0.587
$\eta_c = 0.80$	0.340	0.329	0.375	0.361
$\eta_b = 1.00$	0.463	0.460	0.518	0.520
$\eta_b = 0.80$	0.370	0.378	0.415	0.419
$\eta_e = 1.00$	0.476	0.489	0.539	0.556
$\eta_e = 0.80$	0.355	0.346	0.392	0.381

4.4.3.4 Influence of constant area combustion.   Because of the obvious geometric simplicity, constant area combustion is a realistic candidate for any ramjet or scramjet engine design. The question of immediate interest is whether or not the change from constant pressure combustion will have a dramatic impact on overall efficiency. The answer is obtained, of course, by carrying out the stream thrust analysis with Eqs. (4-45) through (4-47) replaced by Eqs. (4-48) through (4-50). Before any numerical results are presented, however, it is worthwhile to consider the general nature of the flows involved.

In the case of constant *area* heat addition, the governing equations are coupled in such a way that two solutions can occur. They correspond physically to combustor flow with and without the equivalent of a normal shock wave, and entropy considerations forbid the supersonic solution when the combustor entry flow is subsonic. Since scramjet combustor entry flow is supersonic, either solution can occur, depending upon the details of the flow within the combustor. The two solutions also correspond to supersonic and subsonic combustor exit flow. Combining this with the results of Example Case 2.4, it should be easy to understand that the effect of heat addition is to move the exit Mach number of *both* solutions closer to 1. An especially interesting result is that, for the limiting case of thermal choking, both solutions have an exit Mach number of 1. More information on the behavior of constant area heat addition or Rayleigh flow can be found in Ref. 4.6.

In the case of constant *pressure* heat addition, the governing equations are uncoupled, and only one solution occurs. The physical explanation for this is that the presence of normal shock waves would automatically violate the condition of constant pressure flow. The appearance of normal shock waves and dual solutions is therefore forbidden. Nevertheless, as we have already seen in Example Case 2.8, the heat addition can cause the combustor exit flow to be either supersonic or subsonic.

The results of the constant area heat addition calculations are summarized on Fig. 4.10. The engine overall efficiency is evidently strongly dependent upon whether or not a normal shock is present and, as one should expect, is substantially higher when a shock can be avoided. As the cycle static temperature ratio increases, the Mach number entering the combustor decreases. This, in turn, causes the strength of the normal shock wave and the attendant losses to decrease so that there is less difference between the two constant area combustor solutions.

The combustor exit Mach number M_4 does have a supersonic branch and a subsonic branch, in accordance with our expectations. As the cycle static temperature ratio increases, the Mach number entering the combustor decreases which, as we have seen in Example Case 2.4, brings thermal choking closer for constant heat addition. For the case under consideration, thermal choking ($M_4 = 1.0$) occurs for both branches when ψ is somewhat greater than 10, and is therefore outside the expected range of operation.

Returning to the original question regarding overall efficiency, comparison of Figs. 4.9 and 4.10 reveals that, provided there is no shock wave in the combustor, there is little performance difference between constant pressure and constant area heat addition. Their

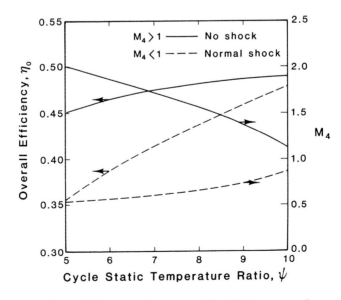

Fig. 4.10 Airbreathing engine overall efficiency and combustor exit Mach number as functions of cycle static temperature ratio, for the case of constant area heat addition.

overall efficiencies at $\psi = 7.0$, for example, are 0.470 and 0.475, respectively. The underlying cause of this difference is that constant pressure heat addition results in a lower T_4 than does constant area heat addition, which causes their thermal efficiencies at $\psi = 7.0$ to be 0.512 and 0.518, respectively. It is certainly comforting to find that the precise details of combustor design have only a minor influence on overall efficiency, because that broadens the usefulness of any conclusions based upon constant pressure design.

4.4.3.5 Influence of fuel mass addition.   In contrast to airbreathing engines that operate at subsonic and slightly supersonic speeds, for which the overwhelming purpose of the fuel is to heat the air, scramjets can derive a considerable fraction of their thrust from the fluxes of mass, momentum, and kinetic energy that accompany the introduction of the fuel into the combustor. This could have been seen as early as Eq. (3-11), if one had known then that V_e approaches V_0 as V_0 becomes large. It is therefore imperative that any realistic performance prediction method incorporate the diverse influences of fuel mass addition.

The results of several variations on the nominal case of this theme are summarized in Fig. 4.11, some of which are quite remarkable. To begin with, even when the fuel flow contributes no axial momentum or kinetic energy whatsoever, the engine overall efficiency increases sharply with fuel/air ratio. For hydrogen fuel, for example, η_o would

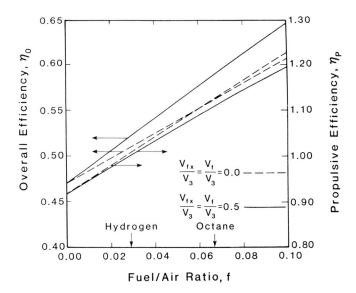

Fig. 4.11 Airbreathing engine overall efficiency and propulsive efficiency as functions of fuel/air ratio and fuel entrance velocities.

increase from 0.470 to 0.510 simply because of the presence of the fuel mass flow. This effect would happen, of course, when additional mass of any sort is introduced into the combustor.

The improved overall efficiency is primarily due to increased propulsive efficiency, as given by Eq. (3-12). As Fig. 4.11 shows, η_p increases from 0.918 to 1.008 for the hydrogen example cited above, while the thermal efficiency actually decreases slightly. This outcome agrees with the intuitively appealing observation that spreading the available energy out across more matter increases the thrust produced, as in the case of bypass turbofan engines.

When the fuel has axial momentum and/or kinetic energy, the benefits are even larger, and for obvious reasons. The data given in Fig. 4.11 demonstrate that overall efficiency increases of an additional several percent are possible, depending upon the speed and direction of the fuel flow expressed as a fraction of the combustor entry velocity. A significant question, therefore, is how large V_{fx}/V_3 and V_f/V_3 are likely to be, especially when the combustor entry velocity is a substantial fraction of the freestream velocity. For hydrogen, happily, the combination of low molecular weight and high temperature (having cooled the engine surfaces) makes fuel entrance velocities of 5000 ft/s (1524 m/s) and higher entirely possible, so that the values used on Fig. 4.11 for V_{fx}/V_3 and V_f/V_0 are feasible. For hydrocarbon fuels, unfortunately, the reverse is true, although this disadvantage is more than made up for by the extra mass flow they provide.

4.4.3.6 Influence of exhaust pressure.   Until this point, it has been assumed that the exhaust flow was *properly expanded,* meaning that $p_{10} = p_0$. The truth is that ramjets and scramjets often operate away from this condition, and scramjets in particular are routinely underexpanded. Stream thrust analysis was therefore performed in order to explore the implications of improper expansion, and the results are presented in Fig. 4.12.

Turning our attention first to the case of ideal or isentropic expansion, where $\eta_e = 1.0$, we find the familiar outcome that the maximum overall efficiency occurs when the exhaust flow is properly expanded. This supports the assertions of Sec. 3.2, and the results based upon them. It is also noteworthy that the falloff of overall efficiency with either overexpansion or underexpansion is extremely gradual, leading to the conclusion that the assertions of Sec. 3.2 need not have been so precise. From the practical standpoint, the outstanding feature of Fig. 4.12 is the enormous range of variation of A_{10}/A_0 with p_{10}/p_0. Since A_{10} must be furnished by engine or vehicle hardware, it represents expensive territory, especially when multiples of an already large A_0 are required. The designer is, therefore, literally driven to operate the exhaust flow slightly underexpanded in order to significantly reduce the cost and weight of the vehicle, while paying a small price in performance.

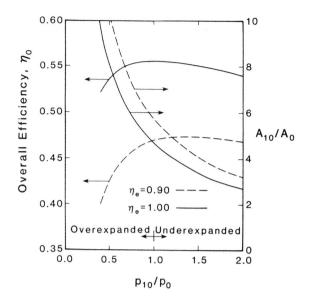

Fig. 4.12 Airbreathing engine overall efficiency and exhaust-to-capture throughflow area ratio as functions of exhaust-to-freestream static pressure ratio and adiabatic expansion efficiency.

When the expansion process is not ideal, the behavior of the engine is similar to the isentropic case, but there are a few important differences, as seen for $\eta_e = 0.90$ in Fig. 4.12. To begin with, as already reported, the overall efficiency is much less than for ideal expansion. Furthermore, the maximum overall efficiency has conveniently moved in the direction of underexpansion, a reflection of the fact that the penalty paid in the expansion process is proportional to the total amount of expansion taking place [see Eq. (4-19)]. Finally, A_{10}/A_0 continues to vary rapidly with p_{10}/p_0, although the levels are quite a bit higher than they were for ideal expansion. The inescapable conclusion is that the motivation to operate the exhaust flow somewhat underexpanded is even stronger when losses are included.

4.4.3.7 Influence of freestream velocity. With regard to scramjet performance, it has been found that the main influence of increasing freestream velocity or flight speed, all other things being equal, is to increase the propulsive efficiency and, therefore, the overall efficiency. The focus here, rather, is upon the engine geometry required to bring about the desired flow conditions.

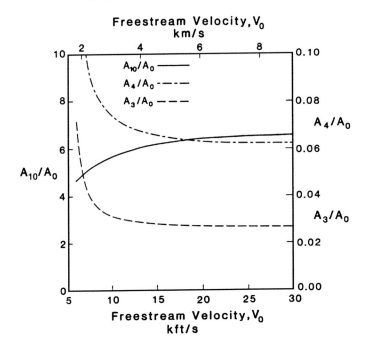

Fig. 4.13 Ratios of combustor entry, combustor exit, and engine exhaust areas to freestream capture area as a function of freestream velocity.

The responses of the three benchmark area ratios to variations of freestream velocity from the nominal case are shown in Fig. 4.13. Perhaps the most fascinating and helpful characteristic of these area ratios is that they change relatively little once a freestream velocity of about 10,000 ft/s (3048 m/s) or a freestream Mach number of about 10 has been reached. This is the airbreathing propulsion analog of the well-known *Mach number independence principle* of external aerodynamics, where it has been found that at sufficiently high Mach numbers many dimensionless aerodynamic quantities become constant.[4.7] This situation suggests that scramjets may need little or no variable geometry in order to do their job.

There are, nevertheless, several downsides that also appear in Fig. 4.13. First, the area ratios change rapidly for freestream velocities below about 10,000 ft/s (3048 m/s), making variable geometry a virtual necessity there. Second, the combustor entry and exit areas are such a small fraction of the capture area that manufacturing tolerance could become a decisive factor, and such unwanted boundary effects as skin friction and heat transfer will grow in importance. Third, even an exhaust system designed to be underexpanded for most of the flight spectrum could easily be operating overexpanded at low flight speeds, with consequent performance penalties.

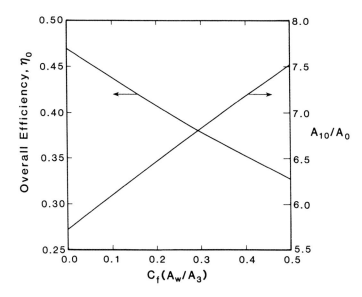

Fig. 4.14 Airbreathing engine overall efficiency and exhaust-to-capture throughflow area ratio as a function of burner effective drag coefficient.

4.4.3.8 Influence of combustor drag. Any drag occurring in the combustor must surely reduce the eventual thrust of the engine. This effect is quantified in Fig. 4.14 for the nominal case and a wide range of burner effective drag coefficients. Although C_f alone is liable to be only about 0.001–0.005 for turbulent flow over smooth, flat walls,[4.6] a wide range of $C_f(A_w/A_3)$ is displayed there in order to accommodate the possible existence of parts that project into the combustor flowpath and cause form drag (such as fuel injectors, mechanical mixing devices, structural members, and joints or steps) or relatively large values of A_w/A_3 resulting from the small height of the combustor passage.

Referring to Fig. 4.14, we see that even such a modest burner effective drag coefficient as 0.1 can reduce overall efficiency by several percent. Moreover, the A_{10}/A_0 required for proper expansion of the exhaust flow also grows rapidly. This information, taken together with Reynolds analogy, which states that heat transfer is proportional to skin friction,[4.6] emphasizes the importance of reducing combustor drag as much as possible.

4.4.4 Composite Scramjet Example Case

In order to both satisfy our curiosity about the possible performance of real scramjets and chronicle the present condition of our predictive techniques, a complete stream thrust analysis was carried out for a typical set of hypersonic engine parameters. The input data and output results (in the order of their appearance in Sec. 4.4.2) are tabulated below without further comment.

Input:

$$\psi = 7.0$$

$$V_0 = 10,000 \text{ ft/s} \quad (3048 \text{ m/s})$$

$$T_0 = 400 \text{ °R} \quad (222 \text{ K})$$

$$f = 0.04$$

$$h_f = 0.0$$

$$T^\circ = 400 \text{ °R} \quad (222 \text{ K})$$

$$\frac{V_{fx}}{V_3} = 0.50$$

$$\frac{V_f}{V_3} = 0.50$$

$$C_f \frac{A_w}{A_3} = 0.10$$

$$\frac{p_{10}}{p_0} = 1.40$$

$$\eta_c = \eta_b = \eta_e = 0.90$$

$$fh_{PR} = 1510 \text{ BTU/lbm} \quad (3510 \text{ kJ/kg})$$

$$R = 1730 \text{ (ft/s)}^2/\text{°R} \quad [289.3 \text{ (m/s)}^2/\text{K}]$$

$$C_{pc} = 0.260 \text{ BTU/lbm} \cdot \text{°R} \quad (1.09 \text{ kJ/kg} \cdot \text{K})$$

$$C_{pb} = 0.360 \text{ BTU/lbm} \cdot \text{°R} \quad (1.51 \text{ kJ/kg} \cdot \text{K})$$

$$C_{pe} = 0.360 \text{ BTU/lbm} \cdot \text{°R} \quad (1.51 \text{ kJ/kg} \cdot \text{K})$$

$$\gamma_c = 1.362$$

$$\gamma_b = 1.238$$

$$\gamma_e = 1.238$$

Output:

		Constant Pressure	Constant Area $(M_4 > 1.0)$
Sa_0	lbf $\cdot$ s/lbm	313	313
	N $\cdot$ s/kg	3070	3070
T_3	°R	2800	2800
	K	1556	1556
V_3	ft/s	8290	8290
	m/s	2530	2530
Sa_3	lbf $\cdot$ s/lbm	276	276
	N $\cdot$ s/kg	2710	2710
$\dfrac{p_3}{p_0}$		258	258
$\dfrac{A_3}{A_0}$		0.0327	0.0327
V_4	ft/s	7730	5940
	m/s	2360	1810
T_4	°R	6730	8090
	K	3740	4500
$\dfrac{A_4}{A_3}$		2.68	1.00
$\dfrac{p_4}{p_0}$		258	1080
Sa_4	lbf $\cdot$ s/lbm	287	258
	N $\cdot$ s/kg	2820	2530
T_{10}	°R	2890	2840
	K	1610	1580
V_{10}	ft/s	11350	11400
	m/s	3460	3470
Sa_{10}	lbf $\cdot$ s/lbm	367	368
	N $\cdot$ s/kg	3590	3610
$\dfrac{A_{10}}{A_0}$		4.74	4.62

Output (cont'd.):

		Constant Pressure	Constant Area $(M_4 > 1.0)$
$\dfrac{F}{\dot{m}_0}$	lbf · s/lbm	60.2	61.5
	N · s/kg	590	603
η_o		0.512	0.524
η_{th}		0.450	0.464
η_p		1.139	1.129
I_{sp}	s	1504	1538

4.5 STATUS ASSESSMENT

During our voyage through this chapter, several independent methods for estimating scramjet engine behavior and measures of performance have been described, developed, and explored. The nature of the progression was from the simple and incomplete but transparent and instructive toward the opposite extreme. The final approach, stream thrust analysis, provides sufficient detail and accuracy to serve as the foundation for the remainder of our work.

While traversing the landscape of hypersonic airbreathing engine performance analysis, the most likely directions for further improvements were discovered. For one, more realistic and accurate models for the functioning of compression and expansion components, as well as an appreciation of the influence of practical considerations upon their design, are needed. This will be the subject of Chaps. 5 and 7. For another, representative air has been taken as far as it can go, and it is now essential to incorporate the actual combustion process in order to provide a correct evaluation of the thermophysical properties of the working medium at any desired point in the flow. This will require an elucidation of several of the quite fascinating processes peculiar to burners, such as fuel injection, mixing, ignition and combustion, their influence on upstream and downstream components, and their control over choking. This will be the subject of Chap. 6. Finally, one could consider *finer-grain* computational schemes, including the one-dimensional differential control volume analysis approach or even two- or three-dimensional computational fluid dynamics (CFD). Although the latter is essential to the design of ramjets and scramjets that must perform properly in actual service, doing CFD is best left to large organizations, although several outstanding examples of its application will be presented later.

The experienced and astute observer will note that flow quantities have not been referenced to total or stagnation pressures and temperatures, such as is done to great advantage in many textbooks on aircraft engines.[4.1,4.2] The main reasons for this are that the stagnation pressures and temperatures of high enthalpy hypersonic flows are no longer related to the static properties by elementary mathematical relationships, and that dealing directly with the static properties is easy enough when there is only a single stream of fluid. And, frankly, this avoids the occurrence of some rather silly numbers.

Nothing should obscure the real accomplishments of this chapter. First, the framework for progress in predicting ramjet and scramjet performance has been established. Second, the predicted values for overall efficiency are more than adequate for the success of the hypersonic cruise and transatmospheric aerospace systems evaluated in Chap. 3. With those encouraging conclusions in mind, we are ready to take a closer look at the component parts.

REFERENCES

[4.1] Oates, G. C., *The Aerothermodynamics of Gas Turbines and Rocket Propulsion,* Revised and Enlarged Edition, AIAA Education Series, Washington, DC, 1988.

[4.2] Mattingly, J. D., Heiser, W. H., and Daley, D. H., *Aircraft Engine Design,* AIAA Education Series, New York, 1987.

[4.3] Reynolds, W. C., and Perkins, H. C., *Engineering Thermodynamics,* McGraw-Hill, New York, 1977.

[4.4] Builder, C. H., "On the Thermodynamic Spectrum of Airbreathing Propulsion," AIAA 1st Annual Meeting, AIAA Paper 64-243, June 1964.

[4.5] Curran, E. T., and Craig, R. R., "The Use of Stream Thrust Concepts for the Approximate Evaluation of Hypersonic Ramjet Engine Performance," Aero Propulsion Laboratory, AFAPL-TR-73-78, Wright-Patterson AFB, OH, July 1973.

[4.6] Shapiro, A. H., *The Dynamics and Thermodynamics of Compressible Fluid Flow,* Ronald Press, New York, 1953.

[4.7] Bertin, J. J., *Hypersonic Aerothermodynamics,* AIAA Education Series, Washington, DC, 1994.

PROBLEMS

4.1 Investigate the burner inlet Mach number situation more thoroughly by plotting the following graphs:

(a) Freestream Mach number M_0 as a function of T_3/T_0 for a burner inlet Mach number M_3 of 0, the upper limit of ramjet operation.

(b) Freestream Mach number M_0 as a function of T_3/T_0 for a burner inlet Mach number M_3 of 1, the lower limit of scramjet operation.

4.2 The combination of Eqs. (4-15) and (4-16) reveals that the airbreathing engine specific impulse I_{sp} is a function of thermal efficiency η_{th}, flight speed V_0, fuel heating value h_{PR}, and fuel/air ratio f.

(a) Assuming a stoichiometric mixture of hydrogen and air, so that $f = 0.0291$ and $h_{PR} = 51,600$ BTU/lbm (120,000 kJ/kg), construct a general performance graph similar to that of Fig. 4.5 showing airbreathing specific impulse as a function of thermal efficiency and flight speed.

(b) Repeat this exercise for any hydrocarbon fuel in Table 3.1.

What conclusions do you draw?

4.3 A contemporary thermodynamic approach to the analysis of steady flow thermodynamic processes is *availability analysis*,[4.3] which uses the steady flow availability function per unit mass

$$b = h - T_0 s$$

a thermodynamic property that depends upon both the state of the working fluid and the static temperature T_0 of the surroundings.

If you treat the engine as a control volume with a fixed amount of heat added, what does availability analysis tell you about engine performance? Can availability analysis be used to prove that maximum engine performance corresponds to minimum h_{10} and s_{10}? If not, why not?

4.4 Test the agreement of Eq. (4-33) plus Eq. (4-28) with your intuitive understanding of thermodynamic closed cycle behavior as follows:

(a) Show that when there is no compression process (i.e., $\psi = 1.0$, the engine produces no thrust. Note in this case that C_{pr} *must* be identical to C_{pb}.

(b) Show that when there is no heat addition, the engine produces a negative thrust or drag (i.e., $V_{10} \leq V_0$).

(c) Show that when the component efficiencies all equal 1.0, the thermodynamic cycle efficiency is given by the expression

$$\eta_{th} = 1 - \frac{C_{pr}}{C_{p0}} \left\{ \left(\psi \cdot \frac{C_{p0} T_0}{\eta_b f h_{PR}} + \frac{C_{p0}}{C_{pb}} \right) \left(\frac{1}{\psi} \right)^{\frac{C_{pc}}{R_c} \cdot \frac{R_e}{C_{pe}}} \right.$$

$$\left. - \frac{C_{p0} T_0}{\eta_b f h_{PR}} \right\}$$

Use this relationship to show that the thermodynamic cycle efficiency equals the *Carnot* value of

$$1 - \frac{1}{\psi}$$

when the thermophysical properties of air are constant. How does the thermodynamic cycle efficiency compare with the Carnot value for representative values of the thermophysical properties of air? Can you explain this result physically?

You will find it helpful in each of these cases to first draw the corresponding T-s diagram and consider it in terms of the discussion of Sec. 4.2.

4.5 Repeat the first law analysis of Sec. 4.3.2 for the turbojet engine, as described in Sec. 4.3.3.1. Show that Eqs. (4-28), (4-32), and (4-33) are unchanged by the exchange of equal work between the compression and expansion processes, provided that η_c and η_e are unchanged. This would mean that the first law analysis is valid even when

$$\frac{T_3}{T_0} > 1 + \frac{\gamma_c - 1}{2} M_0^2$$

4.6 Repeat several of the first law analysis examples of Sec. 4.3.3 in order to confirm and extend the results found there, and to develop your own intuition regarding the behavior of ramjets and scramjets. In addition, test the robustness of the first law approach by varying the assumed thermophysical properties of air both separately and in reasonable combinations and observing the sensitivity of the outcomes.

This exercise will be much more enjoyable if you first write a computer program that automates the calculations.

4.7 Repeat several of the stream thrust analysis examples of Sec. 4.4.3 using HAP(Performance) in order to confirm and extend the results found there, and to develop your own intuition regarding the behavior of ramjets and scramjets. In addition, test the robustness of the stream thrust approach by varying the assumed thermophysical properties of air both separately and in reasonable combinations and observing the sensitivity of the outcomes.

4.8 The relationship between overall and propulsive efficiency and fuel/air ratio on Fig. 4.11 appears to be a straight line. Use the equations of stream thrust analysis to determine whether, or under what conditions, this appearance is correct.

4.9 Prove by mathematical analysis that the *Mach number independence principle,* observed on the basis of calculations in Sec. 4.4.3.7, exists for scramjet engines. Do this for the case of constant pressure combustion and proper expansion of the exhaust flow. What are the limiting values at high M_0 of the area ratios shown on Fig. 4.13?

4.10 The relationship between overall efficiency and burner effective drag coefficient on Fig. 4.14 appears to be a straight line. Use the equations of stream thrust analysis to determine whether, or under what conditions, this appearance is correct.

4.11 Apply the concepts and equations of stream thrust analysis in order to show that the specific axial force (i.e., air mass flow rate specific axial force) exerted on *only* the external and internal *compression* surfaces of the airbreathing engines depicted in Fig. 4.8 acts in the direction of the freestream flow and is given by the expression

$$Sa_3 - Sa_0$$

Calculate the value of the specific axial force for the parameters of Sec. 4.4.3.

4.12 Select a composite scramjet example of your own, as was done in Sec. 4.4.4, and explore with the help of HAP(Performance) the influence of parameters under the control of the designer for both constant pressure and constant area combustion. In particular, find the values of p_{10}/p_0 that yield the highest overall efficiencies.

5
COMPRESSION SYSTEMS
OR COMPONENTS

5.1 INTRODUCTION

The time has come to examine the individual components of the hypersonic airbreathing engine more closely. We will need realistic models of their performance over the expected range of operation, as well as an understanding of their distinctive features or idiosyncracies, in order to identify what will be required for a successful engine design.

This examination will begin with the compression systems or components for the usual reason that it is easier to grasp what happens to the flow by following it along. Compression systems have a number of general characteristics that they happen to have in common with expansion systems. First, the aerothermodynamic processes involved are relatively passive, in the sense that there is no deliberate exchange of energy with the surroundings or release of energy due to combustion. If these do, in fact, happen, they are usually treated as perturbations on the original model. Second, their flows are bounded by surfaces that are both internal and external to the engine flowpath, the latter sometimes being regarded as "belonging to" the vehicle. Third, despite their rather simplistic geometrical appearances, both compression and expansion components exhibit some remarkably complex and intriguing phenomena.

5.2 COMPRESSION COMPONENTS

The function of the compression components (*a.k.a.* compression systems) is to provide the desired cycle static temperature ratio ψ over the entire range of vehicle operation in a controllable and reliable manner with minimum aerodynamic losses (i.e., maximum compression efficiency or minimum entropy increase). Owing to the extraordinarily wide range of flight conditions to be encountered in hypersonic flight, the chosen configuration must provide the means to satisfy several interrelated (and frequently conflicting) design requirements simultaneously. After the general introductory comments that follow, the remaining material of Sec. 5.2 deals with these design requirements more or less in their order of importance, although they must all be attended to before the design is complete.

Readers not currently familiar with the congenial properties of simple two-dimensional oblique compression and expansion waves are encouraged to brush up at this point by means, for example, of Ref. 5.1.

5.2.1 Typical Compression Component Configurations

Before embarking on any compression component analysis, it is important to have a mental picture of the type of geometries most likely to be encountered. One can easily be impressed by human ingenuity when it comes to compression components because so many different types have been invented, each of which has some applications at which it excels. Several of the more common configurations are depicted in Figs. 5.1 through 5.6, although this collection is far from exhaustive. Please bear in mind that, although they are shown here in their two-dimensional shapes, they also have axisymmetric or annular and other multi-dimensional counterparts, and that this introductory discussion centers on the compression caused by shock waves.

The term *spillage* means the difference between the freestream airflow that could pass through the area obtained by projecting the end point of the lip axially to the freestream flow, temporarily referred to as A_1, and the throughflow area of the freestream flow A_0 that actually enters the physical or geometrical opening. The spillage therefore is $A_1 - A_0$.

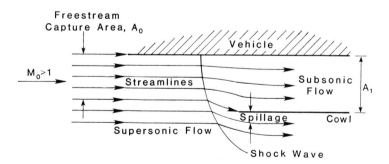

Fig. 5.1 External compression normal shock wave or "pitot" inlet. This device is frequently found on ramjets because it offers reasonable performance for $0 < M_0 < 3$, and the simple, fixed geometry version can be very attractive from the structural standpoint. For higher Mach numbers the normal shock losses become unacceptably high and oblique shock compression becomes necessary. Note that the normal shock wave "warns" the supersonic flow of the blockage awaiting downstream, and allows "spillage" of any excess airflow outside the cowl.

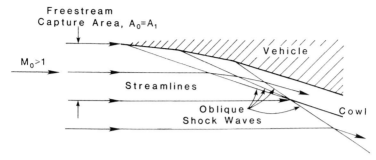

Fig. 5.2 External compression or oblique shock wave compression system. This device is shown at its design point, for which the several oblique shock waves generated by the external compression surface are "focused" on and terminate at the edge or lip of the engine cowl. The internal flow downstream of the final or "terminal" oblique shock wave is uniform and parallel. The uncaptured external flow must adjust its direction due to the presence of the cowl by means of a single oblique shock wave having the same total turning angle as the flow entering the engine.

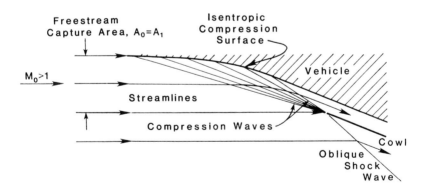

Fig. 5.3 Isentropic external compression system. This device is identical to that of Fig. 5.2, except that the compression surface is continuously curved in such a way that it generates a train of infinitesimal compression waves that cumulatively result in a finite pressure increase, but with no increase in entropy. The uncaptured external flow, however, must still adjust its direction by means of an oblique shock wave.

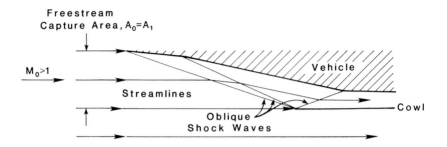

Fig. 5.4 Mixed external and internal compression system. This device incorporates oblique shock waves in various external and internal blends in order to achieve the desired combinations of compression and turning. In comparison to the external compression system of Fig. 5.2, this approach can use multiple internal reflections of weaker shock waves in order to accomplish the same task with less entropy increase, but the overall length must be greater. Mixed compression systems "decouple" the engine cowl angle from the amount of compression and can result in a cowl that is parallel to the freestream flow.

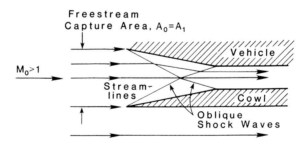

Fig. 5.5 Symmetrical internal compression system. This method employs an adjacent surface in order to generate a "mirror-image" oblique shock wave configuration that produces a uniform and parallel internal flow. This approach obviously shortens the axial length required for a given amount of compression, but requires an additional compression surface and can produce complex flows when operating away from the design point. The axisymmetric version of this type of compression system is known as the "Busemann inlet."

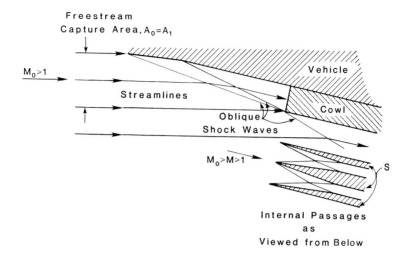

Fig. 5.6   Lateral or sidewall mixed external and internal compression system. This approach combines compression on the external surface with compression generated within closed passages formed by "strut- or pylon-like" structures extending outward from the boundary of the vehicle. In this way the axial length required for a given amount of compression can be reduced and the flow broken into several separate but identical streams, but the crossing oblique shock waves are more difficult to analyze and the internal passage has more surface and corner area.

5.3   COMPRESSION COMPONENT ANALYSIS OVERVIEW

Before proceeding with any quantitative analysis of the behavior of compression components, a few general, clarifying remarks are in order. The notional compression systems pictured in Figs. 5.1 through 5.6 clearly betray the true complexity of the flows encountered in these devices, for a number of reasons, several of which are described below.

First, these devices seldom, if ever, operate exactly at their design points, with the result that extraneous oblique shock and expansion waves flourish and, therefore, that uniform and parallel exit flows would rarely be encountered. This is particularly true for a transatmospheric vehicle, for which the operating range is so large that even a variable-geometry compression system could not cope with the spectrum of possibilities.

Second, the wall friction and heat transfer effects (also known and referred to here as either viscous or boundary-layer effects) lead to two- and three-dimensional distortions of the relatively simple oblique shock and expansion wave-created flowfields. One important

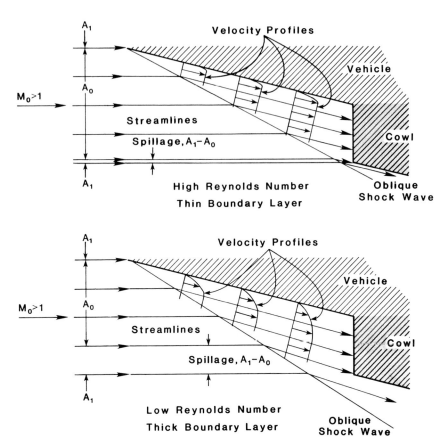

Fig. 5.7 Schematic representation of the behavior of the aerospace vehicle boundary layer upstream of the physical opening of the inlet.

manifestation of the viscous effects is the velocity profile, resulting primarily from the requirement that the relative velocity be 0 at the wall and representing the cumulative effects of skin friction as a boundary-layer momentum deficit. We have already seen in Sec. 2.4.7 that the Reynolds number, the first indicator of boundary-layer effects, will decrease by at least a factor of 10 as a transatmospheric vehicle moves from Earth to orbit. This causes the various boundary-layer thicknesses to increase with flight Mach number and altitude which, in turn, increases both the extent of the flow influenced by viscous effects and the amount of flow displaced away from and thus not available to the physical opening of the inlet. Under some extreme conditions, the outer edge of the boundary layer can extend beyond the cowl lip, so that the entire entering flow is modified. These viscous effects are portrayed in Fig. 5.7.

Another important manifestation of the viscous effects is that the static temperature of the fluid near the wall approaches the stagnation temperature of the freestream flow because it has nearly been brought to rest. As we have seen in Sec. 2.5, at hypersonic velocities this will lead to the dissociation of oxygen and nitrogen and therefore to a change in the chemical composition of the air at the entrance of the combustor. Moreover, the static temperature rise within the boundary layer (or, more precisely, the decrease of availability due to viscous dissipation) both decreases the density and increases the viscosity of the air and therefore has the unfortunate synergistic effect of making the boundary layer grow even faster, roughly in proportion to the square of the Mach number at the outer edge of the boundary layer.[5.2] Ultimately, when the edge Mach number is sufficiently large, the boundary layer grows rapidly enough that the shape of the compression surface is significantly displaced as far as the inviscid external flow is concerned. Under these circumstances, the external pressure distribution is no longer simply "imposed" on the boundary-layer fluid, but instead is altered by the very presence of the boundary layer, and, of course, vice versa. This complex coupling is called *strong viscous interaction* in the open literature.[5.2]

Finally, it must be recognized that many practical vehicle design requirements will prevent the compression system from being either perfectly two-dimensional or axisymmetric, so that the result will at best be a compromise that only approximates one of those ideal extremes. Something even beyond simple wave analysis will therefore be needed for the inviscid flowfield.

This qualitative background should persuade anyone that the flow in the compression component of a hypersonic airbreathing engine combines a wide variety of interlocking physical phenomena, and can only be analyzed properly by experienced professionals using CFD. Several illustrative CFD examples will be found in Sec. 5.8.

The spirit of this textbook, which emphasizes fundamental understanding and realistic estimation, can nevertheless be maintained through the suitable selection of less complex models and analytical tools. For the most part, and unless otherwise stated, several useful approximations will apply to the remainder of this chapter. First, in keeping with the approach of Sec. 2.6, the flow will be treated as one-dimensional in the sense that entrance and exit planes can be found across which the flow properties are uniform. A corollary to this is that the boundary layer will be represented only by its average effect on flow properties. Second, we will represent the air as a calorically perfect gas having the constant properties

$$\gamma_c = \frac{C_{pc}}{C_{vc}} = 1.360$$

and

$$\frac{C_{pc}}{R_c} = \frac{\gamma_c}{\gamma_c - 1} = 3.778$$

during the compression process. Third, heat transfer to or from the wall will be neglected. Fourth, any calculations involving oblique shock waves will be done using HAP(Gas Tables), which can handle either two-dimensional or axisymmetric oblique shock waves, although the emphasis here will be almost entirely upon the former. All results allow the reader to select arbitrary values of γ_c in order to explore the sensitivity of the results to that property of the gas.

5.4 COMPRESSION COMPONENT PERFORMANCE MEASURES

We found in Chap. 4 that the adiabatic compression efficiency η_c exerts a profound influence upon the airbreathing engine overall efficiency η_0 . In fact, we learned there that compression efficiency can "make or break" the performance of an engine as well as its intended aerospace system, which explains why this must be the first consideration of compression components in any study.

Before proceeding with the evaluation of the adiabatic compression efficiency for typical compression component configurations, it is a helpful and revealing exercise to derive and tabulate the relationships between the compression efficiency and other typical one-dimensional performance measures. This compilation will enable us to obtain compression efficiency directly when other measures are specified, as well as to develop an intuitive feeling for the character of the other measures.

Anyone entering the field of hypersonic airbreathing propulsion is liable to be awestruck by the amazing proliferation of parameters that have been formulated for the quantitative evaluation of compression component performance.[5.3] The main value of this taxonomic diversity is that it allows each worker to express performance in terms that also communicate his or her feelings about the important physical phenomena involved. The problem for a textbook is that there are far too many parameters to describe properly. Our solution is to carefully examine and relate to η_c three standard performance measures that fairly represent the entire available spectrum. It is important to recognize that any performance methodology will produce useful results if it is clearly defined and systematically applied. The differences between them are therefore largely questions of taste and tradition, and very much in the eye of the beholder.

5.4.1 Total Pressure Ratio

The *total pressure ratio* across the compression system is defined as the ratio of the total pressure at the entrance of the combustor divided by the total pressure of the freestream flow, and is denoted by the symbol π_c. The total pressure ratio is universally accepted as the meaningful measure of performance for subsonic and supersonic aircraft engine compression systems, and there is an abundance of theoretical and experimental information regarding its behavior in the open literature.[5.4,5.5] It is not as valuable in hypersonic flow because stagnating the flow excites chemical effects that render the total pressure an extraordinarily complicated function of the flow conditions, rather than the simple algebraic formulas that pertain to subsonic and supersonic flow conditions. We, too, must be careful to use our results only where they reasonably apply, or to use them to establish trends. Thus, referring to the results of Example Case 2.1, we see that

$$\pi_c \doteq \frac{p_{t3}}{p_{to}} = \frac{p_3}{p_0} \left\{ \frac{1 + \dfrac{\gamma_c - 1}{2} M_3^2}{1 + \dfrac{\gamma_c - 1}{2} M_0^2} \right\}^{\gamma_c/(\gamma_c-1)} \tag{5-1}$$

Since the total temperature is conserved in such adiabatic flows, then

$$\psi = \frac{T_3}{T_0} = \left\{ \frac{1 + \dfrac{\gamma_c - 1}{2} M_0^2}{1 + \dfrac{\gamma_c - 1}{2} M_3^2} \right\} \tag{5-2}$$

and, therefore,

$$\pi_c = \frac{p_3}{p_0} \left(\frac{1}{\psi} \right)^{\gamma_c/(\gamma_c-1)} \tag{5-3}$$

where the definition of the cycle static temperature ratio ψ of Eq. (4-20) has been used. Since Eqs. (4-21) and (4-23) also apply to the compression process described here, they may be combined to yield

$$\frac{p_3}{p_0} = \left\{ \frac{\psi}{\psi(1 - \eta_c) + \eta_c} \right\}^{\gamma_c/(\gamma_c-1)} \tag{5-4}$$

Finally, Eq. (5-4) may be substituted into Eq. (5-3) to produce the desired results, namely

$$\pi_c = \left\{ \frac{1}{\psi(1 - \eta_c) + \eta_c} \right\}^{\gamma_c/(\gamma_c-1)} \le 1 \tag{5-5}$$

or

$$\eta_c = \frac{\psi - \left(\dfrac{1}{\pi_c} \right)^{(\gamma_c-1)/\gamma_c}}{\psi - 1} \le 1 \tag{5-6}$$

It is interesting to note that these interrelationships are explicitly free of Mach number, even though their range of validity does depend upon the freestream Mach number M_0. The behavior of these equations is portrayed in Fig. 5.8, where it is evident that even modest reductions of adiabatic compression efficiency are associated with enormous decreases in total pressure ratio. The violently nonlinear nature of this relationship, as well as the absolute magnitude of the total pressure ratios (approximately 0.1–0.2 versus the 0.90–0.98 experienced in conventional aircraft), also helps explain why π_c is only occasionally employed by the hypersonic airbreathing propulsion community.

5.4.2 Kinetic Energy Efficiency

The *kinetic energy efficiency* is defined as the ratio of the square of the velocity that the compression component exit flow would achieve if it were isentropically expanded to freestream static pressure to the square of the freestream velocity, and is denoted by the symbol η_{KE}. In the terminology of Fig. 4.2, the mathematical expression for kinetic energy efficiency is

$$\eta_{KE} \doteq \frac{V_X^2}{V_0^2} \tag{5-7}$$

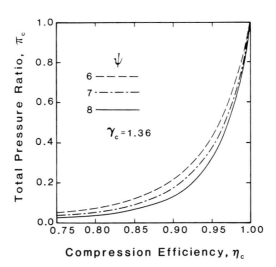

Fig. 5.8 **Compression component total pressure ratio as a function of adiabatic compression efficiency and cycle static temperature ratio.**

A leading virtue of kinetic energy efficiency is the fact that it is referenced to the freestream static conditions, rather than the forbiddingly complex freestream stagnation conditions. The same, of course, can be said for the adiabatic compression efficiency. Kinetic energy efficiency is also intuitively appealing because it is directly related to the preservation of the most important quantity for ramjet or scramjet propulsion, namely the kinetic energy (or velocity or momentum) that will be available to produce thrust. For these reasons, kinetic energy efficiency is frequently encountered in the hypersonic literature.

Combining Eq. (5-7) with Eq. (2-48) in keeping with the compression component assumptions yields

$$\eta_{\text{KE}} = \frac{V_0^2 - 2C_{pc}(T_X - T_0)}{V_0^2} \tag{5-8}$$

which may be combined with Eqs. (2-43) and (2-45) to yield

$$\eta_{\text{KE}} = 1 - \frac{2}{(\gamma_c - 1)M_0^2}\left(\frac{T_X}{T_0} - 1\right) \tag{5-9}$$

Substituting Eq. (4-21) gives the desired results, namely

$$\eta_{\text{KE}} = 1 - \frac{2}{(\gamma_c - 1)M_0^2}(\psi - 1)(1 - \eta_c) \leq 1 \tag{5-10}$$

or

$$\eta_c = 1 - \frac{(\gamma_c - 1)M_0^2}{2}\left(\frac{1 - \eta_{\text{KE}}}{\psi - 1}\right) \leq 1 \tag{5-11}$$

In contrast to the total pressure ratio, these results depend explicitly on the freestream Mach number, although their range of validity is independent of M_0. The behavior of these equations is portrayed in Fig. 5.9, which faithfully reflects the fact that kinetic energy efficiency varies exactly linearly with adiabatic compression efficiency and is very sensitive to freestream Mach number for the range of M_0 encountered in hypersonic flight.

An unfortunate feature of kinetic energy efficiency is that it is very nearly 1, especially for large freestream Mach numbers. This is intrinsic to Eq. (5-10) where, since the cycle static temperature ratio is of the order of 7.0 and the adiabatic compression efficiency is of the order of 0.8, it follows that

$$\eta_{\text{KE}} \sim 1 - \frac{6.7}{M_0^2}$$

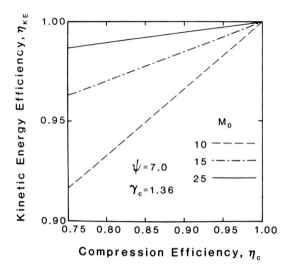

Fig. 5.9 Compression component kinetic energy efficiency as a function of adiabatic compression efficiency and cycle static temperature ratio.

Thus, large differences in adiabatic compression efficiency equate to small differences in kinetic energy efficiency, and kinetic energy efficiency must therefore be quoted to at least three significant figures in order to insure sufficient accuracy.

All of this notwithstanding, kinetic energy efficiency and some of its variants are among the most widely used compression component performance measures in the world of hypersonic airbreathing propulsion.

5.4.3 Dimensionless Entropy Increase

The compression process is inherently irreversible because the viscous forces within the shock waves and boundary layers dissipate availability and generate entropy. The natural compression system performance measure involving entropy is the entropy increase divided by the specific heat of the air at constant pressure, known here as the *dimensionless entropy increase*. This compression component performance measure is included here because it is a universal and fundamental thermodynamic property of the process. Like the adiabatic compression efficiency and kinetic energy efficiency, it is referenced to the freestream static conditions.

A mathematical expression for this term can be obtained by integrating the Gibbs equation, Eq. (2-35), into the form

$$s_3 - s_0 = C_{pc} \ell n \frac{T_3}{T_0} - R_c \ell n \frac{p_3}{p_0} \qquad (5\text{-}12)$$

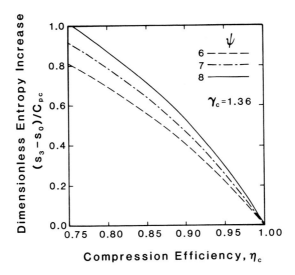

Fig. 5.10 Compression component dimensionless entropy increase as a function of adiabatic compression efficiency and cycle static temperature ratio.

and substituting the definition of cycle static temperature ratio from Eq. (4-20) and the expression for static pressure ratio from Eq. (5-4), in order to produce the desired results, namely

$$\frac{s_3 - s_0}{C_{pc}} = \ell n \left\{ \psi (1 - \eta_c) + \eta_c \right\} \geq 0 \tag{5-13}$$

or

$$\eta_c = \frac{\psi - e^{(s_3 - s_0)/C_{pc}}}{\psi - 1} \leq 1 \tag{5-14}$$

In contrast to the total pressure ratio and kinetic energy efficiency, these relationships are entirely independent of the freestream Mach number. The behavior of these equations is portrayed in Fig. 5.10, where it is evident that the dimensionless entropy increase varies approximately linearly with adiabatic compression efficiency and is only modestly influenced by cycle static temperature ratio over the usual range of interest.

Reflecting on the compression system performance measures discussed above, it is clear that the most easily related are dimensionless entropy increase and adiabatic compression efficiency. One would learn quickly with practice how to use them interchangeably. It is therefore strange but true that, although authors almost always use entropy arguments to explain the qualitative behavior of components or engines, entropy is almost never used in their quantitative analyses. The reasons for this are left to your imagination.

5.4.4 Compression Component Performance Measure Summary

In order to facilitate future compression component analysis, the missing relationships between the four performance measures examined above were derived, and the complete set is presented in matrix form in Table 5.1.

As anticipated by Eq. (2-55), π_c decays exponentially with $s_3 - s_0$. This strong dependence makes π_c less practical as an indicator of the performance of flows having significant entropy increases.

5.5 COMPRESSION COMPONENT PERFORMANCE

We are now in a position to use a variety of analytical tools to make estimates of compression component performance. This approach will at least establish limits for our quantitative performance expectations and reveal the leading trends, and at best provide a reasonably accurate basis for airbreathing engine overall performance analysis.

5.5.1 Compression Component Flowfield Analysis

The underlying foundation of this development is the analysis by means of HAP(Gas Tables) of several *special families* of hypersonic compression systems similar to those already described in Sec. 5.2. Of the universe of possible compression component configurations, the special families chosen for examination have two distinctive features, namely that they are characterized by the *cardinal number* of oblique shock waves available to produce a specified cycle static temperature ratio ψ at a specified freestream Mach number M_0 and, when more than one oblique shock wave is available, all oblique shock waves must provide *equal amounts of geometrical turning* of the flow. As before, they will be evaluated at design point operation, the so-called *shock-on-lip* condition. The resulting configurations and their corresponding performance measures are certainly representative of practical designs, and, as you shall shortly see, include desirable variations for which the cowl is parallel to the freestream flow.

Because of the convenient mathematical properties of oblique shock waves, those families having more than one can have different geometries, depending upon how the oblique shock waves are combined and provided that they do not intersect.[5.1] Please note that we have chosen to begin the compression sequence at the leading edge of the vehicle, which is consistent with the bookkeeping system of the remainder of this textbook and continues the use of freestream

Table 5.1   Interrelationships between the various compression component performance measures. For example, in order to express the compression efficiency η_c in terms of the kinetic energy efficiency η_{KE}, read across the η_c row to the η_{KE} column to find that

$$\eta_c = 1 - \frac{\gamma_c - 1}{2} M_0^2 \left(\frac{1 - \eta_{KE}}{\psi - 1} \right)$$

$\eta_c =$	η_c
$\pi_c =$	$\left\{ \dfrac{1}{\psi\,(1 - \eta_c) + \eta_c} \right\}^{\gamma_c/(\gamma_c - 1)}$
$\eta_{KE} =$	$1 - \dfrac{2}{(\gamma_c - 1)M_0^2} \, (\psi - 1)\,(1 - \eta_c)$
$\dfrac{s_3 - s_0}{C_{pc}} =$	$\ell n \left\{ \psi\,(1 - \eta_c) + \eta_c \right\}$
$\eta_c =$	$\dfrac{\psi - \left(\dfrac{1}{\pi_c}\right)^{(\gamma_c - 1)/\gamma_c}}{\psi - 1}$
$\pi_c =$	π_c
$\eta_{KE} =$	$1 - \dfrac{2}{(\gamma_c - 1)M_0^2} \left\{ \left(\dfrac{1}{\pi_c}\right)^{(\gamma_c - 1)/\gamma_c} - 1 \right\}$
$\dfrac{s_3 - s_0}{C_{pc}} =$	$\dfrac{\gamma_c - 1}{\gamma_c} \ell n \dfrac{1}{\pi_c}$

(continued on next page)

Table 5.1 (continued)

$$\eta_c = 1 - \frac{\gamma_c - 1}{2} M_0^2 \left(\frac{1 - \eta_{KE}}{\psi - 1} \right)$$

$$\pi_c = \left\{ \frac{1}{1 + \frac{\gamma_c - 1}{2} M_0^2 (1 - \eta_{KE})} \right\}^{\gamma_c/(\gamma_c - 1)}$$

$$\eta_{KE} = \eta_{KE}$$

$$\frac{s_3 - s_0}{C_{pc}} = \ell n \left\{ 1 + \frac{\gamma_c - 1}{2} M_0^2 (1 - \eta_{KE}) \right\}$$

$$\eta_c = \frac{\psi - e^{(s_3 - s_0)/C_{pc}}}{\psi - 1}$$

$$\pi_c = exp \left\{ -\left(\frac{\gamma_c}{\gamma_c - 1} \right) \left(\frac{s_3 - s_0}{C_{pc}} \right) \right\}$$

$$\eta_{KE} = 1 - \frac{2}{(\gamma_c - 1)M_0^2} \left\{ e^{(s_3 - s_0)/C_{pc}} - 1 \right\}$$

$$\frac{s_3 - s_0}{C_{pc}} = \frac{s_3 - s_0}{C_{pc}}$$

properties as reference quantities. It is important to recognize that any family of these design point configurations can only be realized along a given flight trajectory with the help of variable vehicle angle of attack and compression component geometry, which presents the designer with a number of exciting challenges.

The families were "designed" by repeatedly applying HAP(Gas Tables) and are illustrated by the selection of configurations presented in Figs. 5.11 through 5.15 below, all of which correspond to a static temperature ratio ψ of 7.0, a ratio of specific heats γ_c of 1.360, capture an equal height of freestream flow, and, except for the streamwise axis of Fig. 5.15, are drawn to the same scale.

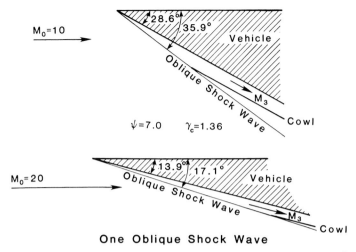

One Oblique Shock Wave

Fig. 5.11 Two external compression systems similar to that of Fig. 5.2, but incorporating *one* oblique shock wave. The vehicle wedge angle and the oblique shock wave angle both diminish with freestream Mach number because the static temperature ratio across an oblique shock wave is proportional to the square of the component of Mach number perpendicular to the wave front. [5.6] This also increases the distance from the leading edge of the vehicle to the leading edge of the cowl.

Two Oblique Shock Waves

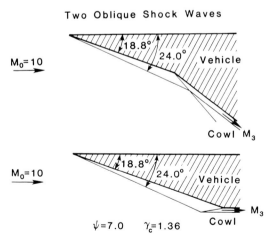

Fig. 5.12 Two compression systems that incorporate *two* oblique shock waves. The upper version is an external compression system, similar to that of Fig. 5.2, while the lower version is a mixed external and internal compression system, similar to that of Fig. 5.4. The lower version has the advantage of providing a cowl that is parallel to the freestream flow and therefore does not disturb the freestream flow, but is longer than the upper version.

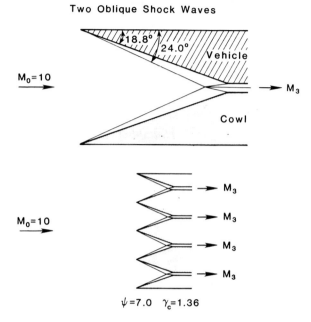

Fig. 5.13 Two more compression systems that incorporate *two* oblique shock waves, as shown in Figs. 5.5 and 5.6.

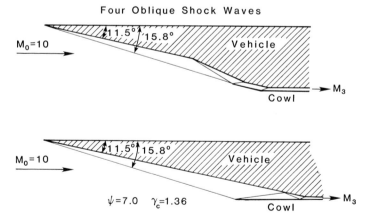

Fig. 5.14 Two mixed external and internal compression systems that incorporate *four* oblique shock waves. Both turn the combustor entrance flow parallel to the freestream, but the upper version, which has more external compression, will generate an oblique shock wave in the freestream because of the drooped cowl lip. The lower version, which has more internal compression, is longer.

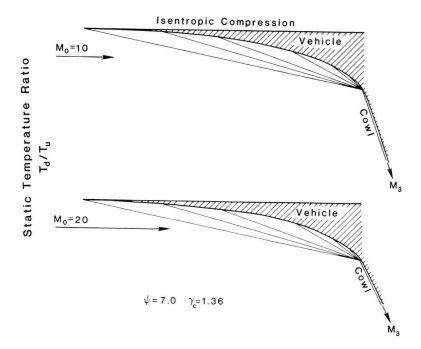

Fig. 5.15 Two isentropic external compression systems, as in Fig. 5.3. Although they appear quite similar, the upper version has had its axial dimensions reduced by a factor of 2, and the lower version has had its axial dimensions reduced by a factor of 4. These illustrations therefore reveal that isentropic external compression is accompanied by very long external compression surfaces and large freestream deflection angles and oblique shock waves created by the cowl. Some or all of the cowl deflection can be avoided if a portion of the isentropic compression is accomplished internally by means of a compression wave train generated from the cowl lip, but the overall length of the compression system will inevitably increase.

The isentropic external compression systems of Fig. 5.15, which correspond to an infinite number of infinitesimal oblique compression waves, were not, of course, "designed" using finite oblique shock wave computations, but were generated instead with the help of the integrated *simple compression wave* relationship between flow deflection angle and local Mach number, and then calculating the length of the Mach line joining the external compression surface to the cowl lip by conserving mass. This procedure is described in greater detail in Refs. 5.1 and 5.5. An alternative but entirely equivalent approach would be to use the "closed form" isentropic ramp solution of Ref. 5.6, which you will find to be more general and slightly more efficient, but considerably less transparent and intuitive.

5.5.2 Adiabatic Compression Efficiency

Each time HAP(Gas Tables) is used to compute the compression system oblique shock wave configuration of a given family (i.e., number of oblique shock waves) for a specified combination of static temperature ratio and freestream Mach number, the results include a determination of all flow properties entering the combustor including, for example, the static pressure ratio and the dimensionless entropy increase. Equation (5-14) can be used to convert dimensionless entropy rise directly to adiabatic compression efficiency. Further, since these performance measures are both referenced to freestream properties, the results are insensitive to freestream Mach number.

The outcome of a systematic exploration of compression system variables over the full expected range of each is presented in Fig. 5.16. The inescapable and remarkable conclusion to be drawn from these numerical data is that adiabatic compression efficiency is almost *independent* of both static temperature ratio and freestream Mach number, but is strongly *dependent* upon the chosen number of oblique shock waves.

This is partly true because the upper and lower bounding lines must be precisely horizontal. On the one hand, the upper lines belong to the isentropic compression family, and therefore must have an adiabatic compression efficiency of exactly 1.0. On the other hand, the lowest lines, which belong to the single oblique shock family, have exactly constant adiabatic compression efficiency because both the static temperature ratio and the dimensionless entropy rise for a single oblique shock depend only upon the square of the component of the Mach number normal to the wave front, $M_0 \sin \theta_0$,[5.7] that is,

$$\frac{s_3 - s_0}{C_{pc}} =$$

$$\ell n \left\{ \left[\frac{(\gamma_c - 1)(M_0 \sin \theta_0)^2 + 2}{(\gamma_c + 1)(M_0 \sin \theta_0)^2} \right] \left[\frac{2\gamma_c(M_0 \sin \theta_0)^2 - (\gamma_c - 1)}{\gamma_c + 1} \right]^{1/\gamma_c} \right\}$$

$$(5\text{-}15)$$

and

$$\psi = \frac{T_3}{T_0} = \frac{\{2\gamma_c(M_0 \sin \theta_0)^2 - (\gamma_c - 1)\} \{(\gamma_c - 1)(M_0 \sin \theta_0)^2 + 2\}}{(\gamma_c + 1)^2(M_0 \sin \theta_0)^2}$$

$$(5\text{-}16)$$

or, using Table 5.1,

$$\eta_c = \frac{\psi - \left\{ \frac{(\gamma_c - 1)(M_0 \sin \theta_0)^2 + 2}{(\gamma_c + 1)(M_0 \sin \theta_0)^2} \right\} \left\{ \frac{2\gamma_c(M_0 \sin \theta_0)^2 - (\gamma_c - 1)}{\gamma_c + 1} \right\}^{1/\gamma_c}}{\psi - 1}$$

$$(5\text{-}17)$$

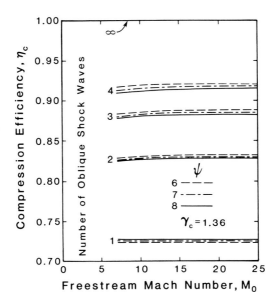

Fig. 5.16 Adiabatic compression efficiency as a function of freestream Mach number, static temperature ratio, and number of oblique shock waves.

which means that there is one adiabatic compression efficiency for each static temperature ratio ψ.

The main conclusions to be drawn for the application of the results portrayed in Fig. 5.16 are self-evident. In view of the cycle estimates presented in Chap. 4, one must employ at least two, and preferably three or four, oblique shock waves in order to achieve adequate compression component performance. Since this is almost always possible in practice, especially because of the presence of the first oblique shock wave emanating from the leading edge of the vehicle and the possibility of at least some isentropic compression, it would seem that the desired adiabatic compression efficiency level of about 0.90 is indeed attainable.

5.5.3 Influence of Boundary-Layer Friction

The foregoing oblique shock wave analyses fail to account for the skin friction that must occur along the external and internal bounding surfaces of the compression component. Since these viscous effects inevitably dissipate the availability and raise the entropy of the air that enters the engine, they also reduce the adiabatic compression efficiency, and therefore must be reckoned with. This cannot be done rigorously within the framework of one-dimensional flow because the

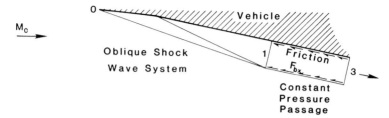

Fig. 5.17 Conceptual analytical model of compression component flow used to estimate the influence of boundary layer friction on adiabatic compression efficiency.

boundary-layer flowfield is inherently two-dimensional (or even three-dimensional). Nevertheless, valuable progress can be made and the necessary insight gained by means of a model that reproduces the overall effects of boundary-layer friction while maintaining the one-dimensional approach.

The conceptual model employed in this analysis is shown in Fig. 5.17, and consists of two separate elements. The first element is the oblique shock wave system, for which skin friction effects will continue to be neglected. Even though a purely external oblique shock wave system is shown for clarity in Fig. 5.17, the approach applies to any blend of external and internal oblique shock wave compression.

The second element is a hypothetical constant pressure passage or duct which is parallel to the flow leaving the oblique shock wave system and in which wall friction is present but shock waves are absent, and for which the flow is assumed to be uniform at the entrance and exit planes. The constant pressure assumption is largely made to simplify the ensuing mathematics, but builds upon our previous findings that the precise details of the flow process have only a slight impact on the resulting performance, as well as the fact that this model is primarily designed to probe the trends and limits of the influence of viscosity, rather than to precisely investigate any particular case. It is important to recognize, then, that the amount of wall friction assigned to the hypothetical passage element of our model can and will be arranged to account for that generated on the actual external surfaces of the compression component as well as on the actual internal surfaces.

Figure 5.18 is intended to further clarify the surrogate flow process for compression component performance estimation by portraying the T-s diagram behavior of the flow of the two elements. The oblique shock wave process, which connects states 0 and 1, will be analyzed, exactly as before, by means of HAP(Gas Tables). The constant pressure frictional process, which connects states 1 and 3, will

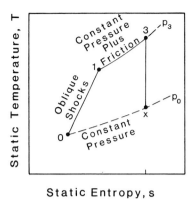

**Fig. 5.18 The _T-s_ diagram portrayal of the compression compo-
nent conceptual analytical model of Fig. 5.17.**

be analyzed by one-dimensional control volume methods. The overall compression process is therefore depicted as a line that connects
states 0 and 3.

The analysis that follows employs the subscript s to denote quantities associated only with the oblique shock wave process. Thus, for
example, p_1 is the static pressure at 1, T_1 is the static temperature
at 1, V_1 is the velocity at 1, η_1 is the adiabatic compression efficiency
of the oblique shock wave process alone, and $\psi_1 = T_1/T_0$. The overall static temperature ratio $\psi = T_3/T_0$ and adiabatic compression
efficiency η_c retain their original meanings.

Recognizing that Eq. (5-4) also applies to the oblique shock process alone, or

$$\frac{p_1}{p_0} = \left\{ \frac{\psi_1}{\psi_1(1 - \eta_1) + \eta_1} \right\}^{\gamma_c/(\gamma_c-1)} \tag{5-18}$$

and that

$$\frac{p_1}{p_0} = \frac{p_3}{p_0} \tag{5-19}$$

then equating Eqs. (5-4) and (5-19) leads to the remarkable result
that

$$\eta_c = \eta_1 \left(\frac{1 - \dfrac{1}{\psi_1}}{1 - \dfrac{1}{\psi}} \right) \tag{5-20}$$

which has the satisfying effect of decoupling the compression efficiency of the oblique shock wave process from that of the duct. This
is particularly attractive because, as a result of our early work, we
are in a position to simply assign realistic values to η_1. Also, since we

treat ψ as an independent parameter, the remainder of this analysis is devoted to determining the missing dependent parameter ψ_1.

For the oblique shock wave process, Eq. (2-48) becomes

$$\psi_1 = 1 + \frac{\gamma_c - 1}{2} M_0^2 \left\{ 1 - \left(\frac{V_1}{V_0} \right)^2 \right\} \tag{5-21}$$

and for the steady-state, one-dimensional, constant pressure flow process of the passage with friction, Eq. (2-29) becomes

$$\dot{m}_0 V_1 + F_{bx} = \dot{m}_0 V_3 \tag{5-22}$$

where F_{bx} is the total force exerted on the walls of the constant pressure passage in the direction of flow, and is due entirely to wall friction. This frictional force can, in turn, be most conveniently represented by the expression

$$F_{bx} = -C_f \cdot \frac{\rho_3 V_3^2}{2} \cdot A_w = -\dot{m}_0 V_3 \left(\frac{C_f}{2} \cdot \frac{A_w}{A_3} \right) \tag{5-23}$$

where C_f must be a suitable average of the skin friction coefficient acting on all the surfaces of interest, and the surface-to-throughflow area ratio A_w/A_3 must be a suitable reflection of the entire compression process. Referencing the skin friction to the combustor entrance velocity V_3 is perfectly reasonable in view of the fact that, for hypersonic airbreathing propulsion, the velocity changes only slightly from its freestream value.

Looking ahead, it is appropriate to note here that the skin friction coefficient C_f is most likely to be found in the range of 0.002–0.005 and is relatively insensitive to Mach number,[5.1] while the surface-to-throughflow area ratio A_w/A_3 is most likely to be found in the range of 5–20. Fortunately, these two parameters are out of synchronization in the sense that, as the vehicle flies to higher Mach numbers along a trajectory of constant q_0, the average skin friction coefficient tends to decrease because the Reynolds number decreases and more of the boundary layer becomes laminar, while the surface-to-throughflow area ratio A_w/A_3 tends to increase. The range of greatest interest for the dimensionless boundary-layer skin friction quantity $(C_f/2)(A_w/A_3)$ is therefore approximately 0.01–0.05, with the most likely values centering on 0.02.

Substituting Eq. (5-23) into Eq. (5-22) yields

$$\frac{V_1}{V_3} = 1 + \frac{C_f}{2} \cdot \frac{A_w}{A_3} \tag{5-24}$$

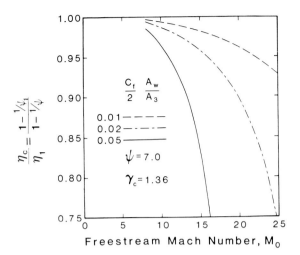

Fig. 5.19 Contribution of boundary-layer friction to adiabatic compression efficiency, from Eq. (5-20), as a function of freestream Mach number and dimensionless skin friction coefficient quantity.

Applying Eq. (2-48) to the overall process leads to the familiar result

$$\left(\frac{V_3}{V_0}\right)^2 = 1 + \frac{2(1 - \psi)}{(\gamma_c - 1)M_0^2} \tag{5-25}$$

Finally, combining Eqs. (5-21), (5-24), and (5-25) produces the desired expression

$$\psi_1 = 1 + \frac{\gamma_c - 1}{2}M_0^2 \left\{1 - \left[1 + \frac{C_f}{2} \cdot \frac{A_w}{A_3}\right]^2 \left[1 + \frac{2(1 - \psi)}{(\gamma_c - 1)M_0^2}\right]\right\} \tag{5-26}$$

which may be used in conjunction with Eq. (5-20) to calculate either the overall adiabatic compression efficiency η_c or only that portion due to boundary-layer skin friction η_c/η_1.

The outcome of calculations based upon this analysis is found in Fig. 5.19. Only the contribution of the frictional losses is shown there because the contribution of oblique shock wave losses η_1 can be considered independently. The results carry several significant messages, the most important of which is that the penalty for boundary-layer friction is extremely sensitive to the prevailing skin friction coefficient. They also show that, because the frictional stresses increase approximately with the square of the flight Mach number (or flight velocity), the penalty is also greatest when the Mach number is large. Taken together, these two observations suggest that, unless the skin friction coefficient is near the low end of its expected range, satisfactorily efficient hypersonic airbreathing propulsion at high Mach numbers will be difficult to achieve. The good news, of course, is that boundary-layer friction should do little to diminish overall scramjet performance for freestream Mach numbers up to about 10 or 15.

5.5.4 Experiential Information

There is a distinct scarcity of experimental compression efficiency data in the open literature available either to compare with theory or to build an extensive design database. There are many obvious reasons for this situation, some of which will be summarized next. On the one hand, high quality hypersonic test facilities are rare, and obtaining the integrated flowfield properties in the narrow exit passages of compression components with intrusive instrumentation is a daunting task (see Chap. 9 for more information regarding facilities and instrumentation). On the other, good experimental data are a precious commodity in this competitive world, and the dissemination of such data are usually prevented by military security classifications and/or industrial proprietary restrictions. Finally, much of the data that appear openly are not complete enough to allow the desired compression efficiency to be pinned down with sufficient accuracy.

We must therefore be resourceful with the information that is presently available in order to establish what level of compression component performance may be achievable in practice. Fortunately, our goal is the relatively modest one of determining general levels and trends, rather than providing the tools necessary for detailed design.

Figure 5.20 portrays the behavior of adiabatic compression efficiency according to a variety of sources. Since the information depicted there is not merely experimental data, but also represents reality as seen by knowledgeable observers, we have chosen to refer to it as experiential data.

The top line shown in Fig. 5.20 builds upon Ref. 5.8, which provides a frequently quoted correlation for kinetic energy efficiency based upon oblique shock wave computations. These computations are quite similar to those of Sec. 5.5.2, but are not identical. Rather than using families of oblique shock waves having different cardinal numbers and equal amounts of geometrical turning, they use an "optimized" set of four oblique shock waves that produces the most efficient compression[5.9] regardless of geometry. Since skin friction forces are neglected, this analysis is referred to as *inviscid*. The resulting correlation, derived for the approximate range $4 < M_0 < 14$,

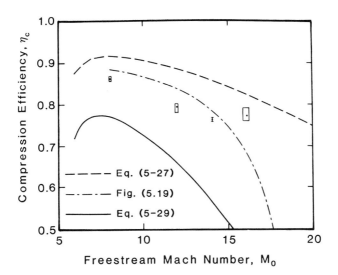

Fig. 5.20 Adiabatic compression efficiency as a function of freestream Mach number taken from several sources. The rectangles display the maximum excursions for the clusters of data points surrounding the averaged points shown for the four freestream Mach numbers of Ref. 5.12. The calculations used ψ = 7.0 and γ_c = 1.36, as necessary.

is

$$\eta_{KE} = 1 - 0.2 \left(1 - \frac{M_3}{M_0}\right)^5 \tag{5-27}$$

Incidentally, the form of this equation agrees with the intuitive notion that the kinetic energy efficiency decreases as the amount of deceleration increases and the amount of compression increases.

The top line of Fig. 5.20 corresponds to a constant static temperature ratio ψ of 7.0, where, according to Eq. (5-2), the combustor entrance Mach number was obtained from

$$\frac{M_3}{M_0} = \sqrt{\frac{2}{(\gamma_c - 1)M_0^2} \left\{ \frac{1}{\psi} \left(1 + \frac{\gamma_c - 1}{2} M_0^2\right) - 1 \right\}} \tag{5-28}$$

and kinetic energy efficiency was converted to adiabatic compression efficiency by means of Eq. (5-11). It should be noted that the "optimum" four oblique shock wave results of Fig. 5.20 fall somewhat short of the four oblique wave results of Fig. 5.16 over the range of validity of the correlation. This should not be taken as astounding because Eq. (5-27) attempts to represent a wide range of freestream Mach numbers and static temperature ratios in a single stroke, and because, as we have seen in Fig. 5.9, adiabatic compression efficiency

Fig. 5.21   Photographs of the inlet model of Refs. 5.10 through 5.12.

is extremely sensitive to small differences in kinetic energy efficiency, especially at the higher freestream Mach numbers.

The bottom line shown in Fig. 5.20 is a frequently quoted correlation for kinetic energy efficiency based upon modifying Eq. (5-27) to include skin friction forces. The resulting "viscous" correlation, also derived for the approximate range $4 < M_0 < 14$, is

$$\eta_{KE} = 1 - 0.4 \left(1 - \frac{M_3}{M_0} \right)^4 \tag{5-29}$$

and the conversion to adiabatic compression efficiency is exactly as previously described.

Taken together, the top and bottom lines of Fig. 5.20 should reasonably be expected to bound reality. In order to test this premise, the lowest line was taken from Fig. 5.19 and multiplied by a constant $\eta_1 = 0.90$ and shown for comparison. This premise is evidently correct.

The remaining information in Fig. 5.20 is *experimental* data. The authors of Refs. 5.10 through 5.12 carried out a remarkable series of truly hypersonic tests in the Cornell Aeronautical Laboratory 48 in (1.2 m) shock tunnel on the inlet model shown in Fig. 5.21. The model design and the data analysis were based on an ingenious application of the principle that compression efficiency can be determined directly from force measurements. The method will be explained in great detail below, but for the time being may be accepted as sound.

A total of 20 data points are presented in Ref. 5.12 and portrayed in Fig. 5.20, clustering around freestream Mach numbers of 8.0 (2 points), 12.0 (10 points), 14.0 (2 points), and 16.0 (6 points). The Reynolds numbers based upon the length of the compression system were in the range of 1.0–3.0 million, indicating that the boundary-layer flow was predominantly laminar.

There are many reasons why these data should not conform to any simple analyses. First, the compression system tested was designed for a freestream Mach number of 16.0, and was therefore running off-design at the other conditions. Second, several variations in minor geometrical details and Reynolds number were tested at each freestream Mach number. Third, various degrees of boundary-layer separation due to oblique shock wave interactions were observed during the tests, and particularly at the design Mach number. Fourth, small corrections to the data were made to account for nonuniform exit profiles and wall heat transfer. Fifth, although the stagnation temperatures were too low for dissociation to occur, the air did not behave as a perfect gas with constant properties. Sixth, the static temperature ratio was not constant in these experiments, but instead was calculated to be as low as 3.5 and as high as 19.9.

All the above notwithstanding, the inescapable visual message of Fig. 5.20 is that the behavior of this "real" compression system is well within our expectations. We therefore conclude from the experiential information that the models of Secs. 5.5.2 and 5.5.3 are fundamentally correct, and furthermore that compression efficiencies in the neighborhood of 0.8 and above are possible for freestream Mach numbers up to about 16.

5.5.5 Determining Compression Efficiency from Global Measurements

The method employed to obtain compression efficiency data in Refs. 5.10 through 5.12 is not only ingenious and sound, as noted above, but it also provides insight into hypersonic flows and an example of the sometimes amazing power of one-dimensional flow analysis. The conceptual one-dimensional flow model to be employed is shown in Fig. 5.22.

The following assumptions are made:

1. The flow into and out of the control volume is steady, one-dimensional, and parallel to the freestream or x direction.

2. All properties of the flow upstream of the control volume are known.

3. The inlet is operating without "spillage," so that the column of airflow captured by the inlet has the same throughflow area as the axial projection of the physical or geometrical opening of the compression system.

4. No air enters or leaves except at the upstream and downstream control volume boundaries.

5. The upstream and downstream throughflow areas are known.

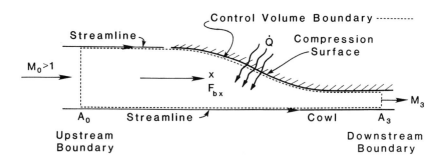

Fig. 5.22 Conceptual one-dimensional flow model of the compression system used to determine adiabatic compression efficiency directly from global force and heat transfer measurements.

6. The air behaves as a calorically perfect gas.

7. The total axial force F_{bx} (pressure and viscous) exerted on the flow by the material boundaries of the compression surface is known. Other than the pressure forces exerted on the upstream and downstream control volume boundaries, this is the total axial force acting on the flow within the control volume.

8. The total rate at which heat is added to the flow at the material boundaries of the compression surface $\dot{Q}$ is known.

Under these conditions, Eqs. (2-28), (2-29), (2-31), and (2-38) can be simplified and combined to yield

$$V_3 = \frac{-b \pm \sqrt{b^2 - 4ac}}{2a} \tag{5-30}$$

where

$$a = \frac{\gamma_c + 1}{2\gamma_c}$$

$$b = -\left(V_0 + \frac{p_0 A_0 + F_{bx}}{\dot{m}_0}\right)$$

$$c = R_c\left(C_{pc}T_0 + \frac{V_0^2}{2} + \frac{\dot{Q}}{\dot{m}_0}\right)$$

The $+$ and $-$ solutions to Eq. (5-30) correspond to supersonic and subsonic downstream velocities, respectively, as we have seen several times previously. For hypersonic scramjet engines we will, therefore, primarily be interested in the supersonic solution. Also,

$$
\begin{aligned}
T_3 &= \frac{1}{C_{pc}}\left(C_{pc}T_0 + \frac{V_0^2}{2} + \frac{\dot{Q}}{\dot{m}_0} - \frac{V_3^2}{2}\right) \\
&= \frac{(\gamma_c - 1)c}{\gamma_c R_c^2} - \frac{V_3^2}{2C_p} \tag{5-31}
\end{aligned}
$$

$$p_3 = \frac{\dot{m}_0 R_c T_3}{A_3 V_3} \tag{5-32}$$

$$\rho_3 = \frac{p_3}{R_c T_3} \tag{5-33}$$

In short, the downstream conditions depend only upon the downstream throughflow area A_3 and the total or *global* quantities F_{bx} and $\dot{Q}$, and *not upon the details of their distribution* along the control volume boundary. This set of equations may be used to determine the downstream conditions, as well as any desired flow property, including compression efficiency. Example 5.1 will demonstrate this for a particular but familiar case.

Example 5.1

Consider the compression system flow illustrated in the sketch below, which is identical to the two oblique shock wave system from the bottom half of Fig. 5.12. There are no friction forces or heat transfer to or from the flow. We wish to find the adiabatic compression efficiency η_c. The upstream conditions are:

$$M_0 = 10.0$$
$$T_0 = 400 \; °R \quad (222 \; K)$$
$$p_0 = 10.0 \; lbf/ft^2 \quad (479 \; N/m^2)$$
$$\gamma_c = 1.36$$
$$R_c = 1716 \; (ft/s)^2/°R \quad (287 \; (m/s)^2/K)$$
$$A_0 = 1.00 \; ft^2 \quad (0.0929 \; m^2)$$

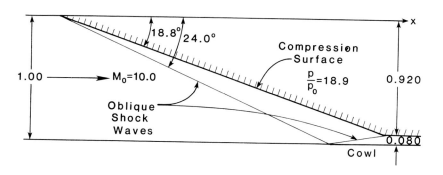

Note that $q_0 = 680 \; lbf/ft^2 \quad (32,600 \; N/m^2)$. The quantitative information shown on the sketch was obtained by means of HAP(Gas Tables). The only force exerted on the flow occurs on the inclined surface and acts in the negative x direction. The value of this force is:

$$F_{bx} = -(18.9)(10.0)(0.920) = -174 \; lbf \quad (-773 \; N)$$

Also:

$$V_0 = M_0\sqrt{\gamma_c R_c T_0} = 9660 \; ft/s \quad (2940 \; m/s)$$

$$\dot{m}_0 = \rho_0 V_0 A_0 = \frac{p_0 V_0 A_0}{R_c T_0} = 4.53 \; lbm/s \quad (2.05 \; kg/s)$$

Substituting the above quantities into Eqs. (5-30) through (5-33), and selecting the supersonic solution to Eq. (5-30), leads to these results:

$$V_3 = 7890 \text{ ft/s} \quad (2400 \text{ m/s})$$
$$T_3 = 2800 \text{ °R} \quad (1560 \text{ K})$$
$$p_3 = 1070 \text{ lbf/ft}^2 \quad (51,200 \text{ N/m}^2)$$

so that Eqs. (4-20) and (5-4) give:

$$\psi = \frac{T_3}{T_0} = 7.00$$

$$\eta_c = \frac{1 - \left(\dfrac{p_0}{p_3}\right)^{(\gamma_c - 1)/\gamma_c}}{1 - \dfrac{1}{\psi}} = 0.828$$

These final two results virtually duplicate those previously obtained, as can be seen on Fig. 5.16.

In the experiments of Refs. 5.10 through 5.12 the total axial force F_{bx} was measured by supporting the compression surfaces on force balances, and physically separating these surfaces from the remainder of the flowpath by means of very low leakage labyrinth seals. The total heat transfer rate $\dot{Q}$ was estimated by extrapolating local measurements taken by thin-film gauges. The compression system model and its internal mechanical configuration are shown in Fig. 5-23.

An interesting philosophical aspect of the analytical process described above is that, by assuming the flow leaving the control volume to be completely uniform, we are also assuming that all possible mixing losses have taken place and that the flow has therefore reached its highest possible entropy. In the real world, significant variations of properties across the flow (or profiles) may persist downstream of the control volume for the *same* F_{bx} and $\dot{Q}$, with a lower "average" entropy than we would calculate with the present model. There are at least two options for proceeding. The first is to attempt to correct the model to more faithfully account for the profiles, as is done in Refs. 5.10 through 5.12. The second is to assume that the mixing will take place eventually and the entropy will inevitably become maximized (or, equivalently, that we know of no Maxwell's demon that will help us deliberately capitalize on the latent potential of the profiles) and thus accept the calculations at face value. In any event,

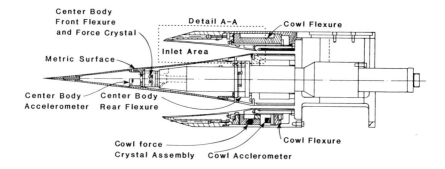

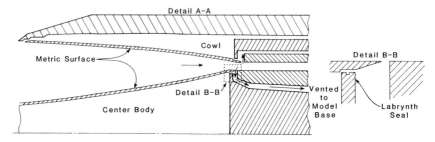

Fig. 5.23 Schematic drawing of the mechanical configuration of the compression system of Refs. 5.10 through 5.12.

there is some comfort in knowing that this time the calculations truly present us with the "worst case."

An interesting practical aspect of the analytical process described above is that it is "robust" in the sense that it is quite insensitive to F_{bx} measurement errors when being used to determine η_c. Example 5.2 will demonstrate this favorable characteristic.

Example 5.2

Repeat Example 5.1, but assume that the magnitude of F_{bx} has been measured 10 percent too high, or 191 lbf (850 N) instead of the correct value of 174 lbf (773 N). Under this condition, the results become:

$$V_3 = 7700 \text{ ft/s} \ (2350 \text{ m/s})$$
$$T_3 = 3030 \ ^\circ\text{R} \ (1680 \text{ K})$$
$$p_3 = 1190 \text{ lbf/ft}^2 \ (56,800 \text{ N/m}^2)$$
$$\psi = 7.57$$
$$\eta_c = 0.827$$

The error in the desired quantity, adiabatic compression efficiency, is merely of the order of 0.1 percent, which can be considered negligible for the purposes at hand.

5.5.6 Influence of Heat Transfer on Compression Efficiency

The global analysis method affords us the almost effortless opportunity to examine how the efficiency of the *nonadiabatic* compression process is affected by the total heat transfer $\dot{Q}$. Since $\dot{Q}$ only appears in Eqs. (5-30) and (5-31) in combination with the freestream total temperature, to wit

$$C_{pc}T_0 + \frac{V_0^2}{2} + \frac{\dot{Q}}{\dot{m}_0}$$

it is both convenient and customary to express the former as a fraction of the latter, namely

$$\frac{\dot{Q}}{\dot{m}_0} = \epsilon \left(C_{pc}T_0 + \frac{V_0^2}{2} \right)$$

so that

$$C_{pc}T_0 + \frac{V_0^2}{2} + \frac{\dot{Q}}{\dot{m}_0} = (1 + \epsilon) \left(C_{pc}T_0 + \frac{V_0^2}{2} \right)$$

This approach lends itself especially well to parametric studies, in which ϵ is varied while everything else is held constant.

Because the stagnation temperature is ordinarily extremely high in hypersonic flight, energy is usually transferred *from* the flowing air to the surroundings by conduction and radiation. Thus, the entering airflow is cooled, not heated, and the algebraic signs of $\dot{Q}$ and ϵ are normally negative. In practice, the magnitude of ϵ is in the range 0.01–0.10. Example 5.3 will illustrate a typical case.

Example 5.3

Repeat Example 5.1, but assume that cooling in the amount $\epsilon = -0.04$ occurs. Use Eq. (4-18) to evaluate the nonadiabatic compression efficiency. Under this condition, the results become:

$$V_3 = 7990 \text{ ft/s} \quad (2440 \text{ m/s})$$
$$T_3 = 2370 \text{ °R} \quad (1320 \text{ K})$$
$$p_3 = 898 \text{ lbf/ft}^2 \quad (43,000 \text{ N/m}^2)$$
$$\psi = 5.93$$
$$\eta_c = 0.837$$

There are many interesting conclusions to be drawn from these results. To begin with, the direction of change of all calculated physical parameters is in agreement with the classical one-dimensional flow results of the Example Cases of Chap. 2 and Ref. 5.1. Furthermore, cooling the flow caused the compression efficiency, as obtained here from the blind application of Eq. (4-18), to increase. This should not be surprising because removing heat from the supersonic flow both decreases the entropy and increases the total pressure. Finally, the change in the compression efficiency due to the cooling is merely of the order of 1.2 percent, which means that $\dot{Q}$ need only be roughly estimated. The small scatter of the data in Fig. 5.20 must be at least partially due to the insensitivity of η_c to errors in the experimental determination of either F_{bx} or $\dot{Q}$.

One should not mistake the higher compression efficiency of Example 5.3 for higher overall performance, because the flow arrives at the downstream boundary with less total enthalpy due to the cooling. In the general case, the velocity obtained by expanding the flow isentropically from Station 3 to atmospheric pressure is given by the ideal velocity expression

$$V_X^2 = V_3^2 + 2C_{pc}T_3 \left\{ 1 - \left(\frac{p_0}{p_3} \right)^{(\gamma_c - 1)/\gamma_c} \right\}$$

so that

$$V_X \text{ (Example 5.1)} = 9380 \text{ ft/s } (2860 \text{ m/s})$$
$$V_X \text{ (Example 5.3)} = 9230 \text{ ft/s } (2813 \text{ m/s})$$

which means that a higher velocity can be generated in the former case despite the lower compression efficiency.

5.6   BURNER ENTRY PRESSURE

We are now able to estimate, without further ado, the static pressure that will exist at the junction or interface between the compression component and the combustion component. This will be done within the framework of one-dimensional, steady flow, and is subject to the usual assumptions, caveats, and limitations.

Although the burner entry pressure p_3 is only one outcome of the aerodynamic behavior of the compression system, it is one of the most important factors involved in the design of a successful combustor. The allowable range of burner entry pressure is, in fact, bounded above and below by considerations of truly basic phenomena.

The burner entry pressure is the highest static pressure to be found within the engine. Since mechanical and thermal loads both increase with p_3, it follows that the combustor is the most challenging structural design, requiring the most advanced materials and cooling techniques. The maximum pressure entering the combustor must therefore be limited to a value that will lead to "reasonable" or "acceptable" weight, complexity, and cost. Although the specific maximum value chosen depends on many considerations, including the mission of the vehicle, the ingenuity of the designer and the availability of materials, the maximum p_3 that will be used here is 10 atm.

Combustion reaction rates increase rapidly (or exponentially) with static pressure and temperature. Unfortunately, the reverse is also true, so that the length of the combustor required to complete the reaction and consume the available fuel will become excessive if p_3 is too low. Although the specific minimum value chosen depends on many considerations, including the type of fuel, the method of fuel injection and the length of the host vehicle, the minimum p_3 that will be used here is the commonly applied rule of thumb of 0.5 atm.

The desired expression for p_3 can be obtained from Eq. (5-4) by replacing p_0 with q_0 from Eq. (2-12), with the result that

$$p_3 = \frac{2q_0}{\gamma_0 M_0^2 \left\{1 - \eta_c \left(1 - \dfrac{1}{\psi}\right)\right\}^{\gamma_c/(\gamma_c - 1)}} \qquad (5\text{-}34)$$

Equation (5-34) may be utilized in a variety of ways in order to examine the behavior of p_3 in flight, several of which are illustrated in Figs. 5.24 through 5.26. Please note that the estimates of ramjet behavior shown there were obtained by setting ψ to its maximum attainable value, namely

$$\psi = 1 + \frac{\gamma_c - 1}{2} M_0^2$$

All other parameters were chosen to reflect their expected ranges.

Referring to Figs. 5.24 through 5.26 collectively, we can see that the burner entry pressure is, fortunately, usually found in the acceptable range, although it tends toward values that are too large at the high freestream Mach number end of ramjet operation, and too small at the high freestream Mach number end of scramjet operation. The latter follows directly by inspecting Eq. (5-34), which reveals that, all other things being equal, the burner entry pressure simply varies inversely with the square of the freestream Mach number. Maintaining complete combustion of the fuel in scramjet operation at high

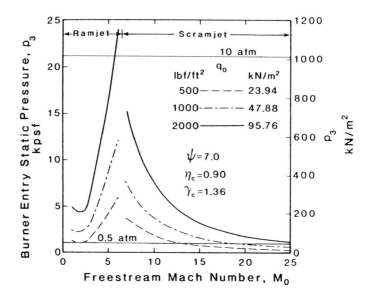

Fig. 5.24 **Burner entry static pressure as a function of freestream Mach number and aerospace vehicle dynamic pressure.**

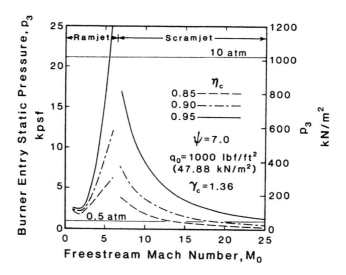

Fig. 5.25 **Burner entry static pressure as a function of freestream Mach number and compression efficiency.**

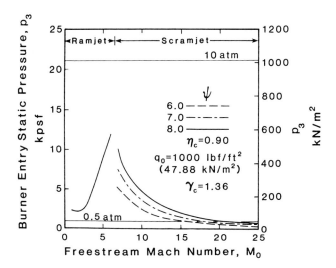

Fig. 5.26   Burner entry static pressure as a function of freestream Mach number and cycle static temperature ratio.

freestream Mach numbers will therefore require higher values of the vehicle dynamic pressure, compression efficiency, and/or cycle temperature ratio. It is, in fact, fairly obvious that at orbital freestream Mach numbers a vehicle dynamic pressure of at least 1000 lbf/ft^2 (47.88 kN/m^2) will be needed in order to provide the conditions for satisfactory combustion.

5.7   COMPRESSION COMPONENT AERODYNAMIC PHENOMENA

There are a number of aerodynamic phenomena that must be considered during the design of compression systems. Although they influence the compression efficiency to differing degrees, they will have to be taken into account for other reasons as well. The majority of these are directly or indirectly related to viscous effects, three-dimensionality, and/or unsteadiness, and therefore require analytical or CFD methods that are beyond the scope of this textbook. Nevertheless, approximate methods will be described that capture the important physics of the phenomena and allow reasonable estimates to be made for the purpose of illustration. Unless otherwise noted, the flows under consideration will be regarded as either one-dimensional or two-dimensional.

5.7.1 Leading-Edge Oblique Shock Wave Geometry

The initial oblique shock wave of the ramjet or scramjet compression system is generated by the leading edge of the vehicle. It is important for the vehicle designer to know what type of initial oblique shock wave is needed as a function of flight condition. We will continue to assume, for consistency, that the geometry can be arranged so that the shock waves are focused on the leading edge of the cowl, the so-called shock-on-lip configuration.

Perhaps the best-known characteristic feature of hypersonic flows is that, for any given body, the shock waves approach or hug the bounding surface more and more closely as the freestream Mach number increases.[5.2] The mental image of a "thin shock layer" leads to a number of hypersonic flow solution techniques, including Newtonian flow theory, probably the most popular of them all. An interesting question to be answered here is whether the oblique shock waves of the compression system obey the same intuitive rules when subject to the additional constraints imposed by the engine.

Figure 5.27 contains flowfield information corresponding to Fig. 5.16 that was also provided by the HAP(Gas Tables) computations. This information reveals that, for constant cycle temperature ratio, the angle between the initial oblique shock wave and the compression surface of the vehicle decreases rapidly as the freestream Mach number increases, regardless of the number of oblique shocks. This angle decreases as the number of oblique shock waves increases simply because they can each be "weaker" and still together achieve the desired cycle temperature ratio.

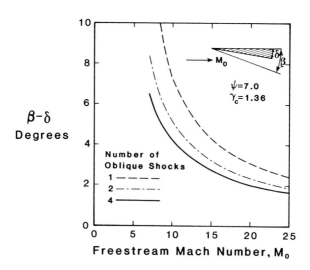

Fig. 5.27   Angle between the initial oblique shock wave and the vehicle compression surface as a function of freestream Mach number.

This agrees completely with the generally observed behavior of hypersonic flows described above. Furthermore, as Fig. 5.27 shows, the angle between the initial oblique shock wave and the vehicle compression surface is only a few degrees at the higher freestream Mach numbers, regardless of the number of oblique shock waves. Nevertheless, the variation of this angle over the hypersonic freestream Mach number range demonstrates that variable geometry is essential to maintaining the shock-on-lip condition.

5.7.2 Shock—Boundary Layer Separation

Oblique shock waves can and do exist throughout ramjets and scramjets. Some are generated by the compression system and not canceled by conveniently located expansion waves, while others originate from various corners and obstacles within the engine flowpath. They may be generated by adjacent vehicle surfaces and propagate onto the exposed engine surfaces, and even into the internal engine flowpath.

Whatever their source, when they impinge upon or originate from a solid boundary, they impose an abrupt, almost discontinuous, increase in pressure on the boundary-layer flow immediately adjacent to the surface. An important property of impinging oblique shock waves is that they are reflected "in kind" (i.e., as oblique shock waves of equal but opposite turning), substantially increasing the pressure rise imposed upon or experienced by the boundary-layer flow.[5.1]

The most violent and unfortunate effect the oblique shock wave pressure rise can have on the boundary-layer behavior is to cause it to separate, as shown schematically in Fig. 5.28. The underlying mechanism for separation is that the cumulative upstream viscous forces have removed the momentum necessary for the boundary-layer flow to overcome the imposed pressure rise. Although reattachment usually eventually occurs, resulting in a finite region of reversed and recirculating flow as shown in Fig. 5.28, there are situations in which reattachment does not take place (see Sec. 5.7.5, for example).

In accordance with intuition, the likelihood of separation is increased by either increasing the "strength" (or pressure rise) of the impinging or originating oblique shock wave, or by "weakening" the boundary layer. The major determinant of boundary layer "strength" seems simply to be whether it is laminar or turbulent.[5.13] Turbulent boundary layers are considerably more resistant to separation than laminar boundary layers because their velocity profiles are "fuller" (i.e., the velocity increases much more rapidly with distance from the wall) and because the intense local turbulent mixing assists the low velocity in overcoming the imposed pressure rise.

These observations have several corollaries. One is that a "worst case" can occur when an impinging oblique shock wave arrives at a location on the boundary from which another oblique shock wave is originating. Another is that removing (by suction) or energizing (by deliberately causing laminar flow to become turbulent or by blowing) the fluid nearest the boundary just prior to the region of imposed pressure rise will also strengthen the flow against separation.

The consequences of separation, particularly in hypersonic flight, are remarkably severe and unpleasant. First, the skin friction drag is replaced by the much more potent pressure or form drag, increasing the local boundary-layer momentum losses by a factor of 2–5.[5.1] This, in turn, both reduces the internal performance of the engine and increases the thickness and distortion of the boundary-layer velocity profile, with the result that more flow blockage occurs and the distribution of flow is harder to predict and control. Second, the separation region itself is unstable and moves rapidly about, causing unsteadiness and uncertainty to propagate downstream.[5.1]

Third, and easily most important, the increased transport of high enthalpy gases from the freestream to the solid boundary increases the wall heat transfer rate adjacent to the separation zone by a factor of 2–5 (see Sec. 9.2.1.3). The resulting *hot spots* enormously complicate the thermal design of the vehicle, particularly since their location and intensity can change with flight condition. Moreover, separation can also cause a laminar boundary layer to become turbulent, so that the heat transfer is increased even for the region well downstream of the attachment point.[5.1] Thus, the uncertainties in the magnitude and location of high wall heat transfer rates associated with shock—boundary layer separation present design challenges of the highest order. A wide range of strategies have been developed for dealing with these phenomena, including protecting the entire surface against the highest anticipated local heat transfer rate, deliberately positioning an oblique shock wave in order to produce a "strong" turbulent boundary layer over most of the compression surface, using suction or blowing in order to prevent separation, and inventing "closed loop" techniques that sense and provide the amount of cooling needed in order to precisely maintain the desired local surface temperature.

Shock—boundary layer separation on hypersonic aerospace vehicles must therefore be avoided to the extent possible. The first step toward this goal is to predict when it will occur, a difficult task even for the most advanced CFD methods. Instead, we will employ results from Ref. 5.13, intended to be suitable for engineering purposes, in order to clarify and quantify this phenomenon.

Using the terminology of Fig. 5.28, reasonable approximations for laminar and turbulent shock—boundary layer separation criteria

Impinging Oblique Shock Wave

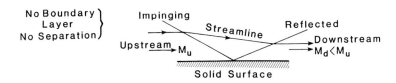

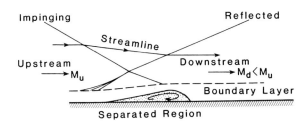

Originating Oblique Shock Wave

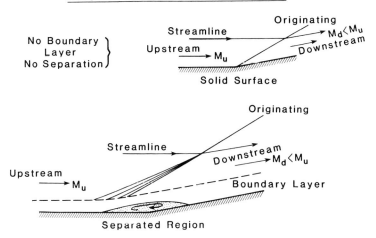

Fig. 5.28   Schematic illustration of boundary-layer separation and reattachment caused by impinging or originating oblique shock waves. Note that the separation or recirculation zone appears to the distant flowfield as a minor perturbation to the geometry of the solid boundary. The downstream compression and expansion waves must therefore eventually coalesce into an oblique shock wave very much like the one expected from the unperturbed boundary.

according to Ref. 5.13 are, respectively,

$$\left(\frac{M_d}{M_u}\right)_{\text{laminar}} < 0.898 \tag{5-35}$$

$$\left(\frac{M_d}{M_u}\right)_{\text{turbulent}} < 0.762 \tag{5-36}$$

Please note that the inequality means that the boundary layer will separate if the downstream Mach number M_d is less than the stated fraction of the upstream Mach number, which means that the deceleration and pressure rise exceed their natural limits.

These two criteria should be employed with the following thoughts in mind. First, although their theoretical basis is broad, they were experimentally confirmed only for the range $1 < M_u < 4$. Second, they apply only to either entirely laminar or entirely turbulent flows. Third, they reflect the experimental observation that separation is negligibly influenced by Reynolds number (or boundary-layer thickness) and, as noted earlier, depends primarily upon whether the flow is laminar or turbulent. Fourth, this is only one of many available approaches to predicting shock—boundary layer separation, and was selected for its instructional value rather than as a recommended design tool.

Inasmuch as we determined in Sec. 4.2.3 that M_3/M_0 is of the order of 0.38, and that this overall thermal compression is largely due to a succession of oblique shock waves, the criteria expressed by Eqs. (5-35) and (5-36) make it abundantly clear that separation is quite likely to come into play. One way to visualize these criteria in action is to apply them to our previously-defined special families of compression systems. This is done by converting the boundary-layer separation criteria into the *maximum* allowable amount of geometrical turning that can be achieved by an *originating* oblique shock wave as a function of the upstream Mach number M_u, and comparing the results with previously obtained designs.

Converting the criteria of Eqs. (5-35) and (5-36) into the maximum allowable amount of geometrical turning can be done either by trial and error using HAP(Gas Tables), or by following Ref. 5.13 and first calculating the maximum allowable static pressure ratio across the oblique shock wave for a given M_u from

$$\frac{p_d}{p_u} = \left\{ (\gamma_c + 1)M_u^2 \left\{ 1 - \left(\frac{M_d}{M_u}\right)^2 \right\} + \left\{ (\gamma_c + 1)^2 M_u^4 \left[\left(\frac{M_d}{M_u}\right)^2 - 1 \right]^2 \right. \right.$$

$$\left. \left. + 4 \left[(\gamma_c - 1) M_u^2 \left(\frac{M_d}{M_u}\right)^2 + 2 \right] \left[(\gamma_c - 1) M_u^2 + 2 \right] \right\}^{1/2} \right\}$$

divided by

$$\left\{ 2 \left(\gamma_c - 1\right) M_u^2 \left(\frac{M_d}{M_u}\right)^2 + 4 \right\}$$

$$(5\text{-}37)$$

and using that value to calculate the maximum allowable geometrical turning from

$$\delta = \tan^{-1} \left\{ \left[\frac{\dfrac{p_d}{p_u} - 1}{\gamma_c M_u^2 - \dfrac{p_d}{p_u} + 1} \right] \left[\frac{2\gamma_c M_u^2 - \left(\gamma_c - 1\right) - \left(\gamma_c + 1\right)\dfrac{p_d}{p_u}}{\left(\gamma_c + 1\right)\dfrac{p_d}{p_u} + \left(\gamma_c - 1\right)} \right]^{1/2} \right\}$$

$$(5\text{-}38)$$

Please note that both methods permit the ratio of specific heats of the gas γ_c to be independently chosen.

The results of this exercise are found in Fig. 5.29, which is rich with worthwhile information. Although it is not surprising that turbulent boundary layers can withstand considerably more turning than laminar boundary layers, it was by no means obvious from the criteria of Eqs. (5-35) and (5-36) that the allowable turning diminishes rapidly with upstream Mach number. Fortunately, the amount of turning required for a given amount of compression also diminishes with upstream Mach number.

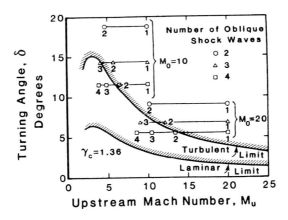

Fig. 5.29 Maximum allowable turning without boundary-layer separation for originating shock waves as a function of upstream Mach number, in accordance with Eqs. (5-35) and (5-36). The discrete symbols are taken from the designs of Sec. 5.5.2 for $\psi = 7.0$.

The discrete symbols appearing on Fig. 5.29 represent the succession of geometrical turns of our earlier designs, for which the first should be ignored because it occurs at the leading edge where no boundary layer exists. As we might expect, more turns having smaller deflections make shock—boundary layer separation less likely, but there are some ominous signs as well. For example, none of these designs avoids the separation of laminar boundary layers, and the two turn designs are inadequate even for turbulent boundary layers. It must therefore be concluded that shock—boundary layer separation is a clear and present danger that must be accounted for in any realistic design. These results suggest that some of the compression be isentropic, as shown in Fig. 5.15.

The criteria of Eqs. (5-35) and (5-36) can be, and often are, presented in other forms in the open literature. Two of these are the maximum allowable dimensionless pressure coefficient

$$\frac{p_d - p_u}{q_u} = \frac{p_u\left(\frac{p_d}{p_u} - 1\right)}{\frac{\gamma_c p_u M_u^2}{2}} = \frac{2\left(\frac{p_d}{p_u} - 1\right)}{\gamma_c M_u^2} \tag{5-39}$$

where p_d/p_u is obtained from Eq. (5-37), and the maximum attainable static temperature ratio from

$$\frac{T_d}{T_u} = \frac{1 + \frac{\gamma_c - 1}{2} M_u^2}{1 + \frac{\gamma_c - 1}{2} M_u^2 \left(\frac{M_d}{M_u}\right)^2} \tag{5-40}$$

These parameters are displayed in Fig. 5.30, the counterpart of Fig. 5.29, where it can be seen that $(p_d - p_u)/q_u$ continuously diminishes as upstream Mach number increases. In contrast, T_d/T_u varies only slightly over the range of interest, asymptotically increasing toward $(M_u/M_d)^2$ as the upstream Mach number becomes indefinitely large. The uniformity of T_d/T_u certainly helps to reach the desired T_3/T_0 with a limited number of separate turns.

5.7.3 The Inlet Flow Starting Process

One of the classical topics in supersonic propulsion that has great meaning for hypersonic propulsion as well is that of the "starting" of inlet flows.[5.1,5.4,5.5] The questions to be answered are how the flow within the compression system can be arranged to change from subsonic to supersonic as the vehicle accelerates, and how efficient compression can be obtained over the entire range of freestream Mach numbers.

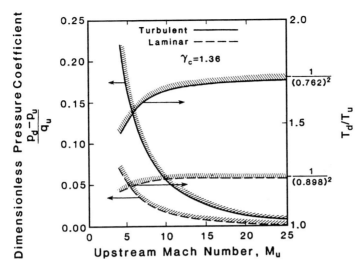

Fig. 5.30 Maximum allowable dimensionless pressure coefficient and maximum attainable static temperature ratio without boundary-layer separation as a function of upstream Mach number, in accordance with Eqs. (5-35) and (5-36).

The key to understanding this critical process will be to recognize the sequence of states involved, as shown in Fig. 5.31. In order to simplify this discussion, the inlet configuration will at first be taken to be of the constant geometry "pitot" variety (see Fig. 5.1), having the minimum throughflow area (usually referred to as the *throat*) within the internal passage, and nothing downstream of the throat to further restrict the amount of flow. Moreover, the flow will be assumed to be steady, one-dimensional, frictionless, and without heat interactions.

Since the starting process takes place at freestream Mach numbers for which chemistry effects are inconsequential and the air can be treated as a calorically perfect gas, the flow therefore acts as one of constant total pressure p_t and total temperature T_t, except for the total pressure losses that take place across any normal shock waves present. This also leads to the great convenience of being able to use the mass flow parameter MFP to analyze the behavior of the flow (see Example Case 2.5).

The sequence of states of the inlet flow starting process will now be described, with the help of Fig. 5.31 for visualization and Fig. 5.32 for quantitative behavior.

State 1. At very low subsonic speeds, the flow is subsonic throughout the inlet. Please note that the capture area A_0 can exceed the inlet area A_1 if air is forcefully drawn into the inlet, as is

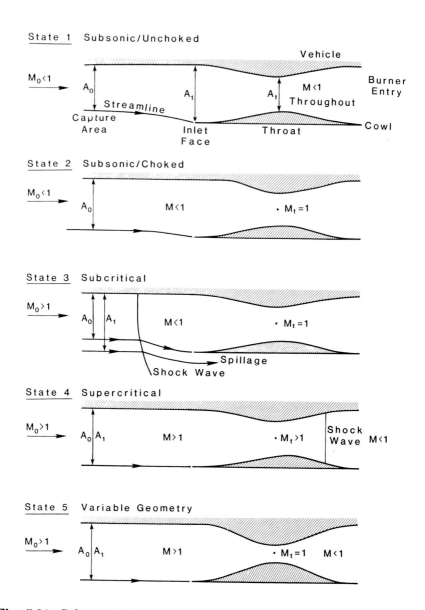

Fig. 5.31 Schematic illustration of states of the inlet flow starting process for a pitot inlet. In keeping with the engine reference station designations of Sec. 4.1, the throughflow area of the entering or "captured" freestream flow is designated A_0, the throughflow area at the beginning of internal compression or "inlet" face area is designated A_1, and the minimum throughflow area at the "throat" is designated A_t.

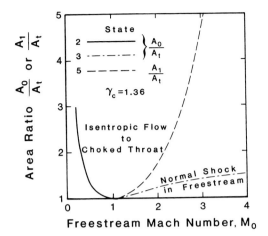

Fig. 5.32 Area ratios as functions of freestream Mach number for several states of the inlet flow starting process.

often the case for turbojet engines. Under this condition, the relationships between the three throughflow areas of interest cannot be determined without additional information.

State 2. At and beyond a sufficiently high subsonic Mach number, depending on the downstream conditions, the velocity at the throat is sonic, and the inlet is said to be *choked,* meaning that the capture area is *controlled* by conditions at the throat. In this state, for which no shock waves can exist, the assumptions above lead to the result that

$$MFP \cdot A = \text{constant} \tag{5-41}$$

or, since the Mach number at the throat is 1, that

$$\frac{A_0}{A_t} = \frac{MFP_t}{MFP_0} = \frac{1}{M_0}\left\{\frac{2}{\gamma_c+1}\left(1 + \frac{\gamma_c-1}{2}M_0^2\right)\right\}^{(\gamma_c+1)/[2(\gamma_c-1)]} \tag{5-42}$$

The application of Eq. (5-42), as displayed in Fig. 5.32, reveals that A_0/A_t decreases with increasing M_0, just as one would expect for converging compressible flows. An especially important and intuitive result is that $A_0 = A_t$ when the freestream flow is sonic. Also observe that A_0/A_t will be greater than the fixed value of A_1/A_t for sufficiently low M_0, while the reverse will be true for sufficiently high, but subsonic, M_0. Finally, since the throat is the first minimum area to be encountered by the flow, the flow remains subsonic everywhere upstream of the throat.

State 3. When the freestream flow is supersonic, approximately normal shock waves form in the oncoming flow in order to "warn" of the flow restriction imposed by the choked throat. These shock waves can be reasonably assumed to take place at the freestream Mach number, and cause the "spillage" (i.e., the mass flow that would pass through the difference between A_1 and A_0 at freestream conditions) necessary to accommodate the requirements of the throat.

This state is also referred to as *subcritical*, and is considered to be undesirable because the inlet is not capturing as much air as the physical opening would otherwise make possible and because the external normal shock wave causes a total pressure loss and distorts the velocity profile.

This state can be analyzed in two steps, the first accounting for the constant area "jump conditions" across the normal shock in the freestream, and the second accounting for the isentropic flow from a point immediately downstream of the shock to the throat. In the first step, the subsonic Mach number M_s immediately downstream of the normal shock is determined from the relationship[5.1]

$$M_s = \left\{ \frac{1 + \dfrac{\gamma_c - 1}{2} M_0^2}{\gamma_c M_0^2 - \dfrac{\gamma_c - 1}{2}} \right\}^{1/2} \qquad (5\text{-}43)$$

which, in the second step, can be inserted in Eq. (5-41) in order to find

$$\frac{A_0}{A_t} = \frac{1}{M_s} \left\{ \frac{2}{\gamma_c + 1} \left(1 + \frac{\gamma_c - 1}{2} M_s^2 \right) \right\}^{(\gamma_c+1)/[2(\gamma_c-1)]} \qquad (5\text{-}44)$$

Although it is not directly obvious from Eqs. (5-43) and (5-44), the results portrayed in Fig. 5.32 make it clear that A_0/A_t increases continuously with freestream Mach number, which means that the freestream shock wave moves steadily "closer" to the inlet face and the spillage decreases. Eventually, when $A_0/A_t = A_1/A_t$ or $A_0 = A_1$, the normal shock wave must be situated precisely at the face of the inlet, and there is no spillage. This condition obviously depends upon the relative size of the inlet area. For example, referring to Fig. 5.32, at a freestream Mach number of 2.00, the normal shock would stand at the face of the inlet if $A_1/A_t = A_0/A_t = 1.23$. This condition is also referred to as *critical*, and represents the upper limit of subcritical operation.

Another interesting feature of these results is the relatively small range of A_0/A_t that can occur in nature. One way to look at this is that the critical Mach number is very sensitive to A_1/A_t.

State 4. Once the freestream Mach number even slightly exceeds the value for which the normal shock wave would stand at the face of the inlet, the shock wave snaps into or is "swallowed" by the inlet, and the flow becomes supersonic from the freestream to the throat and beyond. One way to comprehend this phenomenon is to recognize that, if the normal shock wave remained at the inlet face, A_1/A_t would no longer be large enough to bring the flow back to sonic by the throat, thus violating a fundamental operating principle of the engine. More elaborate calculations show that this would be also true if the shock tried to position itself anywhere within the convergent flowpath.

After the shock wave has been swallowed, the flow is said to be *started* or *supercritical,* which is considered desirable because the physical opening is capturing as much air as possible, and because any normal shock waves that occur will be located within the internal flowpath where they can be controlled. Thus, the critical freestream Mach number is also the starting freestream Mach number.

The position of the normal shock wave in the flowpath now depends on the *back pressure* due to downstream conditions. The Mach number at the throat now depends only upon the freestream Mach number and A_1/A_t, and can be quite supersonic. In the case of scramjet engines, where the flow remains supersonic throughout the engine, the high throat Mach numbers are acceptable. However, in the case of ramjet engines, where the flow must be made subsonic before entering the burner, beginning with normal shock waves located downstream of the throat, these relatively high throat Mach numbers will lead to excessive performance losses.

This can be clearly seen in Fig. 5.33, which presents relevant results for started, fixed geometry pitot inlets at the starting freestream Mach number. This information was generated for each starting freestream Mach number by first finding the corresponding A_1/A_t from Fig. 5.32. Then M_t was found by applying Eq. (5-41) from the inlet face to the throat. Next, the total pressure ratio across the normal shock at M_t was determined from the relationship[5.1]

$$\pi_s = \pi_c = \frac{\left\{ \dfrac{\dfrac{\gamma_c+1}{2}M_t^2}{1+\dfrac{\gamma_c-1}{2}M_t^2} \right\}^{\gamma_c/(\gamma_c-1)}}{\left\{ \dfrac{2\gamma_c}{\gamma_c+1}M_t^2 - \dfrac{\gamma_c-1}{\gamma_c+1} \right\}^{1/(\gamma_c-1)}} \qquad (5\text{-}45)$$

where, for simplicity, it is assumed that this is the only total pressure loss in the compression system. Finally, the static temperature ratio ψ and the adiabatic compression efficiency η_c are obtained, as usual, from Eqs. (5-2) and (5-6), respectively.

It should be obvious by now that these adiabatic compression efficiencies are too low for acceptable engine performance. Furthermore, if the normal shock wave were located downstream of the throat,

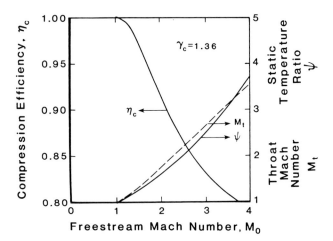

Fig. 5.33 Fixed geometry pitot compression system throat Mach number, static temperature ratio, and adiabatic compression efficiency as functions of the starting freestream Mach number.

the local Mach number would be higher still, the total pressure loss greater, and engine performance poorer. Because of this, inlets will certainly be designed to start at freestream Mach numbers less than 4, thus justifying the calorically perfect gas assumption made at the outset of this analysis. Finally, the performance also worsens as the freestream Mach number increases beyond the starting value because the Mach numbers become higher throughout the flowpath. This situation cries out for remedies.

State 5. One way to reduce the throat Mach number and the attendant losses is to provide variable geometry within the inlet that allows the throat area and the throat Mach number to be reduced once the inlet has started. The ideal limit of this case would be to make the throat flow sonic, and position the normal shock wave an infinitesimal distance downstream, so that the total pressure loss would be negligible. The remainder of the flowpath would then be a subsonic diffuser.

Unfortunately, this situation is unacceptable in two ways. First, deceleration of the flow in the vicinity of sonic velocity is itself inherently unstable. Second, positioning the normal shock too close to the throat invites the possibility that a small disturbance will cause the shock wave to be disgorged back into the freestream, and return to State 3. In practice, therefore, the throat Mach number would always be maintained sufficiently above sonic and the normal shock wave positioned sufficiently far downstream to insure stability.[5.1,5.5] Nevertheless, this condition provides a clearly defined limit from which to reason, and was therefore included in Fig. 5.32 for purposes of elucidation. The limit line was again obtained from Eq. (5-42), which means that States 2 and 5 are merely the subsonic and supersonic branches of the isentropic, choked flow, freestream Mach number—area ratio relationship, respectively.

The obvious conclusion to be drawn from Fig. 5.32 is that a lot of variable geometry would be required in order to approach the ideal limit for a pitot inlet. For example, at a freestream Mach number of 2.00, the A_1/A_t required by this state is 1.72, which is a factor of $1.72/1.23 = 1.40$ larger than that required for starting. This factor obviously grows rapidly with freestream Mach number and reaches 3.22 at $M_0 = 3.00$ and 8.13 at $M_0 = 4.00$.

One may very well ask how the starting process would differ if the compression system included an external compression surface, such as shown in Fig. 5.2, rather than the simple pitot inlet considered above. This would then require the inclusion of the influences of both the oblique shock wave system on any terminal normal shock wave, and of the compression surface boundary layer on the flow within the internal flowpath. These effects clearly require an analysis that is not only much more complex but also specific to a given configuration. The results would therefore be correspondingly less transparent and general. Nevertheless, the qualitative differences between the two cases can and will be described because of the valuable insight obtained.

As far as the oblique shock wave system plus terminal normal shock wave is concerned, the tendency would be for the oblique shock waves to decrease the Mach number just upstream of the normal shock wave, and thus increase the Mach number downstream of the normal shock wave slightly above the value originally given by Eq. (5-43). When substituted into Eq. (5-44), this would then yield a slightly lower value of A_0/A_t than that shown in Fig. 5.32, which means that starting would be delayed to a higher freestream Mach number for an inlet of given A_1/A_t, or, conversely, a smaller A_1/A_t would be needed for starting at a given freestream Mach number. The drag of the boundary layer would also tend to reduce the average Mach number just upstream of the normal shock wave, with similar results.

The other primary influence of the boundary layer is to displace the flowfield away from the compression surface, a phenomenon often picturesquely referred to as *boundary-layer blockage.* The presence of the blockage reduces the throughflow areas available at the inlet face and throat to "effective" sizes that are less than their purely geometric dimensions. Thus, all other things being equal, the freestream capture area and engine thrust are reduced in direct proportion to the effective throat throughflow area. This condition will be greatly aggravated in subcritical operation if the terminal normal shock wave

also separates the boundary layer upstream of the inlet face, causing a substantial increase in the downstream blockage. Provision is
therefore often made to remove (by means of bleeds or diverters) the
boundary layer in the neighborhood of the expected location of the
terminal normal shock in order to avoid that possibility.

Before moving on to the next topic, it should be made clear that
inlet design is a very mature field, and an abundance of reference
material is available to furnish the student and the designer with
whatever level of detail they desire.[5.5,5.14] This is primarily because
many operational aircraft and missiles fly in the supersonic regime,
and practical solutions to the problems they encounter have been
worked out through the years. We have chosen not to dwell on several
of the fascinating phenomena that arise because hypersonic aerospace
systems pass quickly through the supersonic regime, and spend the
vast majority of their time at hypersonic speeds.

5.7.4 Inlet Unstart

Once the inlet is started, it will remain in the preferred supercritical
state unless seriously perturbed. Three types of disturbances that
can separately or in combination cause the inlet to "unstart" and
return to the undesirable subcritical state will be described next.

First, unstart will occur if the freestream Mach number is reduced
sufficiently below the starting value. The required Mach number for
a fixed geometry inlet can be found on Fig. 5.32 by recognizing that
State 5 also represents the maximum possible internal contraction
or A_1/A_t possible for a given freestream Mach number. Therefore,
when A_0/A_t lies above the State 5 line, no isentropic solution can
connect the inlet face to the throat. When this condition is reached,
the flow reverses the procedure described under State 4 and the shock
wave pops out of or is "disgorged" by the inlet, returning the flow
to State 3. Continuing the example used above, Fig. 5.32 reveals
that an inlet that starts at a freestream Mach number of 2.00 (i.e.,
$A_1/A_t = 1.23$) will not unstart until a freestream Mach number
of 1.57 is reached. The difference between starting and unstarting
freestream Mach numbers is a fortunate "hysteresis" that provides
a safe operating margin at the instant of starting. Nevertheless,
even hypersonic aerospace vehicles must eventually decelerate and
land, during which sequence they will experience and must cope with
unstart.

Second, unstart will occur if the flow reaching the inlet face is
sufficiently distorted, even temporarily, by upstream effects that it
can no longer pass through the throat area. These effects could
include total pressure losses, flow angularity, and blockage, and could
be caused, for example, by unexpected changes in angle of attack of

the vehicle, unanticipated boundary-layer separation on the external compression surface, or the ingestion of unusual gases, such as the exhaust of a nearby vehicle. These phenomena are difficult to analyze and predict, but they all occur to some degree, and design margin must be provided in order to avoid their consequences.

Third, unstart will occur if the back pressure is increased to the point that the terminal normal shock wave is moved upstream to the throat. The back pressure will increase, for example, if the chemical energy release in the burner is increased or the exhaust nozzle throat area is decreased. Whether these changes are deliberate or accidental, steady-state or transient, the outcome will be the same. That is, once the shock wave reaches the throat it is unstable in the sense that any infinitesimal disturbance will cause it to be disgorged by the inlet, returning the flow to State 3. Since the unstart in this case is precipitated by downstream conditions, it cannot be depicted on Fig. 5.32. However, the highest total pressure possible for a given inlet can easily be determined by first using Eq. (5-41) to calculate the throat Mach number M_t corresponding to the prevailing values of M_0 and A_1/A_t, and then using Eq. (5-45) to calculate the total pressure ratio across the normal shock at M_t.

Unstart must be avoided at almost any cost. One reason for this is that returning the inlet to the subcritical state deprives the engine of the airflow necessary for thrust, and the normal shock upstream of the inlet face diminishes the total pressure necessary for performance. Another is that unstart is an extremely unsteady and violent phenomenon, in which the swiftly moving shock waves can impose transient loads on the adjacent structure a factor of 10 or more in excess of those found in steady-state. Likewise, the fuel control system may be unable to properly adjust to the rapidly changing burner entry conditions, and allow the burner to overheat or the flame to go out.

5.7.5 Inlet Isolators

As you may very well imagine, the idealized concept of the normal shock wave as a flow discontinuity that is so helpful to the understanding of the general behavior of inlets seldom appears in ramjet and scramjet compression systems. For one thing, a normal shock wave within the internal flowpath is almost certain to cause boundary-layer separation, forcing the normal shock wave to take on an altogether different appearance. For another, the back pressure imposed by the downstream elements may require considerably less compression than would accompany a normal shock wave.

In order to get closer to reality, we will begin by examining the behavior of shock waves in constant area ducts, for which there is an extensive body of experimental and analytical information.[5.15–5.20]

Since these flow devices produce a static pressure rise in a constant area duct, they are also called *constant area diffusers.*

The general features of the two most frequently observed two-dimensional (i.e., no variation into the page) flowfields are portrayed in Figs. 5.34 and 5.35. Although the inlet *i* is often the same as the inlet face (Station 1) and the exit *e* is often the burner entry (Station 3), this may not always be the case, so their designations have been made more general. The back pressure may be the result of chemical energy release in the burner or choking of a downstream area, but may also be caused by such obstructions as fuel injectors or even the blockage due to the presence of the injected fuel. The back pressure imposed in Figs. 5.34 and 5.35 is shown as the exit value associated with a large plenum in order to emphasize the one-dimensional nature of this boundary condition.

The pattern shown in Fig. 5.34, known as a *normal shock train,* tends to occur at lower inlet Mach numbers and thicker inlet boundary layers. The pattern shown in Fig. 5.35, known as an *oblique shock train,* tends to occur at higher inlet Mach numbers and thinner inlet boundary layers. A rough but useful rule of thumb is that the dividing line between normal and oblique shock trains "pseudoshocks" is in the range $2 < M_i < 3$.

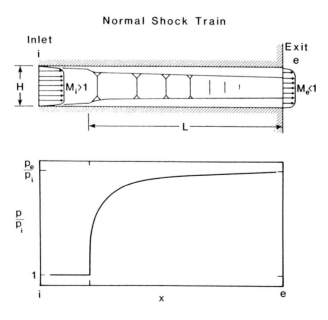

Fig. 5.34 Schematic drawing of the flow pattern and static wall pressure distribution of a complete normal shock train.

An important piece of fundamental background knowledge about shock trains in constant area ducts is that the maximum static pressure rise they can generate is *exactly* the same as the simple normal shock discontinuity would have generated, provided only that the shock train is complete in the sense that the flow properties are uniform across the exit plane and the total wall friction is negligible compared to the inlet stream thrust, both being good first approximations for such flows. This conclusion can easily be reached by observing that the set of control volume conservation equations is identically the same for the simple normal shock discontinuity and the shock trains under these assumptions, so that the solution must also be the same.

Examining the complete normal shock train case of Fig. 5.34 more closely, we see that the upstream or initial normal shock wave "bifurcates" (i.e., decomposes into oblique shock waves near the wall) as it interacts with the separating wall boundary layer. The majority of the possible static pressure increase takes place within this initial region, after which the boundary layer reattaches and grows rapidly, but remains attached. The remaining static pressure increase is associated with a series of "secondary" normal shock waves, which may or may not be bifurcated. The relatively large thickness of the

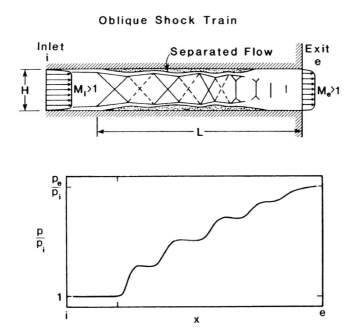

Fig. 5.35 Schematic drawing of the flow pattern and static wall pressure distribution of a complete oblique shock train.

boundary layer "smears out" the effects of the individual secondary shock waves, with the result that the static pressure along the wall increases gradually and continuously.

Examining the complete oblique shock train case of Fig. 5.35 more closely, we see that it starts with an upstream or initial oblique shock wave that separates the boundary layer. This initial wave is reflected from the boundary layer on the opposite wall as an expansion wave, and then continues to propagate down the duct, alternating between compression and expansion while gradually increasing the static pressure. The boundary layer remains separated for a great distance, even though its thickness waxes and wanes in response to the arriving expansion and compression waves. The static pressure at the wall increases in a series of "plateaus," for which the pressure increases in the region between successive compression waves, and levels off or even decreases in the region between successive expansion waves.

The shock train provides the mechanism for a supersonic flow to adjust to a static back pressure higher than its inlet static pressure. If the back pressure should exceed the maximum possible, the whole shock train will be disgorged and the inlet will unstart, just as described above. If, on the other hand, the back pressure is somewhere between the maximum and inlet static pressures, the shock train simply slides or translates downstream almost entirely intact until its exit plane pressure equals the imposed back pressure. This process is sketched in Fig. 5.36, which shows how translating an oblique shock train can allow the back pressure to be matched.

It may at first seem counter-intuitive that shock trains can communicate upstream so freely in highly supersonic flows. However, the inevitable presence of the subsonic portion of the boundary layer always allows information to be transmitted at least a short distance, and that in turn can lead to shock—boundary layer interactions that continue the propagation process. The shock train is merely the equilibrium outcome of the sequence of events.

We may therefore conclude that the shock train is a useful agent of fluid mechanics, one that can carry out the work of a normal shock wave, while providing a cushion when less static pressure rise is needed. Unfortunately, if the constant area duct is not long enough to contain the entire shock train, the leading edge can migrate upstream far enough to unstart the inlet. Since the constant area diffuser must therefore be long enough to contain any upstream influence of the shock train, it is also known as an *isolator*.

The designer must know how long to make the isolator, but predicting the expected length of the shock train is no easy matter because the fluid mechanics involved are so complex. We will attack the problem by means of a correlation that is widely used and quite

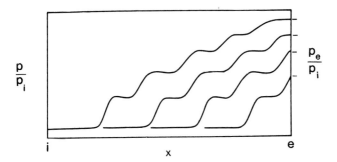

Fig. 5.36 Wall pressure distributions and back pressures corresponding to several axial locations of an oblique shock train.

physically transparent. The work of Ref. 5.19 can be manipulated to yield

$$\frac{L}{H} = \frac{\sqrt{\dfrac{\theta}{H}}}{\sqrt[4]{Re_\theta}} \cdot \frac{\left\{ 50 \left(\dfrac{p_e}{p_i} - 1 \right) + 170 \left(\dfrac{p_e}{p_i} - 1 \right)^2 \right\}}{(M_i^2 - 1)} \tag{5-46}$$

where the nomenclature of Fig. 5.34 has been used. The correlation of Eq. (5-46) reflects the influences of the most pertinent physical parameters, namely the inlet Mach number M_i, the back pressure ratio p_e/p_i, the inlet boundary-layer momentum thickness θ (sometimes referred to as the *confinement* because the blockage of the boundary layer reduces the available duct throughflow area), and the inlet Reynolds number Re_θ (based on the inlet momentum thickness). Please note that these are not, in general, complete shock trains.

Figure 5.37 contains the results of calculations based upon Eq. (5-46) for representative values of the inlet conditions and back pressure ratios. The main conclusions to be drawn from Fig. 5.37 are that the ratio of shock train axial length to duct height L/H can be large compared to 1 when the inlet Mach number is small and the back pressure ratio is large, and that it is easily possible for L/H to equal or exceed 10 for typical flight conditions.

The left-hand boundary shown on Fig. 5.37 corresponds to the maximum static pressure ratio obtainable for a given inlet Mach number, and was obtained by substituting the static pressure ratio for a normal shock wave at the inlet Mach number,[5.1]

$$\frac{p_e}{p_i} = \frac{2\gamma_c}{\gamma_c + 1} M_i^2 - \frac{\gamma_c - 1}{\gamma_c + 1} \tag{5-47}$$

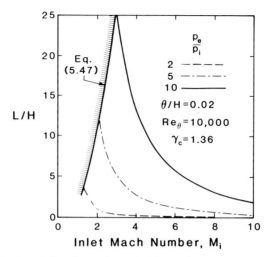

Fig. 5.37   Ratio of shock train axial length to duct height as a function of inlet Mach number and back pressure ratio for an isolator or constant area diffuser with typical inlet conditions.

into Eq. (5-46). It follows that the left-hand boundary is entirely composed of complete shock trains. Conversely, all other points on Fig. 5.37 represent incomplete (or translated) shock trains.

One disconcerting result of this development is that the designer may be required to select a relatively high L/H for the isolator at relatively low supersonic freestream Mach numbers in order to prevent unstart. Then, when the height of the duct is reduced in order to match the freestream capture area at high freestream Mach numbers, the isolator surface area divided by throughflow area may become so large that the internal wall friction losses are unacceptable.

5.7.6 Maximum Contraction Ratio

A final measure of the behavior of compression systems often found in the literature is the maximum practical or "running" value of the freestream area to throat area ratio A_0/A_t or *contraction ratio*. If the compression were entirely isentropic and the flow at the throat were sonic, the maximum contraction ratio would be that given by Eq. (5-42) and shown on Fig. 5.32 as State 5. This value cannot be reached for real inlets because the compression process is neither isentropic nor may the Mach number at the throat be too near to 1.

Once again, the flow is far too complex to provide any simple answers, but a serviceable approximation, based on a great many compression system tests over the range $2 < M_0 < 4$, is that the maximum contraction ratio can be approximated by an isentropic process wherein the freestream flow is compressed to a throat Mach number equal to $M_0/2$.[5.14] This conclusion seems to be rather insensitive to the split between external and internal compression. Beyond this point the total pressure losses of the inlet and/or the possibility of unstart increase precipitously. Applying Eq. (5-41) to this situation shows, for example, that the maximum contraction ratio at $M_0 = 3.00$ is 3.85 and at 4.00 is 7.19, provided that $\gamma_c = 1.36$. These numbers should be compared with those of Fig. 5.32.

Incidentally, the same empirical procedure estimates the total pressure ratio of the compression system as 0.90 of the total pressure ratio of a normal shock wave at a throat Mach number of $M_0/2$, as obtained from Eq. (5-45). Using the compression component performance parameter interrelationships of Table 5.1, this procedure leads to the conclusion that, for example, the adiabatic compression efficiency at $M_0 = 3.00$ is 0.932 and at 4.00 is 0.957, provided that the static temperature ratio ψ corresponds to the flow being brought to rest and that $\gamma_c = 1.36$. These results certainly confirm our adiabatic compression efficiency expectations at the low freestream Mach number end of the scale.

5.8 COMPRESSION COMPONENT CFD EXAMPLES

Thus far the complexities concerning aeropropulsion design and analysis of a compression component have been avoided with several useful approximations. These complexities principally arise because of the following: before the airflow traveling at hypersonic speed enters the burner of an aeropropulsion system, it has to be slowed down to supersonic speed relative to the flight vehicle, with minimum loss of total pressure and within a relatively short distance. This requirement leads to a compression component with mixed compression through mostly oblique shock waves. The interaction among these waves and between the waves and boundary layers produces complex flowfields. Hypersonic boundary layers are relatively thick, owing to high shear stresses that generate high temperatures and low densities. These thick boundary layers significantly affect the inviscid flowfields which in turn influence these layers, causing the hypersonic viscous interaction. Whether the boundary layer entering the internal part of the compression component (referred to here for consistency with the literature as the inlet) is laminar or turbulent is also of significance. Usually, this layer is turbulent. On the cowl surface the flow is laminar. On two vertical surfaces of the inlet, a portion of the flow is turbulent and the rest of the flow is laminar. However, it is conceivable to have fully laminar flow entering the inlet. These laminar flows eventually undergo transition and become turbulent before leaving the inlet. Another complication is the effect of combustion-induced disturbances on the operation of the inlet. These perturbations may cause flow reversal, loss of thrust, inlet unstart, or loss of the propulsion system. Moreover, the physical and chemical states of the flow exiting the compression component and entering the burner determine the performance of the latter.

The only tool that can be used to address all of the complexities of flows in compression components simultaneously is computational fluid dynamics (CFD), provided it is used properly and provided that appropriate allowances are made for its limitations. This power of CFD is due to the following facts: (1) This tool can include simultaneously, at some level, the effects of all known phenomena or the effects of any known phenomenon. (2) Experimental simulations of hypersonic flows cannot include simultaneously all phenomena encountered in flight. These simulations are often limited for determining inlet performance parameters, and they need to be complemented with computational results for computing these parameters. Often the accuracy of the experimentally determined performance parameters, such as air capture ratio and kinetic energy efficiency, is less than that required for design and analysis of inlets for use at high hypersonic Mach numbers.[5.21] Furthermore, the flowfield detail available in CFD results provides an important diagnostic capability for understanding the test data. (3) Computational fluid dynamics tools are the sole source of design and performance data for portions of the flight trajectory that ground tests cannot simulate. Moreover, shock trains in isolators can only be analyzed with CFD. (4) Parametric studies are costly and time-consuming when carried out in test facilities. Computational fluid dynamics tools can compute the complete flowfields and all the performance parameters, and can provide accurate trends when parametric studies are conducted. The purpose of this section is merely to illustrate this power of CFD.

5.8.1 A Study of Sidewall-Compression Inlets

The compression of airflow by compression surfaces takes place in planes normal to these surfaces. The external compression of the flow entering an inlet is achieved essentially in vertical planes by the forebody bow shock and horizontal compression surfaces. In a sidewall-compression inlet consisting of two vertical wedge-shaped surfaces, further compression occurs in horizontal planes, Fig. 5.38. The leading edges of these sidewalls are swept both to reduce the aerothermal loads on these edges and to assist in spilling the flow ahead of the cowl leading edge at low Mach numbers. Primarily, the amount of air captured by the inlet or spilled out depends on the cowl position, the sweep angle of the leading edges of sidewalls, and the freestream Mach number. Unlike the horizontal-wall compression inlets, the sidewall-compression inlets have the potential to provide efficient spillage and good inlet starting characteristics, with their shape and size remaining fixed over a range of Mach numbers. Moreover, the thick boundary layer entering the inlet is less likely to separate because of shock—boundary layer interaction, since further compression of the flow entering the inlet is done in horizontal planes rather than in vertical planes.

The flow in sidewall-compression inlets is fully three-dimensional and more complicated to analyze than that in axisymmetric or horizontal-wall compression inlets. The inlet performance is determined by the amount of air captured or flowing through the inlet and the efficiency of the process used to compress this air. Essentially, this determination requires knowledge of the mass, momentum, and energy of the flow leaving the inlet.

Reference 5.22 reports a computational, parametric study at Mach 10 of the sidewall-compression inlet sketched in Fig. 5.38. This study was used to design a test experiment and to limit the test matrix of this experimental investigation. In this computational-parametric study, two parameters are considered, the leading-edge

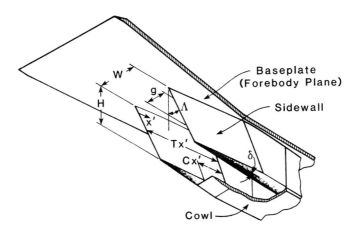

Fig. 5.38 A sketch of sidewall-compression inlet. The inlet side-walls are at a constant height, $H = 7.2$ in (18.3 cm), with a total length of 31 in (78.7 cm). The distance from the sidewall leading edge to the constant-area throat ($T_{x'}$) is 22.8 in (57.9 cm). The side-wall compression angle δ is fixed at 6 deg. The leading-edge sweep varies from 0 to 60 deg. The contraction ratio, that is, the ratio of inlet entrance area to the throat area, is 5. With the inlet height being constant, this ratio is the same as the ratio of the inlet entrance width to the throat gap, W/g. The forebody plate extends 1 in (2.54 cm) ahead of the inlet entrance.

sweep angle and the cowl leading-edge position. The sweep angle is varied between 30 and 60 deg. A no-sweep case is also considered as a reference point. The cowl position is defined by the forward extent of the cowl leading edge ahead of the inlet throat as a percentage of the distance between the constant-area throat and the sidewall leading edge.

The three-dimensional Navier-Stokes equations are solved for laminar flows with a unit freestream Reynolds number of 2.15×10^6/ft $(7.06 \times 10^6$/m$)$. The fluid is assumed to be calorically perfect. The walls are maintained at $T_w = 540$ °R (300 K). The inflow computational boundary is fixed at freestream conditions; that is, no incoming boundary layer is imposed at this computational boundary. Since the flowfield is symmetric, only the flowfield in one half of the inlet is computed. Reference 5.22 reports that increasing the computational grid points by 50 percent in all three spatial directions did not significantly change the integrated performance quantities for the following two extreme cases: 30-deg. sweep with 0 percent cowl position, and 60-deg. sweep with 50 percent cowl position. For example, the momentum-averaged total pressure recovery varied by less than 0.1 percent.

The performance of the inlet is assessed in terms of the percentage mass capture, momentum-averaged exit-throat Mach number, momentum-averaged total pressure recovery, kinetic energy efficiency, area-weighted compression, momentum-averaged static temperature ratio, and adiabatic compression efficiency (Table 5.2). For the selected inlet, the sidewalls could be swept from 0 to 30 deg. without adversely affecting the performance of this inlet when the cowl is positioned at the start of the constant-area throat. The leading-edge sweep promotes the turning of the flow toward the cowl plane. Consequently, the spillage is increased and the mass capture is decreased with increasing sweep angle. For the same sweep angle, the forward motion of the cowl leading edge increases the mass capture. This motion, however, decreases the total pressure recovery for 30-deg. and 45-deg. sweep angles, and increases it for 60-deg. sweep angle. As the leading-edge sweep is increased, the strength of the shock at this edge is decreased, which in turn increases the throat Mach number. Again, a forward cowl position lowers the Mach number because of the shock formed by the cowl leading edge. The kinetic energy efficiency changes in a manner similar to the total pressure recovery. The maximum variation in efficiency is 0.3 percent when the cowl is at 50 percent location. This variation is estimated to change the propulsion-system specific impulse by 1 percent. The area-weighted compression decreases with an increase in leading-edge sweep angle owing to both the weakened swept internal shock waves and the increased spillage. Nevertheless, the adiabatic compression efficiency is almost constant. While no single configuration emerged as clearly optimum, the 45-deg. leading-edge sweep configuration appears to possess the most attractive performance characteristics.[5.22]

Another parametric investigation was also undertaken to understand the effect of Reynolds number on the internal shock structure and on performance parameters.[5.23] The investigation considered an inlet with a contraction ratio of 3, sweep angle of 45 deg., height of 4.0 in (10.2 cm), 9.0 in (22.9 cm) of forebody plate ahead of the entrance plane, and with the cowl leading edge positioned at 0 percent. Two values of Reynolds numbers are considered, one being the same as in the previous study and another being approximately one fourth of that value. Again, the computed results are for laminar flows.

The basic effect of increasing the viscous effects is to effectively increase compression caused by the compressive surfaces, to decrease the mass capture, and to change the internal shock patterns. Table 5.3 summarizes the performance parameters. As can be noticed, there is a significant effect on these parameters.

Table 5.2   Performance parameters for Re = 2.15 x 10^6/ft (7.06 x 10^6/m).

Leading-Edge Sweep (deg.)	% Mass Capture	M_3	π	η_{KE}	p/p_0	T/T_0	η_c
0% Cowl							
0	96.73	5.202	0.443	0.9869	19.91	3.386	0.8902
30	96.05	5.345	0.443	0.9869	18.78	3.249	0.8835
45	94.05	5.449	0.461	0.9876	18.00	3.173	0.8860
60	88.51	5.623	0.499	0.9890	15.66	3.034	0.8920
25% Cowl							
30	98.28	5.227	0.439	0.9867	20.45	3.397	0.8894
45	96.90	5.303	0.455	0.9874	19.95	3.354	0.8928
60	94.03	5.516	0.514	0.9895	18.75	3.201	0.9048
50% Cowl							
30	99.41	5.176	0.431	0.9864	21.12	3.449	0.8890
45	98.26	5.256	0.452	0.9873	20.74	3.411	0.8944
60	96.65	5.507	0.522	0.9898	18.91	3.220	0.9081

Table 5.3   Performance parameters for 0% cowl configuration and 45-deg. leading-edge sweep angle.

Reynolds Number	% Mass Capture	M_3	π	η_{KE}	p/p_0	T/T_0	η_c
0.55×10^6/ft	90.4	5.72	0.448	0.9871	9.70	2.843	0.8601
2.15×10^6/ft	93.4	6.30	0.609	0.9924	9.20	2.621	0.9061

An increase in the boundary-layer thickness causes the movement of the shock impingement on sidewalls forward to the vicinity of the beginning of the constant-area region (Fig. 5.39). This has caused a 3 percent decrease in the mass capture. It is estimated that this decrease corresponds to about 1 percent change in the propulsion-system specific impulse. This change could be significant. The decrease in total pressure recovery, kinetic energy efficiency, and the throat Mach number are explained by the increase in the internal compression resulting from an increase in internal shock strength. At the inlet exit, there are streamwise vortices near the forebody plane and near the cowl surface. The size of these vortices in the crossflow plane depends on the Reynolds number. The axial growth of these vortices is observed in Plate 3 (at end of book). At the entrance of the constant-area section, the differences in the static

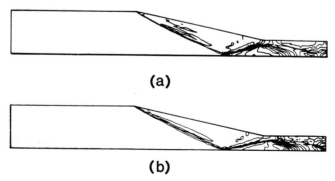

(a)

(b)

Fig. 5.39   Nondimensional static pressure contours in horizontal planes at a vertical length of $H/2$. (a) Re = 0.55 X 10⁶/ft (1.80 X 10⁶/m), and (b) Re = 2.15 X 10⁶/ft (7.05 X10⁶/m) (Ref. 5.24). Only one half of the flow is presented.

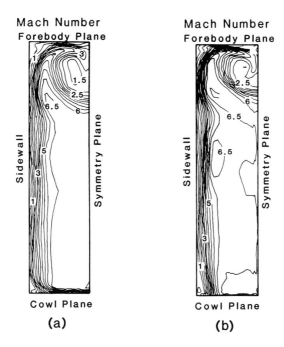

Fig. 5.40   Mach number contours leaving the inlet, (a) *Re* = 0.55 X 10⁶/ft (1.80 X 10⁶/m), and (b) *Re* = 2.15 X 10⁶/ft (7.05 X 10⁶/m) (Ref. 5.24).

pressure contours (Fig. 5.39) owing to the Reynolds number effect are consistent with the differences observed in the oil-flow patterns (Plate 3). The nonuniformity of the flow exiting the inlet, and the streamwise vortices, are evident in the Mach number contours plotted in Fig. 5.40.

5.8.2 Chemical Status of Exit Flow

Chemistry plays an essential role in the operation and performance of aeropropulsion systems, because the chemistry model strongly affects the position and strength of shock waves, the chemical composition of the flow, and the outcome of the chemical reactions. A combination of the following chemical models is used for analyzing the flow through these systems: calorically perfect gas, frozen chemistry, equilibrium, partial equilibrium, finite-rate chemistry, and vibrational excited state. Some chemical reactions are advantageous and some are detrimental. Combustion reactions and recombination reactions release heat that adds to the impulse produced by these systems. Certain detrimental reactions cannot be avoided. These include the dissociation of air on compression surfaces. The question is, which of these chemical models is the appropriate one and for what conditions? Computational fluid dynamics is the only tool that can be used to answer this question and study the role of chemistry in the design of aeropropulsion systems, because the high-temperature effects cannot be accurately simulated in ground-test facilities for speeds (approximately) above Mach 12.

Reference 5.25 discusses the chemistry effects in a generic inlet (Fig. 5.41). Flowfields in this inlet are computed using both equilibrium and finite-rate models at Mach 10 and 13. The upstream conditions are set at $p_0 = 1$ atm and $T_0 = 540$ °R (300 K). Please note that the static pressure is at least two orders of magnitude higher than what it would be for an airbreathing propulsion system at these Mach numbers. No-slip, adiabatic, and noncatalytic conditions are used on inlet walls. Again, note that the adiabatic condition is an idealization that is unlikely to occur in reality.

As the high speed airflow is slowed by the compression process and by the viscous effects, the temperature of the air rises. At both Mach numbers, the equilibrium model predicts a higher level of dissociation of air along walls than those computed with the finite-rate model, because the flow-residence time is too short for the chemical reactions to reach an equilibrium state (Fig. 5.42). The use of the former model leads to conversion of more of the kinetic energy to chemical energy than to thermal energy. Consequently, the static temperature predicted by this model is lower than that predicted by the finite-rate model (Fig. 5.43). Only traces of ions are present at Mach 13 and no ionization is found at Mach 10. This study leads to the conclusion that dissociation effects are significant in the boundary layers of compression components, whereas vibrational effects need to be addressed in the inviscid region. These effects must be appropriately modeled to determine the chemical composition and physical state of the air entering the burner and consequently its performance.

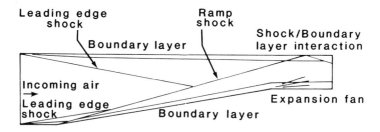

Fig. 5.41 A sketch of physical phenomena in an inlet with a flat-surface cowl and a 10-deg. ramp.

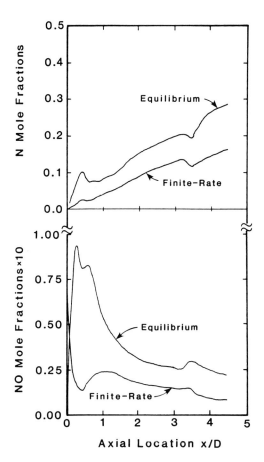

Fig. 5.42   N and NO concentration distributions along the bottom wall, $M_0 = 13$. The horizontal axis represents streamwise locations along the wall normalized by the inlet width.

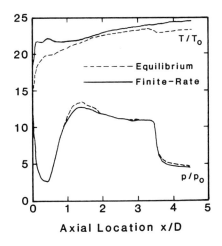

Fig. 5.43   Static temperature and static pressure distributions along the bottom wall, $M_0 = 13$.

5.9 CONCLUSION

Just as the compression system prepares the airflow for the combustion process, the material of this chapter has prepared us for a consideration of the analysis and design of combustors or burners. As we move downstream, we will discover and learn how to deal with the entirely different physical phenomena associated with the mixing and chemical reactions of fuel and air. There is much to look forward to.

REFERENCES

[5.1] Shapiro, A. H., *The Dynamics and Thermodynamics of Compressible Fluid Flow,* Ronald Press, New York, 1953.

[5.2] Anderson, J. D. Jr., *Hypersonic and High Temperature Gas Dynamics,* McGraw-Hill, New York, 1989.

[5.3] Curran, E. T., and Bergsten, M. B., "Inlet Efficiency Parameters for Supersonic Combustion Ramjet Engines," Aero Propulsion Laboratory, APL TDR 64-61, Wright-Patterson AFB, OH, Jun. 1964.

[5.4] Oates, G. C., *The Aerothermodynamics of Gas Turbine and Rocket Propulsion,* Revised and Enlarged Edition, AIAA Education Series, Washington, DC, 1988.

[5.5] Seddon, J., and Goldsmith, E. L., *Intake Aerodynamics,* AIAA Education Series, New York, 1985.

[5.6] Emanuel, G., *Gasdynamics: Theory and Applications,* AIAA Education Series, New York, 1986.

[5.7] NACA Ames Research Staff, "Equations, Tables and Charts for Compressible Flow," NACA Ames Aeronautical Laboratory, NACA Rept. 1135, Moffett Field, CA, 1953.

[5.8] Billig, F. S., Orth, R. C., and Lasky, M., "Effects of Thermal Compression on the Performance Estimates of Hypersonic Ramjets," *Journal of Spacecraft and Rockets,* Vol. 5, No. 9, Sep. 1968.

[5.9] Oswatitsch, K., "Pressure Recovery in Missiles with Reaction Propulsion at High Supersonic Speeds (The Efficiency of Diffusers)," NACA TM 1140, Jun. 1947.

[5.10] Kutschenreuter, P. H. Jr., Balent, R. L., and Richey, G. K., "Design Analysis Summmary of Two Force-Balance Hypersonic Inlet Configurations," Flight Dynamics Laboratory, AFFDL-TDR-64-19, Wright-Patterson AFB, OH, Mar. 1964.

[5.11] Kutschenreuter, P. H. Jr., and Balent, R. L., "Hypersonic Inlet Performance from Direct Force Measurements," *Journal of Spacecraft and Rockets,* Vol. 2, No. 2, Mar. 1965.

[5.12] Balent, R. L., and Stava, D. J., "Test Results of a Supersonic/Hypersonic Combustion Force-Balance Inlet Model," Flight Dynamics Laboratory, AFFDL-TR-66-95, Wright-Patterson AFB, OH, Oct. 1966.

[5.13] Love, E. S., "Pressure-Rise Associated with Shock-Induced Boundary-Layer Separation," NACA TN 3601, Dec. 1955.

[5.14] Mahoney, J. J., *Inlets for Supersonic Missiles*, AIAA Education Series, Washington, DC, 1990.

[5.15] Stockbridge, R. D., "Experimental Investigation of Shock Wave/Boundary-Layer Interactions in an Annular Duct," *Journal of Propulsion and Power*, Vol. 5, No. 3, May 1989.

[5.16] Carroll, B. F., and Dutton, J. C., "Characteristics of Multiple Shock Wave/Turbulent Boundary-Layer Interactions in Rectangular Ducts," *Journal of Propulsion and Power*, Vol. 6, No. 2, Mar. 1990.

[5.17] Lin, P., Rao, G. V. R., and O'Connor, G. M., "Numerical Investigation on Shock Wave/Boundary-Layer Interactions in a Constant Area Diffuser at Mach 3," AIAA Paper 91-1766, Jun. 1991.

[5.18] Lin, P., Rao, G. V. R., and O'Connor, G. M., "Numerical Analysis of Normal Shock Train in a Constant Area Isolator," AIAA Paper 91-2162, Jun. 1991.

[5.19] Waltrup, P. J., and Billig, F. S., "Prediction of Precombustion Wall Pressure Distributions in Scramjet Engines," *Journal of Spacecraft and Rockets*, Vol. 10, No. 9, Sep. 1973.

[5.20] Waltrup, P. J., and Billig, F. S., "The Structure of Shock Waves in Cylindrical Ducts," *AIAA Journal*, Vol. 11, No. 10, Oct. 1973.

[5.21] Van Wie, D., "Techniques for the Measurement of Scramjet Inlet Performance at Hypersonic Speeds," AIAA Paper 92-5104, Dec. 1992.

[5.22] Holland, S, D., "Computational Parametric Study of Sidewall-Compression Scramjet Inlet Performance at Mach 10," NASA TM-4411, Feb. 1993.

[5.23] Holland, S. D., "Reynolds Number and Cowl Position Effects for a Generic Sidewall Compression Scramjet Inlet at Mach 10: A Computational and Experimental Investigation, " AIAA Paper 92-4026, July 1992.

[5.24] Holland, S. D., private communication, 1992.

[5.25] Yu, S.-T., McBride, B. J., Hsieh, K.-C., and Shuen, J.-S., "Numerical Simulation of Hypersonic Inlet Flows with Equilibrium or Finite Rate Chemistry," AIAA Paper 88-0273, Jan. 1988.

[5.26] Billig, F. S., "Research on Supersonic Combustion," *Journal of Propulsion and Power*, Vol. 9, No. 4, 1993.

PROBLEMS

5.1 Derive the following compression component performance interrelationships of Table 5.1, and plot the results in a format similar to Figs. 5.8 and 5.9. You will find that this is most easily accomplished by starting with the interrelationships already developed in Sec. 5.4.

(a) π_c as a function of η_{KE}.

(b) π_c as a function of $(s_3 - s_0)/C_{pc}$.

(c) η_{KE} as a function of $(s_3 - s_0/C_{pc}$.

Which one of these, if any, appears to hold some promise as a simple means for portraying the physical behavior of compression component performance?

5.2 One of the first parameters proposed for use as a compression component performance measure was the *process efficiency*,[5.3] which may be defined using the nomenclature of Fig. 4.2 as

$$K_D \doteq \frac{V_X^2 - V_3^2}{V_0^2 - V_3^2} \leq 1$$

Assuming a constant value of γ_c for the compression process, prove that $K_D = \eta_c$.

5.3 The *static pressure recovery* of a compression system is defined using the nomenclature of Fig. 4.2 as

$$\eta_{pr} \doteq \frac{p_3}{p_{3s}} \leq 1$$

where p_{3s} is the static pressure that would be reached in an isentropic process of the same static temperature ratio ψ as the actual irreversible process.

Assuming a constant value of γ_c for the compression process, prove that $\eta_{pr} = \pi_c$. Why is this true?

5.4 The *polytropic efficiency* is an alternative, and sometimes more intuitively appealing, way to specify compression component performance. Referring to the sketch below, the polytropic efficiency is defined as the adiabatic compression efficiency corresponding to an *infinitesimal* pressure increase at any point along the compression curve of Fig. 4.2, or

$$\varepsilon_c \doteq \frac{dh_s}{dh} = \frac{dT_s}{dT} \leq 1$$

Since ε_c is related to an incremental step, polytropic efficiency is often interpreted as representing a level of technology, and therefore taken to be constant in determining the compression efficiency of the overall compression process.

Assuming constant values of ε_c and γ_c for the compression process and using the usual nomenclature, prove that

$$\eta_c = \frac{\psi^{\varepsilon_c} - 1}{\psi - 1} \le 1$$

Start this derivation by applying the Gibbs equation to the fictitious isentropic process that connects point b to point c in order to find, after dividing by dT, that

$$\frac{dT_s}{dT} = \varepsilon_c = \frac{\gamma_c - 1}{\gamma_c} \cdot \frac{dp/p}{dT/T}$$

Integrate this differential equation and substitute into Eqs. (5-12) and (5-14) in order to obtain the desired result, and then plot η_c as a function of ψ for several constant values of ε_c. Why does η_c increase with ψ for a constant value of ε_c? Why is this the same as the case of compressors for turbojet engines?[5.4]

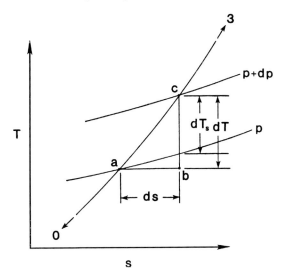

5.5 Use HAP(Gas Tables) to confirm the geometrical turning angles shown on Figs. 5.11 through 5.15, as well as the adiabatic compression efficiencies shown on Fig. 5.16. Create some design point external compression systems of your own and compare the results with those of Figs. 5.11 through 5.15 and Fig. 5.16.

5.6 Convert the information shown on Fig. 5.16 to both η_{KE} and $(s_3 - s_0)/C_{pc}$ as functions of freestream Mach number M_0 using the compression component performance interrelationships of Table 5.1. What conclusions do you draw regarding the behavior of η_{KE} and $(s_3 - s_0)/C_{pc}$ as indicators of compression component performance?

5.7 Use HAP(Gas Tables), HAP(Trajectory), and HAP(Air) to compare the oblique shock wave angle and downstream flow properties for a calorically perfect gas with $\gamma_c = 1.36$ to those of equilibrium air for a typical constant-q_0 trajectory and deflection angles of 5, 10, and 15 deg. How good an approximation is the calorically perfect gas?

5.8 Verify the adiabatic compression efficiency of the single geometric turn, lowest lines of Fig. 5.16 by generating them from Eqs. (5-15) and (5-17). Use this information to plot the geometric turning angle as a function of freestream Mach number M_0 and static temperature ratio ψ.

5.9 Determine for yourself how the performance of normal shock waves compares with that of oblique shock waves by calculating their adiabatic compression efficiencies as functions of freestream Mach number M_0 and θ_0 from Eqs. (5-15) and (5-17).

5.10 Military specification MIL-E-5008B contains the following correlation for the expected inviscid performance of high speed aircraft compression components:

$$\pi_c = \frac{800}{M_0^4 + 935} \qquad (M_0 \geq 5)$$

Compare this correlation with the information presented in Fig. 5.16 by first converting it to adiabatic compression efficiency using the interrelationships of Table 5.1, and then calculating the line that corresponds to $\psi = 7.0$ and $\gamma_c = 1.36$. How do you suppose the correlation was created?

5.11 One of the early methods for estimating viscous compression component performance was[5.3]

$$\eta_{KE} = 0.94 + 0.06 \left(\frac{M_3}{M_0} \right)$$

Compare this correlation with the experiential information of Fig. 5.20 by first converting it to adiabatic compression efficiency using the interrelationships of Table 5.1, and then calculating the line that corresponds to $\psi = 7.0$ and $\gamma_c = 1.36$. Does this information support or oppose our previous conclusions?

5.12   A recent method for estimating viscous compression component performance is[5.26]

$$\frac{A_0}{A_3} = -3.5 + 2.17M_0 - 0.017M_0^2$$

$$\frac{p_3}{p_0} = -8.4 + 3.5M_0 + 0.63M_0^2$$

Assuming that the gas is calorically perfect, use the static pressure mass flow parameter (i.e., Problem 2.12) to show that M_3 can be determined directly from the quadratic equation

$$M_3\sqrt{1 + \frac{\gamma_c - 1}{2}M_3^2} = \frac{p_0}{p_3} \cdot \frac{A_0}{A_3} \cdot M_0\sqrt{1 + \frac{\gamma_c - 1}{2}M_0^2}$$

Calculate the supersonic burner entry value of M_3 for several hypersonic freestream Mach numbers in the range $5 < M_0 < 25$, and use the results to calculate, in turn, ψ from Eq. (5-2), $s_3 - s_0$ from Eq. (5-12), and η_c from Eq. (5-14) in order to demonstrate that this correlation corresponds to an essentially constant compression efficiency $(0.878 < \eta_c < 0.894)$. Use $\gamma_c = 1.36$.

5.13   At hypersonic freestream Mach numbers, Eq. (4-12) shows that

$$\frac{M_3}{M_0} = \sqrt{\frac{T_0}{T_3}} = \sqrt{\frac{1}{\psi}}$$

Combine this result with the correlations of Eqs. (5-27) and (5-29) to demonstrate that these models for η_{KE} do not approach 1 in the limit of high freestream Mach number, and therefore refute Eq. (5-10). How would you reconcile these contradictory conclusions?

5.14   Repeat the development of the influence of boundary-layer friction of Sec. 5.53 using V_1 as the reference value for the skin friction coefficient instead of V_3. How do the results change? Why?

5.15   Introduce the heat addition term $\dot{Q}/\dot{m}_0$ into the analysis of Sec. 5.5.3 via Eq. (5-25) and obtain the corresponding solution for nonadiabatic η_c. How does this result compare with that of the analysis of Sec. 5.5.6?

5.16   Using the model and method of Sec. 5.5.5, derive the expressions necessary to determine the axial force F_{bx} exerted by an isentropic compression system on the oncoming flow with given values of M_0 and ψ. Make a graphical sketch of how F_{bx} varies with M_0 and ψ.

5.17 Using the model and method of Sec. 5.5.5, derive the expressions necessary to determine the axial force F_{bx} exerted by a compression system on the oncoming flow with given values of M_0, and ψ, and η_c. Make a graphical sketch of how F_{bx} varies with η_c.

5.18 Prove that, for flight at a constant geometric altitude, the burner entry static pressure depends on ψ, η_c, and γ_c, but is *independent* of M_0.

5.19 In order to determine the skin friction and/or heat transfer in the burner, it is often useful to have a direct method for calculating the burner entry dynamic pressure $q_3 = \rho_3 V_3^2/2$. For this purpose, show that for constant γ_c and given values of M_0, ψ, and η_c:

$$\frac{q_3}{q_0} = \frac{\rho_3 V_3^2}{2q_0} = \frac{\gamma_c p_3 M_3^2}{2q_0} = \frac{\dfrac{\gamma_c}{\gamma_c - 1}\left\{\psi\left(1 + \dfrac{\gamma_c - 1}{2}M_0^2\right) - 1\right\}}{\dfrac{\gamma_0 M_0^2}{2}\left\{1 - \eta_c\left(1 - \dfrac{1}{\psi}\right)\right\}^{\gamma_c/(\gamma_c-1)}}$$

This can be most conveniently accomplished by using Eq. (5-2) to specify M_3 and Eq. (5-34) to specify p_3. Develop a feeling for the behavior of the burner entry dynamic pressure by reproducing Figs. 5.24 through 5.26 with q_3 as the dependent parameter instead of p_3.

5.20 There are several popular shock—boundary layer separation correlations that are based upon the maximum allowable dimensionless pressure coefficient as defined by Eq. (5-39). Two of them for the turbulent boundary layer case are:

(a) (Gadd, 1953[5.5])

$$\frac{p_d - p_u}{q_u} = \frac{2}{\gamma_c M_u^2}\left\{\left(\frac{1 + \dfrac{\gamma_c - 1}{2}M_u^2}{1 + 0.64\dfrac{\gamma_c - 1}{2}M_u^2}\right)^{\gamma_c/(\gamma_c-1)} - 1\right\}$$

(b) (Reshotko-Tucker, 1955[5.13])

$$\frac{p_d - p_u}{q_u} = \frac{3.2}{8 + (M_u - 1)^2}$$

Compare these with the simple turbulent shock—boundary layer separation criterion of Eq. (5-36) by first combining the correlations above with Eq. (5-39) in order to find the maximum allowable p_d/p_u

for a given M_u, then employing Eq. (5-37) to find the corresponding minimum allowable value of M_d/M_u, and finally plotting the latter as a function of M_u. Is our initial constant M_d/M_u criterion optimistic or pessimistic in comparison to these two correlations?

5.21 Show that the maximum possible value of $A_0/A_t = A_1/A_t$ for which an inlet can be started is given by the expression

$$\frac{A_0}{A_t} = \sqrt{\frac{2\gamma_c}{\gamma_c - 1}} \left(\frac{\gamma_c + 1}{2\gamma_c}\right)^{(\gamma_c+1)/[2(\gamma_c-1)]}$$

This happens because M_s asymptotically approaches a finite lower limit rather than 0 as the freestream Mach number becomes indefinitely large, as given by Eq. (5-43). Substitute the latter result into Eq. (5-44) in order to obtain the equation above. Use this equation to calculate the limiting value of A_0/A_t assuming that $\gamma_c = 1.36$ for comparison with Fig. 5.32.

5.22 The analyses of the inlet flow starting process can also be entirely carried out on the basis of the behavior of total pressure. Even though the calculations tend to be more burdensome, many people find this a more physically intuitive and therefore more appealing approach. As an example, consider the positioning of the external normal shock wave as a function of the freestream Mach number in State 3.

For this situation, the appropriate form of the mass flow parameter relationship is

$$\frac{A_0}{A_t} = \frac{MFP_t \cdot p_{tt}}{MFP_0 \cdot p_{to}} = \frac{MFP(M = 1)}{MFP(M = M_0)} \cdot \pi_s(M = M_0)$$

As the freestream Mach number increases, MFP_t/MFP_0 increases in accordance with Eq. (5-42), while π_s decreases in accordance with Eq. (5-45) (with M_s replaced by M_0). The competition between these conflicting trends must be settled by calculations.

Use this method to reproduce the State 3 line of Fig. 5.32.

5.23 A started inlet is operating supercritically with a normal shock wave located downstream of the throat. The Mach number downstream of the normal shock wave is 0.80, the Mach number at the throat is 1.15, and the ratio of the inlet face area to the throughflow area at the location of the normal shock wave is 1.40. Use the logic and relationships of Sec. 5.7.3 to find:

(a) The Mach number upstream of the normal shock wave.

(b) The ratio of the throughflow area at the throat to that at the location of the normal shock wave.

(c) The existing freestream Mach number.

(d) The starting freestream Mach number.

(e) The minimum possible unstart Mach number.

5.24 An isolator with an inlet Mach number of 5.0 is to be designed to accommodate a shock train with a maximum pressure ratio equal to some fraction of the amount that would be caused by a normal shock wave. If $\theta/H = 0.04$, $Re_\theta = 20{,}000$, and $\gamma_c = 1.36$, calculate and plot the required isolator L/H as a function of the expected pressure ratio.

5.25 An isolator is found with an L/H of 15.0. Assuming that $\theta/H = 0.02$, $Re_\theta = 16{,}000$, and $\gamma_c = 1.36$, find the minimum inlet Mach number for which it could have been designed.

5.26 Using the shock—boundary layer separation criteria of Eq. (5-35) or (5-36) in combination with Eq. (5-43), show that the minimum isolator inlet Mach number that will cause a normal shock wave to separate the wall boundary layer and initiate the shock train is given by the expression

$$M_0^2 = \frac{\frac{\gamma_c - 1}{2}(c^2 + 1) + \sqrt{\left[\frac{\gamma_c - 1}{2}(c^2 + 1)\right]^2 + 4\gamma_c c^2}}{2\gamma_c c^2}$$

where c is the constant of Eq. (5-35) or (5-36). Calculate the inlet Mach numbers corresponding to laminar and turbulent shock—boundary layer separation for $\gamma_c = 1.36$. What do these results tell you about the likelihood of shock trains appearing?

5.27 Use the criteria of Sec. 5.7.6 to determine the entire "running" line of practical inlets and plot it against the information of Fig. 5.32.

5.28 It is often convenient to have available a direct relationship between contraction ratio and compression system performance. Assuming calorically perfect gas behavior, and taking advantage of Eqs. (5-2) and (5-4), show that

$$\frac{A_0}{A_3} = \frac{\rho_3 V_3}{\rho_0 V_0} = \frac{\sqrt{1 + \frac{2(1 - \psi)}{(\gamma_c - 1)M_0^2}}}{\psi\left\{1 - \eta_c\left(1 - \frac{1}{\psi}\right)\right\}^{\gamma_c/(\gamma_c - 1)}}$$

Develop a feeling for the behavior of the contraction ratio by plotting its behavior as a function of freestream Mach number for several combinations of ψ and η_c with $\gamma_c = 1.36$.

6
COMBUSTION SYSTEM
PROCESSES AND COMPONENTS

6.1 INTRODUCTION

The practical problems of employing supersonic combustion are very great: It is necessary to capture a stream tube of supersonic air, inject fuel, achieve a fairly uniform mixture of fuel and air, and carry out the combustion process—all within a reasonable length, and preferably without causing a normal shock within the engine. There is currently no conclusive evidence that these requirements can be met: nevertheless, the present study starts with the basic assumption that stable supersonic combustion in an engine is possible.

> — *Richard J. Weber and John S. McKay, "Analysis of Ramjet Engines Using Supersonic Combustion," NACA TN 4386, NACA Lewis Flight Propulsion Laboratory, Cleveland, OH, Sep. 1958.*[6.1]

In the 30-plus years following the study of Weber and McKay, the requirements of the scramjet burner have been more clearly delineated, as has been shown in the preceding chapters. Airframe structural and heat transfer limitations constrain flight Mach numbers to specific altitudes and corresponding freestream conditions. Cycle efficiency considerations, together with temperature limitations imposed by materials and combustion product gas dissociation, dictate the combustion system entry Mach number and thermodynamic state. As a result, the entry conditions to the combustion system are by now very well defined.

However, the reader may be surprised to learn that there are still serious questions as to whether or not stable supersonic combustion is possible over the required range of burner entry conditions. Our present state of knowledge regarding the design and performance of supersonic combustion systems may be summarized by borrowing a pithy quote from Enrico Fermi, "We are still confused, but at a higher level."

In this chapter, we will first study the fluid mechanics of mixing of fuel with air to flammable proportions, next the chemistry of exothermic chemical reaction between fuel and air after they are mixed, and finally the *aerothermochemistry* (the coupling of finite-rate processes of mixing and chemical kinetics of the combustion process with one-dimensional gasdynamics) of the supersonic flow within the burner.

Along the way, we will look at some typical hardware components and engineering techniques currently being employed to achieve the desired performance.

6.1.1 Combustion Stoichiometry

As background to the main topics of this chapter, it will be helpful to first review and expand on relevant topics from Chaps. 2 and 3.

The stoichiometric or *ideal* fuel/air mixture ratio, f_{st}, introduced in Sec. 3.2.4.1, is of interest because that is the fuel/air ratio which usually results in the greatest liberation of sensible energy from the breaking of molecular bonds. Although there exist general rules of stoichiometry for combustion of arbitrary reactants, we will restrict our attention to fuels of interest for the propulsion of hypersonic aircraft, namely those hydrocarbon fuels listed in Table 3.1: hydrogen H_2, methane CH_4, ethane C_2H_6, hexane C_6H_{14}, and octane C_8H_{18}. The only oxidizer of interest is air, which we assume to be 21 percent oxygen and 79 percent nitrogen by volume, as stated in Sec. 2.5.2.

As stated in Sec. 3.2.4.1, the maximum combustion temperature occurs when hydrocarbon fuel molecules are mixed with just enough air so that all of the hydrogen atoms form water vapor H_2O, and all of the carbon atoms form carbon dioxide CO_2. This particular mixture of fuel and air is represented by a general chemical equation for complete combustion called the *stoichiometric equation*, given by Eq. (3-5) and repeated here for convenience,

$$C_x H_y + \left(x + \frac{y}{4} \right) \left(O_2 + \frac{79}{21} N_2 \right)$$

$$\rightarrow x CO_2 + \frac{y}{2} H_2O + \frac{79}{21} \left(x + \frac{y}{4} \right) N_2 \qquad (3\text{-}5)$$

Note that in Eq. (3-5), the nitrogen N_2 acts merely as an inert *diluent*, absorbing some of the sensible thermal energy released by combustion by virtue of its specific heat capacity.

The stoichiometric fuel/air ratio can readily be determined from the ratio of molar coefficients of the reactants appearing on the left-hand side of Eq. (3-5). The stoichiometric fuel/air ratio expressed as a volume or mole ratio is

$$f_{st} = \frac{1}{\left(x + \frac{y}{4} \right)\left(1 + \frac{79}{21} \right)} = \frac{84}{100(4x + y)} \qquad \text{mols F/mol A} \qquad (6\text{-}1)$$

and the stoichiometric fuel/air ratio on a mass basis is given by Eq. (3-6) from Sec. 3.2.4.1,

$$f_{st} = \frac{36x + 3y}{103(4x + y)} \quad \text{lbm F/lbm A} \tag{3-6}$$

Of course, the units of Eq. (6-1) can be in either lbm-mols or kg-mols, and the units of Eq. (3-6) can be kg F/ kg A as well as lbm F/ lbm A.

When considering off-stoichiometric mixtures of fuel and air, it is conventional to speak of *fuel-rich* and *fuel-lean* mixtures. To quantify this, we define the *fuel/air equivalence ratio*, or simply the *equivalence ratio*, as the ratio of the actual fuel/air ratio to the stoichiometric fuel/air ratio:

$$\phi \equiv \frac{f}{f_{st}} \tag{6-2}$$

The utility of the equivalence ratio ϕ is that it permits us to express either fuel-rich or fuel-lean mixtures by simply multiplying the fuel term in Eq. (3-5) by ϕ:

$$\phi C_x H_y + \left(x + \frac{y}{4} \right) \left(O_2 + \frac{79}{21} N_2 \right) \rightarrow \cdots \text{ products} \tag{6-3}$$

The *complete combustion* assumption behind Eq. (3-5) does not imply that, in actual practice, a stoichiometric mixture of fuel and air will yield only CO_2 and H_2O as combustion products. In reality, the CO_2 and H_2O molecules will dissociate into other molecular fragments at elevated temperature, just as is the case with air. Further, as will be shown presently, it is all too common to have incomplete combustion occur in the very short flow time available in a scramjet combustor, so that the actual gases leaving the combustor may be a mixture of reactants (fuel plus air), dissociated products, and incompletely oxidized fuel molecules. In addition, at elevated temperatures, a very small fraction of the atmospheric nitrogen is in fact oxidized, forming the air pollutant gases nitric oxide NO, nitrous oxide N_2O, and nitrogen dioxide NO_2. These oxidized nitrogen species are collectively referred to as "NO_x."

Finally, off-stoichiometric fuel/air ratios, as characterized by the equivalence ratio, affect the type and distribution of combustion products. As a general guideline, equivalence ratios must be in the range 0.2 to 2 for combustion to occur within a useful timescale. For off-stoichiometric mixtures, as well as for possibly incomplete combustion, Eqs. (3-5) and (6-3) may be generalized to be written as

$$\phi C_x H_y + \left(x + \frac{y}{4}\right)\left(O_2 + \frac{79}{21}N_2\right) \to n_{CO_2}CO_2 + n_{CO} CO + n_{H_2O} H_2O + \cdots$$

$$\cdots + n_{O_2} O_2 + n_O O + n_{NO_2} NO_2 + n_{N_2O} N_2O + \cdots \text{ etc. } (6\text{-}4)$$

where "etc." indicates that the list of possible product gases may be as many combinations of O, H, C, and N atoms as exist in nature. If we denote by "NS" the total number of species that may appear on the right-hand side of Eq. (6-4), the messy right-hand side of Eq. (6-4) can be represented by the notation

$$\phi C_x H_y + \left(x + \frac{y}{4}\right)\left(O_2 + \frac{79}{21}N_2\right) \to \sum_{i=1}^{NS} n_i A_i \qquad (6\text{-}5)$$

where A_i represents the chemical formula of the i-th gas molecule appearing in the NS product gases.

A method for finding the composition of post-combustion gases, as characterized by the set of mole numbers n_i in Eq. (6-5), will be presented later in this chapter, in Sec. 6.3. However, before we consider the chemistry of combustion, we will consider the problem of how the fuel and air can be mixed to suitable equivalence ratios so that combustion can occur.

6.2 FUEL-AIR MIXING

6.2.1 Basic Concepts and Definitions

Gas-phase chemical reactions occur by the exchange of atoms between molecules as a result of molecular collisions. Consequently, fuel and air must be mixed to near-stoichiometric proportions *at the molecular level* before combustion can take place.

To illustrate this point, consider two fluids which are mixed to a very fine macroscopic scale, but still not to the molecular level. Such a mixture is called a *bulk mixture, segregated mixture, macromixture,* or *emulsion.*[6.2] An example is homogenized milk, in which the immiscible fat particles have been shattered by ultrasound to such small size that they are suspended in the milk, giving the outward appearance of being fully mixed. However, we know that molecular mixing has not occurred because, if the milk were left to stand for a long time, the fat particles would eventually rise to the surface and coalesce, forming a layer of cream. The average distance between dispersed fat particles is called the *scale of segregation* of the emulsion, or macromixture.[6.2]

In contrast to a macromixture or emulsion, a mixture which is homogeneous at the molecular level is called a *micromixture.* By definition, the scale of segregation of a micromixture is zero.

To demonstrate these and some other ideas, consider the following "kitchen physics" experiment. Start with a cup of black coffee and a tablespoon of liquid or powdered coffee creamer. The coffee and the creamer are the *mixants.* Carefully place the creamer from the tablespoon onto the coffee surface, taking care to disturb the surface of the coffee as little as possible. Observe how slowly the creamer diffuses into the coffee at the *mixant interface,* leaving the bulk of the coffee and the creamer still largely segregated. Next, take a coffee stirrer or pencil, put it into the coffee near the edge of the cup, and draw it quickly through the creamer just once, stopping the motion of the stirring rod at the other side of the cup. Observe how the motion of the stirring rod induced a pair of counter-rotating vortices, which caused regions of coffee and creamer to be drawn into each other, giving the appearance somewhat of a "jelly roll." In the lexicon of mixing, the action of the stirring rod is called *stirring, bulk mixing,* or *macromixing.* The immediate effect of stirring is both to reduce the *scale of segregation* and to increase the area of the mixant interface, so that the rate of *micromixing* is increased.

Note the cause-and-effect sequence of events: first, the input of mechanical work to the coffee/creamer system by motion of the stirring rod (macromixing), followed by molecular diffusion (micromixing.) This sequence of events characterizes all mixing processes. In steady-state mixing, the initial stirring or macromixing phase is called *near-field mixing,* and the subsequent molecular diffusion or micromixing phase is called *far-field mixing.*

A common mistake in the design of mixing systems is to assume that stirring a fluid automatically causes mixing to occur. However, if two initially segregated mixants such as fuel and air are not present in the region where stirring is made to take place, no mixing whatever results, in spite of the fluid power expended. Swithenbank[6.16] warns against this wasteful expenditure of stirring power in "mixing like with like." This subtle but critically important concept should be borne in mind as we next consider the development of mixing layers in steady flow, with and without stirring imposed by a shear layer.

6.2.2 Fuel-Air Mixing in Parallel Streams

Consider two uniform, parallel flowing streams of air and gaseous fuel issuing continuously from a splitter plate in a constant area, steady-flow mixer, as illustrated in Fig. 6.1. Immediately downstream of the splitter plate, the respective axial velocities are u_1 and u_2, as shown, and the initial scale of segregation is b, the lateral dimension

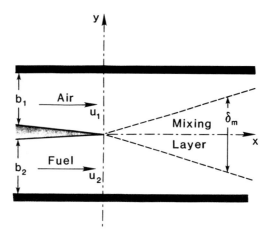

Fig. 6.1   Mixing of parallel streams of air and gaseous fuel in a constant-area duct.

of each *pure* or unmixed stream. For clarity, it is assumed that the pressures and densities of both streams are equal and constant, and that boundary layers on the splitter plate and duct walls can be ignored. The mole fractions of air and fuel in the entry airstream are $(Y_A)_1 = 1.0$ and $(Y_F)_1 = 0.0$, and in the entry fuel stream, $(Y_A)_2 = 0.0$ and $(Y_F)_2 = 1.0$, respectively.

If the two velocities differ, for example if $u_1 > u_2$, a *shear layer* is generated at the interface between the two streams, in which momentum is transported laterally from the faster to the slower stream. A shear layer is merely a special name given to a boundary layer generated when two streams shear against each other, rather than against a wall or bounding surface.

In boundary-layer analysis, velocity profiles are nondimensionalized with respect to the local freestream velocity. Since a shear layer has two, different freestream velocities, it is convenient to nondimensionalize local velocities by some average reference velocity, called the *convective velocity* u_c. One obvious choice is the simple average or mean velocity, $u_c = 0.5(u_1 + u_2)$.

Not only are vorticity and momentum transported laterally within a shear layer, but thermal and mechanical energy as well as mass (molecules) may also be transported laterally as well. If the two streams have different molecular identities, as for example air and fuel as shown, the shear layer is also a *mixing layer,* as denoted by the dashed curves on Fig. 6.1. By analogy with the definition of boundary-layer thickness, the *mixing layer thickness* δ_m is defined as the region within which the mole fractions of air and of fuel differ by one percent or more from their respective values in the unmixed streams.

6.2.2.1 Zero-shear mixing layer.   If the two velocities are equal, no shear stress exists between the two streams, and they simply coflow downstream at a convective velocity $u_c = u_1 = u_2$. Even though there is now no lateral transport of either momentum or vorticity, there is still lateral mass transport due to molecular diffusion at the fuel-air interface. However, note that as soon as diffusion begins to occur, the mixant interface begins to smear out, so that a well-defined mixant interface ceases to exist.

The local rate of molecular diffusion is given by *Fick's law*,[6.3] which states that the time rate of molecular transport of air into fuel (and fuel into air) is proportional to the product of the *interfacial area* and the local *concentration gradient*. The proportionality constant is a molecular property called the *molecular diffusivity*, D_{FA}. The product ρD_{FA} is approximately equal to the molecular viscosity μ for most gases,[6.3] and also like μ, varies approximately with the square root of the absolute temperature, as in Eq. (2-4). The ratio of these two values is called the *Schmidt number*, $S_C \equiv \mu/\rho D_{FA}$.

In equation form, Fick's law for diffusion of air into fuel may be written as

$$j_A = -D_{FA} \cdot \frac{\partial C_A}{\partial y} \tag{6-6}$$

where j_A is the *net molar diffusive flux* (molar flow rate per unit area) of air (lbmolA/ft^2·s, kmolA/m^2·s) in the y direction, C_A is the *concentration* of air (lbmolA/ft^3, kmolA/m^3), and $\partial C_A/\partial y$ is the lateral *concentration gradient*.

The air mole fraction is related to the concentrations of fuel and air by

$$Y_A = \frac{C_A}{C_A + C_F} \tag{6-7}$$

As a result of diffusion of fuel into air (and air into fuel), the mixing layer thickness δ_m grows with downstream distance x approximately as[6.4]

$$\delta_m \approx 8\sqrt{\frac{D_{FA}x}{u_c}} \tag{6-8}$$

and the spatial profile of air mole fraction varies in the x and y directions approximately as[6.4]

$$Y_A = \frac{1}{2}\left\{1 + erf\left(\frac{4y}{\delta_m}\right)\right\} \tag{6-9}$$

where *erf(x)* is the *error function*,[6.5]

$$erf(z) \equiv \frac{2}{\sqrt{\pi}} \int_0^z exp\left(-t^2\right) dt$$

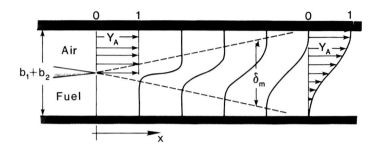

Fig. 6.2 Axial development of cross-stream profiles of air mole fraction Y_A in a zero-shear ($u_1 = u_2$) laminar mixing layer. (Fuel mole fraction profile is $Y_F = 1 - Y_A$.)

The mole fraction profile $Y_A(x, y)$ is illustrated in Fig. 6.2. The fuel mole fraction Y_F profile is the mirror image of the Y_A profile about the x axis: $Y_F = 1 - Y_A$.

It is apparent from Fig. 6.2 that the maximum concentration gradient occurs at $y = 0$. By differentiating Eq. (6-9) and setting $y = 0.0$, the maximum air mole fraction gradient (which is proportional to the air concentration gradient) is shown to change with axial distance as

$$\left. \frac{\partial Y_A}{\partial y} \right)_{y=0} = \frac{4}{\sqrt{\pi}\delta_m} = \frac{1.772}{\delta_m}, \quad \delta_m > 0 \qquad (6\text{-}10)$$

Equations (6-8) and (6-10) show that the maximum mixing rate at any axial location decreases inversely with the square root of x. The greatest concentration gradient, and therefore the greatest diffusive flux, occurs immediately downstream of the splitter plate.

It is evident from Fig. 6.2 that, even when the mixing layer reaches the wall at $x = L_m$, the air mole fraction Y_A still varies from 1.0 at $y = b_1$ to 0.0 at $y = -b_2$. Even though no *pure* fluid exists in the duct beyond $x = L_m$, it is clear that considerable mixing still remains to be accomplished. While the steepest concentration gradient still occurs at $y = 0$, a considerable additional downstream distance is required, perhaps as much as another L_m, to approach complete micromixing.

The total distance L_m required for the mixing layer boundary to reach the walls may be estimated from Eq. (6-8), by setting δ_m equal to $2b$ and solving for $x = L_m$:

$$L_m = \frac{u_c b^2}{16 D_{FA}} \qquad (6\text{-}11)$$

Equation (6-11) can be used to estimate how small the fuel injector height b would have to be in order to reduce L_m to some reasonable distance, say $L_m = 6$ ft (1.8 m). For a flight Mach number $M_0 = 10$, we will use values for inlet compression efficiency and burner entry conditions from the composite scramjet example case of Sec. 4.4.4. Since we assume $u_c = u_1 = u_2$, then the entry fuel and air velocities are 8290 ft/s (2530 m/s). We will assume that $Sc \sim 1$, so that ρD_{FA} is approximately equal to μ determined from Eq. (2-4) at the burner entry static temperature 2800 °R (1556 K) and pressure 4060 lbf/ft² (1.85 $\times 10^5$ N/m²). From Eq. (6-11), the required entry scale of segregation (inlet fuel jet dimension) b is estimated as $b = 0.05$ in (1.2 mm), or $L_m/b \sim 1440$. As mentioned in the preceding paragraph, it may be necessary to double this estimate, in order to allow sufficient convective time for all the lateral concentration gradients to mix out.

From Eq. (6-11), it may be seen that if L_m is to be made as short as possible, it is necessary to decrease u_c, decrease b, increase D_{FA}, or a combination of all three. (Since L_m is also equal to the interfacial area separating the fuel and air streams, the interfacial area cannot be controlled independently.)

As the velocity and temperature of the fuel and air streams are dictated by processes occurring upstream of the burner, there is no chance that we can increase the gas temperature and thereby D_{FA}, or decrease the convective velocity u_c. We can reduce b, the scale of segregation at burner entry, by *manifolding* the fuel and air streams at entry, as illustrated in Fig. 6.3.

For a typical two-dimensional duct burner, the ratio of burner length L to entry duct height H, which we might call the *burner aspect ratio*, is about $L/H \sim 10$. Without manifolding, the inlet scale of segregation b would be equal to the duct height H, so that the burner *mixing aspect ratio* L/b (burner length divided by inlet scale of segregation b) would be simply $L/b = L/H$. The purpose of

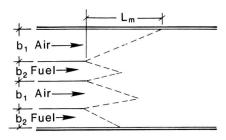

Fig. 6.3 Entry manifolding of fuel and air to shorten the required length to mix, L_m.

manifolding is to reduce b by subdividing the inlet duct height, as
for example $b = H/2$ as shown in Fig. 6.3.

Entry manifolding requires placing struts with internal fuel passages and nozzles into the inlet airstream. As we are dealing with
gas phase reactants, fuel struts spaced only 0.05 in (1.2 mm) apart
would cause far too much blockage of the entry airflow, which would
result in increased internal skin friction and shock wave drag leading
to an unacceptable decrease in overall cycle efficiency. As a practical
matter, these considerations limit the the maximum permitted mixing aspect ratio $(L/b)_{max}$ to about 20, compared to the estimated
value of $L_m/b \sim 1440$ required to achieve complete micromixing in
a zero-shear mixing layer. Clearly, molecular diffusion alone cannot
meet the requirement of rapid lateral mixing in a supersonic flow.

Some possible ways to increase the local rate of mixing are to
somehow increase the mixant interface area, and to increase or
"steepen" the concentration gradients. The obvious (and conventional) way to steepen the concentration gradients is to cause a shear
layer to develop between the two streams, anticipating that the added
lateral transport of x momentum between the two mixant streams
will enhance the growth rate of the mixing layer.

6.2.2.2 Laminar shear/mixing layer. We now consider the case $u_1 >
u_2$, with the convective velocity assumed to be the average of the
two stream velocities, $u_c = 0.5(u_1 + u_2)$. (Later, we will see that
differences in mass density between the two streams require that this
choice be modified.)

Defining the *velocity ratio* $r \equiv u_2/u_1$ and the *velocity difference*
$\Delta u \equiv u_1 - u_2$, the reader can (and should) verify that they are related
to the mean convective velocity $u_c = (u_1 + u_2)/2$ by

$$\Delta u = 2u_c \left(\frac{1-r}{1+r} \right) \qquad (6\text{-}12)$$

Much as with the the mole fraction profiles in the no-shear mixing
layer, the velocity profile in the laminar shear layer is given by[6.6]

$$\frac{u}{u_c} = 1 + \left(\frac{1-r}{1+r} \right) erf\left(\frac{4y}{\delta} \right) \qquad (6\text{-}13)$$

where δ is the local shear layer width which, for laminar flow, is
found to grow with x in the same manner as the zero-shear, diffusive
mixing layer,

$$\delta = 8\sqrt{\frac{\nu x}{u_c}} = 8\sqrt{\frac{2\nu x}{\Delta u} \left(\frac{1-r}{1+r} \right)} \qquad (6\text{-}14)$$

where ν is the *kinematic viscosity*, $\nu \equiv \mu/\rho$.

Since the Schmidt number is approximately unity in gas flows (that is, $\rho D_{FA} \simeq \mu$),[6.3] a negligible increase in growth rate for δ_m is observed in a laminar shear layer. We shouldn't be too surprised at this outcome, however, since in laminar flows, *all* lateral transport is by molecular processes. In order to significantly increase the lateral transport in the shear layer, it is necessary to increase Δu until the flow in the shear/mixing layer is no longer laminar. Again, this is consistent with what we know from boundary-layer theory, namely that "energizing" the boundary layer to maintain flow attachment in the presence of adverse pressure gradients requires either that a laminar flow be tripped to induce turbulence, or that a vortex generator be installed on the wall to reach out into the local freestream in order to "grab" some high momentum freestream fluid and draw it into the boundary layer.[6.6]

6.2.2.3 Turbulent shear/mixing layer.   As we further increase the velocity difference Δu between the two streams, the flow at any downstream station eventually undergoes transition from laminar flow. When this happens, a dramatic change in the structure of the shear layer results: the previously time-steady shear layer becomes unstable, and large vortices are periodically formed between the two streams. The result of this *Kelvin-Helmholtz instability* can be seen in the formation of large ocean waves as a result of sustained wind shear at the air-water interface. These large vortex structures act somewhat like "roller bearings" to accommodate the imposed velocity difference. This phenomenon is illustrated schematically in Fig. 6.4. Since this complex, nonsteady flow is quasiperiodic in nature, it is very difficult to represent by instantaneous streamlines or path lines. Consequently, schematics such as Fig. 6.4 are sometimes referred to by researchers as "cartoons."

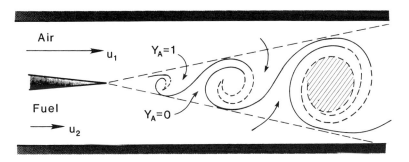

Fig. 6.4 Formation of vortex structures in a transitional shear layer, for $u_1 > u_2$. Dashed curves at mixant boundaries indicate molecular diffusion. Crosshatched area represents fully micromixed region.

Now we're getting somewhere! As a result of the formation of vortices, higher speed fluid rolls up or *entrains* the slower fluid into large-scale "jelly-roll" structures, which are periodically formed downstream of the splitter plate. The net effect, within each large-scale vortex, is to stretch the interface between the unmixed fluids. This stretching not only increases the interfacial area, but also locally steepens the concentration gradients and simultaneously reduces the local scale of segregation—here, the thickness of the "jelly-roll" layers within the vortex structure. Fortuitously, *all* of these effects act to reduce the convective time and axial distance required for mixing to be completed.[6,7]

The crosshatched area shown in the third vortex structure in Fig. 6.4 represents the fully micromixed region. Of course, molecular diffusion occurs continuously at the fuel-air interface immediately after the splitter plate, and the fully micromixed end state is simply the cumulative result of interfacial diffusion.

Note that if the fluids in the two streams were immiscible, the jelly roll structures would still form, causing the interface to stretch. The interface area would still grow, and the scale of segregation would still decrease, just as described previously. However, since the two fluids are assumed immiscible, their bimolecular diffusivity D_{12} is zero, and therefore *micromixing* cannot occur. Interestingly, the mixant interface would remain well defined in spite of the continuous reduction in scale of segregation. (In immiscible liquid flows, surface tension at the interface would cause the interface to break up into spherical droplets. Since the gaseous phase is conventionally defined by the absence of such interfaces, a *gaseous emulsion* is a contradiction in terms and cannot exist at all in nature. We invoke the concept here merely as a limiting condition in order to emphasize the role of molecular diffusion in mixing.)

The sequential events which ensue following the formation of the large vortices act in concert to accelerate the end-state, diffusive micromixing by enhancing various terms in Fick's law. Those steps are, in order of occurrence:

1. Shear stress, which arises because $u_1 > u_2$, causes the periodic formation of large vortices.

2. The vortex sheet between the two streams rolls up and engulfs fluid from both streams, and stretches the mixant interface.

3. Stretching of the mixant interface increases the interfacial area and simultaneously steepens the local concentration gradients along the entire interface.

4. Molecular diffusion occurs at the stretched mixant interface, causing the interface to smear out and eventually disappear in the fully micromixed state.

The preceding four steps have been purposely oversimplified, to illustrate both the importance and the sequential nature of each process step in getting from an initially segregated macromixture to a fully micromixed state, so that the fuel and air molecules can react. In fact, when the two coflowing streams are *very* strongly sheared (that is, $u_1 \gg u_2$), the large-scale vortices formed at the splitter plate immediately begin to break down into smaller-scale vortices, in parallel with Steps 3 and 4 above, thus further reducing L_m, the length required for micromixing. This apparently chaotic distribution of vortex sizes is what is usually referred to as *turbulence*. In addition to becoming turbulent, the large vortex structures are often observed to *pair*, a process in which two vortices are mutually drawn into each other's structure to form a single, larger vortex.

Of course, no gain is without cost. While we may continue to increase the strength of the shear layer in order to speed up the entrainment-stretching-mixing process, we must remember that a turbulent shear layer or boundary layer transports momentum and vorticity as well as molecules of fuel and air. When this occurs in a boundary layer, the result appears as increased drag on the walls. However, inside a shear/mixing layer, where there are no walls to exhibit drag, the result is simply increased viscous dissipation of mechanical energy to thermal energy, with accompanying irreversible entropy increase, total pressure loss, and ultimately decreased cycle efficiency. In addition, as we will see presently, continued increase of the velocity difference Δu (thereby reducing the velocity ratio $r = u_2/u_1$) in order to strengthen or *pump* the shear layer soon leads to compressibility effects which have a very negative impact on the shear layer growth rate.

6.2.3 Quantitative Measures of Local "Goodness" of Mixing

The presentation so far has focused on a Lagrangian or *following-the-motion* description of the formation, engulfment, and diffusive processes occurring within each vortex structure. However, all of the instruments employed for diagnostic analysis or process monitoring are fixed in laboratory or device coordinates. Consequently, it is necessary to focus our attention on the Eulerian or *fixed-in-space* perspective, and to see how measurements obtained by fixed-in-space instruments can be interpreted in terms of the following-the-motion structure of the flow.

Imagine that a gas sampling probe is inserted into the shear layer of Fig. 6.4, at a fixed point $(x, y) = (4b, 0)$, approximately in the center of the crosshatched region which just happens to be at the $(4b, 0)$ location at the instant represented by the *snapshot* or *freeze-frame* of Fig. 6.4. As the flow is nonsteady in nature, we further assume that

the sampling probe has sufficiently fast transient response characteristics to track the rapid changes in concentration as the partially
segregated jelly-roll structures wash past the fixed sampling site.

The instantaneous values of air mole fraction Y_A sensed by the
probe would vary between 0 and 1 as patches of air, fuel, or mixed air
and fuel (hatched region) pass over the probe. Figure 6.5(a) shows
how the continuous signal output from the sampling probe might
appear. Take a moment to visualize the relationship between the
partially segregated, jelly-roll structure shown frozen in time in Fig.
6.4 with the continuous time record of instantaneous local values of
Y_A at $(4b, 0)$, as shown in Fig. 6.5(a).

If the sensor probe output of Fig. 6.5(a) were printed from a
chart recorder for a sufficiently long time, we could obtain an estimate of the Eulerian time-mean value of Y_A , $\langle Y_A \rangle$, at the point
$(4b, 0)$. However, modern gas sampling systems incorporate signal
processing electronics and software which operate on the incoming
data stream to produce a continuously updated record similar to Fig.
6.5(b), which is output to an oscilloscope or computer screen.

Figure 6.5(b) is a plot of the *probability density function* (PDF)
for $Y_A(t)$, denoted $P(Y_A)$. The PDF for $Y_A(t)$ is defined implicitly as
follows: $P(Y_A)dY_A$ is the fraction of elapsed time (over a sufficiently
long sampling time) during which Y_A is within the range $(Y_A, Y_A +
dY_A)$.

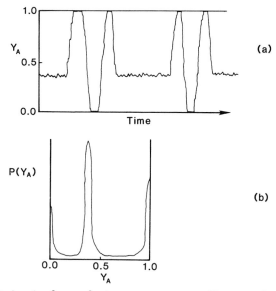

Fig. 6.5   Outputs from fast-response sampling probe located at
(x,y) = $(4b,0)$ in Fig. 6.4, measuring $Y_A(t)$.   (a) Raw data record
of $Y_A(t)$. (b) Probability density function (PDF) for $Y_A(t)$.

Since Y_A can only have values between 0 and 1, it follows that

$$\int_0^1 P(Y_A)dY_A = 1 \qquad (6\text{-}15)$$

The utility of the PDF for Y_A is that the long-time-*mean* value of Y_A can be continuously computed and output by the signal processing circuitry, by utilizing the basic definition of the mean of a distribution. Consequently, the time-mean value of $Y_A(t)$, $\langle Y_A \rangle$, is defined as the first moment of $P(Y_A)$ about the origin:

$$\langle Y_A \rangle \equiv \int_0^1 Y_A P(Y_A)dY_A \qquad (6\text{-}16)$$

When patches of pure fuel or air are present, "spike" or *Dirac delta functions* appear in the PDF for Y_A in the vicinity of 0 and 1, as shown in Fig. 6.5(b). (A *Dirac delta function* $\delta_D(z)$ is merely a useful notation for representing the area under a spike function: by definition, $\delta_D(z) = 1$ when $z = 0$, and is zero for all nonzero values of z.)

The area under $P(Y_A)$ in the vicinity of $Y_A = 1$ is called the *intermittency* of air, i_A, at the measured station. (Put another way: if the signal probe detects the presence of pure air 27 percent of the time, $i_A = 0.27$ at that location.)

We can also find the *variance* (square of the standard deviation) of Y_A, $g(Y_A)$, defined as the second moment of $P(Y_A)$ about the mean value of Y_A:

$$g \equiv \int_0^1 [\langle Y_A \rangle - Y_A]^2 \, P(Y_A)dY_A \qquad (6\text{-}17)$$

Take another moment to think about the relationship between the *raw data* record of Fig. 6.5(a) and the signal-processed output record, the PDF for Y_A represented in Fig. 6.5(b). In particular, think about what the two records would look like if the two fluids were immiscible: since Y_A could then assume values of *only* 0 or 1, the raw signal would jump discontinuously back and forth between 0 and 1 whenever a jelly-roll layer passed over the probe. Since the fraction of time Y_A can assume values *other* than 1 or 0 is precisely nil, the resulting PDF for Y_A would be a pair of Dirac delta or spike functions located at $Y_A = 0$ and at $Y_A = 1$, weighted by the intermittencies at $Y_A = 0$ and 1:

$$P(Y_A) = (1 - i_A)\,\delta_D(Y_A) + i_A\,\delta_D(Y_A - 1) \qquad (6\text{-}18)$$

The no-micromixing PDF for Y_A of Eq. (6-18) could be called (at least, in Texas) a "hook-'em-horns" distribution. Notice once again that *no micromixing* occurs in an emulsion.

In contrast to the no-micromixing PDF for Y_A of Eq. (6-18), the presence of the central "hump" in the PDF for Y_A illustrated in Fig. 6.5(b) is direct evidence that significant micromixing has occurred in some patches of the fluid flowing past the sampling station at $(4b, 0)$—in the case illustrated by Fig. 6.4, the crosshatched region within each large-scale vortex structure.

Still considering the shear/mixing layer of Fig. 6.4, note that if Y_A were to be continuously monitored at a location very far downstream of $x = L_m$, the raw data signal would be a "flatliner" and the resulting PDF for Y_A would be a single Dirac delta function located at $\langle Y_A \rangle$, $\delta_D(Y_A - \langle Y_A \rangle)$, and the resulting variance $g(Y_A)$ would be exactly 0.

We are now ready to formally define a local, quantitative measure of "goodness" or completeness of mixing: The *intensity of segregation* (or *segregation index* or *unmixedness*) I_s may be defined as the ratio of the variance to the square of the mean of one of the mixants. For the air mole fraction,

$$I_s = \frac{g(Y_A)}{\langle Y_A \rangle^2} \qquad (6\text{-}19)$$

Ideally, I_s should be defined so that $I_s = 0$ in a region where molecular homogeneity has been achieved, and $I_s = 1$ for a completely unmixed (fully segregated) region. I_s of Eq. (6-19) satisfies the fully mixed limit, but satisfies the fully segregated limit only when $i_A = 0.5$. Further, Eq. (6-19) is obviously meaningful only in regions where $\langle Y_A \rangle > 0$. (In practice, I_s is defined by more complex definitions than Eq. (6-19) in order to exactly satisfy the ideal criteria given.)

The complement of I_s is sometimes referred to as the *contact index*, $CI \equiv 1 - I_s$. The term *contact* suggests that molecules are able to freely contact or collide when they are fully micromixed ($CI = 1$), and that they are unable to contact (and therefore unable to react chemically) when the macromixture is fully segregated ($CI = 0$).

Finally, it should be noted that the intensity of segregation I_s, as defined by Eq. (6-19), is only a *local*, time-averaged measure of goodness or completeness of mixing. We will presently define a suitable *global* value of goodness of mixing, spatially averaged over the lateral direction, and varying only in the downstream or axial direction.

6.2.4 Time-Averaged Characteristics of a Turbulent Shear Layer

In Sec. 6.2.2, it was shown that the growth rates of mixing layers and shear layers are essentially identical for time-steady, laminar flows.

Until about 1970, it was commonly (and mistakenly) assumed that the *time-mean* growth rates of turbulent shear layers and turbulent mixing layers were essentially the same, as well.

Prior to 1970, research in fluid mechanics was aimed principally at the prediction of aerodynamic drag, which required analysis of momentum transport between bounding walls and the fluid flow within the boundary layer. The aerodynamic drag on a wall is sensitive only to the momentum transported by the collective action of many molecules acting as a continuum, and does not depend at all on the molecular identity of those molecules. In other words, if we think of a boundary layer in one sense as a mixing layer, where the mixants are low momentum flux fluid and high momentum flux fluid, it doesn't matter whether or not the cross-stream mixing is accomplished by macromixing of fluid particles ("chunks" of molecules), or micromixing due to the momentum transport by individual molecules. Consequently, the standard approach to this problem has been to "Reynolds-average" (time-average) the partial differential equations of conservation of mass, momentum, and energy, collectively referred to as the *Navier-Stokes equations.* In this approach, instantaneous velocities appearing in the Navier-Stokes equations are formally replaced by the sum of a time-mean and a transient or fluctuating component of velocity. The equations are then time-averaged, and in conventional notation, *overbars* are placed over each resulting time-averaged term, to indicate that time-averaging has been performed. The resulting equations are called the *Reynolds equations.*

As a result of the time-averaging process, a new term called the *Reynolds stress* appears in the Reynolds-averaged momentum equation. This term represents the time-mean shear stress in the fluid resulting from time-mean turbulent transport of fluctuating momentum flux, and must be modeled. Postulating theoretical or empirical models for the Reynolds stress term is referred to as *closure* of the time-averaged momentum equation.[6.6]

In the process of Reynolds-averaging the Navier-Stokes equations, no reference whatever is made to the complex vortical structure of the flowfield, as illustrated in Fig. 6.4. Of the wealth of statistical information concerning the flowfield structure represented by measured PDF's—defined in Sec. 6.2.3 for Y_A, but equally well defined and measurable for velocity components as well—only the first moments of the PDF's of flow variables, namely the time-mean values, are represented in the Reynolds equations.

By 1970, it had become apparent that, since chemical reaction within turbulent shear/mixing layers depends critically upon *micromixing* within the layer, it was necessary to investigate the structure of the shear layer, which had been largely ignored up to that time. Professor J. E. Broadwell of CalTech summarized the dilemma with the rueful observation, "It appears that we put the overbars in too soon."

Professor Broadwell's observation can be illustrated by considering a simplified version of the chemical-kinetic rate law (to be treated more carefully in Sec. 6.3), for chemical reaction between fuel and air, in which the instantaneous rate of reaction R of fuel and air is taken to be proportional to the product $Y_A Y_F$, or $R = k Y_A Y_F$, where k is a constant of proportionality (k actually depends on both pressure and temperature). In the example case of a hypothetical emulsion represented by the "hook-'em-horns" $P(Y_A)$ of Eq. (6-18), we know that chemical reaction is impossible, as fuel and air molecules never coexist at any point in space. The Reynolds decomposition represents the instantaneous values for Y_A and Y_F as $Y_A = \langle Y_A \rangle + Y_A'$ and $Y_F = \langle Y_F \rangle + Y_F'$, where the prime denotes the *fluctuation*, defined as the difference between the time-mean value and the instantaneous value. Thus, the instantaneous reaction rate is given by $R = k(\langle Y_A \rangle \langle Y_F \rangle + Y_A' \langle Y_F \rangle + Y_F' \langle Y_A \rangle + Y_A' Y_F')$. If this expression is time-averaged, there results $\bar{R} = k(\overline{\langle Y_A \rangle \langle Y_F \rangle} + \overline{Y_A' Y_F'}$, since the time-averaged two middle terms are 0. The term $\overline{Y_A' Y_F'}$ is the analogue of the Reynolds stress in the Reynolds-averaged momentum equation.

If we further assume for *simplicity* that the air intermittency $i_A = 0.5$, then $\overline{\langle Y_A \rangle \langle Y_F \rangle} = \langle Y_A \rangle \langle Y_F \rangle = (0.5)(0.5) = 0.25$, and since the fluctuations Y_A' and Y_F' are negatively correlated (that is, Y_A' equals $+0.5$ whenever $Y_F' = -0.5$, and vice versa), then $\overline{Y_A' Y_F'} = (0.5)(-0.5) = -0.25$. Therefore, $\bar{R} = k(0.25 + [-0.25]) = 0.0$, which is correct. However, if we had naively assumed that the time-average reaction rate was proportional to the product of the time-average mole fractions, we would have calculated $\bar{R} = k\langle Y_A \rangle \langle Y_F \rangle = k(0.25) > 0$, which is incorrect. This erroneous result would be obtained as a result of having "put the overbars in too soon."

As a background to the post-1970 research in turbulent mixing layers, we will first consider Prandtl's 1926 approximate solution for the time-mean growth rate and velocity profiles in a turbulent shear layer, as summarized in Chap. XXIV of Schlichting.[6.6] This solution assumes constant and equal densities, constant-pressure flow, and uniform and time-steady inlet axial velocities $u_1 > u_2$. Overbar notation is not employed: the same notation will be used for time-averaged values as we used previously to represent instantaneous values in laminar shear layers.

In Prandtl's solution, the effective or virtual turbulent kinematic viscosity is modeled by means of the mixing length assumption.[6.6] The resulting time-mean velocity profile is given approximately by

$$\frac{u}{u_c} = 1 + \left(\frac{1-r}{1+r}\right) \left\{ 3\left(\frac{y}{\delta}\right) - 4\left(\frac{y}{\delta}\right)^3 \right\} \qquad (6\text{-}20)$$

where δ is the local shear layer width, which varies as

$$\delta = 6B^2 \left(\frac{1-r}{1+r} \right) x \qquad (6\text{-}21)$$

and where δ is the ratio of the Prandtl mixing length ℓ_m to the shear layer width, $B = \ell_m / \delta$. The ratio B is the only empirical constant in Prandtl's solution, and can only be determined from experimental data.[6.6]

Note that according to Eq. (6-21), turbulent shear layers grow linearly with x, rather than as the square root of x, as was the case for laminar shear layers, Eq. (6-14).

In an excellent recent (1991) survey of post-1970 research on turbulent mixing layers,[6.7] Dimotakis shows that the time-mean shear layer growth rate for constant and equal densities is given by

$$\delta = C_\delta \left(\frac{1-r}{1+r} \right) x \qquad (6\text{-}22)$$

where values of C_δ reported from different experiments and investigators vary from 0.25 to 0.45. Note that Eqs. (6-22) and (6-21) are of the same form, and differ only in the representation of the proportionality constant.

6.2.4.1 Density effects on shear layer growth.   We now consider the more general case where $\rho_1 \neq \rho_2$, which requires us to define another parameter s, the mass density ratio:

$$s \equiv \frac{\rho_2}{\rho_1} \qquad (6\text{-}23)$$

Turning our attention once again to Fig. 6.4, note that as fluids 1 and 2 are entrained, stagnation points arise between the two relative flows, in the region between two adjacent vortex structures. Still assuming constant (but not equal) densities, the stagnation pressure may be expressed in the convective reference frame as[6.7]

$$p_1 + \frac{1}{2}\rho_1 (\Delta u_1)^2 = p_2 + \frac{1}{2}\rho_2 (\Delta u_2)^2 \qquad (6\text{-}24)$$

where $\Delta u_1 = u_1 - u_c$, and $\Delta u_2 = u_c - u_2$. If we also assume that $p_1 = p_2$, then Eq. (6-24) can be solved for the density ratio s,

$$s^{1/2} = \frac{\Delta u_1}{\Delta u_2} = \frac{u_1 - u_c}{u_c - u_2} \qquad (6\text{-}25)$$

which can in turn be solved for the convective velocity u_c:

$$u_c = \frac{u_1 + s^{1/2} u_2}{1 + s^{1/2}} \qquad (6\text{-}26)$$

Equation (6-26) is the corrected convective velocity of the vortex structures, with density differences in the two streams taken into account, but for incompressible flow. Note that for equal densities, $s = 1$, Eq. (6-26) reverts to the arithmetic mean value assumed in Sec. 6.2.2.

For compressible flows, Eq. (6-24) must be replaced by equating the total (isentropic stagnation) pressures in the convective reference frame, calculated from familiar one-dimensional gasdynamic relations as

$$\left(1 + \frac{\gamma_1 - 1}{2} M_{c1}^2\right)^{\frac{\gamma_1}{\gamma_1 - 1}} = \left(1 + \frac{\gamma_2 - 1}{2} M_{c2}^2\right)^{\frac{\gamma_2}{\gamma_2 - 1}} \qquad (6\text{-}27)$$

where M_{c1} and M_{c2} are the *convective Mach numbers*, defined by

$$M_{c1} \equiv \frac{u_1 - u_c}{a_1} \quad \text{and} \quad M_{c2} \equiv \frac{u_c - u_2}{a_2}$$

where a_1 and a_2 are the sonic speeds in inlet streams 1 and 2, respectively.

Since u_c is algebraically implicit in Eq. (6-27), it must be found iteratively, or by using a root-finder or "solve" function in a software package or scientific programmable calculator. However, note that if the specific heat ratios are the same in both streams, and if both streams are at the same temperature, u_c is given exactly by the incompressible expression Eq. (6-26).

The density-corrected expression for shear layer growth rate[6.7] is considerably more complex than Eq. (6-22):

$$\frac{\delta}{x} = C_\delta \left(\frac{1 - r}{1 + s^{1/2} r}\right) \left(\frac{1 + s^{1/2}}{2}\right)$$

$$\times \left\{1 - \frac{\left(1 - s^{1/2}\right) / \left(1 + s^{1/2}\right)}{1 + 1.29 \left(1 + r\right) / \left(1 - r\right)}\right\} \qquad (6\text{-}28)$$

Note that Eq. (6-28) reduces to Eq. (6-22) when $s = 1$.

6.2.4.2 Compressibility effects on shear layer growth. Up to this point, attention has been focused on the growth of constant (but not necessarily equal) density shear layers. Equations (6-21), (6-22), and (6-28) show the very strong, positive effect on shear layer growth of decreasing the velocity ratio $r = u_2/u_1$. Before we can see how far we can carry this strategy, it is necessary to consider Mach number effects on shear layer growth rate.

At the present time, researchers have investigated Mach number effects on shear layers only as high as $M_{c1} = 2$.[6.7] The results are somewhat alarming. At values of M_{c1} as low as 0.2, marked suppression of the shear layer growth rate begins to occur, and by $M_{c1} = 0.8$ and greater, the shear layer growth rate is suppressed to only 0.2 of the incompressible value given by Eq. (6-28)! While the reasons for this discouraging finding are currently under investigation, it is speculated that the principal suppression mechanism is the stabilization of the modes of fluid instability that are responsible for the formation of the large vortex structures. The precise physical reason for this mode stabilization and the low growth rate of other modes is an open question. A recent model ascribing importance to *sonic eddies* is one attempt to find a physical interpretation.[6.8]

For design purposes, Dimotakis[6.7] proposes an empirical compressibility correction factor:

$$f(M_{c1}) = 0.2 + 0.8e^{-3M_{c1}^2} \tag{6-29}$$

Equation (6-29) is to be used in the following way: For any given set of inlet velocities, temperatures, and densities, first calculate the $M_{c1} = 0$ shear layer growth rate from Eq. (6-28), and then multiply that value by the correction factor of Eq. (6-29) to obtain the correct growth rate $\delta(x)$ for $M_{c1} > 0$.

6.2.5 Mixing in a Turbulent Shear/Mixing Layer

By inspection of Fig. 6.4, it is apparent that, at any axial station, micromixing is not complete throughout the shear layer, which is indicated by the dashed lines. In fact, for the first one or two vortices sketched in Fig. 6.4, it may be seen that comparatively little micromixing has yet occurred. The distance downstream of the splitter plate at which a significant amount of mixed fluid is first present is called the *mixing transition point*.[6.7] The mixing transition occurs approximately at the point where

$$\frac{(u_1 - u_2)\delta}{\nu} \approx 10 \tag{6-30}$$

where ν is a representative average value of the molecular kinematic viscosity within the shear layer.

Equation (6-30) was determined for $M_{c1} = 0$; it is not known at present what effect if any $M_{c1} > 0$ may have on the estimate.[6.7] By invoking the $M_{c1} = 0$ shear layer growth rate, Eq. (6-22), the mixing transition can be estimated to occur approximately at

$$x_m \cong \frac{20\nu u_c}{C_\delta (\Delta u)^2} \qquad (6\text{-}31)$$

Following the mixing transition, the time-averaged micromixing layer is observed to grow approximately as a constant fraction of the shear layer,

$$\frac{\delta_m}{\delta} \approx 0.49 \qquad (6\text{-}32)$$

The estimate Eq. (6-32) is believed to apply without regard to density or compressibility effects, and is independent of the location of the mixing transition point, x_m.[6.7] The estimates of Eqs. (6-31) and (6-32) are illustrated in Fig. 6.6.

Combining Eqs. (6-28) and (6-29), together with Eq. (6-32), results in an equation for the axial growth rate of the turbulent mixing layer:

$$\frac{\delta_m}{x} = 0.49 f(M_{c1}) C_\delta \left(\frac{1 - r}{1 + s^{1/2} r} \right) \left(\frac{1 + s^{1/2}}{2} \right)$$

$$\times \left\{ 1 - \frac{\left(1 - s^{1/2}\right) / \left(1 + s^{1/2}\right)}{1 + 1.29(1 + r)/(1 - r)} \right\} \qquad (6\text{-}33)$$

Equation (6-33) can now be used to obtain a new estimate of the distance required to mix, for comparison with our estimate from Sec. 6.2.2.1 of $L_m/b \sim 1500$ to 3000 for a laminar, no-shear mixing layer. As before, we assume burner entry conditions of 2800 °R

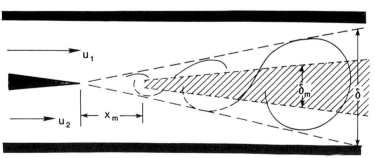

Fig. 6.6 Growth of time-averaged shear layer thickness δ, and of mixing layer thickness δ_m, following mixing transition.

(1556 K), air velocity u_1 = 8290 ft/s (2530 m/s), and fuel (H_2)/air velocity ratio $r = u_2/u_1$ = 0.5. Assuming the air and fuel are at the same temperature and pressure, the density ratio s is given by Eq. (6-23) as 0.0693. Equation (6-27) in turn gives the convective velocity u_c = 7426 ft/s (2266 m/s) and convective Mach numbers M_{c1} and M_{c2} = 0.331 and 0.359, respectively, so that Eq. (6-29) gives a compressibility factor $f(M_{c1})$ = 0.776. Finally, the mixing layer growth rate is estimated from Eq. (6-33) to be $\delta_m/x \sim 0.0241$ to 0.0433, depending on the value of C_δ in Eq. (6-22). Assuming as in Sec. 6.2.2.1 that $x = L_m$ when $\delta_m = 2b$, we estimate $L_m/b \sim$ 46 to 83. The required entry scale of segregation for an L = 6 ft (1.8 m) combustor is therefore increased to b = 0.9 to 1.5 in (2.2 to 4 cm). While this is a considerable improvement over the no-shear mixing layer, even a one-inch entry scale of segregation is undesirably constrictive. For $(L/b)_{max} \sim 20$, b should be at least 3.6 in (9.1 cm). In addition, there are other problems with shear layer mixing, as we will soon see.

6.2.5.1 Heat release effects on the mixing layer. Two principal effects of heat release on the mixing layer have been observed, both theoretically and experimentally. First, at any axial location, both the shear layer and the included mixing layer occupy a greater volume fraction of the channel, due to the volume dilation resulting from the temperature rise due to combustion at essentially constant pressure. The second effect is to *reduce the rate of growth* of the mixing layer.

For equal mass densities in both streams (which is far from the case when H_2 is the fuel), Dimotakis[6.7] suggests that the estimate for δ_m/x in Eq. (6-33) should be multiplied by $(1 - C_q q)$, where $q = \Delta\rho/\rho_0$ is the ratio of the positive difference between mass density within the product layer (i.e., the burning or burned mixing layer) and the freestream mass density $\rho_0 = \rho_1 = \rho_2$, and where $C_q \sim 0.05$.[6.7] No similar quantitative estimates are presently available for H_2/air mixing and combustion, but both experimental and CFD modeling investigations of mixing and combustion of H_2/air show a marked decrease in growth rate of the burning mixing layer, compared to a nonreacting mixing layer with the same entry conditions.[6.9, 6.10] However, this effect diminishes with increasing Mach number, as the increment in total enthalpy due to heat release becomes a smaller fraction of the total enthalpy.

6.2.5.2 Gas composition within the mixing layer. It might be naively assumed that the mole fraction of air within the mixing layer is simply proportional to the entry mass flow rates of air and fuel into the mixer, namely $\dot{m}_{A,0} = \rho_A b_A u_A$ and $\dot{m}_{F,0} = \rho_F b_F u_F$, respectively. At some distance far downstream, conservation of mass certainly requires that the mixture fuel/air ratio must eventually reach the limit

$$f_0 \equiv \lim_{x \to \infty} f = \frac{\dot{m}_{F,0}}{\dot{m}_{A,0}} = \begin{cases} b_2 sr/b_1, & u_1 = u_A > u_f = u_2 \\ b_1/b_2 sr, & u_1 = u_F > u_A = u_2 \end{cases} \tag{6-34}$$

with a corresponding *overall equivalence ratio* ϕ_0 obtained from Eqs. (3-6) and (6-2) as

$$\phi_0 = \frac{f_0}{f_{st}} = f_0 \left[\frac{103(4x+y)}{36x+3y} \right] \tag{6-35}$$

However, in the mixing near-field, where we expect that combustion will begin and therefore where the value of the local fuel/air ratio is very important, the local equivalence ratio ϕ_m is *not* equal to the far-field limit ϕ_0 of Eq. (6-35). If we again look at Figs. 6.1 through 6.4 and think about it, we should realize that the interaction between the two streams is strictly local, and should not be influenced by the lateral extent of either mixant. Rather, the mixture composition is determined by the *volumetric entrainment ratio* E_v, defined as the volume ratio of high speed fluid to low speed fluid entrained within each vortex structure—that is, wrapped up into each vortex or jelly-roll structure.[6.7] The volumetric entrainment ratio E_v is observed to depend on the convective reference frame velocity ratio and on the *large structure spacing-to-position ratio* ℓ/x [6.7] as

$$E_v \cong \left(\frac{u_1 - u_c}{u_c - u_2} \right) \left(1 + \frac{\ell}{x} \right) \tag{6-36}$$

At any instant, the large-structure spacing ℓ is simply the distance between the centers of two adjacent vortices, as depicted in Fig. 6.4. The spacing-to-position ratio (ℓ/x) is estimated by the relation

$$\frac{\ell}{x} \approx 0.68 \left(\frac{1-r}{1+r} \right) \tag{6-37}$$

Dimotakis[6.7] argues that, since ℓ is observed to grow proportionally to δ, the compressibility correction factor $f(M_{c1})$ of Eq. (6-29) can be applied to Eq. (6-37) as well, in order to correct for Mach number effects on the entrainment ratio. Equations (6-36) and (6-37) can be combined to give the estimate, good for different densities and for $M_{c1} > 0$,

$$E_v \cong \left(\frac{u_1 - u_c}{u_c - u_2} \right) \left[1 + 0.68 f(M_{c1}) \left(\frac{1-r}{1+r} \right) \right] \tag{6-38}$$

The corresponding mass-basis and mole-basis entrainment ratios E_m and E_n are given by

$$E_m = \frac{\rho_1}{\rho_2} E_v = \frac{E_v}{s} \quad \text{and} \quad E_n = \frac{W_2}{W_1} E_m \qquad (6\text{-}39)$$

where W_1 and W_2 are the mean molecular weights of the gases in streams 1 and 2, respectively.

When air is the high speed fluid, E_m of Eqs. (6-38) and (6-39a) estimates the mass-basis air/fuel ratio within the mixing layer. For any $C_x H_y$ fuel, Eq. (6-39) can be combined with Eqs. (3-6) and (6-2) to give an estimate for the equivalence ratio within the mixing layer ϕ_m:

$$\phi_m \cong \begin{cases} \dfrac{1}{E_m}\left[\dfrac{103(4x + y)}{36x + 3y}\right] , & u_1 = u_a > u_f = u_2 \\[4mm] E_m\left[\dfrac{103(4x + y)}{36x + 3y}\right] , & u_1 = u_f > u_a = u_2 \end{cases} \qquad (6\text{-}40)$$

Equations (6-38) to (6-40) will now be used to estimate the equivalence ratio in a mixing layer ϕ_m, using the same assumptions and conditions used in our previous turbulent mixing layer growth estimates: $r = 0.5$, $f(M_{c1}) = 0.776$, and $u_c = 7426$ ft/s (2266 m/s). For H_2 fuel, $x = 0$ and $y = 2$ in Eq. (6-40a). The resulting estimate for the equivalence ratio in the mixing layer is $E_m = 4.46$ lbm A/lbm F, and $\phi_m = 7.69(!)$. This is very bad news, as we must have an equivalence ratio in the range 0.2 to 2.0 for combustion to occur promptly, rapidly, and efficiently.

The problem stems from the large difference in molecular weight between air and H_2 (14.42:1). In the present example, since $\rho_A > \rho_F$ and $u_A > u_F$, the mixing layer entrains 4.46 times more mass of air than of H_2. However, Eq. (6-40a) shows that we must have $E_m = 34.48$ lbm A/lbm F in order to achieve $\phi_m = 1$.

Can anything be done to increase E_m nearly eightfold? Equations (6-38) and (6-39) show that E_m can be increased by reducing either $s = \rho_2/\rho_1$ or $r = u_2/u_1$, or both. But u_1 and ρ_1 are fixed by thermodynamic cycle considerations, so s and r can be lowered *only* by reducing ρ_2 and/or u_2. To put it another way: since we cannot drive the shear layer to entrain more air, we must try to force it to *entrain less hydrogen*. However, mass conservation requires that $\dot{m}_F = b_2 \rho_2 u_2$ must remain constant to satisfy the desired ϕ_0, so ρ_2 can be reduced only if u_2 and/or b_2 is increased accordingly. We can't increase u_2 very much, as that tends to cancel the effect on E_m of reducing ρ_2, and increases in b_2 are constrained by the aircraft fuselage envelope. It's a classic "catch-22"! We can tweak these values to reduce ϕ_m to perhaps 5 or 6, but in the end we are forced to conclude reluctantly that, for H_2 fuel, *the stoichiometric requirements for optimal combustion cannot be met by shear layer mixing.*

Of course, if we had the luxury of an infinitely long mixer/combustor, the far-field mixing would eventually have to yield the overall equivalence ratio ϕ_0 given by Eq. (6-35), in order to satisfy global mass conservation. After the fuel-rich mixing layer has entrained all of the fuel at the rate given by Eqs. (6-38) and (6-39), the remaining unmixed airstream must still be mixed into the fuel-rich mixing layer until $\phi_m \rightarrow \phi_0$ as $x \rightarrow \infty$. Regrettably, there is at present no theory to describe how the flow mixes from the near-field stirring or macromixing to the far-field, turbulence-dominated micromixing, to (if given sufficient time) the overall equivalence ratio ϕ_0 given by Eq. (6-35). As a result, we have to abandon further analysis of the internal structure of the near-mixing region and focus attention on empirical global measures of mixing.

6.2.6 Axial Mixing "Efficiency"

At the end of Sec. 6.2.3, we promised to define a suitable global value of "goodness" of mixing, spatially averaged over the lateral direction, and varying only in the axial direction. The mixing layer growth rate δ_m defined by Eq. (6-33) would appear to satisfy this requirement to some extent. However, as Fig. 6.7 illustrates, when the entry scales

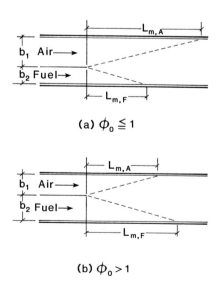

(a) $\phi_0 \leqq 1$

(b) $\phi_0 > 1$

Fig. 6.7 Growth of mixing layer when one mixant is depleted before the other. (a) Fuel-lean case. (b) Fuel-rich case.

of segregation b_1 and b_2 (duct height, jet diameter, etc.) are not equal, it is possible for one mixant to be depleted ("mixed out" into the mixing layer), while the other mixant remains partially unmixed. As we have seen, the equations defining δ_m are concerned only with near-field mixing, and do not address the case of far-field mixing.

Anderson[6.11] defines an empirical, one-dimensional measure of the degree of mixing completeness which takes into account both near- and far-field mixing. This measure is termed the *mixing efficiency* $\eta_M(x)$, and is defined for overall fuel-lean mixing at any axial station x as "the amount of fuel that would react if complete reaction occurred without further mixing divided by the amount of fuel that would react if the mixture were uniform."[6.11]

In the language of this chapter, for $\phi_0 \leq 1$, $\eta_M(x)$ is the fraction of fuel which is micromixed with air at any axial location x. For overall fuel-rich mixtures, $\phi_0 > 1$, $\eta_M(x)$ is defined as the fraction of *air* which is micromixed at the axial location x. We can say that $\eta_M(x)$ measures the completeness of micromixing of the *minor mixant*, defined as either fuel or air, depending on whether the overall equivalence ratio ϕ_0 is less than or greater than 1.0, respectively.

Although not stated explicitly in the above definition of $\eta_M(x)$, it is implicit in its definition that, as long as $\eta_M < 1$, fuel and air are micromixed *in stoichiometric proportion* within the mixing layer; that is, $\phi_m = 1.0$ is assumed.

Strictly speaking, as pointed out by Swithenbank,[6.16] the term "mixing efficiency" should be reserved for a cost/benefit ratio of mixing, where the "cost" is the mechanical energy drawn from the air or fuel stream expended for stirring, and the "benefit" is either increased rate of growth of the near-field mixing layer or reduced axial distance required for near-complete micromixing. Nonetheless, we will follow the convention of referring to $\eta_M(x)$ as the "mixing efficiency."

We have seen that $\delta_m(x)$ represents the axial growth rate of the mixing layer from the fluid mechanical perspective, but does not address the composition within the mixing layer. The mixing efficiency η_M differs from $\delta_m(x)$ in that $\eta_M(x)$ represents the rate at which *only one* of the mixants, the minor mixant, is mixed into the mixing layer. However, the assumption that $\phi_m = 1$ within the mixing layer implicitly defines the mixing rate of the major mixant as well. The $\phi_m = 1$ assumption is in stark contradiction to Eqs. (6-35) through (6-39) which show that, at least for near-field mixing in turbulent shear layers, ϕ_m is not equal to 1.0 in general. In spite of this contradiction, we will utilize the mixing efficiency $\eta_M(x)$ for our purposes, simply because it is the only such measure available for describing mixing in systems other than parallel-injection mixing layers.

Data from experimental measurements of supersonic mixing between parallel coflowing streams of hydrogen and air[6.9] can be represented approximately by the empirical relations

$$
\eta_{M_{0^\circ}} \cong \begin{cases} x/L_{m,F} , & \phi_0 \leq 1 \\ x/L_{m,A} , & \phi_0 > 1 \end{cases} \tag{6-41}
$$

where $L_{m,F}$ and $L_{m,A}$ are the axial distances required for all of the fuel or air, as appropriate, to be mixed to stoichiometric proportion, as illustrated in Fig. 6.7.

Since the micromixed portion of fuel and air are assumed to be in stoichiometric proportion as long as $\eta_M < 1$, the distance L_m required to achieve $\eta_M = 1$ is greatest when the overall equivalence ratio ϕ_0 is unity, as can be seen from Fig. 6-7. The axial distance at which the minor mixant is depleted is given approximately by the empirical relations[6.12]

and

$$
\frac{L_{m,F}}{b} \cong 0.179 C_m\, e^{1.72\phi_0} , \qquad \phi_0 \leq 1
$$

$$
\frac{L_{m,A}}{b} \cong 3.333 C_m\, e^{-1.204\phi_0} , \qquad \phi_0 > 1 \tag{6-42}
$$

where $b = (b_1 + b_2)$ is the sum of the entry scales of segregation for both streams, and where the mixing constant C_m in Eq. (6-42) is reported as varying from 25 to 60.[6.12] Note that these values for C_m are of the same order of magnitude as the estimates of $L_m/b \sim 46$ to 83 obtained in Sec. 6.2.5.

Finally, with regard to the mixing scenarios depicted in Fig. 6.7, the reader may wonder how to represent the mixing rate of the major mixant following depletion of the minor mixant. Lacking guidance from the formal definition of $\eta_M(x)$, one researcher suggests "the major constituent stream is assumed to continue to mix at the rate it was mixing when the minor constituent ran out, until it too is depleted."[12] This assumption can be represented by

$$
L_{m,0} = \begin{cases} L_{m,F} + \dfrac{1 - \phi_0}{\phi_0 (d\eta_M/dx)_{x=L_{m,F}}} , & \phi_0 \leq 1 \\[3mm] L_{m,A} + \dfrac{\phi_0 - 1}{(d\eta_M/dx)_{x=L_{m,A}}} , & \phi_0 > 1 \end{cases} \tag{6-43}
$$

where $L_{m,0}$ is the axial distance required to fully micromix *both* streams. Even though Eqs. (6-42) show that the required axial distance $L_{m,F}$ or $L_{m,A}$ to deplete the minor mixant is decreased for off-stoichiometric mixtures, Eq. (6-43) shows that the length required to fully micromix *both* streams $L_{m,0}$ remains approximately equal to C_m , irrespective of ϕ_0.

6.2.7 Mixing with Normal Fuel Injection

In this chapter, we have learned much about the internal structure of turbulent shear/mixing layers, and have learned that, for H_2/air, it is essentially impossible to achieve near-stoichiometric mixtures in the near-field. When this fact was realized, supersonic combustion researchers turned their attention to injecting the fuel at right angles to the flow, in the hope that reducing the velocity ratio $r = u_2/u_1$ to 0 would maximize the mixing layer performance measures given by Eqs. (6-22), (6-33), and (6-38). However, the apparent benefit of $r = 0$ on those estimates does not tell the whole story, as we shall see.

Figure 6.8 illustrates the significant fluid mechanical features of fuel injected from a normal jet into a supersonic crossflow.[6.13] The most significant feature is that the supersonic crossflow is displaced by the fuel jet exactly as if a cylindrical rod were inserted into the freestream from the wall at a right angle. As a result, a detached normal shock wave forms just upstream of the jet, causing the upstream wall boundary layer to separate. In addition, a bluff-body wake region is formed immediately downstream of the jet core. The recirculating flow in the wake acts exactly as the subsonic flame stabilization zone in a gas turbine combustor primary zone, or the wake of the flameholding gutter in ramjet combustors and turbojet afterburners.

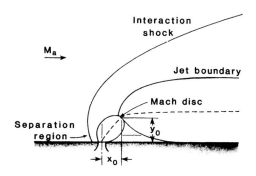

Underexpanded jet injected normally into a supersonic freestream

Fig. 6.8   Schematic of flowfield in normal fuel injection (Refs. 6.13 and 6.14).

The overall effect of the normal jet on the flowfield is to anchor the mixing layer firmly to the jet core, reducing the mixing transition distance x_m (see Fig. 6.6) not only to 0, but actually to a small negative distance, as the mixing and flameholding zone extends slightly upstream of the jet via the separated boundary layer.

Unfortunately, the significant gain in near-field mixing completeness is not without cost. The detached normal shock wave causes a severe local loss of total pressure which, together with the boundary-layer separation total pressure loss, leads to a decrease in overall cycle efficiency.

Experimental measurements of the axial variation of the mixing efficiency $\eta_M(x)$ for normal injection of hydrogen into a crossflowing airstream[6.14] can be represented approximately by an empirical relation,

$$\eta_{M_{90^\circ}} = \left\{ \frac{x}{L_m} + \frac{1}{(50 + 1000\alpha)} \right\}^\alpha \qquad (6\text{-}44)$$

where α is a fit parameter which varies from $\alpha = 0.17$ for "widely spaced" injectors to $\alpha = 0.25$ for "closely spaced" injectors, and where L_m is the axial distance required for the minor mixant to be depleted while mixing to stoichiometric proportion, as given by Eq. (6-41).

Comparing values of $\eta_M(x)$ for parallel and normal injection at the same value of x/L_m, determined from Eqs. (6-41) and (6-44), the normal jet mixing layer can be seen to grow initially (small x) at a significantly greater rate than the parallel jet mixing layer. It should be especially noted that the presence of the recirculating wake region around the jet results in η_M being much greater than 0 at $x = 0$: specifically, Eq. (6-44) gives values of $\eta_M(0) = 0.24$ and 0.40 for closely and widely spaced injectors, respectively.

However, it is of even greater significance that the value of L_m which is to be used in Eq. (6-44) is the same value given by Eq. (6-42a). While normal injection "jump-starts" the mixing and causes a greater initial growth rate of η_M, the far-field mixing is apparently somewhat weakened in such a way that approximately the same axial distance is required for the minor mixant to be fully ingested into the mixing layer as with parallel fuel injection! Apparently, the near-field mixing is dominated by the stirring or macromixing driven by the large-scale vortices generated by the jet/freestream interaction, whereas the far-field mixing depends as before only on the small-scale turbulence within the plume/mixing layer, such that the far-field mixing essentially has no "memory" of the near-field stirring. The transition from near-field to far-field mixing takes place about 10 to 20 jet diameters downstream.[6.15]

It is of course possible to inject the fuel into the cross stream at angles between 0 and 90 deg., in order to enhance the near-field mixing, while holding down the severity of shock losses from the impingement of the fuel jet on the freestream. To reasonably good approximation, so-called *vectored jet* injection can be modeled by linearly varying the fit parameter α in Eq. (6-44) between $\alpha = (0.17$ to 0.25) for 90-deg. injection and $\alpha = 1$ for parallel fuel injection.[6.12]

6.2.8 Axial Vortex Mixing

It is useful at this point to briefly review the spatial coordinate system introduced in Fig. 6.1. The x direction is axially downstream, while the y coordinate denotes the lateral "up" direction. The z coordinate (not shown in Fig. 6.1) represents the horizontally lateral "out of the page" direction.

It is now known that the flow instabilities induced in both parallel fuel injection and normal fuel injection give rise principally to *lateral* or *spanwise* vorticity—that is, vorticity vectors aligned in the z direction for parallel fuel injection, and in the y direction for normal fuel injection—and only secondarily to axial or streamwise vorticity (vorticity vectors aligned in the x direction.) This can be visualized by recalling that the vortices formed in the planar shear layer of Fig. 6.4 act somewhat like "roller bearings" rotating about the z axis to accommodate the velocity difference between the two streams. Similarly, recall that jets injected normal to the wall act much like cylindrical rods, as illustrated in Fig. 6.8. Flow over cylinders in a crossflow causes periodic vortex shedding into the cylinder wake, with the vortex axes aligned parallel to the symmetry axis of the cylinder[6.6] which, in the coordinate system of Fig. 6.1, is in the y-direction. Given the general finding that mixing from vectored fuel injection is costly in terms of total pressure loss, and that the dominant sense of the vorticity which is responsible for the near-field stirring is predominantly lateral vorticity, the next thing to try is stirring with axial vortices.

6.2.8.1 Axial vorticity in the fuel stream.
In 1968, Swithenbank and Chigier postulated that "substantial increase in mixing rates can be obtained by applying a swirling motion to the fuel jet." [17] It was well known in 1968 that this was true for subsonic mixing and combustion, but there was very little experimental verification to support their assertion. Two experiments reported in 1972–73 found little or no mixing enhancement by swirling the fuel jet as it issued parallel to a supersonic airstream. However, more recent investigations reported in 1989–90 supported Swithenbank and Chigier's assertion. In a 1992 research paper,[6.18] Naughton and Settles review the historical research mentioned here, and state "increases in mixing rate of up to 60% are possible through the addition of streamwise vor-

ticity" using the supersonic, swirled injection fuel nozzle shown in Fig. 6.9. While this improvement in mixing rate over parallel injection is no greater than for normal fuel injection, it is anticipated, based on theoretical considerations as yet unsubstantiated, that the same degree of mixing enhancement can be achieved with less mixing power expended, and therefore lower total pressure loss, than with normal fuel injection. To this end, note the sweepback angle of the fuel strut in Fig. 6.9 which, together with the lenticular shape of the strut cross section (not shown), is designed to minimize shock losses in the flow.

Another method of inducing axial vorticity in the parallel-flow fuel jet includes inducing axial vorticity through secondary flows which arise when supersonic fuel flows through a converging tapered slot jet, which features an elliptic-to-conical duct transition just before sonic injection of the fuel into the parallel airstream.[6.19]

6.2.8.2 Axial vorticity in the airstream.

A variety of mixing devices have been proposed, all with the basic objective of converting a fraction of the flow energy in the air stream into tangential kinetic energy, in the hope that the resulting axial vortex will sweep through and entrain an unswirled, parallel-injected fuel jet. These devices have been termed *hypermixers.*

One class of such devices which has been studied intensively in recent years is an array of wall-mounted ramps of various configurations, as illustrated in Figs. 6.10 and 6.11.

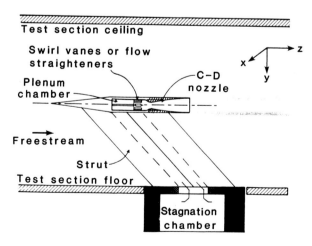

Fig. 6.9 Vortex injector for imparting swirl to fuel jet in supersonic mixing (Ref. 6.18).

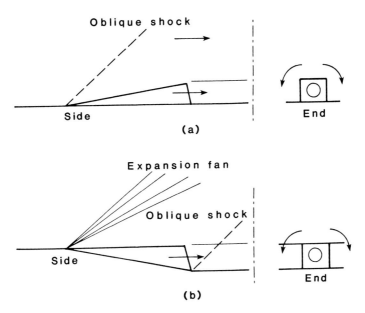

Fig. 6.10   Geometry of wall-mounted, unswept ramp fuel injectors. (a) Raised ramp.   (b) Relieved or exposed ramp.   Side elevation: view is along z axis.   End elevation: view is upstream, along x axis.

In configuration (a) of Fig. 6.10, an oblique shock stands at the base of the ramp where it rises from the wall. The compressed air above the ramp spills over the sides into the lower pressure along the ramp sidewall, forming a counter-rotating pair of axial vortices in the sense shown in the end view of Fig. 6.10. The design goal of this configuration is for the axial vortices to entrain the central fuel jet, ultimately leading to downstream mixing.

In configuration (b) of Fig. 6.10, the wall is turned away from the flow, while the top surface of the ramp remains in the plane of the upstream wall. When the wall has relieved far enough to expose the fuel jet in the downstream end of the ramp, the wall turns back into the flow until it is again parallel with the plane of the upstream wall. In this case, a Prandtl-Meyer expansion fan is anchored at the upper edge of the inclined plane, causing a pressure difference between the flow on the upper ramp surface and the expanded flow along the sidewalls of the ramp. The design goal is also to form counter-rotating pairs of axial vortices having the same sense of rotation as in configuration (a). However, configuration (b) is significantly different from (a) in that the fuel jet now has to pass through a planar oblique shock anchored at the lower edge of the inclined ramp, where the wall is turned back into the freestream. If the fuel jet is gaseous

H_2, its mass density is much less than that of the adjacent air, due to the $\sim 14{:}1$ difference in molecular weight of fuel and air. This fact, which is responsible for the poor mixing in parallel fuel injection, may possibly be used to our advantage. In addition to the axial vorticity generated by the flow spilling over the shoulders of the relieved ramp, additional axial vorticity is generated due to *baroclinic torque* at the mixant interface between the fuel jet and the airstream. This occurs because a strong spatial gradient of mass density exists at the mixant interface, with the gradient direction radially outward from the surface of the cylindrical mixant interface. At the same time, a strong pressure gradient exists, with direction normal to the surface of the planar oblique shock, through which the fuel jet must pass. Wherever the density and pressure gradients are not collinear, baroclinic torque is generated. The pointwise rate of generation of baroclinic torque in the vicinity of the fuel/air interface is given by[6.20]

$$\rho \frac{D}{Dt}\left(\frac{\omega}{\rho}\right) = \frac{1}{\rho^2}\,\nabla\rho \times \nabla p \qquad (6\text{-}45)$$

From a CFD modeling study comparing near-field mixing for a round, parallel-injected jet with and without passage of the jet through an oblique shock (but without a ramp), Drummond concluded there was a significant increase in the near-field fuel/air mixing and subsequent combustion in the case of the shocked jet compared to the unshocked jet.[6.21] However, it is not clear at this time whether the gain in mixing enhancement due to the vorticity generated by baroclinic torque is sufficient to justify the additional loss of total pressure caused by the oblique shock, through which *all* of the airstream must pass.

Further, the effectiveness of unswept ramps as axial vorticity mixers is questionable, as the mixing which does occur may be due principally to stirring by *lateral* vortices generated in the wake flow, downstream of the ramp face in the vicinity of the jet exit plane as shown in the "end" view of Fig. 6.10, and only secondarily to the weak axial vortices generated by spillage over the ramp sidewalls.[6.22]

In order to strengthen the formation of axial vorticity, sweepback has been added to both raised and relieved ramps, as illustrated in Fig. 6.11. This configuration has been shown to be superior to the corresponding unswept ramp configurations for enhancement of near-field mixing.[6.7, 6.16]

Other strategies for inducing axial vorticity into the airstream include inserting various configurations of vortex generators into the freestream, including "micropylons"[6.23] and swept delta-wing tab mixers[6.22, 6.24] An interesting hybrid scheme, which induces axial vorticity in both the fuel and air streams, uses serrated or fluted

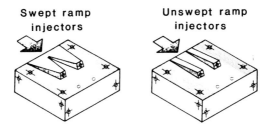

Fig. 6.11 Perspective view of swept and unswept ramp fuel injectors.

downstream edges on the fuel/air splitter plate, somewhat like the corrugated edge of a pie crust.[6.25] The laterally undulating surface introduces alternately raised and relieved ramps into both streams as they flow past the corrugated splitter plate. While this fluted or lobed mixer arrangement has been found very effective in subsonic mixing, both for bypass air mixing and noise reduction in turbojet engine exhaust nozzles, there is as yet no experimental confirmation of its effectiveness in supersonic flows. While holding promise for increased near-field mixing, there is at present no theory to predict the effectiveness of any of these devices for enhanced fuel/air mixing at reasonable cost of total pressure loss.

6.2.8.3 A mixing efficiency model for axial vortex mixers.   In a modeling study of chemical kinetics in scramjet burners, Jachimowski suggests an exponential function representation for the mixing efficiency,[6.26] which can be adapted for present purposes as

$$\eta_M = \frac{1 - e^{-Ax/L_m}}{1 - e^{-A}} \tag{6-46}$$

where A is a fit parameter which varies in the range $(1,5)$ to represent increasingly effective near-field mixing,[6.27,6.28] and where L_m is the distance for minor mixant depletion as given by Eqs. (6-42). Values of $A = 1.77$ and 3.4 may be used in Eq. (6-46) to represent axial mixing efficiency of unswept and swept 10-deg. raised ramp mixers, as illustrated in Figs. 6.10a and 6.11, based on results of a CFD comparison of swept ramp, unswept ramp, and 30-deg. vectored wall injection mixers.[6.27] In a subsequent study by the same researchers,[6.28] $A = 4.9$ fits the mixing efficiency for a swept ramp mixer at a higher freestream Mach number. Both studies concluded that the swept ramp is a more effective mixer than either vectored wall injection or unswept ramps, and that the unswept ramp performed so poorly that it should not be considered further as a candidate mixing device.

The mixing efficiency model for normal injection, Eq. (6-44), together with the linear variation scheme for wall injection between 0 and 90 deg. as given at the end of Sec. 6.2.7, did not agree well with the mixing efficiency data in Ref. 6.23 at an overall equivalence ratio $\phi_0 = 3.0$.

Equation (6-46) has much to recommend it as a generic model for hypermixers or axial vortex enhanced mixing systems, compared to Eqs. (6-41) and (6-44) for parallel and normal jet mixing, respectively. Because of the nature of far-field mixing by small-scale turbulence, complete micromixing is achieved only in the asymptotic limit of infinite x, which is well represented by the decaying-exponential term in Eq. (6-46). As better information becomes available in the future, from both experimental and CFD modeling efforts, Eq. (6-46) will be a useful form for incorporating these data in scramjet combustor design studies.

6.2.9 Summary of Fuel-Air Mixing

In this chapter, we have drawn extensively from Dimotakis' summary[6.7] of more than twenty years of research on mixing in turbulent shear/mixing layers. The most significant finding of this research for supersonic combustion is that shear between two coflowing streams cannot achieve near-field mixing to near-stoichiometric proportions of air with hydrogen, the fuel of choice for hypersonic propulsion. It now appears that lateral or spanwise vorticity, whether generated by shear layers, lateral fuel injection, or separated base/wake flows, is unlikely to produce the required H_2/air mixing at a reasonable cost in total pressure loss. Regrettably, although the need for stirring by axial vorticity was recognized in 1968,[6.17] the subsequent body of reported research on this subject is miniscule compared to that on mixing in shear layers.

However, the twenty years of intense research on shear/mixing layers has not been wasted with regard to supersonic combustion for hypersonic propulsion. Shear layer research has clarified the concepts that must be utilized and the questions that must be addressed with respect to the effectiveness and energy efficiency of axial vortex mixing. Much more basic research remains to be done with respect to mixing in axial vortices. CFD studies will greatly assist the design of experiment and interpretation of experimental results, but CFD alone cannot supply all the answers. CFD researchers are the first to recognize and state that the quality and reliability of CFD analysis are in direct proportion to the adequacy of the included physical submodels, which still remain to be discovered. "GIGO" (garbage in, garbage out) still applies.

In the interim, we will have to make do with empirical measures of mixing effectiveness, as represented by the mixing "efficiency" η_M.

In the remainder of this chapter, we will next explore how the interaction of mixing and chemistry determines a corresponding combustion "efficiency" η_b, and finally how the combination of mixing, chemistry, and aerothermodynamics of the bulk flow interact to determine the performance of the combustion system.

6.3 COMBUSTION CHEMISTRY

In Sec. 2.5, you were introduced to the phenomenon of molecular dissociation of air at elevated temperature, and how to determine its equilibrium composition at given values of temperature and pressure by using the software program HAP(Air). In the present section, we are concerned not with the equilibrium dissociation of molecules of gaseous mixtures at prescribed values of pressure and temperature, but rather with finding the temperature and composition that results from the burning of gaseous hydrocarbon fuels (including hydrogen H_2) with air, at a given pressure. It is this release of sensible thermal energy in an exothermic chemical reaction which provides the energy required by the thermodynamic propulsion cycle, as discussed in Sec. 3.2.5. Given sufficient time for the chemical reactions to fully occur, the end-state products of combustion will be in chemical equilibrium, and the corresponding equilibrium temperature is called the *adiabatic flame temperature* (AFT). However, it is usually the case that there is inadequate time for either complete mixing or complete combustion to occur. The resulting *nonequilibrium* state of the gas can be determined only from considering *chemical kinetics,* the study of finite-rate chemical reactions. Each of these topics will be considered in turn in this section.

6.3.1 Equilibrium Concepts and Definitions

The basic problem of thermodynamics is the determination of the equilibrium state that eventually results after the removal of internal constraints in a closed composite system.

— *H. B. Callen, Thermodynamics*[6.29]

To illustrate the above statement from Ref. 6.29, consider a *composite system* consisting of two rigid gas bottles, one containing oxygen O_2 and the other nitrogen N_2. Each is in internal equilibrium, but we would like to know the properties of the mixture of O_2 and N_2 that would result if we were to interconnect the two bottles. The composite system is assumed to not leak any gases to or from the surroundings, nor to exchange either heat or work with the surroundings—hence the composite system is called a *closed system*.

In the case of intermixing of O_2 and N_2, we know from experience that the final equilibrium state will be a mixture of the two gases, having final values of pressure and temperature lying somewhere between the respective initial property values in the bottles prior to mixing. However, if the two tanks initially contained gases which are mutually *hypergolic* (react spontaneously and exothermically upon contact), the final equilibrium state will be at the adiabatic flame temperature (AFT), which is considerably greater than that of either of the two initially separated gases. (In a closed system, as described here, the equilibrium pressure would also be very much greater than that of either initially separated gas.) Moreover, the final composition will no longer be simply a mixture of the two initial gases and possibly their dissociation products, but will include one or more different gases which are the *products of combustion* of the exothermic chemical reaction. The software program HAP(Equilibrium) calculates the AFT and thermodynamic state and composition of the equilibrium products of combustion.

Before we address the details of how HAP(Air) and HAP(Equilibrium) calculate the thermodynamic equilibrium of chemically reacting gases, we will first consider the mechanical equilibrium of an idealized ball bearing rolling around in a rigid, shallow bowl. We know from common experience, as well as from the study of dynamics and mechanics of solids, that the ball-bowl system is in mechanical equilibrium only when the ball is at rest in the bottom of the bowl. We arrive at this conclusion formally by assuming that the ball and bowl are a *conservative system*; that is, that wind drag and rolling friction can be neglected, so that if the ball is in motion it will continue in motion indefinitely, with the sum of its kinetic energy and gravitational potential energy remaining constant. Trying to place the ball at rest anywhere but in the bottom of the bowl would result in the ball moving spontaneously toward the bottom of the bowl.

In the language of thermodynamics, we say that in the absence of viscous or frictional processes, there is no *dissipation of mechanical energy* (gravitational potential and kinetic energy) *to thermal energy* (internal energy of ball and bowl, heat transfer between ball and bowl and surroundings.) Such nondissipative motion is described in thermodynamic parlance as an *isentropic process.* The corresponding thermodynamic statement of the ball-bowl mechanical equilibrium is:

> For a collection of identical ball-plus-bowl systems having the same entropy, but having different mechanical energies, the system in equilibrium is that which has the least mechanical energy.

6.3.2 Thermodynamic Equilibrium of Ideal Gas Mixtures

The equivalent statement for thermodynamic equilibrium of a mixture of gases is:

> For a collection of identical closed thermodynamic systems having the same entropy and volume but having different internal energy, the system in thermodynamic equilibrium is that system having the least internal energy.[6.29]

By "identical" is meant having the same number of atoms of each chemical element, but having different numbers of molecules which are made up of the various atoms present. The equilibrium postulate may be stated in mathematical form as

$$E_{eq} = \min \{E(S, V, \{N_i, i = 1, NS\})\} \quad \text{for constant } S, V \quad (6\text{-}47)$$

where E is the internal energy, S is the entropy, V is the volume, and N_i is the number of molecules (or moles) of each of the NS gases present.

There is a practical problem with utilizing Eq. (6-47) to determine the thermodynamic equilibrium state, namely that the constraints S and V are both *extensive* (additive) properties. While we can easily measure and constrain the volume, we cannot as a practical matter either measure or constrain the entropy. (Ask your laboratory technician if you can check out an entropimeter!) Happily, thermodynamics provides a way to reformulate Eq. (6-47) in terms of familiar and more readily measurable and controllable *intensive* thermodynamics properties, namely pressure and temperature. We start with the *fundamental equation*[6.29] of the system of interest, which appears on the right-hand side of Eq. (6-47),

$$E = E (S, V, \{N_i, i = 1, NS\}) \quad (6\text{-}48)$$

It is desired to perform some mathematical operation on Eq. (6-48) which will replace the extensive properties S and V with the intensive properties p and T as independent variables. The desired operation is called a partial Legendre transformation,[6.29] in which the entropy S and volume V are formally replaced with their respective partial derivatives of E with respect to S and to V, as follows:

$$L[E]_{S, V} \equiv E - S \left(\frac{\partial E}{\partial S}\right)_{V, N_i} - V \left(\frac{\partial E}{\partial V}\right)_{S, N_i} \quad (6\text{-}49)$$

Further, we know from thermodynamic theory that the partial derivatives in Eq. (6-49) are identically equal to T and $(-p)$, respectively:

$$T \equiv \left(\frac{\partial E}{\partial S}\right)_{V,N_i} \quad \text{and} \quad -p \equiv \left(\frac{\partial E}{\partial V}\right)_{S,N_i} \qquad (6\text{-}50)$$

With the substitution of Eqs. (6-50) into Eq. (6-49), there results

$$L[E]_{S,V} = E - ST - V(-p) \qquad (6\text{-}51)$$

A partial Legendre transformed fundamental equation such as Eq. (6-51) is called a *thermodynamic potential* function, and the particular one defined by Eq. (6-51) is called the *Gibbs potential* function, or simply the *Gibbs function*:

$$G(T, p, \{N_i, i = 1, NS\}) \equiv E - TS + pV = H - TS \qquad (6\text{-}52)$$

where the enthalpy $H = E + pV$ has been introduced for convenience. The equilibrium postulate can now be restated as

> For a collection of identical thermodynamic systems having the same temperature and pressure but different values of Gibbs function, the system in thermodynamic equilibrium is that system having the least value of the Gibbs function.[6.29]

As before, all of the systems are identical in that they all have the same number of atoms of each chemical element present, but they differ from each other in the variety and number of ways the atoms present may be combined into molecules. In mathematical form, the equilibrium requirement is restated as

$$G_{eq} = \min \{G(T, p, \{N_i, i = 1, NS\})\} \quad \text{for constant } T, p \qquad (6\text{-}53)$$

For a multicomponent mixture of ideal gases, the mixture Gibbs function per unit mass is obtained by summing over the partial molal Gibbs function of each of the gases in the mixture,

$$g = \sum_{k=1}^{NS} n_k g_k \qquad (6\text{-}54)$$

where n_k are the *mass-specific mole numbers* of the k-th gas present (units: lbm-moles of gas k/lbm of mixture, or kg-moles k/kg mixture); and where

$$g_k \equiv h_k - T s_k = h_k - T s_k^0 + RT \log \frac{n_k}{n_m} + RT \log \frac{p}{p_0} \qquad (6\text{-}55)$$

is the *partial molal Gibbs function* of the k-th species.

In Eq. (6-55), $n_m \equiv \sum_{i=1}^{NS} n_i$ is the sum of the mole numbers, p and T are the pressure and temperature of the mixture, and the subscripts and superscripts 0 (zero) denote values at 1 atmosphere pressure. The one-atmosphere enthalpy and entropy for ideal gases are given by

$$h_k = (\Delta h_{f_k}^0)_{298} + \int_{298}^{T} C_{p_k} \, dT'$$

and

$$s_k^0 = (s_k^0)_{298} + \int_{298}^{T} C_{pk} \frac{dT'}{T'} \qquad (6\text{-}56)$$

where $(\Delta h_{f_k}^0)_{298}$ is *the enthalpy of formation* of the k-th gas, which is the sum of the molecular bond energy and the sensible thermal energy at 298 K (25 °C), and where $(s_k^0)_{298}$ is the absolute entropy of the k-th gas at the standard reference state of 1 atmosphere and 298 K (25 °C).

Table 6.1 lists the enthalpies of formation for some of the reactant and product species of interest in airbreathing propulsion combustion. Note that the enthalpy of formation at 25 °C or 298 K for H_2, O_2, and N_2 is equal to 0. That is because, by definition, the datum state for enthalpy of formation is 0 for the elements "in their naturally occurring allotropic form" at the standard reference state of 1 atmosphere and 298 K.

The equilibrium composition at given p and T, $\{n_i^{eq}, i = 1, NS\}$, is that set of mole numbers which satisfies the minimum Gibbs function requirement

$$dg = \sum_{i=1}^{NS} \left. \frac{\partial g}{\partial n_i} \right)_{T,p,n_{k \neq i}} dn_i = 0 \qquad (6\text{-}57)$$

subject to the atom-number constraint

$$\sum_{k=1}^{NS} a_{ik}^L n_k - b_i^* = 0, \qquad i = 1, NLM \qquad (6\text{-}58)$$

where b_i^* is the number of kg-atoms (or lbm-atoms) of element i present per unit mass (kg or lbm) of mixture, a_{ik}^L is the number of i-th atoms in a molecule of the k-th gas, and NLM is the number of distinct chemical elements present in the mixture.

Because of the materials and dissociation temperature limitation and the need to maintain sufficiently high combustor pressures to enable chemical reactions to occur rapidly, ionization of combustion product gases is neglected.

Table 6.1   Enthalpy of formation $(\Delta h^0_{f_k})_{298}$ for some reactant and product species of interest in airbreathing propulsion combustion (Ref. 6.30).

		BTU/lbmol	kJ/kmol
Methane	CH_4	−32,192	−74,877
Ethane	C_2H_6	−36,413	−84,695
Hexane	C_6H_{14}	−71,784	−166,964
Octane	C_8H_{18}	−89,600	−208,403
Carbon monoxide	CO	−47,520	−110,527
Carbon dioxide	CO_2	−169,181	−393,503
Atomic hydrogen	H	93,717	217,979
Hydrogen	H_2	0	0
Water vapor	H_2O	−103,966	−241,818
Hydrogen peroxide	H_2O_2	−58,518	−136,108
Hydrogen peroxyl	HO_2	899	2,091
Atomic oxygen	O	107,139	249,197
Oxygen	O_2	0	0
Hydroxyl	OH	16,967	39,463
Atomic nitrogen	N	203,200	472,629
Nitrogen	N_2	0	0
Nitrous oxide	N_2O	35,275	82,048
Nitric oxide	NO	38,817	90,286
Nitrogen dioxide	NO_2	14,228	33,093

The algorithm utilized by the programs HAP(Air) and HAP(Equilibrium) for solving Eqs. (6-57) and (6-58) was developed in the 1960's by engineers at the NASA-Lewis Research Center.[6.30] The method of Lagrange multipliers is used to generate a coupled set of $(NLM + 1)$ nonlinear algebraic equations, which are in turn solved iteratively by an accelerated Newton-Raphson method. Since $NLM = 4$ (for C, H, O, and N) is a very small number, the equilibrium solution is obtained very quickly. A complete derivation of the computational algorithm can be found in Ref. 6.31.

6.3.2.1   Adiabatic flame temperature (AFT).   Using the algorithm described in Sec. 6.3.2, the software program HAP(Air) determines the composition of a gas mixture (in this case, air) whenever the temperature and pressure are different from near-standard conditions, where air is composed only of undissociated oxygen and nitrogen. This scenario is referred to by Gordon and McBride as "the (T,p) problem" of chemical equilibrium.[6.30]

Combustion of fuel and air presents a somewhat different problem, however. Typically, while we know the initial composition and state of the fuel/air mixture and at what pressure burning will occur, we do not know what the temperature will be after combustion. However, if it is assumed that combustion occurs without either heat or work interaction with the surroundings, then the enthalpy of the (final) products will be the same as the (initial) reactants, and that value is known. In this case, it is necessary to specify the final *enthalpy*, rather than the unknown final *temperature*, of the equilibrium products. For this reason, the determination of the *adiabatic flame temperature* AFT is referred to as "the (h, p) problem."[6.30]

When molecular collisions result in the exchange of atoms between molecules, the number of molecules of each kind changes. Exothermic reactions result in the release of chemical bonding energy, which appears as sensible thermal energy. These two kinds of energy associated with each molecule appear in Eq. (6-56), repeated here for convenience:

$$h_k = (\Delta h_{f_k}^0)_{298} + \int_{298}^{T} C_{p_k} dT' \tag{6-56}$$

The mass-specific enthalpy of a mixture of gases is given by

$$h = \sum_{k=1}^{NS} n_k h_k \tag{6-59}$$

and for the particular mixtures representing the reactants (fuel plus air) and products, that is, those gases appearing on the left-hand side and on the right-hand side of Eq. (6-5), respectively,

$$h_R = \sum_{k=1}^{NS} (n_k)_R h_k \quad \text{and} \quad h_P = \sum_{k=1}^{NS} (n_k)_P h_k \tag{6-60}$$

If the reactants are ignited and allowed to burn to the final equilibrium state without heat being added or removed during the process, the final equilibrium temperature is the adiabatic flame temperature, as mentioned previously. For example, consider a case where the reactants are initially at 298 K. After combustion occurs, resulting in an adiabatic process which releases sensible thermal energy, the adiabatic flame temperature is found by solving the algebraically implicit equation

$$h_P = h_R \tag{6-61}$$

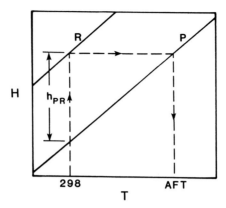

Fig. 6.12 Enthalpy-temperature diagram illustrating relationship between enthalpies of reactants h_R and products h_P, and between initial reactants temperature (example shown as 298 K) and the equilibrium adiabatic flame temperature (AFT).

Figure 6.12 illustrates the solution of Eqs. (6-60) and (6-61). Note that at the initial temperature, assumed here to be 298 K, h_P is less than h_R. This is because the principal product molecules have larger negative values of enthalpy of formation than reactant molecules, as can be seen from Table 6.1.

If the fuel/air reactant mixture temperature is initially 298 K, as illustrated in Fig. 6.12, and if the fuel and air are in stoichiometric proportion so that $\phi = 1.0$, then the stoichiometric AFT is found from the solution of Eq. (6-61). If the products are subsequently cooled at constant pressure until the mixture temperature is reduced back to 298 K, then the amount of heat removed is called the *heating value* or *heat of reaction* h_{PR}, as mentioned in Sec. 3.2.5. Since the end states of the overall process (adiabatic burning followed by cooling) are both at the reference temperature 298 K, the relationship between the enthalpies of formation of each species and the heating value of the fuel are determined from an energy balance for the overall process,

$$h_{PR} = (h_R)_{298} - (h_P)_{298} = \sum_{i=1}^{NS}(n_i)^R \left(\Delta h_{f_i}^0\right)_{298} - \sum_{i=1}^{NS}(n_i)^P \left(\Delta h_{f_i}^0\right)_{298}$$

$$(6\text{-}62)$$

The values of h_{PR} given in Table 3.1 were determined for *complete combustion* (no dissociation of products), so they are all somewhat greater than the equilibrium values of h_{PR} calculated by the (h,p) option of HAP(Equilibrium).

If the fuel and air are initially at a temperature greater than 298 K, the solution of Eq. (6-61) results in a proportionately greater

AFT, as can be seen from tracing out the path on the H-T diagram, Fig. 6.12. For example, for stoichiometric H_2-air reactants at 298 K (25 °C) and 1 atm, HAP(Equilibrium) gives a constant-pressure adiabatic flame temperature of 2382.76 K. If the reactant temperature is raised to 500 K, the new AFT is calculated to be 2480.22 K. (Why does a 202 °C increase in reactant temperature cause less than 100 °C increase in AFT?)

The algorithm used in HAP(Equilibrium) for solving the (h,p) problem is the same as that for the (T,p) problem described in Sec. 6.3.2, except that Eq. (6-61) is added to the problem as an additional constraint. As a result there are now $(NLM+2)$ simultaneous equations to be solved, rather than the $(NLM+1)$ equations required for the (T,p) problem.

6.3.3 Chemical Kinetics

As has been already been pointed out, it is usually the case in supersonic combustion that insufficient time is available for the exothermic combustion reactions to reach equilibrium. Consequently, it is necessary to consider the rate at which chemical reactions proceed.

For purposes of mathematically modeling finite-rate chemical kinetics for homogeneous gas-phase chemical reaction, it is assumed that a great many individual, reversible reactions of the form CO + OH $\rightarrow$ CO_2 + H occur. By convention, species appearing on the left-hand side of each such reaction are called *reactants*, and those on the right-hand side are called *products*. Note that the example reaction, being reversible, could just as well have been written CO_2 + H $\rightarrow$ CO + OH, in which case CO_2 and H would be called reactants, rather than CO and OH. With a suitable collection of such reactions, it is possible to approximately describe the time rates of change of all species. Such a set of reactions is referred to as a *reaction mechanism*. Table 6.2 is such a mechanism for combustion of gaseous hydrogen H_2 with air.

For a closed thermodynamic system (fixed amount of mass) at constant pressure, the system of ordinary differential equations which describe the isobaric "batch reaction" time rate of change of the i-th chemical species is given by[6.32]:

$$\frac{dn_i}{dt} = f_i(n_k, T) \qquad i, k = 1, NS \qquad (6\text{-}63)$$

where

$$f_i = -\rho^{-1} \sum_{j=1}^{JJ} \left(\alpha'_{ij} - \alpha''_{ij} \right) (R_j - R_{-j}) \qquad (6\text{-}64)$$

Table 6.2   Hydrogen-air combustion mechanism (Ref. 6.26). The symbol "M" stands for *third body,* meaning any species acting as a gas-phase catalyst. Units: s, gmol, cm^3, K.

j	Reaction							A_j	B_j	E_j/R
1.	H_2	$+ O_2$		$= OH$	$+ OH$			1.70E13	0	24,157
2.	H	$+ O_2$		$= OH$	$+ O$			2.60E14	0	8,455
3.	O	$+ H_2$		$= OH$	$+ H$			1.80E10	1.00	4,479
4.	OH	$+ H_2$		$= H_2O$	$+ H$			2.20E13	0	2,592
5.	OH	$+ OH$		$= H_2O$	$+ O$			6.30E12	0	549
6.	H	$+ OH$	$+ M$	$= H_2O$	$+ M$			2.20E22	-2.00	0
7.	H	$+ H$	$+ M$	$= H_2$	$+ M$			6.40E17	-1.00	0
8.	H	$+ O$	$+ M$	$= OH$	$+ M$			6.00E16	-0.60	0
9.	H	$+ O_2$	$+ M$	$= HO_2$	$+ M$			2.10E15	0	-503
10.	HO_2	$+ H$		$= O_2$	$+ H_2$			1.30E13	0	0
11.	HO_2	$+ H$		$= OH$	$+ OH$			1.40E14	0	543
12.	HO_2	$+ H$		$= H_2O$	$+ O$			1.00E13	0	543
13.	HO_2	$+ O$		$= O_2$	$+ OH$			1.50E13	0	478
14.	HO_2	$+ OH$		$= H_2O$	$+ O_2$			8.00E12	0	0
15.	HO_2	$+ HO_2$		$= H_2O_2$	$+ O_2$			2.00E12	0	0
16.	H	$+ H_2O_2$		$= H_2$	$+ HO_2$			1.40E12	0	1,812
17.	O	$+ H_2O_2$		$= OH$	$+ HO_2$			1.40E13	0	3,221
18.	OH	$+ H_2O_2$		$= H_2O$	$+ HO_2$			6.10E12	0	720
19.	H_2O_2	$+ M$		$= OH$	$+ OH$	$+ M$		1.20E17	0	22,899
20.	O	$+ O$	$+ M$	$= O_2$	$+ M$			6.00E17	0	-906
21.	N	$+ N$	$+ M$	$= N_2$	$+ M$			2.80E17	-0.75	0
22.	N	$+ O_2$		$= NO$	$+ O$			6.40E9	1.00	6,300
23.	N	$+ NO$		$= N_2$	$+ O$			1.60E13	0	0
24.	N	$+ OH$		$= NO$	$+ H$			6.30E11	0.50	0
25.	H	$+ NO$	$+ M$	$= HNO$	$+ M$			5.40E15	0	-600
26.	H	$+ HNO$		$= NO$	$+ H_2$			4.80E12	0	0
27.	O	$+ HNO$		$= NO$	$+ OH$			5.00E11	0.50	0
28.	OH	$+ HNO$		$= NO$	$+ H_2O$			3.60E13	0	0
29.	HO_2	$+ HNO$		$= NO$	$+ H_2O_2$			2.00E12	0	0
30.	HO_2	$+ NO$		$= NO_2$	$+ OH$			3.40E12	0	-260
31.	H	$+ NO_2$		$= NO$	$+ OH$			3.50E14	0	1,500
32.	O	$+ NO_2$		$= NO$	$+ O_2$			1.00E13	0	600
33.	NO_2	$+ M$		$= NO$	$+ O$	$+ M$		1.16E16	0	66,000

In Eq. (6-64), R_j and R_{-j} $(j = 1, JJ)$ are modified Arrhenius expressions for the forward and reverse rates of the j-th reaction,

$$R_j = k_j \Pi_{k=1}^{NS}(\rho n_k)^{\alpha'_{kj}} \qquad (6\text{-}65)$$

and

$$R_{-j} = k_{-j} \Pi_{k=1}^{NS}(\rho n_k)^{\alpha''_{kj}} \qquad (6\text{-}66)$$

In Eqs. (6-63) to (6-66), n_i is the mass-specific mole number of the i-th species $(i = 1, NS)$, T is the temperature, ρ is the mixture mass density, α'_{ij} and α''_{ij} are the stoichiometric coefficients of species i $(i = 1, NS)$ in reaction j $(j = 1, JJ)$ as a reactant and as a product species, respectively; k_j and k_{-j} are the forward and reverse *rate constants* in the modified Arrhenius rate expressions for R_j and R_{-j}, which in turn are the forward and reverse rates of the j-th reaction $(j = 1, JJ)$. NS is the total number of distinct chemical species in the gas mixture, and JJ is the total number of independent chemical reactions. For example, $NS = 13$ and $JJ = 33$ for the hydrogen/oxygen reaction mechanism in Table 6.2.

For adiabatic batch reaction, the equation for conservation of static enthalpy, Eqs. (6-57) through (6-59), constitutes an algebraic constraint on Eqs. (6-63) to (6-66):

$$h \equiv \sum_{i=1}^{NS} h_i n_i = h_0 = \text{constant} \qquad (6\text{-}67)$$

where h_i is the molal enthalpy of the i-th species, defined in Eq. (6-56), and h_0 is the mass-specific enthalpy of the mixture.

The mass density ρ in Eqs. (6-64) through (6-66) is determined from the temperature and pressure by the equation of state for an ideal gas, $\rho = p/(RT n_m)$, where R is the universal gas constant, and n_m is the sum of the mole numbers, $n_m \equiv \sum_{i=1}^{NS} n_i$.

6.3.3.1 Kinetic rate constants and the equilibrium constant.
The forward rate constant k_j in Eq. (6-65) is usually prescribed by three empirical constants, A_j, B_j, and E_j, in the form

$$k_j = A_j T^{B_j} exp\left(\frac{-E_j}{RT}\right) \qquad (6\text{-}68)$$

as for example in the hydrogen-air reaction mechanism presented in Table 6.2.

The reverse rate constant k_{-j} for the j-th reaction in the mechanism *can* be prescribed by means of three additional fit coefficients, A_{-j}, B_{-j}, and E_{-j}, but it is conventional to calculate the reverse rate constant k_{-j} from the fact that, at chemical equilibrium, the forward rate R_j of every reaction must be equal to its reverse rate R_{-j}. By setting Eq. (6-65) equal to Eq. (6-66), there follows

$$k_{-j} = k_j \frac{\Pi_{k=1}^{NS}(\rho n_k)^{\alpha'_{kj}}}{\Pi_{k=1}^{NS}(\rho n_k)^{\alpha''_{kj}}} = k_j \frac{[RT]^{\sum_{i=1}^{NS}(\alpha''_{ij}-\alpha'_{ij})}}{K_{p_j}} \qquad (6\text{-}69)$$

where K_{p_j} is the *equilibrium constant* for reaction j, which can be evaluated directly from thermochemical properties of the individual species in each j-th reaction,[6.29, 6.30]

$$K_{p_j} \equiv exp\left[\sum_{i=1}^{NS}(\alpha'_{ij} - \alpha''_{ij})\frac{g_i^0}{RT}\right] = \Pi_{i=1}^{NS}\left(\frac{p_i}{p_0}\right)^{(\alpha''_{ij}-\alpha'_{ij})} \qquad (6\text{-}70)$$

where $g_i^0 \equiv h_i - Ts_i^0$ is the temperature-dependent part of the partial molal Gibbs function of the i-th species, as in Eq. (6-55), p_i is the partial pressure of the i-th species in the equilibrium mixture of gases $p_i = n_i p/n_m$, p is the mixture pressure, and $p_0 = 2116$ psf (101 kPa) is the reference pressure of 1 standard atmosphere.

Since the left-hand side of Eq. (6-70) is a function only of temperature-dependent thermodynamic data, K_{p_j} (or log K_{p_j}) for many reactions have been calculated and tabulated. In addition to using K_{p_j} for determining reverse rate constants as in Eq. (6-69), tabulated values of K_{p_j} can also be equated to the right-hand side of Eq. (6-70) and used to independently verify the equilibrium values of species partial pressures predicted by HAP(Equilibrium) for those species which appear as reactants and products in the j-th reaction. More details on the derivation and other applications of equilibrium constants can be found in any undergraduate thermodynamics textbook, such as Ref. 6.29.

It is helpful for purposes of discussion to restate the expression for net species production, Eq. (6-64) together with Eqs. (6-65) and (6-66), as a difference between two positive-definite terms:

$$f_i = Q_i - D_i \qquad (6\text{-}71)$$

where

$$Q_i \equiv \rho^{-1}\sum_{j=1}^{JJ}\left(\alpha'_{ij}R_{-j} + \alpha''_{ij}R_j\right) \qquad (6\text{-}72)$$

and

$$D_i \equiv \rho^{-1}\sum_{j=1}^{JJ}\left(\alpha'_{ij}R_j + \alpha''_{ij}R_{-j}\right) \qquad (6\text{-}73)$$

In Eqs. (6-71) to (6-73), Q_i and D_i represent the gross rates of production and destruction of the i-th species, respectively, due to the contributions of all the JJ forward and reverse reactions in the assumed or prescribed mechanism.

6.3.4 Physical and Computational Scenario for Isobaric Batch Reaction

Consider an initially quiescent micromixture of fuel and air in a shock tube. At time 0, a shock wave passes quickly through the mixture, rapidly raising the pressure and temperature well above the ignition limits. The subsequent events leading to release of sensible thermal energy occur in three distinctly different chemical-physical periods, as illustrated in Fig. 6.13. These three periods or regimes are called the *induction, heat release,* and *equilibration* regimes.

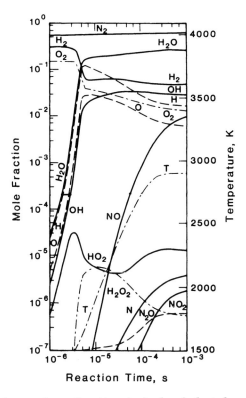

Fig. 6.13 Regimes of combustion in isobaric batch reaction of stoichiometric ($\phi = 1$) hydrogen and air. Initial conditions: $p = 2$ atm, $T = 1500$ K. Reaction mechanism includes 30 reactions and 15 species (Ref. 6.32).

The *induction* period is the time interval immediately following some form of homogeneous bulk ignition. In the homogeneous (completely micromixed) case under consideration, ignition occurs due to shock compression, whereas in the scramjet engine, ignition occurs only after fuel and air are micromixed to flammable proportions $(0.2 < \phi < 2.0)$, as explained in Sec. 6.1.1. During the induction period, the mole numbers of *intermediates* or chain carriers, such as O, H, OH, HO_2, and H_2O_2, increase by many orders of magnitude from near-zero values in the uncompressed mixture, or in the unmixed streams in the scramjet case. During this period, the species production terms Q_i in Eq. (6-71) are very large compared to the destruction terms D_i, so that f_i is large and positive. Also, the coupling with the enthalpy equation (6-67) is very weak, so that the induction process is essentially isothermal as well as adiabatic; no sensible energy is released. When the intermediate species have reached some critical value of concentration sufficient to begin to react with fuel and oxygen molecules, the process of releasing sensible thermal energy can begin. Therefore, the induction period ends when the mixture temperature begins to rapidly increase. In the hydrogen/air example illustrated in Fig. 6.13, the induction time (sometimes called the *ignition delay time*) is about 2×10^{-4} s. Many researchers have proposed empirical equations for induction times for stoichiometric mixtures of hydrogen and air, such as [6.33]

$$t_{\text{ind}} \approx 4.5 \times 10^{-9} \left(\frac{p_0}{p} \right) exp \left(\frac{10^4}{T} \right) \qquad (6\text{-}74)$$

where t_{ind} is in s, p is in the same units as p_0, the pressure of a standard atmosphere, and T is in Kelvins. Note that, for the mixture pressure and initial temperature of Fig. 6.13, Eq. (6-74) estimates $t_{\text{ind}} = 1.768 \times 10^{-6}$, in good agreement with Fig. 6.13.

During the heat release period, very rapid changes in temperature and species mole numbers occur. During this period, the species equations and the energy conservation equations are all very strongly coupled. The heat release period ends when the reaction intermediates have all passed their peak values, at about 1×10^{-5} s in Fig. 6.13.

The equilibration period begins when all species mole numbers begin a decaying-exponential approach toward their respective equilibrium values. The equilibration process does not have a clearly defined termination, due to the asymptotic nature of the approach to the chemical equilibrium state. However, since equilibrium values of temperature and species concentration can be determined in advance by a Gibbs function minimization scheme, as shown in Sec. 6.3.2, the end of the equilibration period can be defined as the time at which all of the mole numbers and the temperature are within (say)

1 percent of their chemical equilibrium values, at about 2×10^{-4} s in Fig. 6.13.

Note that Fig. 6.13 is presented on log-log coordinates. This choice of scale tends to obscure the many orders-of-magnitude variation in mole fractions of the various species. In addition, note that each of the three sequential combustion periods requires an order-of-magnitude longer time than the preceding period. The implications of this slowing down of chemical reaction rates are of obvious concern to supersonic combustion, just as is the slowing down of the mixing process from the rapid near-field stirring to the slower far-field micromixing.

6.4 COMBINED MIXING AND CHEMICAL KINETICS

In the preceding section, we studied the sequence of chemical-kinetic events which occur in homogeneous (fully micromixed) batch reaction at constant pressure, in order to clarify the differences between the processes of induction, heat release, and equilibration. We now wish to consider the combined, simultaneous effects of both finite-rate micromixing, as represented by the model equations for the mixing parameter Eqs. (6-41) through (6-44) and (6-46), and finite-rate chemical kinetics as represented by Eqs. (6-63) through (6-67). We will do this by considering the effects of these combined processes in a finite control volume, as illustrated in Fig. 6.14.

Whereas Eq. (6-63) represents the rate of change of mole numbers of the i-th species in a *closed* thermodynamic system, we are concerned here with the rate of chemical reaction in a steady-state,

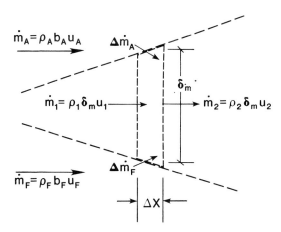

Fig. 6.14 Control volume for steady-flow analysis of simultaneous mixing and chemical reaction within a mixing layer.

steady-flow *open* thermodynamic system or control volume having multiple inlet streams, as described in Sec. 2.6.2.

The expression for mass conservation within the c.v. (control volume) of Fig. 6.14 is given by Eq. (2-28),

$$\dot{m}_1 + \Delta \dot{m}_F + \Delta \dot{m}_A = \dot{m}_2 \qquad (6\text{-}75)$$

where the subscripts 1 and 2 denote the upstream and downstream faces of the c.v., respectively, and where $\Delta \dot{m}_F$ and $\Delta \dot{m}_A$ are the mass inflow rates due to in-mixing from the fuel and air streams. Since $\eta_M(x)$ is assumed to be given, the mass flow rates in each of the three streams $(F, A, 1)$ at any axial location x can be calculated at once, as given in Table 6.3.

Assuming constant pressure for the steady-flow mixing and reaction process and neglecting shear stresses at c.v. boundaries, conservation of linear momentum, Eq. (2-29), gives

$$\dot{m}_1 u_1 + \Delta \dot{m}_F u_F + \Delta \dot{m}_A u_A = \dot{m}_2 u_2 \qquad (6\text{-}76)$$

from which the exiting convective velocity u_2 from the mixing layer c.v. is

$$u_2 = \frac{\dot{m}_1 u_1 + \Delta \dot{m}_A u_A + \Delta \dot{m}_F u_F}{\dot{m}_1 + \Delta \dot{m}_A + \Delta \dot{m}_F} \qquad (6\text{-}77)$$

Note that the convective velocity u_2 in this c.v. analysis is different from that for the mean velocity of shear layer vortex structures u_c given by Eq. (6-27).

Conservation of i-th species mole numbers is given by

$$\dot{m}_1 n_{i,1} + \Delta \dot{m}_F n_{i,F} + \Delta \dot{m}_A n_{i,A} + f_i \Delta x \delta_m = \dot{m}_2 n_{i,2} \,,$$

$$i = 1, NS \qquad (6\text{-}78)$$

Table 6.3 Axial variation of mass flow rates for the three streams illustrated in Fig. 6.15. $L_{m,F}$ and $L_{m,A}$ are given by Eq. (6-42).

	$\phi_0 \leq 1$ $(0 \leq x \leq L_{m,f})$	$\phi_0 > 1$ $(0 \leq x \leq L_{m,A})$
$\dot{m}_F(x)/\dot{m}_{A,0}$	$f_{st}\phi_0\left(1 - \eta_M(x)\right)$	$f_{st}\left(\phi_0 - \eta_M(x)\right)$
$\dot{m}_A(x)/\dot{m}_{A,0}$	$1 - \phi_0\eta_M(x)$	$1 - \eta_M(x)$
$\dot{m}_1(x)/\dot{m}_{A,0}$	$\phi_0\left(1 + f_{st}\right)\eta_M(x)$	$\left(1 + f_{st}\right)\eta_M(x)$

where $\Delta x \delta_m$ is the volume of the c.v., $n_{i,F}$ and $n_{i,A}$ are the mole numbers of i-th species in the fuel and air streams, respectively, $n_{i,1}$ is the mole number of i-th species inflowing from the upstream control volume, and f_i is the net volumetric production rate of i-th species as given by Eq. (6-64).

Neglecting the rate of dissipation of mechanical energy to thermal energy within the control volume, the adiabatic constraint is given by conservation of energy as in Eq. (2-31),

$$\dot{m}_1 h_{t,1} + \Delta \dot{m}_F h_{t,F} + \Delta \dot{m}_A h_{t,A} = \dot{m}_2 h_{t,2} \qquad (6\text{-}79)$$

where the total enthalpy in each of the c.v. three inflow streams and the single outflow stream is given by the sum of the static enthalpy and kinetic energy,

$$h_{t,k} \equiv \frac{u_k^2}{2} + \sum_{i=1}^{NS} h_i n_{i,k} , \quad k = 1, A, F, 2 \qquad (6\text{-}80)$$

Note that, although the total enthalpy is assumed to be conserved throughout the processes of mixing and chemical reaction, the static temperature within the mixing layer will rise as exothermic chemical reaction results in sensible "heat release." If we had specified an axial distribution of cross-sectional area $A(x)$, the static pressure would change as well. In the interests of simplicity, however, we here assume constant pressure, and allow the cross-sectional area to passively adjust to accommodate the resulting density changes. The interaction between $A(x)$, $p(x)$, and $T(x)$ will be considered in detail in Sec. 6.5.

The nature of chemical reaction within the mixing layer is such that all of the chemical reaction regimes described for batch reaction, namely induction, heat release, and equilibration, occur *in parallel* within the control volume shown in Fig. 6.14. At the same time as gases flowing in from the upstream mixing layer continue to burn, fresh fuel and air enter the c.v. from the remaining unmixed streams and are mixed in. Consequently, the outflow from the c.v. is a micromixture of unreacted fuel and air, reaction intermediates such as H, O, and OH, and products of reaction resulting from combustion completed somewhere upstream. This model for steady-flow, steady-state, simultaneous inflow/outflow, micromixing and chemical reaction within a control volume is referred to in chemical engineering parlance as a *continuous stirred-tank reactor* (CSTR), and in the combustion literature as either a *well-stirred reactor* (WSR) or a *perfectly-stirred reactor* (PSR) model.[6.34]

6.4.1 "One-Dimensionalized" Axial Variation of Properties

Although Eqs. (6-75) through (6-80) were all written for coflowing but crossmixing one-dimensional streams, the c.v. calculations result in three different axial distributions of every property or variable of interest, namely axial velocity $u(x)$, temperature $T(x)$, and mole numbers $\{n_i(x),\ i = 1, NS\}$. However, we need axial profiles of only one, not three, of each property for subsequent inclusion in one-dimensional aerothermodynamic analysis. This is done in a very straightforward way by mass-flow-averaging the axial fluxes of momentum, species, static enthalpy, and energy.[6.12, 6.27] For conservation of mass,

$$\dot{m}_\sigma \equiv \dot{m}_A + \dot{m}_F + \dot{m}_2 = \sum_{k=A,F,2} \dot{m}_k = \text{constant} \qquad (6\text{-}81)$$

where, at any axial station, $\dot{m}_2$ denotes the outflow from the mixing layer c.v., and $\dot{m}_A$ and $\dot{m}_F$ denote the *remaining* mass flow rates in the air and fuel streams, which have been depleted from their combustor entry values $\dot{m}_{A_0}$ ($= \dot{m}_0$) and $\dot{m}_{F_0}$ by the sum of the mixing inflows $\Delta \dot{m}_A$ and $\Delta \dot{m}_F$ to all of the upstream control volumes, respectively.

Using the summation notation of Eq. (6-81), the mass-averaged velocity $\tilde{u}$ ("u-tilde") is determined from the sum of the axial momentum fluxes,

$$\dot{m}_\sigma \tilde{u} = \sum_{k=A,F,2} \dot{m}_k u_k \qquad (6\text{-}82)$$

Similarly, the mass-averaged mole number of the i-th species $\tilde{n}_i$ is determined for each of the NS species,

$$\dot{m}_\sigma \tilde{n}_i = \sum_{k=A,F,2} \dot{m}_k n_{i,k}\ , \qquad i = 1, NS \qquad (6\text{-}83)$$

With the mass-averaged axial velocity $\tilde{u}$ determined from Eq. (6-82), the mass-averaged static enthalpy $\tilde{h}$ is given by

$$\dot{m}_\sigma \left(\tilde{h} + \frac{\tilde{u}^2}{2} \right) = \sum_{k=A,F,2} \dot{m}_k \left(h_k + \frac{u_k^2}{2} \right) \qquad (6\text{-}84)$$

where $(h_k = \sum_{i=1}^{NS} h_i(T_k)n_i, k,\ k = A, F, 2)$ is the mass-specific static enthalpy in each of the three streams, and $\tilde{h} = \sum_{i=1}^{NS} h_i(\tilde{T})\tilde{n}_i$ is the static enthalpy based on mass-averaged mole numbers, at the as yet undetermined static temperature $\tilde{T}$. Since Eq. (6-84) is algebraically implicit in the unknown temperature $\tilde{T}$, a solver or root finder is required to determine $\tilde{T}$ at each axial station.[6.27, 6.34]

Finally, the mass-averaged total enthalpy $\tilde{h}_t$ and corresponding total temperature $\tilde{T}_t$ are determined by

$$\tilde{h}_t = \sum_{i=1}^{NS} h_i\left(\tilde{T}_t\right)\tilde{n}_i = \sum_{i=1}^{NS} h_i\left(\tilde{T}\right)\tilde{n}_i + \frac{\tilde{u}^2}{2} \qquad (6\text{-}85)$$

where it has been assumed that the mass-averaged composition is "frozen" at the values $\{\tilde{n}_i,\ i = 1, NS\}$ given by Eqs. (6-83). As with Eq. (6-84), a solver or root finder must be used to evaluate the algebraically implicit total temperature $\tilde{T}_t$ in Eq. (6-85).

6.4.2 Axial Combustion "Efficiency"

There are many ways to define a measure of completeness of combustion, or combustion "efficiency." For example, if we were primarily concerned with what fraction of the fuel burns to completion, an appropriate measure would be the fraction of some atomic component in the fuel stream (say, hydrogen H) which is converted to some product species (say, water vapor H_2O) in the incompletely micromixed and incompletely reacted product stream. However, our primary concern here is with the amount of chemical "heat release" which contributes to propulsion. As shown in Eq. (2-104), the equivalent heat "added" to the gases in the burner changes both the static temperature and the flow velocity, and so is reflected in the increase of total temperature. On the other hand, it is the *static* temperature, not the total temperature, on which both the rate (chemical kinetics) and extent (chemical equilibrium) of heat release depends. Consequently, the most appropriate measure of combustion efficiency $\eta_b(x)$ at any axial station is the ratio of the rise in mass-averaged static temperature from burner inlet to the static temperature rise which would have resulted if all the fuel and air at that axial station were completely micromixed and burned to chemical equilibrium. In equation form,

$$\eta_b(x) \equiv \frac{\tilde{T}(x) - \tilde{T}_\sigma}{\tilde{T}_{AFT,\sigma} - \tilde{T}_\sigma} \qquad (6\text{-}86)$$

where $\tilde{T}_\sigma$ is the mass-averaged static temperature of both fuel and air streams at burner entry, as determined by Eqs. (6-81) through (6-84), and $\tilde{T}_{AFT,\sigma}$ is the adiabatic flame temperature obtained as described in Sec. 6.3.2.1, where the reactant composition and static enthalpy are the mass-averaged values for the fuel and air streams at burner entry.

The difference between a combustion efficiency defined in terms of static temperature rise as defined by Eq. (6-86), and one based on the rise in *total* temperature, is negligible. This is because our working assumption of frictionless, constant-pressure flow requires also that the flow velocity remain constant, as given by Eq. (2-107), so that changes in static temperature are proportional to changes in total temperature as seen from Eq. (2-104). In fact, for assumed constant C_p's, the changes are identical.

Note the similarity of the definition of combustion efficiency given by Eq. (6-86) to that of the mixing efficiency η_M defined in Sec. 6.2.6. As we have seen, combustion cannot take place until micromixing has first occurred, so that $\eta_b(x)$ *must always be less than, or at most equal to,* $\eta_M(x)$ at the same x. Thus, no matter how fast the chemical kinetics may be—in the infinitely fast limit, chemical equilibrium would occur instantly—the combustion process in a scramjet burner is said to be *mixing limited.*

Figure 6.15 illustrates dramatically how greatly $\eta_b(x)$ can lag behind $\eta_M(x)$ when the entry static temperature of the reactants is so low as to cause significant ignition delay. The case illustrated is from supersonic H_2/air combustion experiments performed at NASA-Lewis in 1973,[6.35] as analyzed in Ref. 6.12. Parallel streams of air and fuel enter the burner in the configuration of Fig. 6.1, with airstream duct height $b_1 = 8.9$ cm and H_2 stream duct height $b_2 = 0.4$ cm. In this case, the airstream is vitiated air, in which some H_2 has already been burned in order to raise the static temperature, and additional O_2 introduced to make up for the oxygen used in burning the H_2. The composition of the vitiated airstream is O_2, N_2, and H_2O with mole fractions of 0.203, 0.438, and 0.359, respectively, with static pressure 1 atm, static temperature 1270 K, and $M = 2.44$. The H_2 fuel stream enters at $M = 1$, static pressure 1 atm, and total temperature 300 K. From these data, using equations and software described previously in this chapter, it can be shown that the overall equivalence ratio is 0.83, and the induction or ignition delay time inferred from the induction distance of 0.17 m from Fig. 6.15 is about 10^{-4} s. This estimate of ignition delay time agrees well with Eq. (6-74), from which the exponential sensitivity of the ignition delay time on the reactant temperature is apparent. Note also from Fig. 6.15 that, even after the first reactants to mix are ignited, the continued addition of "cold" reactants to the mixing layer causes η_b to grow at a slower rate than η_M.

6.5   AEROTHERMODYNAMICS OF THE COMBUSTION SYSTEM

We will consider the combustion system as consisting of two components, the inlet isolator and the combustor or burner, as illustrated in the schematic Fig. 6.16. Since fuel and air are still mixing and burning as they flow supersonically out of the burner and into the

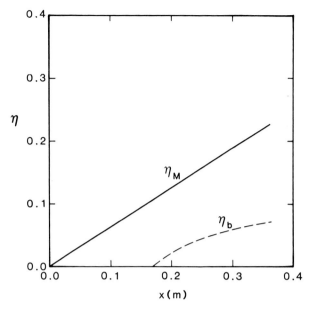

Fig. 6.15 Axial growth of mixing efficiency η_M and combustion efficiency η_b for supersonic combustion of vitiated air and parallel-injected hydrogen. Burner entry conditions as described in text. Data from Ref. 6.35 as analyzed in Ref. 6.12.

expansion system, the thrust nozzle could also logically be regarded as part of the combustion system. However, we adopt the viewpoint that the function of the combustion system is to cause the fuel and air streams to begin to mix and start to burn, and that the design of the expansion system must take into account the degree of incompleteness of mixing and reaction in the gas stream exiting the burner, as a too-rapid drop in pressure will inhibit further mixing and "freeze" the reaction chemistry. Conversely, while thrust is generated in any burner with an expanding area ratio, the burner is not usually regarded as part of the expansion system.

In Fig. 6.16, engine reference Stations 3 and 4 designate burner entry and exit, respectively, consistent with the designations in Table 4.1 and Fig. 4.1. Station 2 will be used to designate entry to the isolator. Station 3, which designates both isolator exit and burner entry, is defined as the axial location *of the most upstream fuel injector.* Stations *u* and *d* designate the upstream and downstream limits or "ends" of a positive or adverse axial pressure gradient, respectively, and Station *s* designates the upstream "end" of the negative or favorable pressure gradient which extends through the remainder of the burner and right on through the expansion system. As will be shown presently, Station *s* is also the location of the lowest Mach number in the combustion system.

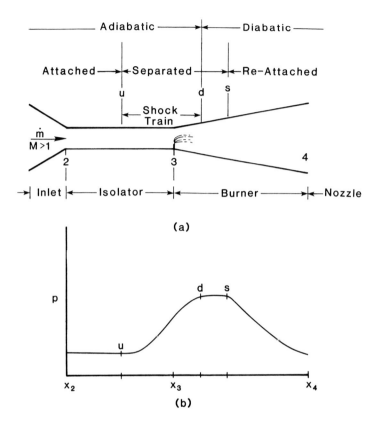

Fig. 6.16 (a) Designation of axial locations for combustion system geometry (Stations 2, 3, and 4) and axial variation of static pressure within the combustion system (Stations *u*, *d*, and *s*). (b) Typical axial distribution of wall pressure for scramjet mode operation.

Before getting deeply into analytical detail, we will review certain relevant material from earlier chapters, introduce some new concepts, define some new terms, and establish some cause-and-effect relationships which will help guide us through some rather tricky phenomena.

6.5.1 The Dual-Mode Combustion System

The aerothermodynamics of ramjets and scramjets was introduced in Sec. 2.6.5, and an H-K diagram for the air being processed in both ramjets and scramjets was presented in Figs. 2.19 and 2.20. As was discussed in Sec. 4.2.3, the conflicting requirements of high cycle thermal efficiency and dissociation of the working fluid at excessively high static temperatures dictate that the combustion process must be subsonic (ramjet) for flight Mach numbers less than about 5, and supersonic (scramjet) for M_0 greater than about 7.

A "pure" ramjet engine, which operates at supersonic flight speeds but with subsonic combustion, requires two area constrictions (physical throats), as illustrated in Fig. 1.18. As described in Sec. 1.4.1, the first throat, at the outlet from the inlet diffuser, is required to stabilize the final, normal shock wave in the area expansion downstream of the throat (sometimes called the *transition section* or *trans-section*), in order to deliver subsonic flow to the burner. The second throat, downstream of the burner, is required to accelerate the subsonic flow to supersonic velocity in the expansion nozzle. It is important to note that the flow is *choked* ($M = 1$) only in the second throat. The choking condition determines the static pressure at burner entry, which appears as a subsonic *back pressure* to the trans-section and which in turn determines the location and strength of the normal shock in the trans-section.

Unlike the ramjet engine depicted in Fig. 1.18, the "pure" scramjet engine of Fig. 1.19 has no physical throat. As explained in Sec. 1.4.2, since the Mach number never drops to or below unity in a "pure" scramjet, there is no need for either an upstream or a downstream throat.

To avoid having to carry two different engines for ramjet and scramjet operation, we would like to be able to operate in *either* ramjet or scramjet mode using only the "no-throat" geometry of the scramjet, Fig. 1.19. In other words, we would like to be able to have subsonic flow in the burner *without area constriction either upstream or downstream of the burner,* as such constrictions would limit the mass flow rate at higher flight Mach numbers, when supersonic combustion is required. Of course, while it is conceivable to design "rubber" or variable-throat inlet compression and expansion system geometries to accomplish this, it is obviously highly impractical.

To satisfy this design goal, Curran and Stull proposed in 1963, and patented in 1969, the concept of a *dual mode combustion system*,[6.36] in which both subsonic and supersonic combustion can be made to occur within the same scramjet engine geometry. The first experimental demonstration of ramjet mode operation in the open literature was reported by Billig in 1966.[6.37, 6.38]

6.5.1.1 Ramjet mode (subsonic combustion).   In ramjet mode, the flow must be subsonic at burner entry. The transition from supersonic flow to subsonic flow is accomplished in the dual-mode engine by means of a constant-area diffuser called an *isolator,* the characteristics of which have been described in Sec. 5.7.5. In order that the burner entry flow be subsonic, the flow must be *choked* ($M = 1$) somewhere downstream, which causes a large back pressure p_3 at burner entry. This back pressure causes a *normal shock train* to form in the isolator, just upstream of Station 3. As long as the back pressure p_3 does not exceed the isolator's ability to contain the normal shock train, the isolator performs the same functions as the variable-area diffuser or trans-section of Fig. 1.18. The function of the second ramjet throat—to choke the flow and thereby fix the burner entry back pressure p_3, and to accelerate the subsonic flow through $M = 1$ to supersonic velocities in the nozzle—is provided for in the dual-mode burner by means of a *choked thermal throat*, which is brought about by choosing the right combination of area distribution $A(x)$ and fuel-air mixing and combustion, as represented by the total temperature distribution $T_t(x)$.[6.39, 6.40] We will show in some detail in Sec. 6.5.4.4 how this is accomplished. Although not shown on Fig. 6.16, an asterisk * will be used to designate the axial location of the choked thermal throat, whenever one exists.

6.5.1.2 Scramjet mode (supersonic combustion).
In scramjet mode, since there is no need for a physical throat either upstream or downstream of the burner, and the flow is supersonic at burner entry, there is apparently no need for an inlet isolator. Indeed, none is shown in the schematic Fig. 1.19. However, even though the flow is *ideally* neither choked nor subsonic anywhere within the engine, it frequently happens that if the area increase in the burner is not sufficient to relieve the *thermal occlusion* resulting from heat "addition" to a supersonic stream, an adverse pressure gradient arises. This effect can be seen for frictionless constant-area heat addition (Rayleigh flow) by substituting Eq. (2-105) into Eq. (2-103). If the pressure rises too abruptly within the burner, the boundary layer will separate. The resulting pressure rise propagates freely upstream through the separated boundary layer, even though the *confined core flow* remains supersonic.[6.41] Unless the upstream migration of the pressure rise is contained, the engine inlet will unstart. Happily, we get "double duty" from the isolator, which not only can contain the normal shock train required for subsonic burner entry in ramjet mode, but also can contain an *oblique shock train* with a *supersonic* confined core outflow, as described in Sec. 5.7.5, which provides the necessary adiabatic pressure rise in the confined core flow to match the pressure rise resulting from heat addition in the burner, and thereby prevents unstart of the engine inlet.

6.5.1.3 Transition from scramjet to ramjet mode.
An example process path for shock-free supersonic combustion on H-K coordinates is shown in Fig. 6.17. This example process path was calculated for a variable-area scramjet process, using the software program HAP(Burner), which uses analytical methods to be described in Sec. 6.5.4.

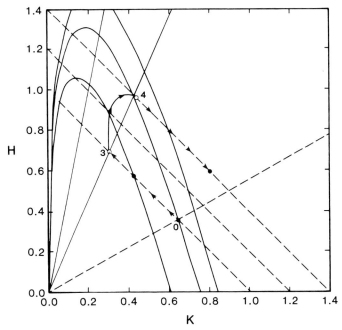

Fig. 6.17. Supersonic combustion process path on H-K coordinates. Example case illustrated: Scramjet mode with shock-free isolator, $M_2 = M_3 = 1.5$. $A(x)/A_3 = 1 + \chi$, where $\chi \equiv (x - x_3)/(x_4 - x_3)$, and $\theta = 2$, $x_i = x_3$, and $\tau_b = 1.4$ in Eq. (6-91).

The shape of the burner process path in Fig. 6.17 is interesting. At burner entry, where the heat "addition" rate dT_t/dx is greatest, the process path is close to the desirable constant-p/constant-K process path, as the pressure rise due to heat "addition" is counteracted by the relief due to increasing area. However, as dT_t/dx tapers off toward burner exit, the relieving effect of increasing area now dominates the occlusion due to heat addition, so the pressure starts to decrease (K increases), and the burner process path approaches in turn a constant-M path, then a constant-T path, and finally approaches a constant-T_t (adiabatic) process path near burner exit. Note that, in this example, the burner Mach number passes through a minimum $M_s = 1.33$, which is well above Mach 1. The axial locus of the minimum Mach number, denoted Station s in Fig. 6.16, is called the *thermal throat* of the burner, by analogy to the physical throat of a converging-diverging nozzle. Also note that, just as with a physical throat, it is possible for a thermal throat to exist without being choked.

In the burner process path illustrated in Fig. 6.17, the minimum Mach number $M_s = 1.33$ is not sufficiently less than $M_3 = 1.5$ to cause the boundary layer to separate, so the isolator is shock-free.

However, note that the burner area ratio $A_4/A_3 = 2$ is too great to maintain the exit pressure p_4 close to p_3. This drop-off in static pressure is accompanied by a decrease in static temperature and an increase in Mach number, compared to the ideal constant-P heat addition process. By reducing the mean temperature at which heat is added in the burner, the cycle thermal efficiency is reduced, as discussed in Sec. 4.2. In addition, as we have learned in this chapter, supersonic mixing is inhibited at higher Mach numbers, and chemical kinetic rates are strongly proportional to both static pressure and temperature. Thus for the heat addition process shown in Fig. 6.17, compared to the desirable constant-P process, burner residence time is decreased, the fuel-air mixing rate is depressed, combustion reactions tend to "freeze" before the desired amount of heat has been released, and the mean temperature at which heat is added is reduced, even for the same $\tau_b = T_{t4}/T_{t3}$. Clearly, it is very important to maintain the design pressure in the combustion system whenever possible.

In order to make the burner outlet pressure p_4 closer to the inlet value p_3, the burner area ratio could be reduced from 2 to, say, 1.73, and the calculation repeated. The result is plotted as process path B on Fig. 6.18. In this case, the Mach number at the thermal throat, $M_s = 1.19$, is still not sufficiently less than $M_3 = 1.5$ to separate the boundary layer. However, it is getting closer to unity. Also, the exit pressure p_4 hasn't been raised that much, and is still less than p_3.

As the burner area ratio is further decreased to 1.57, resulting in process path C in Fig. 6.18, the thermal throat moves increasingly closer to the $M = 1$ ray, as now $M_s = 1.03$. However, the flow is still not quite choked at the thermal throat, and p_4 is still less than p_3.

A further decrease in the area ratio to $A_4/A_3 = 1.55$ *does* cause the flow to choke at the thermal throat. As a result, the flow must now be subsonic at burner entry, so that M_3 and Φ_3 can be made sufficiently low to allow more "room" for the desired heat addition, as described in Sec. 2.6.5. A normal shock train forms in the isolator to provide the required pressure and subsonic flow at burner entry. The resulting heat addition process is represented as path D in Fig. 6.18. The dual-mode combustion system is now operating in ramjet mode, with *subsonic* flow into and *supersonic* flow out of the burner.

There are other ways besides changing the engine geometry to transition from scramjet to ramjet mode and back, for example by keeping $A(x)$ fixed while increasing the fuel flow rate to "add" more heat, thus changing $T_t(x)$ until the increase in τ_b is sufficient to cause thermal choking. This process can be reversed by reducing τ_b until the flow "un-chokes" and supersonic flow is re-established at burner entry. A detailed analysis of this and other methods of mode transition will be given in Sec. 6.5.5.3.

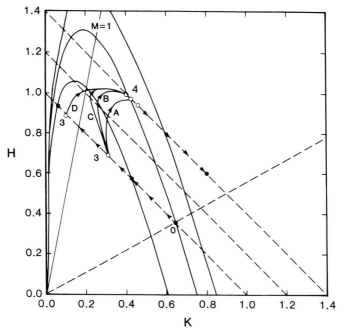

Fig. 6.18 Scramjet-to-ramjet mode transition by reducing burner area ratio A_4/A_3, for same $T_t(x)$ as in Fig. 6.17. Path A is that illustrated in Fig. 6.17 for $A_4/A_3 = 2.0$, path B for $A_4/A_3 = 1.73$, path C for $A_4/A_3 = 1.57$, path D for $A_4/A_3 = 1.55$.

6.5.2 Cause and Effect Within the Dual-Mode System

Table 6.4 summarizes the axial locations of regions of rising and falling static pressure $p(x)$ in relation to the burner and isolator, for the three different cases of burner-isolator interaction, each of which will be described in detail. It is important to study Table 6.4, together with the descriptions of each case or scenario which follow, in order to grasp the cause-and-effect relations or *causal chain* between heat addition, area change, thermal choking, and flow separation in each of the three cases. Without such understanding, the formal aerothermodynamic analysis which follows will make little sense.

The first case shown in Table 6.4 is supersonic combustion with no shock train in the isolator. An example of this case was shown in Fig. 6.17. At some point downstream of fuel injection at Station 3, enough fuel and air have mixed and ignited so that heat "addition" begins, which is designated as Station u. If the area does not increase enough to accommodate the heat release, the pressure begins to rise at Station u and reaches a maximum at Station d, which in this case coincides with Station s. The pressure decreases after Station s,

Table 6.4 Relative axial location of rising (*u-d*) and falling (*s-4*) pressure gradients for different combustion system modes and type of shock train. Axial station designations are as illustrated in Fig. 6.16 and as described in text. *M* = 1 at Stations & and *.

System mode	Shock train	Axial location of pressure gradients					Separated	Subsonic	Adiabatic
scramjet	none	2	u	3	$d = s$	4	no	no	2-u
	oblique	2	u	3	$d \qquad s$	4	u–s		2–d
ramjet	normal	2	u	$d = 3$ &	s	* 4		& – *	

where the effect of increasing area overcomes that of increasing total temperature. In this case, the adverse pressure gradient is not great enough to separate the boundary layer. Since in this case the flow is everywhere supersonic and attached, no back pressure $p_3 > p_2$ arises at burner entry, so no pre-combustion shock train is formed in the isolator, and the isolator is said to be *shock-free*. The only effect of the isolator on the engine is the generation of some skin friction drag. If the area increase provides so much relief that the pressure does not rise at all, as for example in Fig. 6.17, then Stations u, d and s are all coincident.

The second case shown in Table 6.4 differs from the first, "shock-free" scramjet case in that the adverse pressure gradient due to heat addition is now so great that boundary layer separation does occur. This case is illustrated in Fig. 6.16b. When the supersonic flow entering the burner encounters the blockage presented by the separated flow near the walls, it is forced into a confined core flow through an oblique shock train which adiabatically compresses the flow until its pressure is equal to the peak pressure p_s in the burner. The resulting oblique shock train is located partially in the isolator and partially in the burner, as was stated in Sec. 5.7.5, in response to varying imposed back pressures, "the shock train simply slides or translates downstream almost entirely intact until its exit plane pressure equals the imposed back pressure." Stations u and d now designate the upstream and downstream end planes of positive pressure gradient, but in this case the pressure rise is not due *directly* to heat release, but rather is caused by the *adiabatic* compression

within the oblique shock train. In this case, heat "addition" does
not begin until Station d, which is the plane at which the back pressure due to heat release in the burner is matched or supported by the
adiabatic pressure rise from the pre-combustion shock train. From
Station d to s, the pressure is constant while heat is added to the confined core flow within the burner. The pressure is constant because
the maximum pressure p_s is transmitted freely upstream through the
separated region near the walls, where it presses laterally inward (radially inward for cylindrical geometry) on the confined, supersonic
core flow. The flow reattaches at Station s, where the effect of relief
due to area increase is greater than that of the occlusion due to heat
release, so that the pressure begins to decrease, as indicated by the
$p(x)$ curve of Fig. 6.16b. It is helpful to bear in mind that while
Stations 2, 3 and 4 are fixed "hardware" locations, Stations u, d and
s may translate upstream or downstream in response to variations
in flight Mach number, altitude and engine operating conditions. In
the example case illustrated in Fig. 6.16b, heat "addition" does not
begin immediately following fuel injection at Station 3, but is delayed
due to mixing transition and/or chemical-kinetic induction delay. In
some cases, Stations d and 3 may be coincident. In fact, it is even
possible that Station d may occur *upstream* of Station 3, for example
if the fuel injector is a normal jet, with a recirculation or backflow
of fuel upstream of the fuel jet core, as represented by the mixing
parameter $\eta_{M_{90^\circ}}$ of Eq. (6-44).

The last case shown in Table 6.4 is ramjet mode, for which *subsonic* flow is required at burner entry. This case is distinctly different
from the preceding two scramjet cases in which the flow is everywhere
supersonic, except of course in the separated flow regions adjacent to
the walls in the isolator and burner. In order to have subsonic flow
into the burner, the flow must be *thermally choked* somewhere in the
burner, at a station which will be designated by an asterisk *. In this
case, the resulting back pressure is transmitted upstream *through the
attached, subsonic bulk flow*, rather than selectively through the separated flow region near the walls, as in scramjet mode. As a result,
the back pressure at burner entry is impressed more or less uniformly
across the entire cross-section plane of the burner, which causes a
normal shock train, rather than an oblique shock train, to form in
order to compress the confined core flow. The flow in the confined
core passes from supersonic to subsonic flow within the normal shock
train, at a location designated by an ampersand & in Table 6.4. As
in the scramjet mode with oblique shock train case, the pressure rise
begins in the isolator at Station u, and reaches a maximum at Station d where the pressure matches the back pressure p_3 and where
heat "addition" begins. Note especially that in this subsonic flow
case, Stations d and 3 are coincident. Because heat addition in a

subsonic stream causes the pressure to *decrease*, as can be seen from Eqs. (2-103) and (2-105) for constant-A heat addition, the confined core flow reattaches quickly, so that the axial distance between $d = 3$ and s is very short. Note that in this case, all of the normal shock train u-d is contained in the isolator, but the confined, subsonic core flow is still separated as it enters the burner.

6.5.3 Control Volume Analysis of the Isolator

Now for some analysis! In Sec. 5.7.5, it was pointed out that if wall friction is neglected, and if the flow is attached at both isolator entry Station 2 and exit Station 3, then conservation of mass, momentum and energy admit only two exit states: the same state as at entry ("shock-free" flow), and the state immediately downstream of a normal shock with approach Mach number $M_2 > 1$, given by Eq. (5-47). For any given $M_2 > 1$, these two limiting isolator exit pressure limits will be designated as $p_{3min} = p_2$ and $p_{3\,max} = p_{2y}$. A back pressure p_3 less than p_2 will never occur, as the flow at isolator entry is always supersonic. A back pressure p_3 greater than p_{2y} will cause the engine inlet to unstart. *For any other back pressure p_3 between these two limits, the exit flow from the isolator must be a confined core flow surrounded by a region of separated flow.*

We would like to calculate the isolator exit/burner entry Mach number M_3 in the confined core flow, for any imposed back pressure p_3 in the admissible range $[p_2, p_{2y}]$. Fig. 6.19 shows the finite control volume for this analysis.

Since wall friction is neglected and the isolator area is constant, no axial forces act on the surfaces of the control volume except at the entry and exit planes, so the axial stream impulse function $I = pA + \dot{m}u$ is necessarily equal at Stations 2 and 3, whether or not the flow is attached or separated at Station 3. We will assume that the pressure p_3 is uniform across the exit cross-section plane at Station 3. While this is a very bad assumption for supersonic *expansion,*[6.39, 6.40]

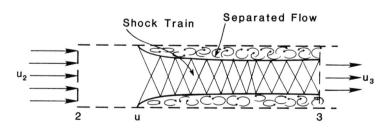

Fig. 6.19   Control volume for analysis of isolator with confined or separated flow at exit.

it is a reasonably good assumption for supersonic *compression*,[6.41] as the pressure in the separated flow region presses inward on the confined core flow. At the isolator exit plane, I_3 is given by the sum of two terms, one for the confined core and one for the region of separated flow. Since the mean axial velocity in the separated region is either zero or negligible, the impulse function at isolator exit is given by

$$I_3 = I_{\text{sep}} + I_{\text{core}}$$

$$= [p_3 (A_2 - A_{3c}) + 0] + (p_3 A_{3c} + \dot{m} u_3) = p_3 A_2 + \dot{m} u_3 \quad (6\text{-}87)$$

where the subscript "c" for "core" emphasizes that A_{3c} represents the cross-sectional area *of the confined core flow* at Station 3, and *not* of the entire cross section $A_3 = A_2$. The "c" subscript has not been applied to u_3, T_3, and M_3 because they are all *intensive properties* of the confined core flow at Station 3. Although the term A_{3c} cancels in Eq. (6-87), this is the area which *must* be used with the core flow velocity u in the mass conservation equation $\dot{m} = \rho A_c u$ wherever the flow is separated.

The velocity u_3 of the confined core at isolator exit can be determined immediately by equating Eq. (6-87) to I_2. Since the flow in the isolator is adiabatic, the energy equation can be solved for the core flow static temperature T_3. With u_3 and T_3 known, the Mach number M_3 is given by

$$M_3 = \left\{ \frac{\gamma_b^2 M_2^2 \left(1 + \frac{\gamma_b - 1}{2} M_2^2\right)}{\left(1 + \gamma_b M_2^2 - \frac{p_3}{p_2}\right)^2} - \left(\frac{\gamma_b - 1}{2}\right) \right\}^{-1/2} \quad (6\text{-}88a)$$

Equation (6-88a) can be solved algebraically for the back pressure ratio p_3/p_2,

$$\frac{p_3}{p_2} = 1 + \gamma_b M_2^2 - \gamma_b M_2 M_3 \sqrt{\frac{1 + \frac{\gamma_b - 1}{2} M_2^2}{1 + \frac{\gamma_b - 1}{2} M_3^2}} \quad (6\text{-}88b)$$

For a given isolator entry Mach number M_2 and back pressure ratio p_3/p_2 imposed by the burner, Eq. (6-88a) gives the resulting isolator exit Mach number M_3. Note that M_3 decreases approximately linearly with increasing back pressure p_3, from $M_3 = M_2$ when $p_3/p_2 = 1$ (scramjet mode with shock-free isolator) to the lowest, subsonic value $M_3 = M_{2y}$, which is the post-normal shock Mach number corresponding to the normal shock pressure ratio p_{2y}/p_2 given by Eq. (5-47).

For any pressure ratio p_3/p_2 between the shock-free and normal-shock limits, the ratio of the confined core area A_{3c} to the isolator cross-sectional area A_2 may be determined immediately from conservation of momentum as

$$\frac{A_{3c}}{A_2} = \frac{1}{\gamma_b M_3^2}\left[\frac{p_2}{p_3}\left(1 + \gamma_b M_2^2\right) - 1\right] \tag{6-89}$$

where M_3 is given by Eq. (6-88a).

Equation (6-89) is plotted on Fig. 6.20 for an isolator entry Mach number $M_2 = 2$ and ratio of specific heats $\gamma_b = 1.4$. Note that the confined core area A_{3c} is a minimum for a back pressure p_3 corresponding to a supersonic exit Mach number of the confined core. Generally speaking, pressure ratios corresponding to M_3 less than and greater than 1 correspond to the formation of normal-shock and oblique-shock pre-combustion shock trains, respectively.

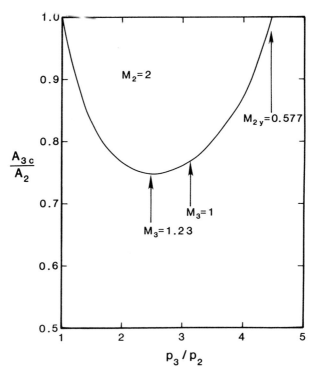

Fig. 6.20   Variation of confined flow area fraction with imposed back pressure ratio, for isolator entry Mach number M_2 = 2 and ratio of specific heats γ_b = 1.4.

There is evidence from CFD modeling,[6.42] confirmed by experiment,[6.43] that the nature of the separated flow is quite different in the two cases. In the case of scramjet mode with oblique-shock train, the region outside the confined supersonic core is a fully recirculating flow, as represented in Fig. 5.35. This is an important consideration, as the wall-mounted "hypermixer" fuel injector/mixers described in Sec. 6.2.8.2 and shown in Fig. 6.11 can function as designed only if the local flow is supersonic, not detached and recirculating!

In the ramjet mode with normal-shock train case, the flow outside the core apparently alternately separates and reattaches in the lateral vicinity of each normal shock of decreasing strength.[6.43] As a result, the structure of the flow outside of the confined core is somewhat like a "Swiss cheese," having pockets of separated, recirculating flow embedded in a low axial momentum, very thick boundary layer. However, because of the on-average low axial momentum flux in the off-core region, the separated flow model of Fig. 6.19 and Eqs. (6-87) to (6-89) gives a very good representation of the mean flow behavior of the subsonic outflow from the isolator.[6.41]

Although the flow is adiabatic within the isolator, the change in thermodynamic state is caused by dissipative processes within that part of the shock train contained within the isolator, between stations (u-3) in the scramjet mode with oblique-shock case, and between stations (u-d) in the ramjet mode with normal-shock train case. Adiabatic flows having other dissipative mechanisms result in pressure rise-Mach number relations which are different from Eq. (6-88). Two notable examples are the adiabatic, constant-area flow with wall friction of Example Case 2.9 called *Fanno flow*,[6.39,6.40] and flow through a single oblique shock wave, for which the relation between pressure rise ratio and entry and exit Mach numbers is given by Eq. (5-37). These three cases are illustrated on temperature-entropy coordinates in Fig. 6.21, from which it can be seen that the isolator does not act exactly as a thermodynamically equivalent oblique shock wave, even when it contains the upstream end of an oblique-shock train, but is significantly more dissipative than a single oblique shock wave for the same pressure rise ratio p_3/p_2.

The axial distribution of pressure and confined core area within the isolator cannot be predicted by one-dimensional analysis, due to the complex three-dimensional character of the normal or oblique shock wave interactions within the shock train, as illustrated in Figs. 5.34 and 5.35. However, Waltrup and Billig[5.19] have experimentally measured the axial variation of wall pressures generated by confined normal and oblique shock trains, including the overall axial length of the shock train $L = x_d - x_u$, for a variety of isolator entry conditions.

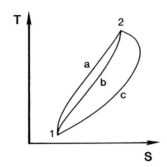

Fig. 6.21 Temperature-entropy diagram showing entry and exit end states for flow through (a) an oblique shock, (b) a constant-area isolator, and (c) constant-area duct with wall friction (Fanno flow), all having the same shock-free and normal-shock limiting end states.

For purposes of preliminary design, the Waltrup-Billig correlation, Eq. (5-46), can be used for sizing the length of the inlet isolator.

6.5.4 One-Dimensional Flow Analysis of the Burner

As we have seen, the isolator is a strictly passive component, caught as it were between the proverbial rock (the inlet compression system) and a hard place (the burner). The sole function of the isolator is to prevent inlet unstart, by providing sufficient additional adiabatic compression above its entry pressure p_2 to match or support whatever back pressure p_3 the burner may impress upon it. We now turn our attention to the *active* component of the combustion system, the burner, in order to analyze and quantify the processes which determine the magnitude of the back pressure p_3. It should be kept in mind that there are two very different physical mechanisms which cause a back pressure $p_3 > p_2$ to arise at burner entry: in scramjet mode, thermal occlusion unrelieved by area expansion which causes *unwanted* flow separation, and in ramjet mode, the *required* thermal choking which insures that the flow into the burner will be subsonic.

We consider the simple case of one-dimensional flow of an ideal gas with constant specific heat ratio γ_b. It is difficult to recommend a single, representative value for γ_b because the choice of fuel, changes in chemical composition, molecular weight and temperature during fuel-air mixing and subsequent combustion all cause γ_b to decrease from $\gamma_c = 1.36$ at isolator entry, as explained in Sec. 5.3, to as low as $\gamma_e = 1.24$ at burner exit, where the mixture of fuel, air and combustion products flows from the burner into the expansion system. Consequently, the reader must exercise good judgment in selecting values for γ_b, as well as for the gas constant R_b, which will be representative of the flow within the particular component, or for the specific conditions being analyzed at the moment. For our part, we will state what values of γ_b and R_b are assumed.

6.5.4.1 Generalized one-dimensional flow.

For the case of frictionless flow without mass addition, but with change in both cross-sectional area A and total temperature T_t due to heat "addition," the governing *ordinary differential equation* (ODE) for axial variation of Mach number is given by[6.39,6.40]

$$
\frac{dM}{dx} = M \left(\frac{1 + \dfrac{\gamma_b - 1}{2} M^2}{1 - M^2} \right) \left\{ -\left(\frac{1}{A} \frac{dA}{dx} \right) + \frac{(1 + \gamma_b M^2)}{2} \left(\frac{1}{T_t} \frac{dT_t}{dx} \right) \right\}
$$

(6-90)

For purposes of brevity and clarity, we have omitted the additional terms in Eq. (6-90) for effects of wall friction, drag of internal struts or fuel jets, and of mass addition due to fuel injection. All of the omitted effects are secondary in importance compared to the strong interaction between axial variation of cross-sectional area $A(x)$ and of total temperature $T_t(x)$,[6.37] and can easily be included in Eq. (6-90) if desired.[6.39,6.40]

When $M > 1$, it can be seen from inspection of Eq. (6-90) that *occluding* the flow, either by decreasing A or by increasing T_t, causes M to decrease in the axial direction. Conversely, *relieving* the flow, either by increasing A or decreasing T_t, will cause M to increase. However, in the dual-mode combustion system, $A(x)$ will never decrease with x, as there is no physical throat, and $T_t(x)$ will never decrease, because exothermic combustion can only "add" heat to the flow. Consequently, Eq. (6-90) can be used to determine how much increase of $A(x)$ is required to accommodate increases of $T_t(x)$ in order to control the axial variation of Mach number, pressure, and other thermodynamic properties and flow variables.

When both the $A(x)$ and $T_t(x)$ terms are present as in Eq. (6-90), only a very few closed form, integral solutions are known to exist, for very simple algebraic forms for $A(x)$ and $T_t(x)$. Consequently, it is necessary in general to use approximate methods to solve the governing ODE. Shapiro refers to this method as *generalized one-dimensional flow analysis*.[6.39,6.40]

For present purposes, we assume that $A(x)$ and $T_t(x)$ in Eq. (6-90) are prescribed or given a priori. In the language of Chap. 8 of Shapiro,[6.39] $A(x)$ and $T_t(x)$ are chosen as independent variables, and their coefficients in Eq. (6-90) are the *influence coefficients* of the respective independent variables on the single dependent variable M. Of course, while $A(x)$ may be prescribed by design, we do not actually know in advance what $T_t(x)$ will result for a given set of flow entry boundary conditions, as $T_t(x)$ depends on the rate of combustion heat release, which in turn is determined by the finite-rate processes of mixing and chemical reaction. However, we do know that the fundamental physics and chemistry of the finite-rate processes of mixing and exothermic chemical reaction dictate the *general form* of $T_t(x)$. As we have already seen in this chapter, for supersonic combustion (scramjet mode), both mixing and chemical heat release rates are greatest at onset, and relax asymptotically toward their respective fully-mixed and chemical equilibrium (or kinetically frozen) zero-rate values with infinite convective time or distance. As a result, the axial total temperature *gradient* dT_t/dx is greatest shortly after ignition, usually near burner entry, and decreases monotonically to its least value at burner exit, as for example in Fig. 6.15. For a wide variety of scramjet mixers, and to represent both mixing transition delay and induction or ignition delay, $T_t(x)$ can be usefully represented in nondimensional form by a *rational function* (ratio of polynomial functions) given by

$$\tau(x) = 1 + (\tau_b - 1) \left\{ \frac{\theta\chi}{1 + (\theta - 1)\chi} \right\}, \quad \theta \geq 1 \qquad (6\text{-}91)$$

where $\tau(x) \equiv T_t(x)/T_{t2}$, $\chi \equiv (x - x_i)/(x_4 - x_i)$, x_i is the axial location at which supersonic combustion or heat "addition" begins ($i = u$ when the isolator is shock-free, otherwise $i = d$, as shown in Table 6.4), θ is an empirical constant of order 1 to 10 which depends on the mode of fuel injection and fuel-air mixing, and $\tau_b = T_{t4}/T_{t2}$ is the overall total temperature rise ratio in the burner, which is determined from considerations of inlet fuel manifold aspect ratio, fuel type, fuel injection/mixing system, and chemical kinetics, as described in previous sections in this chapter. As a result, in the spirit of preliminary design, $T_t(x)$ can be varied systematically within realistic limits, in order to explore the effects of various fuel injection and mixing strategies and devices on scramjet mode burner performance. In ramjet mode, the physical mechanisms of fuel-air mixing and flameholding are very different from scramjet mode, and have not been dealt with in this chapter. However, we can approximate $T_t(x)$ for ramjet mode subsonic combustion by assuming $x_i = x_3$ and $\theta = 40\text{--}50$ in Eq. (6-91).

To determine a unique solution to Eq. (6-90), it is necessary to specify the burner entry Mach number M_3, as well as the *forcing functions* $A(x)$ and $T_t(x)$. With the given M_3 as an *initial condition*, the ODE Eq. (6-90) can be solved by a straightforward step-by-step or "marching" method, starting from burner entry and numerically integrating Eq. (6-90) for $M(x)$ along the burner axis, right up to burner exit. This is easily done with a standard ODE-solver algorithm such as a fourth-order Runge-Kutta method, as is used in the software program HAP(Burner). As long as $M(x)$ remains well above 1 in the burner, the ODE Eq. (6-90) never becomes singular due to the $(1 - M^2)$ term in the denominator, and as long as the integration step lengths Δx are kept small enough to insure the cumulative numerical error stays within reasonable bounds, which is handled automatically in HAP(Burner), the calculation proceeds quickly, accurately and without difficulty.

By recording the intermediate values of $M(x)$ as Eq. (6-90) is solved step-by-step from burner entry to exit, the axial distribution of Mach number, $M(x)$, can be obtained as a discrete set of values distributed along the axis of the burner. The axial distributions of all other flow variables and thermophysical properties of interest in the burner can then be determined from the point-by-point solution set $M(x)$, together with the corresponding values of the prescribed functions $A(x)$ and $T_t(x)$, by means of a collection of what Shapiro terms "useful integral relations"[6.39]:

$$T(x) = T_2 \cdot \frac{T_t(x)}{T_{t2}} \left[\frac{1 + \left(\frac{\gamma_b - 1}{2}\right) M_2^2}{1 + \left(\frac{\gamma_b - 1}{2}\right) M^2(x)} \right] \tag{6-92}$$

$$p(x) = p_2 \cdot \frac{A_2}{A_c(x)} \cdot \frac{M_2}{M(x)} \sqrt{\frac{T(x)}{T_2}} \tag{6-93}$$

$$P_t(x) = P_{t2} \cdot \frac{p(x)}{p_2} \left[\frac{T_2}{T(x)} \frac{T_t(x)}{T_{t2}} \right]^{\gamma_b/(\gamma_b - 1)} \tag{6-94}$$

$$u(x) = u_2 \cdot \frac{M(x)}{M_2} \sqrt{\frac{T(x)}{T_2}} \tag{6-95}$$

Equations (6-92) through (6-95) are simply combinations of the mass flow parameter equation Eq. (2-83) with the ideal gas equation of state Eq. (2-38), the definition of the Mach number Eq. (2-43), and the definitions of total temperature and total pressure, Eqs. (2-50) and (2-54), respectively. Note that in Eq. (6-93) the term $A_c(x)$ has the subscript "c" for "core" as a warning when using these integral equations to calculate properties in the separated flow between Stations u and s, in the two cases in Table 6.4 which have pre-combustion shock trains.

The following additional equations are useful for plotting the burner process path on H-K coordinates:

$$H(x) = \frac{T(x)}{T_{t2}} \quad \text{and} \quad K(x) = \left(\frac{\gamma_b - 1}{2}\right) M(x)^2 H(x) \qquad (6\text{-}96)$$

The impulse or stream thrust function $I(x)$ can be evaluated directly from its definition, Eq. (2-60). In its nondimensional form, the stream thrust function Φ may be evaluated in terms of $\tau(x)$ and $M(x)$ from Eq. (2-127), or in terms of H and K from Eq. (2-63), by

$$\Phi = \left(\frac{\gamma_b - 1}{\gamma_b}\right) \frac{A}{A_c} \frac{H}{\sqrt{2K}} + \sqrt{2K} \qquad (6\text{-}97)$$

The method of generalized one-dimensional analysis described in this section is utilized in the program HAP(Burner), which was used to generate Fig. 6.17.

6.5.4.2 Frictionless constant-area burner.
When only one of the two forcing functions $A(x)$ and $T_t(x)$ is active, Eq. (6-90) can be integrated in closed form to give an integral, algebraic relation between $M(x)$ and the single forcing function. These special cases are referred to by Shapiro as "simple types" of compressible flow.[6.39] For example, when $T_t(x)$ is held constant so that $dT_t/dx = 0$ while only $A(x)$ is allowed to vary, Eq. (6-90) can be formally integrated to give the familiar $(A/A*)$ vs. M algebraic equation for isentropic flow with area change.[6.39,6.40] If $A(x)$ is held constant so that $dA/dx = 0$ but $T_t(x)$ is allowed to vary, Eq. (6-90) reduces to Eq. (2-106), which can be integrated in closed form to obtain the algebraic equation for frictionless, diabatic, constant-A (Rayleigh) flow:

$$M(x) = \sqrt{\frac{1 + \gamma_b \Omega + (\gamma_b + 1)\sqrt{\Omega}}{1 - \gamma_b^2 \Omega}} \qquad (6\text{-}98)$$

where

$$\Omega \equiv 1 - \tau(x) \left[\frac{2(\gamma_b + 1)M_3^2 \left(1 + \dfrac{\gamma_b - 1}{2} M_3^2\right)}{(1 + \gamma_b M_3^2)^2} \right]$$

All other variables of interest can be determined along x as before, with Eqs. (6-92)-(6-97). These equations can be evaluated with the Rayleigh flow option in HAP(Gas Tables), or as a special case in HAP(Burner).

Historically, most analyses of supersonic combustion have assumed constant-A combustion, as the constant-A geometry is both easy to fabricate and to analyze by naive application of Eqs. (6-98) and (6-92) through (6-97). (By "naive" is meant ignoring the possibility of flow separation.) However, as can be seen from Eqs. (2-103) and (2-105), an adverse pressure gradient is created whenever heat is added to a frictionless, supersonic flow in a constant-A burner, and if the pressure rise is too great, it will cause separation of the boundary layer. It is important to note that cause-and-effect relationships are somewhat different in the combustion system. In the inlet compression system, the causal chain is: (1) the supersonic flow is turned through some angle by the wall, (2) the turning flow is compressed through an oblique shock wave, and (3) if the oblique shock pressure rise exceeds an empirically observed threshold value, such as Eqs. (5-35) through (5-37), the boundary layer separates. In the case of a frictionless constant-A burner, the causal chain is: (1) the pressure rises as a result of heat "addition" to the supersonic flow, and (2) if the pressure rise exceeds some threshold value, the boundary layer separates, so that (3) the oncoming supersonic flow is turned into itself by the effective area blockage of the separated flow near the wall, and is compressed into a confined core flow through an oblique shock train, until the confined core flow pressure matches the pressure in the region of separated flow in the burner.

6.5.4.3 Frictionless constant-pressure burner.

In Example Case 2.8, the one-dimensional aerothermodynamic equations for frictionless heat addition at constant pressure were introduced. For a constant-p process with prescribed $T_t(x)$, the axial variation of Mach number is given by Eq. (2-112),

$$M(x) = \frac{M_3}{\sqrt{\tau(x)\left(1 + \frac{\gamma_b - 1}{2}M_3^2\right) - \left(\frac{\gamma_b - 1}{2}\right)M_3^2}} \qquad (6\text{-}99)$$

and the area distribution required to maintain constant pressure for any given $T_t(x)$ is given by Eq. (2-120) as

$$A(x) = A_3\left[\tau(x)\left(1 + \frac{\gamma_b - 1}{2}M_3^2\right) - \left(\frac{\gamma_b - 1}{2}\right)M_3^2\right] \qquad (6\text{-}100)$$

Equation (6-100) shows that the area distribution $A(x)$ required to maintain constant pressure is simply a scaled multiple of $\tau(x)$, as for example the rational-function shape of $\tau(x)$ in Eq. (6-91), a shape which would be very difficult to fabricate for any *fixed* set of values for x_i, τ_b, M_3 and $\theta > 1$ in Eqs. (6-91) and (6-100), and even more difficult when we realize that x_i, τ_b, M_3 and θ will all change in response to varying altitude and flight Mach number. Obviously, short of having a "smart," infinitely variable geometry or "rubber" burner, it is virtually impossible to maintain *exactly* constant pressure in any burner, even more so in a burner of fixed geometry. However, in a two-dimensional planar geometry for the burner, it is possible to admit *some* degree of variable geometry by hinging one or both of the burner walls so that A_4/A_3 can be varied in flight, within reasonable limits. Consequently, Eq. (2-120), which is just Eq. (6-100) evaluated at burner exit, still remains important for sizing the overall burner area ratio A_4/A_3 required to maintain *equal* entry and exit pressures. In addition, Eq. (6-100) will be useful for analyzing the constant-p confined core flow in a dual-mode combustion system when operating in scramjet mode.

To sum up: If it were practical to do so, we would design a "smart, rubber" burner for supersonic combustion which would always operate at constant-p, so that there would be no need for an isolator. However, for reasons of manufacturing feasibility, active cooling of burner walls and off-design engine operation, we will have to settle for a more-or-less straight-walled burner, with $A(x)$ linear in x for planar geometry (quadratic in x for axisymmetric geometry) and with overall area ratio A_4/A_3 sized, or varied in flight, to achieve at best *nearly equal* pressures at burner entry and exit. In addition, in order to maximize combustion efficiency at burner exit, an "early" heat-release distribution such as Eq. (6-91) with as large a value of θ as possible is desirable, which will very often cause flow separation due to the large adverse pressure gradient resulting from a high rate of heat release unrelieved by an equivalent area increase. Consequently, an isolator is necessary to prevent inlet unstart when the engine is operating in scramjet mode with oblique shock train.

6.5.4.4 Establishing a choked thermal throat.
As was stated in Sec. 6.5.1.1, in order to operate the dual-mode combustion system in ramjet mode, since there is no physical throat between the burner and the expansion system, the required choking must be provided within the burner by means of a *choked thermal throat,* which can be brought about by choosing the right combination of area distribution $A(x)$ and fuel-air mixing and combustion, as represented by the total temperature distribution, $T_t(x)$.[6.39, 6.40] The locus of a choked thermal throat on T-s, h-s, or H-K coordinates is referred to as a *critical point.*[6.39, 6.40] We will now see how this is accomplished.

It is instructive to think of the sum of the area and total temperature terms in curly brackets in Eq. (6-90) as an equivalent physical area, or *effective area,* defined by

$$\left(\frac{1}{A_{\text{eff}}}\frac{dA_{\text{eff}}}{dx}\right) \equiv \left(\frac{1}{A}\frac{dA}{dx}\right) - \frac{(1+\gamma_b M^2)}{2}\left(\frac{1}{T_t}\frac{dT_t}{dx}\right) \tag{6-101}$$

We can think about the *effective* area distribution in the familiar way we have learned to think about the *physical* area distribution for frictionless, adiabatic (isentropic) flow with area change. We know that the only way to accelerate an initially subsonic flow through $M = 1$ to supersonic velocity is to pass the initially subsonic flow through a converging-diverging nozzle, in which $A(x)$ goes through a minimum called a throat. This fact is represented mathematically in Eq. (6-90) by the requirement that, as M goes to unity, the singularity due to the $(1 - M^2)$ term in the denominator must be offset by dA_{eff}/dx in the numerator of Eq. (6-101) going to 0 at exactly the same location x where $(1 - M^2)$ goes through 0—that is, through the throat of a converging-diverging *effective* area distribution. Since $A(x)$ and $T_t(x)$ are given functions, a zero-solver or root-finder can be used with Eq. (6-101) to find the value of x at which *both* $M = 1$ *and* $dA_{\text{eff}}/dx = 0$:

$$\left(\frac{1}{A}\frac{dA}{dx}\right)_* - \frac{(1+\gamma_b)}{2}\left(\frac{1}{T_t}\frac{dT_t}{dx}\right)_* = 0 \tag{6-102}$$

where the asterisk $*$ denotes the choked thermal throat or *critical point,* at which $M = 1$.

It is thus possible to determine from Eq. (6-102) the value of x_* where a critical point *can* occur in the burner, for any given pair of functions $A(x)$ and $T_t(x)$. However, solving Eq. (6-102) for x_* does not tell us if a critical point *will* occur. In other words, Eq. (6-102) gives the sufficient conditions for a critical point to exist at x_*, but it does not give the *necessary* conditions. It is important to note that if, for a particular $A(x)$ and $T_t(x)$, there is no solution to Eq. (6-102)— that is, if there is no value of x_* between x_3 and x_4 which satisfies Eq. (6-102)—then *that burner cannot operate in ramjet mode.*

Having determined the axial location of the critical point x_* from Eq. (6-102), if one exists, Eq. (6-90) can be solved *backward* from the critical point to burner entry. The problem is well posed, since $A(x)$ and $T_t(x)$ are given, and the required initial condition for M is $M(x_*) = 1$. However, the dilemma presented by the $(1 - M^2)$ term in the denominator of Eq. (6-90) has yet to be resolved. Even though we know that $(dA_{\text{eff}}/dx)_* = 0$ at the same axial location as $M = 1$, it can be seen by inspection of Eq. (6-90) that $(dM/dx)_* = 0/0$ is algebraically indeterminate.

In Chap. 8 of Ref. 6.39, Shapiro shows that *l'Hôpital's rule* can be used to evaluate Eq. (6-90) at the critical point:

$$\left(\frac{dM}{dx}\right)_* = \frac{1}{4}\left\{-\Omega \pm \sqrt{\Omega^2 - 4\Psi}\right\} \tag{6-103}$$

where

$$\Omega \equiv \gamma \left(\frac{1}{A}\frac{dA}{dx}\right)_*$$

and

$$\Psi \equiv (\gamma_b - 1)\left(\frac{1}{A}\frac{dA}{dx}\right)_*^2 - (1+\gamma_b)\left(\frac{1}{A}\frac{d^2A}{dx^2}\right)_* + \frac{(1+\gamma_b)^2}{2}\left(\frac{1}{T_t}\frac{d^2T_t}{dx^2}\right)_*$$

Note that, depending on the sign of the square root term in Eq. (6-103), $(dM/dx)_*$ can be either positive or negative at the critical point, just as is the case for isentropic nozzle flow. To find the required subsonic entry Mach number, the root having the positive sign in Eq. (6-103) is selected, and the ODE solver is marched *upstream* from x_* to x_3. The resulting subsonic entry Mach number is *unique* for any given $A(x)$ and $T_t(x)$, and will be designated M_{3p}, with the subscript "3p" to indicate the selected "plus" sign in Eq. (6-103), as illustrated in Fig. 6.22.

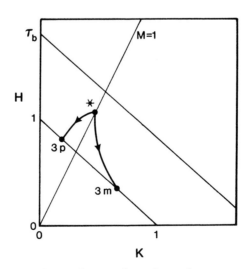

Fig. 6.22 Locating the unique subsonic and supersonic burner entry States 3p and 3m, respectively, required for a choked thermal throat (critical point) to exist at $x = x_*$, for specified $A(x)$ and $T_t(x)$. x_* is determined from Eq. (6-102).

The calculation is then repeated with the negative-sign root in Eq. (6-103) selected. The resulting, unique *supersonic burner* entry Mach number is labeled "$3m$," as before to indicate the choice of "minus" sign in Eq. (6-103). This supersonic solution branch is also shown in Fig. 6.22.

When a choked thermal throat is known or assumed to exist, the corresponding (to M_{3p} and M_{3m}) subsonic and supersonic burner entry pressures p_{3p} and p_{3m}, which the isolator "sees" as back pressures, are determined by substituting M_{3m} or M_{3p} into Eq. (6-88b) to obtain the corresponding back pressure ratio p_3/p_2. It is important to note that these are unique states, which for the given $A(x)$ and $T_t(x)$ satisfy not only the numerical solution to Eqs. (6-90) and (6-103) for M_{3p} and M_{3m}, but also satisfy conservation of the particular values of mass, momentum and energy at isolator entry Station 2. Note that the confined core area A_{3c} at burner entry is given for both cases by Eq. (6-89), and that all other properties of the confined core flow at isolator exit/burner entry are determined as usual by Eqs. (6-92) through (6-97). To complete the calculation of properties for the rest of the burner downstream of the critical point, the ODE Eq. (6-90) is solved for $M(x)$ by marching *downstream* from x_* to x_4, starting with $M = 1$ and choosing the positive-sign root for $(dM/dx)_*$ in Eq. (6-103). As usual, Eqs. (6-92) through (6-97) are used to determine all other property values of interest.

6.5.5 System Analysis of Isolator-Burner Interaction

Having developed the necessary analytical tools for calculating the aerothermodynamic behavior of the isolator and burner, we are now ready to analyze the interaction between these two components of the dual-mode combustion system, for each of the three cases in Table 6.4 and described in Sec. 6.5.2.

6.5.5.1 Scramjet with shock-free isolator. In the first scramjet case shown in Table 6.4, there is really no interaction at all between the burner and the isolator. As there is no pressure feedback from the burner, the aerothermodynamic state of the inflowing air is unaltered between Stations 2 and 3, still assuming frictionless flow, of course. The calculation of axial variation of all properties within the burner is carried out by the direct marching solution of Eq. (6-90) from Station 3 to Station 4, together with Eqs. (6-92) through (6-97), as described in Sec. 6.5.4.1. An example calculation for this case was illustrated in Fig. 6.17.

6.5.5.2 Scramjet with oblique shock train. Given the entry state to the isolator T_{t2}, p_2 and $M_2 > 1$, and given the functions $A(x)$ and $T_t(x)$, it must first be determined whether or not the static pressure rise resulting from thermal occlusion in the burner will separate the boundary layer. If the flow is predicted to separate, then the state of the confined core flow at burner entry must be determined. This is accomplished in two steps:

1.   First, assume that the flow in the isolator is supersonic and shock-free throughout, so that $M_3 = M_2$, and solve Eq. (6-90) for the given functions $A(x)$ and $T_t(x)$ to determine the axial variation of Mach number, $M(x)$. Identify the axial location where the greatest static pressure occurs, Station s in Fig. 6.16 and Table 6.4. Further assume that the same empirical criterion for boundary layer separation applies as in the analysis of the inlet compression system, namely Eq. (5-36) for a turbulent boundary layer. *If M_s is less than $0.762\, M_2$, the boundary layer is assumed to separate.*

2a. If the boundary layer does *not* separate, the flow is scramjet mode with shock-free flow in the isolator, which is analyzed as described in the preceding Sec. 6.5.5.1.

2b. If the boundary layer *does* separate in the burner, the flow is assumed to internally adjust itself in such a way that the heat added *to the separated core flow* in the burner (process *d-s* in Fig. 6.16 and Table 6.4) occurs *at constant pressure equal to the maximum pressure $p_s = p(x_s)$ in the burner, as determined in Step 1.*

The back pressure at Station d, where the pressure rise through the oblique shock train matches the pressure of the separated flow, is therefore determined by the requirement $p_d = p_s$, and the complete thermodynamic state of the separated core flow at Station d, as well as the axial variation of all properties between Stations d and s, is obtained from the constant-p solution for $M(x)$, Eq. (6-99), together with Eqs. (6-92) through (6-97). Between Stations d and s, the confined flow does not conform to the burner area $A(x)$, but rather forms its own axial area variation $A_c(x)$ as given by Eq. (6-100). The confined core flow is assumed to re-attach immediately downstream of the axial location of maximum pressure at Station s, due to the establishment of a favorable pressure gradient at that location.

6.5.5.3 Constant-area scramjet with oblique shock train.   Application of the preceding two-step analysis for a constant-A burner is illustrated on both H-K and T-s coordinates in Fig. 6.23. To keep the analysis simple, it has been assumed in Fig. 6.23 that the heat addition process begins immediately at burner entry, so that Station d is co-incident with Station 3. Note that, in a constant-A burner, Station s will *always* be coincident with Station 4, as the first area relief encountered by the flow is at burner exit/expansion system entry.

From the T-s diagram of Fig. 6.23b, it can be seen that the irreversible, adiabatic entropy increase along path u-d within the oblique shock train is exactly offset by the reduced reversible, diabatic entropy increase along path d-s, as the same amount of heat is added at a higher mean temperature along path d-s as is added along the assumed attached-flow path 2-4:

$$(\Delta s_{2-3})\underset{\substack{\text{adiabatic,}\\\text{shock train}}}{\text{irreversible}} + (\Delta s_{3-4})\underset{\substack{\text{const-}p\\\text{heating}}}{\text{reversible}} = (\Delta s_{2-4})\underset{\substack{\text{const-}A\\\text{heating}}}{\text{reversible}} \qquad (6\text{-}104)$$

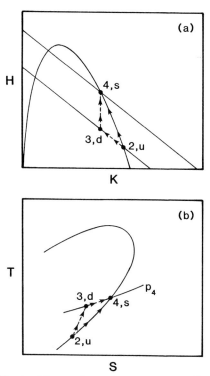

Fig. 6.23 **(a) Effect of boundary layer separation on isolator and burner entry and exit states, for constant-A burner without wall friction or mass addition.   Dotted lines indicates state of separated core flow within both burner and isolator. (b) Temperature-entropy diagram of processes illustrated in (a). Path u-d (2-3) represents irreversible adiabatic change of state in the oblique shock train. Path d-s (3-4) represents constant-p, reversible heat addition to the confined core flow within the constant-A burner.**

This result is consistent with the experimental observation that a scramjet combustion system adjusts itself in such a way as to minimize the entropy rise in the burner.[6.37]

6.5.5.4 Constant-area ramjet with normal shock train. In the constant-A burner of the preceding Sec. 6.5.4.2, if τ_b is increased so that the flow is at *incipient* thermal choking, that is if "just one more millijoule" would cause the flow to choke, the isolator still contains an oblique shock train, which provides a confined core flow with supersonic burner entry Mach number $M_3 < M_2$, shown as point $3a$ on Fig. 6.24. The oblique shock train plus constant-p heat addition process is represented as path 2-$3a$-4 in Fig. 6.24.

The addition of that "just one more millijoule" causes thermal choking to occur, which forces an abrupt change to a fully *subsonic*

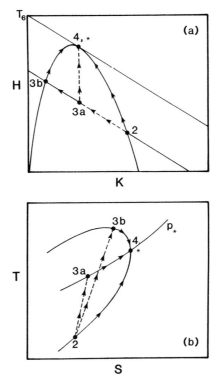

Fig. 6.24   (a) H-K diagram for thermally choked flow in constant-A burner. Path 2-$3a$-4∗ is for supersonic combustion with separated flow in both isolator and burner. Path 2-$3b$-4∗ is for normal-shock equivalent in isolator, all-subsonic, attached flow throughout burner. (b) T-s diagram illustrating processes in (a).

burner entry flow with $M_3 < 1$, shown as point 3b on Fig. 6.24. However, because the heat is now "added" subsonically, a *favorable* pressure gradient is established immediately, so that the flow is attached everywhere within the burner, and Stations d, 3, and s are now essentially coincident. From the process paths in T-s coordinates, Fig. 6.24b, it can be seen that the sum of the irreversible entropy rise in the shock train and the reversible entropy rise due to heat "addition" are equal along both paths, since the same amount of heat represented by τ_b is "added" at a higher mean temperature along the subsonic path $3b$-4_*:

$$
\left(\Delta s_{2-3a}\right)_{\substack{\text{irreversible}\\\text{adiabatic,}\\\text{oblique}\\\text{shock train}}} + \left(\Delta s_{3a-4_*}\right)_{\substack{\text{reversible}\\\text{const-}p\\\text{heating}}} =
$$

$$
\left(\Delta s_{2-3b}\right)_{\substack{\text{irreversible}\\\text{adiabatic,}\\\text{normal}\\\text{shock train}}} + \left(\Delta s_{3b-4_*}\right)_{\substack{\text{reversible}\\\text{const-}A\\\text{heating}}} \qquad (6\text{-}105)
$$

Since the flow is attached (unseparated) at the subsonic State $3b$, that state must have the same value of stream thrust function Φ as the isolator entry State 2. Consequently, the normal-shock train in the isolator is established at its *normal shock* limit of operation, that is, $M_{3b} = M_{2y}$ and $p_{3b} = p_{2y}$. As a result, any further increment in τ_b unstarts the engine inlet, and any decrement in τ_b causes reversion to scramjet operation with burner entry State $3a$.

In a constant-A burner, for any given isolator entry Mach number $M_2 > 1$, there is only one, unique value of τ_b which will cause thermal choking, which is given by the Rayleigh flow relation[6.39, 6.40]

$$
\tau_{b*} = \frac{\left(1 + \gamma_b M_2^2\right)^2}{2(\gamma_b + 1)M_2^2 \left(1 + \dfrac{\gamma_b - 1}{2}M_2^2\right)} \qquad (6\text{-}106a)
$$

or conversely,

$$
M_2 = \sqrt{\frac{1 + \gamma_b\left(\dfrac{\tau_{b*} - 1}{\tau_{b*}}\right) + (\gamma_b + 1)\sqrt{\dfrac{\tau_{b*} - 1}{\tau_{b*}}}}{1 - \gamma_b^2\left(\dfrac{\tau_{b*} - 1}{\tau_{b*}}\right)}} \qquad (6\text{-}106b)
$$

is the supersonic, constant-A entry Mach number for which any given $\tau_b = \tau_*$ will choke the burner. If τ_b is held constant and M_2 varied, then for M_2 greater than that given by Eq. (6-106b), all-supersonic, nonchoked heat "addition" occurs in the constant-A burner, with or without flow separation. For M_2 *less* than that given by Eq. (6-106), the burner entry Mach number M_3 abruptly jumps to a subsonic Mach number $M_{2y} < 1$ corresponding to a normal shock from M_2, and any *further* reduction in M_2 unstarts the engine inlet.

Within the limits of our assumption of one-dimensional frictionless flow with no mass addition, we conclude that constant-A burners have no operating margin at all for transitioning stably back and forth between modes. If a constant-A burner were to be specified for a dual-mode system, the inlet isolator would either have to be replaced by an increasing-area diffuser ("trans-section"), or enough air would have to be bled from the boundary layer near the isolator exit to give the same increasing-area effect as a trans-section to the remaining (not bled) airflow.

Clearly, a constant-A burner with a constant-A isolator is not a good combination for a dual-mode combustion system. Happily, a burner with *increasing* $A(x)$ provides the needed margin to allow a useful operating range for mode transition when coupled with a constant-A isolator.

6.5.5.5 Variable-area ramjet with normal shock train.

In Sec. 6.5.4.4, we showed how to find the two unique subsonic and supersonic entry Mach numbers M_{3p} and M_{3m}, respectively, required by a variable-area burner when a choked thermal throat or sonic point exists at the axial location x_* determined from Eq. (6-102). The process paths resulting from this calculation were shown on H-K coordinates in Fig. 6.22. The *two* Mach numbers M_{3p} and M_{3m} were shown to be unique for a given pair of functions $A(x)$ and $T_t(x)$, just as the *single* M_2 of Eq. (6-106) is unique for a given $\tau_b = \tau_{b*}$ in a constant-A burner. For isolator entry Mach numbers M_2 greater than M_{3m}, all-supersonic, nonthermally choked heat "addition" occurs, with or without flow separation. There is now a narrow but useful range of supersonic values of M_2 *less than* M_{3m} for which thermally choked, ramjet mode operation is possible. We will next show how this range can be determined.

As we are assuming frictionless flow, the only forces acting on the burner walls are static pressure forces. Consequently, axial change in the impulse function $I \equiv pA(1 + \gamma_b M^2)$, as defined in Example Case 2.2, is given by

$$dI = pdA \qquad (6-107)$$

Solving the I-definition equation explicitly for p and substituting into Eq. (6-107), there results

$$\frac{dI}{I} = \frac{1}{(1 + \gamma_b M^2)} \frac{dA}{A} \left(= \frac{d\Phi}{\Phi} \right) \qquad (6-108)$$

Consider the integral of Eq. (6-108) along the two process paths shown in Fig. 6.22. Along the subsonic path $*$-$3p$, M is always less than 1, while M is always greater than 1 along the supersonic path $*$-$3m$. Consequently, it can be seen from Eq. (6-108) that, at the same axial location x along both process paths, and therefore at the same A and dA/dx, dI/dx is always greater on the subsonic path. Consequently, the integral of Eq. (6-108) from $3p$ to $*$ is also greater than that from $3m$ to $*$, so that I_{3p} *is always less than* I_{3m} *for axially-increasing* $A(x)$. If we characterize constant-Φ isolines on H-K coordinates as "arches," as for example in Fig. 2.20, then the subsonic point $3p$ on Fig. 6.22 lies *inside* the constant-Φ isoline ("arch") which passes through the supersonic point $3m$. Conversely, point $3m$ lies *outside* the constant-Φ isoline which passes through point $3p$. An alternative, physical explanation of this result may be stated as follows: Since p is always greater along the subsonic integration path $*$-$3p$ than along the supersonic path $*$-$3m$, then the integral of pdA, and therefore the change in stream thrust I from Eq. (6-107), is also greater along the subsonic path $*$-$3p$. Since I_* is greater than both I_{3m} and I_{3p} (because $dA/dx > 0$), then I_{3p} is necessarily less than I_{3m}.

Unstart will occur only when the isolator entry Mach number $M_2 > 1$ is sufficiently low that $\Phi_2 = \Phi_{3p}$, that is, when the unique-for-$A(x)$-and-$T_t(x)$ subsonic burner entry Mach number $M_{3p} < 1$ corresponds to the outflow from a normal shock wave with a supersonic, upstream Mach number $M_{2x} > 1$ given by[6.39, 6.40]

$$M_{2x} = \sqrt{\frac{M_{3p}^2 + \dfrac{2}{\gamma_b - 1}}{\dfrac{2\gamma_b M_{3p}^2}{\gamma_b - 1} - 1}} \qquad (6\text{--}109)$$

Thus, the particular burner specified by any given $A(x)$ and $T_t(x)$ can operate stably in ramjet mode for supersonic isolator entry Mach numbers M_2 in the range $[M_{2x}, M_{3m}]$, where M_{2x} is given by Eq. (6-109) and M_{3m} is determined as described in Sec. 6.5.4.4 and illustrated in Fig. 6.22.

Before leaving the subject of subsonic heat "addition" in ramjet mode, it must be emphasized once again that the physical mechanisms of fuel-air mixing, flameholding and combustion are very different in subsonic and supersonic combustion. This fact presents some peculiarly difficult design problems for mode transition in dual-mode combustion systems. In ramjet mode, local recirculation or *backmixing*[6.31] must be provided, not only to mix the fuel and air, but also to locally backmix some burned gases into the unburned fuel-air mixture in order to provide a pilot flame or flameholder wake region.

This recirculation "eddy" might be the wake of a normal fuel jet, as described in Sec. 6.2.7, or possibly an array of V-gutters or similar bluff-body flameholders which might be dropped into the burner for ramjet mode operation and retracted for scramjet operation. In either case, it is certain that the resulting $T_t(x)$ will be *very different* for ramjet and scramjet modes, as was described in the paragraph surrounding Eq. (6-91).

6.5.6 Interpretation of Experimental Data

In the preceding Sec. 6.5, a procedure was developed for designing an isolator-burner system "from the inside out," by assuming various algebraic equation models for $A(x)$ and $T_t(x)$, then numerically solving Eqs. (6-90) together with Eqs. (6-92) through (6-97) for $M(x)$ and related properties. Due to very high velocities and temperatures, test conditions in experimental supersonic combustion burners are extremely hostile to intrusive thermocouple or pitot probes or rakes. While nonintrusive (optical) instruments are playing an increasingly important role in experimental scramjet burner research, measurement of static pressure along the burner walls is relatively straightforward, accurate, and reliable. Consequently, it is of interest to see how much knowledge concerning processes within the combustion system can be determined "from the outside in" from measurement of static pressures along the burner walls.

Figure 6.16b can be thought of as an idealized plot of measured wall static pressures $p(x)$ at various axial stations in a scramjet burner. The flow and thermodynamic state of the gas at entry is known, as is the geometric area distribution $A(x)$. It is desired to determine the axial distributions of $M(x)$ and $T_t(x)$ from the known $A(x)$ and measured $p(x)$ data, in order to determine both the overall heat release and the axial distribution of heat release.

The first step is to inspect the $p(x)$ data to identify the axial location of adverse or favorable pressure gradients with respect to Station 3, where the fuel is first injected, and identify which of the three cases of Table 6.4 is indicated by the data. Choices are then made for the axial station locations x_u, x_d and x_s, defined as appropriate to the case identified from Table 6.4 and accompanying text. It is emphasized that the data reduction analysis which follows is only meaningful if the axial location of regions of separated flow are recognized as such and separation is taken into account.

The next step is to "smooth" the $p(x)$ data by curve-fitting in the sense of least-squares. Following a recommendation of Waltrup and Billig,[5.19] $p(x)$ between u and d can be represented by a cubic polynomial,

$$\frac{p(x)}{p_u} = 1 + \left(\frac{p_d}{p_u} - 1\right)(3 - 2\chi)\chi^2 \ , \quad \text{where } \chi \equiv \frac{x - x_u}{x_d - x_u} \quad (6\text{-}110)$$

In the interval $[x_d, x_s]$, $p(x) = p_s = p_d = $ constant.

Whereas any smoothly decreasing function *could* be used to fit the $p(x)$ data from x_s to x_4, Billig[6.37,6.38] recommends a particularly useful function, the "polytropic process" relationship $pA^n = $ constant, where the exponent n is determined from the pressure data on the interval $[x_s, x_4]$ by

$$n = -\frac{\ell n\,[p(x_s)/p(x_4)]}{\ell n\,[A(x_s)/A(x_4)]} \tag{6-111}$$

It is certainly not obvious that Eq. (6-111) will fit any arbitrary set of $p(x)$ data, but experience shows that this is so, although it may be necessary to adjust the end-state values $p(x_s)$ and $p(x_4)$ a bit in Eq. (6-111) to obtain the best fit in the sense of least-squares to all of the intermediate $p(x)$ data.

Since it is assumed that only pressure forces act on the duct walls, a differential change in stream thrust function is given by Eq. (6-107), so that between any two axial locations x_i and x_e in the entire data range $[x_2, x_4]$,

$$I(x_e) = I(x_i) + \int_{x_i}^{x_e} p(x')\frac{dA(x')}{dx'}dx' \tag{6-112}$$

It is important to recognize that, since Eq. (6-112) is based on pressure at the walls, Eq. (6-112) is valid whether the flow is separated or attached. It is in this sense that we are analyzing "from the outside in."

The evaluation of the definite integral in Eq. (6-112) is straightforward:

1. If $A_e = A_i$, then $I_e = I_i$ follows at once.

2. If $p(x)$ is locally fitted by the cubic Eq. (6-110) and $A(x)$ is linear, then dA/dx is constant and factors, leaving just the cubic polynomial integrand which integrates to a quartic expression.

3. If $A(x)$ is quadratic in x (as in a straight-walled conical geometry), then dA/dx is linear, and the integrand is a fourth-order polynomial which integrates to a fifth-order polynomial for $I(x)$.

4. If $A(x)$ is variable in the range $[x_d, x_s]$, then the constant pressure term factors, and the definite integral is evaluated as $p_d[A(x_e) - A(x_i)]$.

5. In the range $[x_s, x_4]$, because of the choice of curve-fit function pA^n with n determined from Eq. (6-111), the definite integral

in Eq. (6-112) is evaluated as

$$\frac{p(x_e)A(x_e) - p(x_i)A(x_i)}{1 - n} \, , \quad n \neq 1 \qquad (6-113)$$

With $I(x)$ thus determined for all x, the Mach number $M(x)$ is obtained from the definition of the impulse function as defined in Example Case 2.2, $I \equiv pA(1 + \gamma_b M^2)$, as

$$M(x) = \sqrt{\frac{1}{\gamma_b} \left[\frac{I(x)}{p(x)A(x)} - 1 \right] \frac{A(x)}{A_c(x)}} \qquad (6-114)$$

Once again, $M(x)$ from Eq. (6-114) must be interpreted as the Mach number *of the separated core flow* within the separated flow range $[x_u, x_s]$ in the last two cases in Table 6.4, in which a pre-combustion shock train exists.

All other properties of interest may be determined for the appropriate case from the "useful integral relations," Eqs. (6-92) through (6-97). In the scramjet with shock-free isolator case, as the flow is attached everywhere, all pressure rise and fall is due to the interaction of heat "addition" and area increase, so that $T_t(x)$ is determinate immediately from Eqs. (6-92) and (6-93). In the two cases with shock trains originating in the isolator, the data treatment is different in regions of separated and attached flow. In the *adiabatic, separated* flow interval $[x_u, x_d]$, $T_t = T_{t2}$ is constant and known, so that the confined core area $A_c(x)$ is evaluated from Eqs. (6-92), (6-93), and Eq. (6-114). In the *diabatic, attached* flow interval $[x_s, x_4]$, $A(x)$ is given, and $T_t(x)$ is evaluated from Eqs.(6-92) and (6-93). Thus all properties are determinate from the measured and smoothed $p(x)$ and the known state of the air at isolator entry Station 2, everywhere *except in the subinterval* $[x_d, x_s]$, within which the flow is both separated and diabatic, so that both $A_c(x)$ and $T_t(x)$ are unknown. In this interval, any simple, smooth function could be used to patch $T_t(x_d) = T_{t2}$ to $T_t(x_s)$, from which the remaining properties could be determined *approximately* within $[x_d, x_s]$. To patch smoothly (matching the first derivatives) at Station s, the differential forms of the conservation equations and of the $pA^n = $ constant relation may be combined to give

$$\frac{dT_t}{dx} = \frac{T_t}{A} \left[\frac{1 + \dfrac{n(1 - M^2)}{\gamma_b M^2}}{1 + \dfrac{\gamma_b - 1}{2} M^2} \right] \frac{dA}{dx} \qquad (6-115)$$

which can then be evaluated at Station s.

Having determined the axial distribution of $T(x)$ from the $p(x)$ data and the known or given isolator entry conditions, and having determined the adiabatic flame temperature T_{AFT} for the given overall equivalence ratio ϕ_0, Eq. (6-86) can be used to determine the axial distribution of combustion efficiency $\eta_b(x)$. Recall that the AFT can be calculated by means of the software program HAP(Equilibrium), using the (h, p) option as described in Sec. 6.3.2.1, with the analyzed burner entry air static temperature T_3 and pressure p_3 (together with the static temperature of the fuel if it is known) as inputs. If desired, the axial distribution of total temperature $T_t(x)$ can be curve-fit by means of Eq. (6-91) to determine the best-fit value of the empirical constant θ.

6.5.6.1 Billig's experimental wall pressure measurements.
Figures 6.25 through 6.27 present experimental pressure measurements from hydrogen-air combustion in laboratory scramjet burners, reported by Billig in Refs. 44 and 45.

From the schematic Fig. 6.25a, it is apparent that a variable-area diffuser was used instead of a constant-area isolator. It is apparent from the relative axial location of the constant-pressure plateau in Fig. 6.25b that the combustion system is in scramjet mode with an oblique shock train. Application of Eqs. (6-110) through (6-114), together with Eqs. (6-92) through (6-97), gives results summarized in Table 6.5.

Note in Table 6.5 the rise in static temperature in the precombustion shock train at burner entry 2-3 or u-d. An H-K diagram of the analyzed process, constructed by using Eqs. (6-96), is shown in Fig. 6.25c. Note that the process path 2-3 in the isolator and the constant-p heat "addition" in the confined flow within the constant-A portion of the burner look exactly like Fig. 6.23a. In the variable-area portion of the burner, from Station s to Station 4, the H-K process path asymptotes rapidly to the $\tau_b = 1.65$ adiabat.

Table 6.5 Analysis of wall pressure data of Fig. 6.25. Overall H_2-air equivalence ratio $\phi_0 = 0.50$, assumed $\gamma_b = 1.31$.

Station	x (in, cm)	T_t (°R, K)	T (°R, K)	u (ft/s, m/s)	M
$2 = u$	-8.16 (-20.7)	4109 (2283)	1570 (872)	6081 (1853)	3.23
3	0.0	4109 (2283)	2510 (1394)	4822 (1470)	2.03
d	2.04 (5.18)	4109 (2283)	2562 (1423)	4822 (1470)	1.97
s	12.00 (30.48)	6644 (3691)	5098 (2832)	4743 (1446)	1.40
4	35.00 (88.90)	6813 (3785)	3934 (2186)	6476 (1974)	2.17

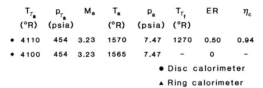

T_{T_a} (°R)	p_{T_a} (psia)	M_a	T_a (°R)	p_a (psia)	T_{T_f} (°R)	ER	η_c
● 4110	454	3.23	1570	7.47	1270	0.50	0.94
◆ 4100	454	3.23	1565	7.47	–	0	–

● Disc calorimeter

▲ Ring calorimeter

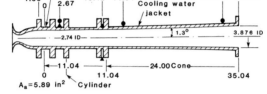

(a)

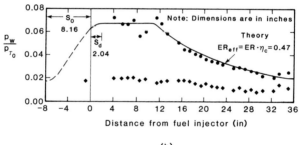

(b)

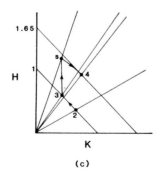

(c)

Fig. 6.25 Experimental wall pressure measurements in a cylinder-cone H₂-air scramjet combustor, Fig. 8 of Ref. 6.44. (a) Schematic of burner. (b) Axial pressure distribution. (c) H-K diagram of process path within burner.

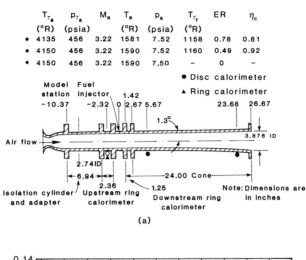

T_{T_a}	p_{T_a}	M_a	T_a	p_a	T_{T_f}	ER	η_c
(°R)	(psia)		(°R)	(psia)	(°R)		
• 4135	456	3.22	1581	7.52	1158	0.78	0.81
▪ 4150	456	3.22	1590	7.52	1160	0.49	0.92
♦ 4150	456	3.22	1590	7.50	–	0	–

(a)

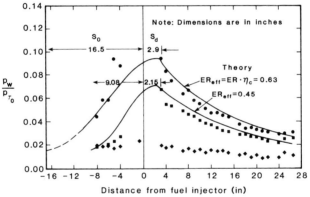

(b)

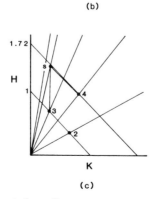

(c)

Fig. 6.26   Experimental wall pressure measurements from H_2-air combustion in a short cylinder-cone scramjet combustor, Fig. 9 of Ref. 6.44. (a) Schematic of burner. (b) Axial pressure distribution. (c) H-K diagram of process path within burner.

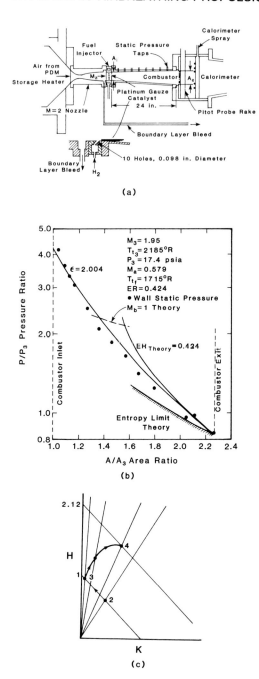

Fig. 6.27   Experimental wall pressure data in a hydrogen combustor test apparatus. (Figs. 7 and 8 of Ref. 6.45).   (a) Schematic of burner.   (b) Axial pressure distribution.   (c) *H-K* diagram of process path within burner.

While the burner satisfies the design requirement $p_4 \sim p_2$, 95 percent of the heat "addition" occurs at constant-p along path 3-s, where the pressure is more than three times greater than p_2 or p_4.

Table 6.6 summarizes results from integral analysis using Eqs. (6-110) through (6-112) for the wall pressure data (upper trace) of Fig. 6.26.

Note that in Fig. 6.26, there is considerably more heat "added" than in Fig. 6.25, so that the minimum Mach number at the thermal throat is not far from choking. Note also the two "outlying" pressure data points at about $x = -5$ in (-12.7cm), which Billig did not include when curve-fitting the shock train pressure rise. Because a normal fuel jet injection system was used, it is entirely possible that combustion in the separated flow region around the fuel jet may have penetrated that far upstream of the fuel injector. If interpreted this way, the data would show $x_d = -5$ in (-12.7cm). Table 6.6 and Fig. 6.26c were constructed assuming $x_d = 0$ in on Fig. 6.26b.

Figure 6.27 is the test data from what is believed to be the first experimental demonstration of ramjet mode operation in a dual-mode combustion system.[6.45] Results from applying Eqs. (6-110) through (6-114) to these data are summarized in Table 6.7.

There are some very interesting features in the data and analyzed results summarized in Table 6.7. First, note that the burner entry state corresponds to State 2y, the normal shock limit for inlet unstart. These data were collected as part of a series of experiments during which Billig first recognized the need for an inlet isolator.[6.38] In the case represented in Fig. 6.27, boundary layer bleed was used, in the absence of an isolator, to stabilize the normal shock at burner entry. The region of flow separation is very short, as the pressure begins to drop immediately downstream of the stabilized normal shock wave at burner entry. Note also that the flow passes smoothly through the critical point as it accelerates from subsonic entry to supersonic flow at burner exit.

Table 6.6 Analysis of wall pressure data of Fig. 6.26. Overall H_2-air equivalence ratio $\phi_0 = 0.78$, assumed $\gamma_b = 1.34$.

Station	x (in, cm)	T_t (°R, K)	T (°R, K)	u (ft/s, m/s)	M
$2 = u$	-16.56 (-42.06)	4138 (2299)	1581 (878)	6088 (1856)	3.22
3,d	0.0	4138 (2299)	3015 (1675)	4036 (1230)	1.55
s	2.88 (7.32)	7312 (4062)	6188 (3438)	4036 (1230)	1.08
4	26.64 (67.66)	7237 (4020)	4114 (2286)	6727 (2050)	2.21

6.5.6.2 Billig's "entropy limit" for nonreacting flow. In Figs. 6.25 and 6.26, a lower $p(x)$ data trace is shown. These data were collected to calibrate the wall pressure measuring instruments in a nonreacting flow having the same nominal entry Mach number $M_2 = 3.2$. Since no fuel was injected, the only occluding mechanisms present were area change and wall friction. Note that there is a noticeable pressure rise in the constant-A section due to wall friction alone, which reminds us that while wall friction may be a secondary effect compared to that of heat "addition," it is by no means negligible.

A comparison of the nonreacting $p(x)$ data with the reacting $p(x)$ data shows that both traces become parallel near burner exit, Station 4. Since the nonreacting cases are adiabatic throughout, $dT_t/dx = 0$ for all x. In the reacting, diabatic cases, however, dT_t/dx approaches 0 asymptotically at burner exit, due to the decaying rate of far-field mixing and kinetic "freezing" of the chemical reactions. Thus, the condition $(dT_t/dx)_4 \rightarrow 0$ may be termed the "adiabatic limit" for heat release. Since wall friction has been neglected, the term "isentropic limit" or, as Billig has termed it, the "entropy limit,"[6.45] is appropriate.

The "entropy limit" value of the exponent n in the data-fitting expression $pA^n =$ constant may be evaluated by setting $dT_t/dx = 0$ in Eq. (6-115) and evaluating at Station 4. By setting the numerator on the right-hand side of Eq. (6-115) to 0, the "entropy limit" value of n, denoted n_e, is found to be

$$n_e = \frac{\gamma_b M_4^2}{M_4^2 - 1} \tag{6-116}$$

It is interesting to note that the polytropic process relation pA^n, with $n = n_e$ given by Eq. (6-116), also describes the variation of pressure with area and Mach number in isentropic flow with area change.

Table 6.7 Analysis of wall pressure data of Fig. 6.27. Overall H_2-air equivalence ratio $\phi_0 = 0.424$, assumed $\gamma_b = 1.31$.

Station	x (in, cm)	T_t (°R, K)	T (°R, K)	u (ft/s, m/s)	M
2	0.00	2186 (1214)	1320 (733)	3410 (1039)	1.95
$3 = d$	11.00 (27.95)	2186 (1214)	2066 (1148)	1272 (388)	0.58
s	12.00 (30.48)	2304 (1280)	2183 (1213)	1272 (388)	0.57
$*$	18.55 (47.12)	4112 (2284)	3506 (1948)	2851 (869)	1.00
4	34.00 (86.35)	4752 (2640)	3245 (1803)	4501 (1372)	1.64

6.6   BURNER COMPONENT CFD EXAMPLES

From the preceding discussions, we can conclude that in the burner there is a strong interaction and synergism between the fuel, fuel injectors, and the burner configuration, with a number of issues related to each one of them. Temperature, kinetics, and ignition are issues associated with the fuel. The injection scheme, mixing enhancement and control, axial momentum, and thermal protection are problems related to fuel injectors. Entrance flow conditions, area ratio and distribution, length, wall friction and heat transfer, mixing, turbulence, and chemistry are concerns regarding burner configurations. These issues are addressed by burner designers to attain the highest performing, lightest, lowest cost, and most durable and reliable burner. On each of these design requirements, different priorities are placed for different applications.

Of all the components of a hypersonic propulsion system, the burner is the least understood in terms of achieving desired design requirements. Just from the point of view of performance, burner designs differ at low, moderate, and high hypersonic, flight Mach numbers. To understand burner performance throughout the Mach number range, an extensive, systematic research effort is required. Research approaches based on testing and on analytical studies have some limitations. On the one hand, testing burners at simulated hypersonic flight speeds in test facilities pose three challenges: realistic simulation of the flow entering the burner, proper instrumentation and measurements to provide the burner flowfields and performance, and accurate quantification of uncertainties in measured and derived quantities. On the other hand, valid fluid dynamics models and uncertainties in them are needed for properly analyzing a burner's flowfields and for determining the burner's performance. In both these activities, computational fluid dynamics (CFD) is an extremely useful tool.

As we have seen, a significant issue of burner design is the burner length required for complete mixing of air and fuel at the molecular level to produce the desired mixture ratio of these two, while a parcel of air is inside the burner. This issue is significant because the specific impulse produced by the burner decreases rapidly and the takeoff gross weight of the flight vehicle equipped with this burner increases quickly as the distance for complete mixing increases. The mixing efficiency is, in part, determined by the injector design. A study of the mixing of fuel with air requires tracking of the fuel, modeling of turbulence, finite-rate chemical kinetics in the entering boundary layers at moderate and high flight Mach numbers, and simulation of combustion. To begin with, such studies are conducted with "cold flows," that is, with nonreacting flows. Later the effect of finite-rate chemistry is investigated. Another simplification is made concerning the turbulence model. Frequently, first laminar flows are studied and then turbulence models are introduced. Moreover, separate studies are necessary at low, moderate, and high hypersonic flight Mach numbers. Following these investigative steps, we present below a few CFD examples, comparing the effectiveness of ramp injectors (Fig. 6.11) and flush-wall injectors (Fig. 6.8) for mixing.

In Ref. 6.46, a numerical study of mixing enhancement in a burner with ramp injectors (Fig. 6.11) at a low-hypersonic flight Mach number is reported. The burner entry airflow and the injector exit hydrogen-flow are, respectively, at Mach 2.0 and at Mach 1.7. Each ramp is approximately 2.76 in (7.0 cm) long with a rectangular base, 0.6 in (1.52 cm) by 0.5 in (1.27 cm). These ramps are inclined at 10.3 deg. to the burner wall. The sidewalls of the unswept ramps are aligned with the entry flow, whereas the swept ramps are swept back at an angle of 10 deg. The fuel equivalence ratio is set at 1.8. Wall temperatures are held constant at 1350 °R (750 K). The circular injector orifice is approximately modeled with a Cartesian computational grid system. Computations are carried out for the following conditions: (1) laminar and cold flow, (2) turbulent and cold flow, and (3) turbulent and reacting flow. Near-field computations are conducted with the full Navier-Stokes equations, whereas far-field computations are done with parabolized Navier-Stokes equations.

Figure 6.28 shows for case (1) the spanwise transport of hydrogen. The swept ramp causes hydrogen to travel an appreciably greater distance than the unswept ramp. Moreover, the swept ramp helps

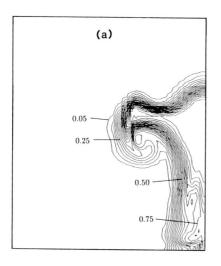
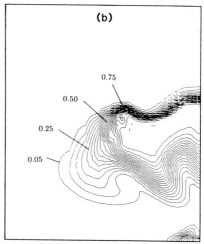

Fig. 6.28 Cross-stream hydrogen mass fraction contours for (a) unswept and (b) swept ramp at a distance of 5.2 in (13.2 cm) beyond the end of the ramps, assuming laminar flow [case (1)].[6.47]

to transport hydrogen completely off the lower wall of the burner. These related observations are explained by the fact that higher levels of streamwise vorticity are introduced by the swept ramps than by those with unswept ramps. Figure 6.29 for case (2) again confirms that swept ramps enhance the mixing of fuel with air. However, turbulent plus molecular diffusive processes produce less steep gradients of hydrogen mass fractions than those developed by the molecular viscosity alone.

A measure of mixing effectiveness at molecular levels is given by the mixing efficiency η_M. This parameter is computed at each crossflow plane and is presented in Fig. 6.30. The swept ramp is more effective for mixing, both in the near- and far-field. Turbulence greatly enhances far-field mixing. The near-field mixing is primarily controlled by large-scale, counter-rotating vortices with streamwise vorticity. These vortices distort the core and disrupt the outer regions of the injected stream of hydrogen and transport it in the spanwise direction. The far-field is largely controlled by small-scale, turbulent diffusive processes. And combustion slightly improves mixing in the far field. Please note that a heuristic argument presented previously concluded that an effect of heat release during combustion is to make the mixing layer to occupy a greater volume fraction at any axial location and to reduce the rate of growth of this layer. The length to achieve a desired mixing level and consequently the length of the burner are shorter with swept ramps than with unswept

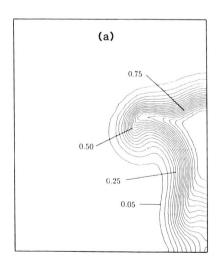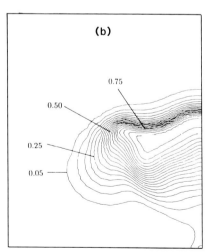

Fig. 6.29   Cross-stream hydrogen mass fraction contours for (a) unswept and (b) swept ramp at a distance of 5.2 in (13.2 cm) beyond the end of the ramps, assuming turbulent flow [case (2)].[6.47]

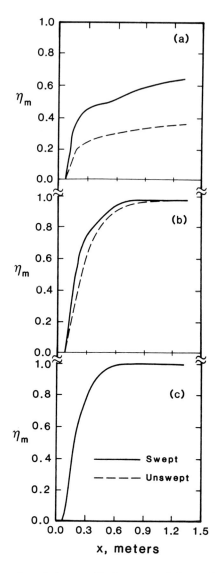

6.30 Variation of mixing efficiency with axial distance for unswept and swept ramps and for (a) case (1), (b) case (2), and (c) case (3).[6.27] Please note that, for case (3), the reacting flow is computed using a finite-rate, two-step combustion model. In computing mixing parameters, molecular hydrogen and total atomic hydrogen are used for nonreacting and reacting flows, respectively. Moreover, the flow conditions considered in these cases correspond to those likely to occur at low flight Mach numbers.

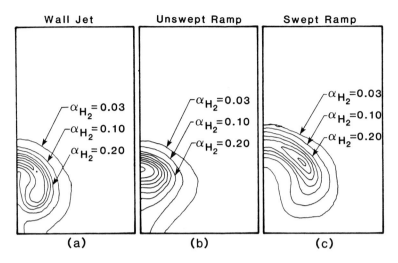

Fig. 6.31 Cross-stream hydrogen mass fraction contours for (a) wall jet, (b) unswept ramp, and (c) swept ramp at a distance of 1.97 in (5.0 cm) downstream of the injection location, assuming cold, turbulent flow.[6.27]

ramps. Although case (1) does not represent a realistic situation, it does help to provide some understanding of mixing phenomenon in the near field.

The mixing effectiveness of a flush-wall injector (Fig. 6.8) is compared with that of ramp injectors in Fig. 6.31. In this study, all flow and shape conditions are the same as those reported previously, except that the fuel equivalence ratio is 1.2 and wall temperatures are held at 1800 °R (1000 K). The flush-wall injector provides a 30-deg. downstream-directed hydrogen jet with equivalent injection conditions.[6.27] The observed differences between hydrogen mass fractions owing to the sweeping of the ramps are similar to those noticed in Fig. 6.29. In the case of the wall jet, there is appreciable distortion of the core of the jet and little distortion of the outer envelope of this jet. The swept ramp has produced the strongest streamwise vortices, the unswept ramp the weakest, and the wall jet has developed intermediate strength vortices. The corresponding mixing efficiencies are presented in Fig. 6.32. Again, the swept ramp turns out to be superior, the wall jet is a close second, with the unswept ramp providing relatively inefficient mixing. Please note that the far-field mixing efficiency values for the swept and unswept ramps differ much more at the fuel equivalence ratio of 1.2 (Fig. 6.32) than at 1.8 (Fig. 6.30b) with other conditions, except the wall temperatures, being the same.

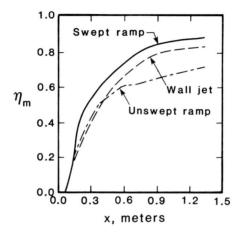

Fig. 6.32 Variation of mixing efficiency with axial distance for unswept and swept ramps and for wall jet with turbulent, cold flow conditions, at a low flight Mach number.

Computations with the swept ramp are also done at moderate and high flight Mach numbers, and those with the 30-deg. wall jet are carried out at high flight numbers.[6.28] As an example of a moderate flight Mach number, the following conditions are considered. For $M_0 = 13.5$, the burner and injector inflow conditions are the following: $p_3 = 19.73$ lbf/in^2 (136 kPa), $T_3 = 2416$ °R (1342 K), $M_3 = 4.1$ (Ref. 6.48), $p_i = 307.5$ lbf/in^2 (2120 kPa), $T_i = 1267$ °R (704 K), $M_i = 1.7$, and ϕ (fuel equivalence ratio) $= 1.0$. The burner wall is maintained at 2459 °R (1366 K). As an example of a high Mach number, the following conditions are studied. For $M_0 = 17$, the burner inflow conditions are the following: $p_3 = 2.39$ lbf/in^2 (16.5 kPa), $T_3 = 3760$ °R (2089 K), and $M_3 = 5.75$ (Ref. 6.48). The ramp injector conditions are $p_i = 17.69$ lbf/in^2 (122 kPa), $T_i = 342$ °R (190 K), $M_i = 1.7$, and $\phi = 1.0$. The wall injector conditions are $p_i = 97.32$ lbf/in^2 (671 kPa), $T_i = 439$ °R (244 K), $M_i = 1.0$, and $\phi = 3.0$. The burner wall is maintained at 540 °R (300 K). The mixing parameters are computed assuming turbulent, cold flow.

The study of swept ramp injectors shows that the mixing of hydrogen with air is slightly worse at $M_0 = 13.5$ than at $M_0 = 17$ (Fig. 6.33), mainly because of different jet-to-freestream conditions. The near-field mixing with the flush-wall injector is better than that with the swept ramp, at high Mach number. In the far-field, the swept ramp and flush wall injectors are comparable. Please note that these conclusions need to be used with caution since some of the injector inflow conditions and the burner conditions are not the same.

Plate 4 (at end of book) shows, for $M_0 = 13.5$, downstream the evolution of (1) the region containing hydrogen in a turbulent, cold flow simulation with air as the burner inflow fluid and (2) of the region containing water in a turbulent, reacting flow simulation with oxygen as the inflow fluid. All other conditions are identical. Pressure contours on some of the computational planes are also shown. These pictorial presentations of burner flows speak for the power of CFD eloquently. The mushroom-shaped, mixing and reacting zones downstream of the ramp are caused by the interaction of injected hydrogen with streamwise vortical flows.

These examples suggest that estimates of performance quantities, such as the mixing efficiency, as well as the visualization of the highly complex fluid dynamics occurring in realistic burners throughout the Mach number range of airbreathing, hypersonic propulsion systems are only feasible with CFD tools in the absence of necessary test facilities and instrumentation. Furthermore, these examples demonstrate the current state of affairs for just one aspect of designing burners, namely, effective mixing of air and fuel. Although the results are qualitative, they are extremely useful for indicating trends, although these trends need to be verified by independent studies. Moreover, the computational accuracy of these results, and the validity of the turbulence and chemistry models used have yet to be established. Ultimately, the credibility of performance estimates can only be established during flight tests, by measuring quantities for deriving these estimates, along with the determination of uncertainties in measurands and derived estimates.

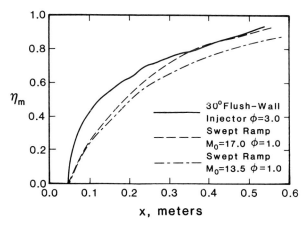

Fig. 6.33   Variation of mixing efficiency with axial distance for unswept and swept ramps and for wall jet with turbulent, cold flow conditions, at moderate and high flight Mach numbers.[6.28]

6.7 CLOSURE

For preliminary design of a scramjet combustion system, the "generalized one-dimensional flow analysis" method of Sec. 6.5.4 is a straightforward *synthesis* procedure for finding the right combination of $A(x)$ and $T_t(x)$ "from the inside out" to achieve a desired burner performance. The method is especially advantageous for designing dual-mode burners, because of its capability of independently locating a potential critical point prior to solving the ordinary differential equation for M, thus avoiding iteration of the numerical solution to locate the critical point. When the design calculations show that flow separation is expected, adjustments can be made to the required burner entry states, which can then be compared with the admissible range of isolator exit states to see whether or not the isolator can contain the pre-combustion shock train.

For determining burner performance "from the outside in" from experimental wall pressure data, the integral analysis method of Sec. 6.5.6 is an easily applied, powerful *analysis* method. In particular, when a critical point occurs in dual-mode combustion, the integral analysis method can locate the choked thermal throat explicitly, whereas an attempt to replicate the $p(x)$ data by trial-and-error application of the generalized one-dimensional flow analysis method would lead to numerical difficulties owing to the $(1 - M^2)$ term in the denominator of Eq. (6-90).

At the beginning of this chapter, it was stated that serious doubts still exist concerning whether or not stable and efficient supersonic combustion is possible over the required range of burner entry conditions. We will conclude this chapter with a recap of some of the more perplexing issues concerning a dual-mode combustion system, and will list some other issues which were omitted from or oversimplified in this chapter.

1. Having shown the impracticality of fabricating variable-in-flight or "rubber" geometry which would be desirable for constant-p burner operation, the flow separation caused by heat addition in a constant-A burner causes constant-p heat addition to occur anyway. Ironically, the self-adjusting area $A_c(x)$ of the confined core provides just the "rubber" geometry required, although heat addition occurs at a higher constant pressure than either burner entry or exit.

2. Although flow separation in a constant-A burner leads to increased loss of total pressure in the isolator shock train, this loss is offset by an accompanying decrease of total pressure loss within the burner, due to the increased mean temperature and reduced burner Mach number. This suggests the possibility of

designing a fully separated, constant-A burner, but with reduced pressure rise from the inlet compression system, to compensate for the additional compression within the isolator.

3. While the constant-A burner design of item 2 might work for all-supersonic combustion, it has no operational margin in the ramjet (subsonic burner entry) mode to be a useful dual-mode engine design, due to the fact that a thermally choked, constant-A burner requires a normal shock upstream, which is precisely at the limit of the constant-A isolator. To stabilize normal shocks of greater strengths, boundary layer bleed could be used near isolator exit,[6.45] or else the isolator could be replaced with a conventional diverging transition section.[6.36]

4. The swept-ramp "hypermixer" was designed to enhance fuel-air mixing by the generation of axial vorticity in the airstream from a system of interacting oblique shock waves and Prandtl-Meyer expansion waves. However, neither shock waves nor P-M expansions can be present in a zero-Mach number, separated flow. Consequently, experimental measurement of nonreacting mixing performance of these devices is irrelevant to their installed performance in a combustion system operating with separated flow.

5. Normal injection of sonic fuel jets into a supersonic crossflow causes detached normal shocks and wakes to form around the cylinder of the jet itself. These negative effects may be significantly ameliorated if the jet must first penetrate a region of separated flow before interacting with the supersonic core. As is the case for swept-ramp "hypermixers," cold flow or nonreacting mixing measurements of normal fuel jets mixing into a supersonic crossflow are probably irrelevant to their installed performance.

Some important issues that were inadequately addressed, or not at all, include:

1. The effect of total pressure losses from all sources within the combustion system is very important to overall cycle efficiency, and was not dealt with adequately in this chapter.

2. Chemical kinetics was shown to exert a very large influence on combustion efficiency, especially at lower flight Mach numbers. While a conceptual and computational framework for performing a mixing-plus-chemistry calculation was presented in Sec. 6.4, a software program to actually do the calculations is not included in the HAP software package.

3. Wall friction, wall heat transfer, and mass and momentum effects of fuel injection were all omitted from consideration, on the grounds that their effects are of secondary importance to $A(x)$ and $T_t(x)$. However, these effects are not so small as to be totally neglected in detailed analysis or design.

4. Standing shock systems within both the isolator and burner are very important in all internal processes, but except for empirical treatment of gross effects such as shock trains and flow separation, were necessarily omitted from the simplified, quasi-one dimensional formulation.

5. In the dual-mode burner, the requirements of subsonic flameholding are very different from those of supersonic flameholding, so that $T_t(x)$ could not in reality be constant during ramjet-scramjet mode transition. Omitted from this chapter were descriptions of the different locations and types of mixers, flameholders and fuel injection systems that will be needed in the each of the two modes. In addition, no mention was made of the possibility of multiple fuel injection sites, or of partially premixing the fuel and air in the isolator, to satisfy some of the problems arising at very high Mach numbers, where convective residence times in the burner are very short.

It is clear that we do not yet have all of the answers to the puzzle of optimal design of scramjet and dual-mode combustion systems. However, it is hoped that the persistent reader has gained enough appreciation for the nature of the unresolved problems and unanswered questions to be inspired to join the search.

REFERENCES

6.1 Weber, R. J., and McKay, J. S., "Analysis of Ramjet Engines Using Supersonic Combustion," NACA TN-4386, 1958.

6.2 Brodkey, R. S., *The Phenomena of Fluid Motions,* Addison-Wesley, New York, 1967.

6.3 Bird, R. B., Stewart, W. E., and Lightfoot, E. N., *Transport Phenomena,* John Wiley, New York, 1960.

6.4 Pai, S.-I., *The Fluid Dynamics of Jets,* D. van Nostrand, New York, 1954.

6.5 Abramovitz, M., and Stegun, I., *Handbook of Mathematical Functions,* Dover Publications, 9th printing, 1972.

6.6 Schlichting, H., *Boundary Layer Theory,* 7th Edition, McGraw-Hill, 1979.

6.7 Murthy, S. N. B., and Curran, E. T., (eds.), *High-Speed Flight Propulsion Systems,* AIAA Progress in Astronautics and Aeronautics Series, Vol. 137, Washington, DC, 1991.

[6.8] Breidenthal, R. E., "Sonic Eddy – A Model for Compressible Turbulence," *AIAA Journal,* Vol. 30, No. 1, Jan. 1992.

[6.9] Beach, H. L., "Supersonic Mixing and Combustion of a Hydrogen Jet in a Coaxial High-Temperature Test Gas," AIAA Paper 72-1179, Nov. 1972.

[6.10] Planche, O. H., and Reynolds, W. C., "Heat Release Effects on Mixing in Supersonic Reacting Free Shear Layers," AIAA Paper 92-0092, Jan. 1992.

[6.11] Anderson, G. Y., "An Examination of Injector/Combustor Design Effects on Scramjet Performance," *Proceedings of the 2nd International Symposium on Air Breathing Engines,* Sheffield, England, Mar. 1974.

[6.12] Pulsonetti, M. V., Erdos, J., and Early, K., "An Engineering Model for Analysis of Scramjet Combustor Performance with Finite Rate Chemistry," AIAA Paper 88-3258, 1988.

[6.13] Billig, F. S., Orth, R. C., and Lasky, M., "A Unified Analysis of Gaseous Jet Penetration," *AIAA Journal,* Vol. 9, No. 6, June 1971.

[6.14] Rogers, R. C., "Mixing of Hydrogen Injected from Multiple Injectors Normal to a Supersonic Airstream," NASA TN D-6476, Sep. 1971.

[6.15] Hollo, S. D., McDaniel, J. C., and Hartfield, R. J., Jr., "Characterization of Supersonic Mixing in a Nonreacting Mach 2 Combustor," AIAA Paper 92-0093, Jan. 1992.

[6.16] Hussaini, M. Y., Kumar, A., and Voight, R. G., *Major Research Topics in Combustion,* ICASE/NASA Langley Research Center Series, Springer-Verlag, New York, 1992.

[6.17] Swithenbank, J., and Chigier, N. A., "Vortex Mixing for Supersonic Combustion," *Twelfth Symposium (International) on Combustion,* The Combustion Institute, Pittsburgh, 1969.

[6.18] Naughton, J. W., and Settles, G. S., "Experiments on the Enhancement of Compressible Mixing via Streamwise Vorticity," AIAA Paper 92-3549, July 1992.

[6.19] Gutmark, E., Schadow, K. C., Parr, T. P., Parr, D. M., and Wilson, K. J., "Combustion Enhancement by Axial Vortices," *AIAA Journal of Propulsion and Power,* Vol. 5, No. 5, Sept. 1989.

[6.20] Waitz, I. A., Marble, F. E., and Zukoski, E. E., "Vorticity Generation by Contoured Wall Injectors," AIAA Paper 92-3550, June 1992.

[6.21] Drummond, J. P., "Mixing Enhancement of Reacting Parallel Fuel Jets in a Supersonic Combustor," AIAA Paper 91-1914, Jan. 1992.

[6.22] Martin, J. T., Mausshardt, S. L., and Breidenthal, R. E., "Incompressible Mixing from a Ten Degree Ramp and a Delta Wing Vortex Generator," unpublished manuscript, 1992.

[6.23] Avrashkov, V., Baranovsky, S., and Levin, V., "Gasdynamic Features of Supersonic Kerosene Combustion in a Model Combustion Chamber," AIAA Paper 90-4568, 1990.

[6.24] Broadwell, J. E., "Delta Wing Nozzle Assembly for Chemical Lasers," U.S. Patent No. 4,466,100, Aug. 14, 1984.

[6.25] McVey, J., and Kennedy, J., "Flame Propagation Enhancement Through Streamwise Vorticity Stirring," AIAA Paper 89-0619, 1989.

[6.26] Jachimowski, C. J., "An Analytical Study of the Hydrogen-Air Reaction Mechanism With Application to Scramjet Combustion," NASA TP-2791, 1988.

[6.27] Riggins, D. W., and McClinton, C. R., "Analysis of Losses in Supersonic Mixing and Reacting Flows," AIAA Paper 91-2266, 1991.

[6.28] Riggins, D. W., McClinton, C. R., Rogers, R. C., and Bittner, R. D., "A Comparative Study of Scramjet Ignition Strategies for High Mach Number Flows," AIAA Paper 92-3287, 1992.

[6.29] Callen, H. B., Thermodynamics, John Wiley, 1960.

[6.30] Gordon, S., and McBride, B., "Computer Program for Calculation of Complex Chemical Equilibrium Compositions," NASA SP-273, 1971.

[6.31] Pratt, D. T., "Calculation of Chemically Reacting Flows with Complex Chemistry," Studies in Convection, Vol. 2, B. E. Launder (ed.), Academic Press, New York, 1977.

[6.32] Radhakrishnan, K., and Pratt, D. T., "Fast Algorithm for Calculating Chemical Kinetics in Turbulent Reacting Flow," Combustion Science and Technology, Vol. 58, pp. 155–176, 1988.

[6.33] Korobeinikov, V. P., "The Problem of Point Explosion in a Detonating Gas," Astronautica Acta, Vol. 14, No. 5, 1969.

[6.34] Pratt, D. T., "Mixing and Chemical Reaction in Continuous Combustion," Progress in Energy and Combustion Science, Vol. 1, pp. 73–86, 1976.

[6.35] Burrows, M. C. and Kurkov, A. P., "Analytical and Experimental Study of Supersonic Combustion of Hydrogen in a Vitiated Airstream," NASA TMX-2828, 1973.

[6.36] Curran, E. T., and Stull, F. D., "The Utilization of Supersonic Combustion Ramjet Systems at Low Mach Numbers," Aero Propulsion Laboratory, RTD-TDR-63-4097, Jan. 1964.

[6.37] Billig, F. S., "Combustion Processes in Supersonic Flow," AIAA Journal of Propulsion and Power, Vol. 4, No. 3, May 1988.

[6.38] Billig, F. S., "Research on Supersonic Combustion," Journal of Propulsion and Power, Vol. 9, No. 4, July 1993.

[6.39] Shapiro, A. H., The Dynamics and Thermodynamics of Compressible Fluid Flow, Volume I, Ronald Press, New York, 1953.

[6.40] Zucrow, M. A., and Hoffman, J. D., Gas Dynamics, Volume I, John Wiley, New York, 1976.

[6.41] Shchetinkov, E. S., "Piecewise-One-Dimensional Models of Supersonic Combustion and Pseudo Shock in a Duct," Combustion, Explosion and Shock Waves, Vol. 9, No. 4, 1975, pp. 409-417.

[6.42] Lin, P., Rao, G.V.R., and O'Connor, G. M., "Numerical Investigation on Shock Wave/Boundary Layer Interactions in a Constant Area Diffuser at Mach 3," AIAA Paper 91-1766, 1991.

[6.43] Elmquist, A. R., "Evaluation of a CFD Code for Analysis of Normal-Shock Trains," AIAA Paper 93-0292, 1993.

[6.44] Billig, F. S., Dugger, G. L. and Waltrup, P. J., "Inlet-Combustor Interface Problems in Scramjet Engines," *Proceedings of the 1st International Symposium on Airbeathing Engines,* Marseilles, France, June 1972.

[6.45] Billig, F. S., "Design of Supersonic Combustors Based on Pressure-Area Fields," *Eleventh Symposium (International) on Combustion,* The Combustion Institute, Pittsburgh, 1967.

[6.46] Riggins, D. W., Mekkes, G. L., McClinton, C. R., and Drummond, P. J., "A Numerical Study of Mixing Enhancement in a Supersonic Combustor," AIAA Paper 90-0203, Reno, NV, Jan. 1990.

[6.47] Drummond, J. P., Carpenter, M. H., and Riggins, D. W., "Mixing and Mixing Enhancement in Supersonic Reacting Flowfields," *High-Speed Flight Propulsion Systems,* Murthy, S. N. B., and Curran, E. T., (eds.), AIAA Progress in Astronautics and Aeronautics Series, Vol. 137, Washington, DC, 1991.

[6.48] Riggins, D. W., private communication, Dec. 1992.

PROBLEMS

6.1 Could the flow depicted in Fig. 6.4 be properly termed either laminar or turbulent? Give arguments for both answers. (Sec. 6.2.2.3)

6.2 PDF measurements taken in a fuel-air shear layer near the edge of the splitter plate, where negligibly little micromixing has yet taken place, show an intermittency for air of $i_A = 0.3$. Find: (a) The intermittency of fuel, i_F, (b) the time-mean values $\langle Y_A \rangle$ and $\langle Y_F \rangle$, and (c) the variance g for Y_A, all at the same location. (Sec. 6.2.3)

6.3 From Eqs. (6-21) and (6-22), estimate the Prandtl mixing length ℓ_m in the shear layer, expressed as a fraction of the local shear layer width d. Are these estimates physically plausible? (Sec. 6.2.4)

6.4 Show that, if both mixant streams have the same temperature $T_1 = T_2$ and the same specific heat ratio $\gamma_1 = \gamma_2$, Eq. (6-27) gives the same expression for u_c as the incompressible relation Eq. (6-24). (Sec. 6.2.4.1)

6.5 On p. 406 of Ref. 7, Drummond et al. give the following entry conditions for parallel stream mixing of hydrogen and air: H_2: 1 atm, 293 K, 1953 m/s; Air: 1 atm, 2000 K, 1297 m/s. Estimate the equivalence ratio within the resulting mixing layer. (Sec. 6.2.5.2)

6.6   From a number of recent technical papers, it appears that inlet manifolding should be limited to a maximum mixing aspect ratio $(L/b)_{max} \sim 20$, where L is the burner length and b is the burner inlet scale of segregation. For this mixing aspect ratio, use Eqs. (6-41), (6-44) and (6-46) to estimate the mixing efficiency η_M at burner outlet for parallel injection, normal injection, and swept ramp mixers. Assume the entry fuel and air mass flow rates are in overall stoichiometric proportion. (Secs. 6.2.6 through 6.2.8)

6.7   From the values of enthalpy of formation given in Table 6.1, find the enthalpy of reaction or heating value h_{PR} in BTU/lbm fuel, for all five fuels listed (four hydrocarbons plus hydrogen.) For any mission-required total combustion energy (BTU or kJ), which fuel would weigh the least? Which would require the least fuel tank volume? (Sec. 6.3.2)

6.8   A 1980 technical paper gives forward and reverse rate data for Reaction No. 1 of Table 6.2 as $A_j = 1.7 \times 10^{13}$, $B_j = 0$, $E_j/R = 24,232$, $A_{-j} = 5.7 \times 10^{11}$, $B_{-j} = 0$, $E_{-j}/R = 14,922$. Using HAP(Equilibrium) to find the equilibrium combustion products from stoichiometric hydrogen/air burning at 1 atm and combustor entry temperature 1000 K, show that the 1980 rate data do or do not satisfy the equilibrium requirement $R_j = R_{-j}$. (Sec. 6.3.3.1)

6.9   At 2000 K, the equilibrium constant for the reaction $CO_2 \rightarrow CO + \frac{1}{2}O_2$ has a value $\log_e K_p = -6.635$. Using HAP(Equilibrium), find the equilibrium composition of combustion products at 2000 K and 2.5 atm, for any hydrocarbon fuel and air at any equivalence ratio between 0.2 and 2. Show that the mole fractions of CO, CO_2 and O_2 predicted by HAP(Equilibrium) satisfy Eq. (6-70). (Sec. 6.3.3.1)

6.10   Verify the assertion in Sec. 6.4.2 that, if it is assumed that the scramjet burner velocity is constant, the combustion efficiency defined in terms of static temperature changes, Eq. (6-86), is equal to the combustion efficiency defined in terms of total temperature changes. If variable specific heats were assumed, which of the two combustion efficiencies would be greater?

6.11   Re-define the combustion efficiency, Eq. (6-86), in terms of total temperature changes, and show that Eq. (6-91) can be rewritten as

$$\eta_b(x) = \frac{\tau_b - 1}{\tau_{AFT} - 1}\left[\frac{\theta\chi}{1 + (\theta - 1)\chi}\right]$$

6.12 Use Eqs. (6-88) and (6-89) to find the confined core flow area fraction A_c/A_2 at burner entry Station 3, for the two scramjet mode cases of Figs. 6.25 and 6.26. Utilize the system data given in the two figures and in the corresponding Tables 6.5 and 6.6.

6.13 Using the generalized one-dimensional method of Sec. 6.5.4.1, with the help of the Design option in HAP(Burner), design a dual-mode (isolator plus burner) combustion *system* to meet the combustion system requirements assumed in the composite scramjet example case of Sec. 4.4.4, and in Sec. 6.2.2.1 and 6.2.5. Note that Station 3 in Sec. 4.4.4 must be interpreted as entry to the combustion system, corresponding to Station 2 in Chap. 6. Your design should include an estimate of the minimum length required for the isolator if required, and your selection of injection/mixing system should be fully justified. Sketch the layout of the isolator-burner system to scale, and make plots of axial variation of $M(x)$, $T_t(x)$, $T(x)$, $A_c(x)/A_2$ (only if flow separation occurs), and $p(x)$. Use the following design assumptions or guidelines:

    i. Assume a planar burner geometry with constant-A isolator and linearly increasing area burner to obtain *equal* pressures at entry and exit $p_2 = p_4$.

    ii. The burner should be no more than 10 ft (3.05 m) in length, and the mixing aspect ratio L/b should be no more than 20.

    iii. Assume infinite-rate chemistry: that is, assume that "mixed is burned," so the mixing efficiency $\eta_M(x)$ and combustion efficiency $\eta_b(x)$ are identical.

    iv. Neglect effects of wall friction and mass addition.

6.14 The same as Problem 6.13, except assume a constant-A burner, which will necessarily result in $p_4 > p_2$.

6.15 Use the methods of Sec. 6.5.6, with the help of the Analysis option in HAP(Burner), to analyze the solution to either of the two design problems 6.13 or 6.14 as if it were experimental data. Plot the aerothermodynamic paths in H-K coordinates, as in part (c) of Figs. 6.25 through 6.27.

7
EXPANSION SYSTEMS
OR COMPONENTS

7.1   INTRODUCTION

The function of the expansion systems or expansion components is to provide acceleration of the flow from the burner-exit static pressure to local atmospheric or freestream static pressure over the entire range of vehicle operation in a controllable and reliable manner with maximum performance (i.e., maximum expansion efficiency or minimum entropy increase and aimed in the desired direction). Expansion components are somewhat less fearful in terms of the number of requirements to be met and the catastrophes that can accompany design mistakes than compression components. Nevertheless, they have their own quirks and can exert an enormous influence on overall engine performance, so they must also be treated with respect and care.

Method of characteristics design techniques for ordinary two-dimensional exhaust nozzles form the basis for the design of ramjet and scramjet exhaust systems. Therefore, readers not currently familiar with these procedures are encouraged to brush up on them at this point by means, for example, of Refs. 7.1 and/or 7.2.

7.2   TYPICAL EXPANSION COMPONENT CONFIGURATIONS

Before embarking on any expansion component analysis, it is important to have a mental picture of the type of geometry most likely to be encountered in practice. The highly integrated scramjet engine expansion systems found on hypersonic vehicles are actually very clever variations on the classical exhaust nozzle theme. It is therefore appropriate to begin by developing an understanding of how classical exhaust nozzles would be configured for typical scramjet operating conditions.

7.2.1 Ideal Hypersonic Nozzle Configurations

We begin our exploration of expansion systems by considering the properties of an idealized family of two-dimensional exhaust nozzles. Figure 7.1 contains a fairly accurate portrayal of such an "ideal" two-dimensional exhaust nozzle designed for a uniform entry Mach number of 1.5 and perfectly expanded to freestream static pressure at an exit Mach number of 3.0. The main purpose of this drawing is to accent the key features of both the design philosophy and the method of analysis that is employed. Thus, for example, these design Mach numbers were chosen from the low speed end of the flight trajectory, or, as you shall soon see, the proportions of the drawing would have made a clear visualization impossible. Please refer repeatedly to Fig. 7.1 when considering each of the following points. The term *exhaust nozzle* will be reserved in what follows only for this conventional type of *fully confined* expansion system.

First, the exhaust nozzle is two-dimensional or planar rather than axisymmetric or circular, so the flow properties do not vary into the page. This choice is often made because circular nozzles are comparatively heavy and do not lend themselves easily to variable geometry. For these reasons, plus the need for tight integration, most hypersonic vehicle designs also use two-dimensional expansion.

Second, the flow at the exhaust nozzle entry is supersonic, rather than the sonic or choked throat condition of conventional convergent-divergent nozzles. This facilitates the analysis a great deal because it conveniently avoids the mathematical complexities connected with starting the supersonic flow analysis from the transonic flow of the throat region. In short, the entire flowfield under consideration is governed by differential equations that are of the hyperbolic, rather than elliptic or parabolic, type.

Third, the design is based on the assumption that the exhaust stream can be modeled as the isentropic flow of a calorically perfect gas. This allows the exhaust nozzle to be designed using the transparent and straightforward method of characteristics. The most important attributes of the characteristics themselves are that they propagate at the Mach wave angle (or Mach angle) relative to the local flow, that they can only be generated (or absorbed) by the changing slope of the bounding surface, and that, once generated, they carry with them a fixed amount and direction of turning of the local flow. It will become evident as we proceed that the method of characteristics plays essentially the same role for exhaust system analysis that the oblique shock wave analysis did for compression systems.

Fourth, the design produces uniform, parallel flow at the desired exit Mach number because that maximizes the resulting thrust.

Fifth, the design is a *minimum length* exhaust nozzle. This is achieved by placing sharp corners that generate centered simple or Prandtl-Meyer expansion fans at the nozzle entry, and absorbing or canceling each of the characteristics upon their arrival at the opposite nozzle boundary by means of a change in slope of equal magnitude and sign. One useful conclusion to be drawn from this description is that the amount of flow turning caused by the initial expansion

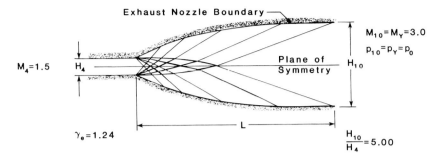

Fig. 7.1 Schematic diagram of an ideal, minimum length, two-dimensional exhaust nozzle designed by means of the method of characteristics for a uniform entry Mach number of 1.5 and a uniform exit Mach number of 3.0.

fan must be precisely equal to the amount of opposite flow turning along the remainder of the boundary, and therefore that each must equal one half of the total flow turning angle associated with the expansion process. Note should also be taken that this approach guarantees that there will be no compression waves anywhere in the flowfield, so that the assumed isentropic flow will be realized.

In addition to presenting a beautiful and meaningful picture of the flowfield, the method of characteristics allows the most important quantities of the flowfield behavior and exhaust nozzle geometry to be either directly calculated or reasonably estimated. Before summarizing the governing equations, it is beneficial, with the help of Fig. 7.2, to describe the various zones involved in the analysis. It should be apparent that the plane midway between the upper and lower boundaries is a plane of symmetry for the entire flowfield.

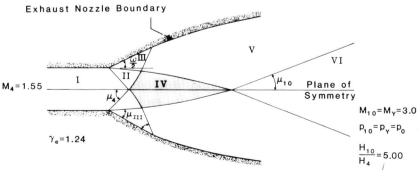

Figure 7.2 Schematic diagram of the entrance region of the exhaust nozzle of Fig. 7.1.

Zone I is the uniform, supersonic, combustor exit or nozzle entry flow having a known Mach number M_4 and a corresponding Mach wave angle μ_4

Zone II is the centered simple or Prandtl-Meyer expansion fan radiating along straight lines from the sharp expansion corner. The design amount of flow turning is one half of the total flow turning angle ω associated with the expansion from M_4 to M_Y . The angular extent of the expansion fan is determined by the upstream or *leading-edge* Mach wave angle μ_4 and the downstream or *trailing-edge* Mach wave angle μ_{III} as well as the corner flow turning $\omega/2$. The characteristics do not intersect or "cross" in Zone II, greatly simplifying the analysis there.

Zone III is a uniform flow zone where the conditions are the same as along the trailing edge of the expansion fan. The exhaust nozzle boundary is a straight line there because the expansion waves or characteristics from the opposite side have yet to arrive. The Mach wave angle relative to the straight boundary μ_{III} must have the same magnitude at the beginning and end of Zone III.

Zone IV is the only place where the characteristics cross. The intersections cause the characteristics to curve or bend, and require a more elaborate (yet still fairly elementary) analysis than the simple expansion fan. The single most important quantity produced by this analysis is the axial location of the termination of Zone IV, because that determines the overall length of the exhaust nozzle. The flowfield of Zone IV also provides the "initial conditions" for Zone V along their common boundary.

Zone V is a region in which the characteristics emanating from the boundary shared with Zone IV no longer cross. As each successive characteristic of Zone V reaches the exhaust nozzle boundary, it is absorbed by turning the boundary an amount equal to the amount of flow turning carried by the characteristic. Thus, the flow in Zone V determines the shape of the nozzle boundary.

Zone VI has uniform flow at the desired nozzle exit Mach number M_Y and pressure $p_Y = p_0$.

The curvature of the characteristics back toward the plane of symmetry in Zone IV is essential to the formation of the final characteristic that joins Zones V and VI, and therefore to the eventual closure of the nozzle. This can be seen especially clearly in the example of Fig. 7.3, which represents the entrance region for higher entry and exit Mach numbers than Figs. 7.1 and 7.2. As the entry and exit Mach numbers increase, the flow turning angle of the sharp expansion corner $\omega/2$ can exceed the Mach wave angle μ_{III} of Zone III, with the result that the trailing edge of the initial expansion fan actually slopes *away* from the plane of symmetry. This creates the striking "tulip" appearance of this type of entrance region that is a trademark of hypersonic nozzles.

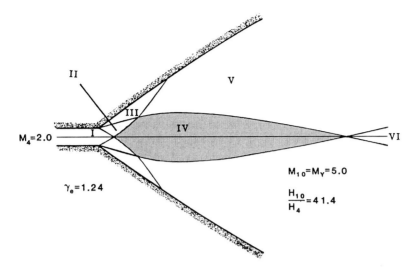

Fig. 7.3 Schematic diagram of the entrance region of an ideal, minimum length, two-dimensional exhaust nozzle designed by means of the method of characteristics for a uniform entry Mach number of 2.0 and a uniform exit Mach number of 5.0.

7.2.2 Hypersonic Expansion System Configurations

All that remains in order to "invent" the modern hypersonic expansion system is to recognize that the same result could be achieved for either the top or bottom half of the entry flow if the plane of symmetry were replaced by a physical surface that reflects the arriving characteristics. The reflecting surface must extend, in fact, only from the entry to the end of Zone IV because no characteristics reach the plane of symmetry beyond that point and because the flow has already reached the freestream static pressure everywhere along the final characteristic. Since this reflecting surface tends to be relatively short and usually contains a hinge for rotation, it is often referred to as a *flap* (or *splitter*).

The resulting configurations can be easily visualized by covering up the top or bottom half of Figs. 7.1 through 7.3. They really do resemble the depictions of hypersonic vehicle expansion systems found in Plate 2 (at end of book), Fig. 1.20, and elsewhere, where most of the expansion surface is also the underside of the aft end of the vehicle. Based on their appearance, they are also variously known as *single-sided nozzles*, *free expansion nozzles*, *unconfined nozzles*, and *expansion ramps*.

An agreeable feature of this type of expansion system is that the ratio of surface area to entry throughflow area remains about the same as for the original closed exhaust nozzle. That is, if we ignore the relatively minor contribution of the flap area, both the surface area and entry throughflow area are cut in two. On the other hand, the ratio of length to entry height is double that of a closed nozzle because the length is unaffected by halving the entry height.

7.3 IDEAL EXPANSION COMPONENT ANALYSIS

The expansion components, like their compression component siblings, encounter a staggering variety of operating conditions and have few, if any, geometrical degrees of freedom at their disposal. We shall begin our exploration of their ability to meet this challenge as we did for compression components, namely by investigating the properties of a wide but relevant range of design point expansion components.

With the discussion and assumptions of Sec. 7.2.2 in mind, the analytical relationships of Ref. 7.1 available, and referring to Fig. 7.2 for guidance and nomenclature, the important properties of the ideal, minimum length, two-dimensional exhaust component design point geometry can be calculated by following the sequence of steps delineated below. The expansion fan portion of HAP(Gas Tables) greatly facilitates the actual calculations. The various heights that define these expansion systems are, for consistency, referenced to one half the entry height of their closed nozzle counterparts.

1. Zone I Mach wave angle

$$\mu_4 = \arcsin \frac{1}{M_4} \tag{7-1}$$

2. Total method of characteristics flow turning from M_4 to M_Y

$$\omega = \sqrt{\frac{\gamma_e + 1}{\gamma_e - 1}} \left\{ \arctan \sqrt{\frac{\gamma_e - 1}{\gamma_e + 1} \left(M_Y^2 - 1 \right)} \right.$$

$$\left. - \arctan \sqrt{\frac{\gamma_e - 1}{\gamma_e + 1} \left(M_4^2 - 1 \right)} \right\}$$

$$- \left\{ \arctan \sqrt{M_Y^2 - 1} - \arctan \sqrt{M_4^2 - 1} \right\} \tag{7-2}$$

3. Zone III Mach number for flow turning of $\omega/2$ (requires iterative solution)

$$
\frac{\omega}{2} = \sqrt{\frac{\gamma_e + 1}{\gamma_e - 1}} \left\{ \arctan \sqrt{\frac{\gamma_e - 1}{\gamma_e + 1} \left(M_{III}^2 - 1 \right)} \right.
$$

$$
\left. - \arctan \sqrt{\frac{\gamma_e - 1}{\gamma_e + 1} \left(M_4^2 - 1 \right)} \right\}
$$

$$
- \left\{ \arctan \sqrt{M_{III}^2 - 1} - \arctan \sqrt{M_4^2 - 1} \right\} \quad (7\text{-}3)
$$

4. Zone III Mach wave angle

$$
\mu_{III} = \arcsin \frac{1}{M_{III}} \quad (7\text{-}4)
$$

5. Zone III static pressure (constant total pressure flow)

$$
\frac{p_{III}}{p_4} = \left\{ \frac{1 + \dfrac{\gamma_e - 1}{2} M_4^2}{1 + \dfrac{\gamma_e - 1}{2} M_{III}^2} \right\}^{\gamma_e/(\gamma_e - 1)} \quad (7\text{-}5)
$$

6. Zone VI Mach wave angle

$$
\mu_{10} = \arcsin \frac{1}{M_Y} \quad (7\text{-}6)
$$

7. Ratio of expansion component exit height to entry height (constant total temperature and total pressure flow) from the mass flow parameter MFP

$$
\frac{H_{10}/2}{H_4/2} = \frac{H_{10}}{H_4} = \frac{M_4}{M_Y} \left\{ \frac{1 + \dfrac{\gamma_e - 1}{2} M_Y^2}{1 + \dfrac{\gamma_e - 1}{2} M_4^2} \right\}^{(\gamma_e + 1)/2(\gamma_e - 1)} \quad (7\text{-}7)
$$

8. Ratio of Zone VI axial length to entry height

$$
\frac{L_{VI}}{H_4/2} = \frac{1}{\tan \mu_{10}} \cdot \frac{H_{10}}{H_4} \quad (7\text{-}8)
$$

9. Expansion component static pressure ratio (constant total pressure flow)

$$\frac{p_Y}{p_4} = \frac{p_0}{p_4} = \left\{\frac{1 + \frac{\gamma_e - 1}{2}M_4^2}{1 + \frac{\gamma_e - 1}{2}M_Y^2}\right\}^{\gamma_e/(\gamma_e - 1)} \tag{7-9}$$

The results of expansion component design point calculations based on Eqs. (7-1) through (7-9) are presented in Figs. 7.4 through 7.7. The value of $\gamma_e = 1.24$ used throughout these analyses reflects the fact that the exhaust flow has a larger number of molecular degrees of freedom energized than air under normal conditions, and is consistent with the thermodynamic cycle values of Chap. 4 and the discussions of chemistry of Chap. 2.

Table 7.1 contains information that can be quite helpful in interpreting these results, namely typical exhaust component operating condition examples corresponding to three very different freestream Mach numbers. This information was obtained from simple scramjet cycle calculations using nominal component efficiency values. Special note should be taken of the fact that the expansion component static pressure ratio is not an independent variable, but is largely prescribed by Eq. (5-4). If we assume that $\psi = 7.0$, $\eta_c = 0.90$, and $\gamma_c = 1.36$, and that combustion takes place at constant static pressure, then $p_4/p_0 = 264$. This value would be somewhat lower if ψ and/or η_c were lower, and somewhat higher if the combustion takes place at constant area. Thus, the examples of Table 7.1 are based on $p_4/p_0 = 264$, and the likely range for this parameter is 100–1000.

Table 7.1   Three typical exhaust component operating condition examples. These case designations will be used in the examples and discussions that follow.

Case	M_0	M_4	M_Y	V_Y/V_0	V_4/V_Y
A	8.0	1.5	5.0	1.30	0.55
B	10.0	2.0	5.5	1.25	0.70
C	18.0	3.0	6.5	1.20	0.80

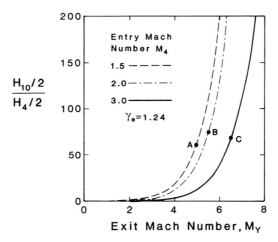

Fig. 7.4 Ratio of exit height to entry height for ideal design point expansion components as a function of exit Mach number and entry Mach number. The three prominent dots signify the example cases of Table 7.1.

The aggregate impression created by the calculated results lies somewhere between interesting and astounding because the geometrical and flow quantities become ever more sensitive to M_0 as M_0 increases.

The ratio of the expansion system exit area or height to the entry area or height is shown in Fig. 7.4. The exit height is not only a large multiple of the entry height, but the ratio increases rapidly with exit Mach number. This remains true even if we pay attention only to the prominent dots that designate the probable operating points of the example cases. The expansion system must therefore provide a sizable H_{10}/H_4 as well as some variable geometry.

The ratio of the axial length of Zone VI to entry height is shown in Fig. 7.5. The overall length of the expansion system exceeds this by the distance from the entry to the end of Zone IV, so this ratio may be taken as a conservative estimate of the overall length. Even so, L_{VI} must be such an enormous multiple of the entry height that it is hard to imagine a vehicle that could provide this much expansion surface length.

The total method of characteristics flow turning angle is shown in Fig. 7.6. The flow turning at the initial simple expansion fan corner is one half this value. These results suggest that some variable geometry will be necessary, especially because the properties of high Mach number flow are quite sensitive to amount of flow turning.

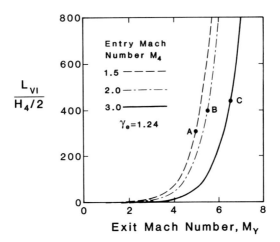

Fig. 7.5 Ratio of Zone VI axial length to entry height for ideal design point expansion components as a function of exit Mach number and entry Mach number. The three prominent dots signify the example cases of Table 7.1.

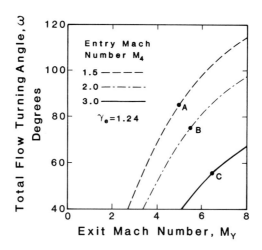

Fig. 7.6 Total method of characteristics flow turning angle for ideal design point expansion components as a function of exit Mach number and entry Mach number. The three prominent dots signify the example cases of Table 7.1.

Finally, the ratio of static pressure of Zone III to entry static pressure is shown in Fig. 7.7. This reveals that the vast majority of the entry static pressure is removed by the initial simple expansion fan, which suggests, in turn, that most of the available axial thrust is realized in the early part of the expansion system.

This possibility can be explored for ideal nozzles by means of stream thrust analysis. Imagine a one-dimensional nozzle being cut off at a location x corresponding to any axial Mach number larger than M_4. The axial Mach number of this approach can be roughly equated to the average Mach number existing at any axial station of the confined nozzle. Recognizing that the total axial force on the internal nozzle surface is the difference between the local stream thrust function Sa_x and the entry stream thrust function Sa_4, and that the maximum possible force on the internal nozzle surfaces is the difference between the exit stream thrust function Sa_Y and the entry stream thrust function Sa_4, then the ratio $(Sa_x - Sa_4)/Sa_Y - Sa_4)$ indicates what fraction of the available stream thrust has been gained. Since the stream thrust function can be written

$$Sa = \dot{m}\sqrt{\frac{RT_t}{\gamma}} \cdot \frac{1}{M} \cdot \frac{1 + \gamma M^2}{\sqrt{1 + \dfrac{\gamma - 1}{2}M^2}} \tag{7-10}$$

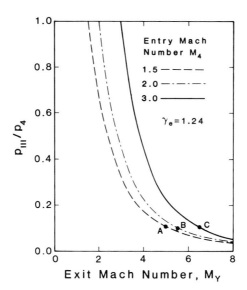

Fig. 7.7   Ratio of Zone III static pressure to entry static pressure for ideal design point expansion components as functions of exit Mach number and entry Mach number. The three prominent dots signify the example cases of Table 7.1.

(this can be most easily done with the help of the solution to Problem 2.12), then,

$$\frac{Sa_x - Sa_4}{Sa_Y - Sa_4} = \frac{\dfrac{1 + \gamma_e M_x^2}{M_x\sqrt{1 + \dfrac{\gamma_e - 1}{2}M_x^2}} - \dfrac{1 + \gamma_e M_4^2}{M_4\sqrt{1 + \dfrac{\gamma_e - 1}{2}M_4^2}}}{\dfrac{1 + \gamma_e M_Y^2}{M_Y\sqrt{1 + \dfrac{\gamma_e - 1}{2}M_Y^2}} - \dfrac{1 + \gamma_e M_4^2}{M_4\sqrt{1 + \dfrac{\gamma_e - 1}{2}M_4^2}}} \qquad (7\text{-}11)$$

The ratio of local height to entry height can be obtained by substituting M_x for M_Y in Eq. (7-7).

The results of the stream thrust analysis are presented in Fig. 7.8 for the example case B. The other cases are not included there only because they fall virtually upon the same line. The prominent dot designates the flow conditions that would exist in Zone III for a minimum length nozzle designed for example case B. This information certainly confirms the suspicion that most of the thrust is generated early in the expansion process. One is therefore driven to the encouraging conclusion that a partial expansion can indeed recover most of the available thrust, and that the enormous heights and lengths of Figs. 7.4 and 7.5 may not be necessary. When this is coupled with the visual observations based on Figs. 7.1 and 7.3 that the height and length of the entrance region are relatively small multiples of

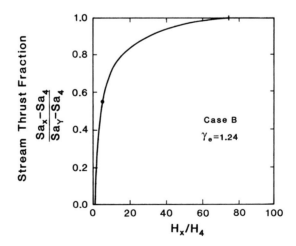

Fig. 7.8 Fraction of the maximum attainable stream thrust obtained as a function of the one-dimensional axial Mach number for example case B. Here the prominent dot signifies the point at which the one-dimensional axial Mach number equals M_{III}.

the entry height (only a factor of 10 or so, as may be confirmed by estimates based on angles available from the design calculations), it appears that expansion components having reasonable performance and practical size are possible.

Anyone applying this design procedure should be made aware that several of the calculated quantities are quite sensitive to the selected value of γ_e. For example, in the case of a minimum length nozzle designed for $M_4 = 2.0$ and $M_Y = 5.0$, reducing γ_e from 1.24 to 1.20 will increase H_{10}/H_4 from 41.4 to 61.8, $L_{VI} / H_4/2$ from 203 to 302, and ω from 68.6 to 75.5, and will reduce p_{III}/p_4 from 0.0127 to 0.0109.

7.4 EXPANSION COMPONENT PERFORMANCE

The function of the expansion component is to produce thrust, or to capitalize on the thrust potential of the flow leaving the combustor. Thus, the exhaust system performance measures must eventually relate to the total amount of thrust or the fractional amount of the potential thrust that is realized.

The one-dimensional analysis approach can still bear fruit despite the unconfined nature of the exhaust system flowfield and the fact that the exhaust system is seldom, if ever, operated precisely at its design point. In order to understand this, it will be helpful to refer to Figs. 7.9 and 7.10, which contain schematic illustrations of typical exhaust system off-design operation flowfields. Off-design conditions are generally said to be either *overexpanded* (i.e., p_4/p_0 is less than the value required by the design H_{10}/H_4) or *underexpanded* (i.e., p_4/p_0 is greater than the value required by the design H_{10}/H_4), in accordance with traditional nozzle terminology. Although there are other types of off-design circumstances for exhaust systems, such as having the correct expansion ratio but the wrong entry Mach number, the present example serves well for illustrative purposes.

The top of Figure 7.9 shows a typical exhaust system operating at its design point and therefore resembles the ideal examples of Sec. 7.3. It also contains the nomenclature and symbology used in the rest of this discussion which, for clarity, have been omitted from the lower portions of Fig. 7.9. This includes, in particular, the slip line (denoting the parallel direction and equal static pressure but unequal velocity boundary between the internal engine exhaust flow and the external vehicle flow), the control volume to be used for determining the total internal force on the exhaust system, and the corner expansion fan Mach waves and their reflections existing at the design point. The two lower drawings indicate only the principal expansion and compression wave *perturbations* to the internal and external flowfields for overexpanded and underexpanded conditions.

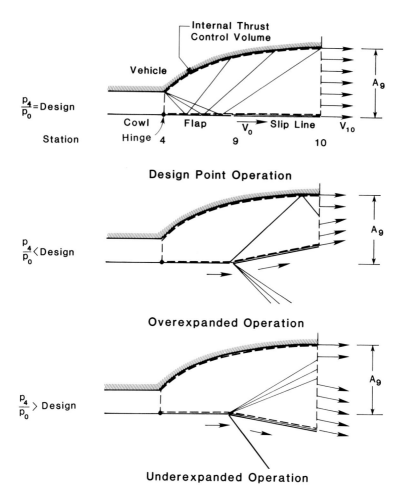

Fig. 7.9   Schematic illustrations of design point, overexpanded, and underexpanded exhaust system operation. The radiating bundles of light lines denote expansion fans and the single heavy lines denote oblique shock waves.

The control volumes are carefully selected to facilitate the calculation of the total force or *gross thrust* due to the pressure and frictional forces acting only on the *internal surfaces* of the expansion system, namely those of the vehicle, cowl, and flap. For mathematical simplicity, it is assumed that the desired thrust direction is aligned with the axial or x direction and the flow leaving the combustor, and is perpendicular to the throughflow areas A_4 and A_{10}. This assumption is not only reasonable, but geometrical corrections are easily made when necessary. Viscous forces acting on the slip line are also neglected. The gross thrust F_{eg} is, then, equal to the integral of the sum of the pressure forces and momentum fluxes acting in the thrust direction on the portion of the control volume other than the surface of the exhaust system (i.e., the slip line and the entry and exit planes). The mathematical expression for this is

$$F_{eg} = \int_{\text{exit plane}} (p+\rho u^2)\, dA_x + \int_{\text{slip line}} p\, dA_x + \int_{\text{entry plane}} (p+\rho u^2)\, dA_x \quad (7\text{-}12)$$

where the term dA_x denotes the axial projection of the differential area, and is counted as positive when the outwardly directed normal is in the positive x direction, and the far right-hand integral is presumed known and may be written as $-\dot{m}_4 S a_4$.

This expression becomes tractable if the exhaust system is sufficiently long that the integral of the static pressure acting on the *external* portion of the control volume may be taken to be $p_0 A_9$, where the area obtained by projecting the end point of the flap axially to the exit plane at any operating condition is temporarily referred to as A_9. Uniform static pressure is, after all, what the flow and the designer are both trying to achieve. The impact of any error in this assumption is greatly diminished for highly supersonic flows by the fact that, as the stream thrust function reveals, the ratio of momentum to pressure contributions is of the order of γM^2. We will use this assumption in our analysis, largely because it is realistic and convenient, but it must be checked in practice and, if found wanting, corrections *must* be made. Finally, since no external mass flow crosses the slip line, the mass flow leaving at the exit plane is equal to that leaving the combustor, and the axial momentum is synonymous with the mass average axial velocity of that flow as it crosses the exit plane. The foregoing allows Eq. (7-12) to be written

$$F_{eg} = \dot{m}_4 u_{10} + p_0 A_9 - \dot{m}_4 S a_4 \quad\quad\quad (7\text{-}13)$$

where u_{10} is the mass average axial component of velocity of the flow. Equation (7-13) makes it clear that maximizing u_{10} is the first priority of the exhaust system designer.

When operating in the overexpanded condition, the adjustment to the static pressure mismatch at the end of the flap is made by two waves of equal flow deflection angles, an oblique shock wave in the exhaust stream and an expansion fan in the external stream. The oblique shock wave can reduce the value of u_{10} in the exit plane in three ways. First, the entropy increases and the total pressure decreases each time the exhaust flow traverses the wave. When the deflection is very large, the oblique shock wave can partly transform into a normal shock (or "barrel") wave, with significantly greater losses. Second, the oblique shock wave can separate the boundary layer on the surface of the expansion ramp, creating additional losses. Third, the exit flow will have some angularity (i.e., not be aligned everywhere with the axial or thrust direction) unless the oblique shock wave happens to intersect the slip line and the exit plane simultaneously.

When operating in the underexpanded condition, the adjustment to the static pressure mismatch at the end of the flap is again made by two waves of equal flow deflection angles, an expansion fan in the exhaust stream and an oblique shock wave in the external stream. The control volume analysis is superfluous in this situation because the pressure and friction force distributions on the entire internal surface are independent of the degree of underexpansion, and so, therefore, is the gross thrust. A control volume analysis would reveal that the potential for additional gross thrust due to the greater V_{10} is lost due to angularity.

Exhaust system designers therefore sometimes employ variable geometry in order to recapture some of the gross thrust lost to angularity. Figure 7.10 shows how this might be done by providing the flap with a hinge at the entry plane, and illustrates only the condition in which the flap is positioned so that the slip line is aligned with the desired thrust direction at the end or tip of the flap.

In the overexpanded case the flap must be closed in order to improve exit plane angularity. For the internal flow, the oblique shock wave generated at the hinge line and the weaker reflection of the original expansion fan waves from the internal flap surface combine to increase the static pressure just upstream of the flap tip. For the external flow, the expansion fan generated at the hinge line reduces the static pressure just upstream of the flap tip. There is usually some position of the flap for which the slip line is pointed in the proper direction. This is obviously a tradeoff situation because any gross thrust gain due to reduction of angularity and elimination of the original flap tip oblique shock wave will be partially offset by the losses of the new hinge line oblique shock wave, and the lowered pressure on the external surface of the flap creates a drag or installation penalty for the vehicle.

In the underexpanded case the flap must be opened in order to improve exit plane angularity. For the internal flow, the expansion fan generated at the hinge line and the stronger reflection of the original expansion fan waves from the internal flap surface combine to decrease the static pressure just upstream of the flap tip. For the external flow, the oblique shock wave generated at the hinge line increases the static pressure just upstream of the flap tip. There is usually some position of the flap for which the slip line is pointed in the proper direction. This is also a tradeoff situation because

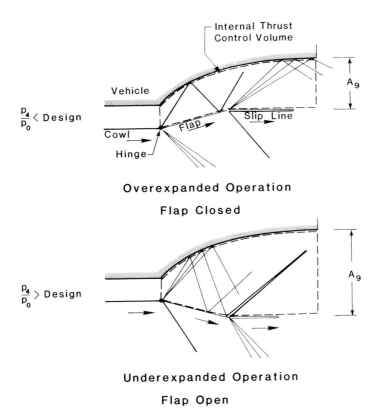

$\dfrac{p_4}{p_0} < $ Design

Internal Thrust
Control Volume

Vehicle

A_9

Cowl

Flap

Slip Line

Hinge

Overexpanded Operation

Flap Closed

$\dfrac{p_4}{p_0} > $ Design

A_9

Underexpanded Operation

Flap Open

Fig. 7.10 Schematic illustrations of overexpanded and underexpanded exhaust systems using a hinged flap to reduce angularity losses. The symbols and nomenclature are the same as Fig. 7.9.

any gross thrust gain due to reduction of angularity will be partially offset by the losses of the new flap tip oblique shock wave, and the increased pressure on the external surface of the flap creates a drag or installation penalty for the vehicle.

Note should be taken that the assumption of uniform static pressure around the external control volume is much more reliable when a flap is available to provide more nearly correct expansion. Also, another possible advantage of the hinged flap would be to provide a controllable minimum area for choking in the ramjet mode of operation, provided that it can be rotated sufficiently to close.

The foregoing discussion leads to the conclusion, embodied by Eq. (7-13), that the exit plane mass average axial velocity of the internal flow is the quantity of greatest importance to the exhaust system designer. The means to describe and obtain estimates for u_{10} are therefore the next topics for study.

7.5 EXPANSION COMPONENT PERFORMANCE MEASURES

The magnitude of the mass-average total exit velocity V_{10} will first be treated by the methods made familiar in Sec. 5.4, and the effects of the angularity and chemistry of the exhaust flow will be considered later. The losses accounted for in these analyses include only the dissipation due to shock waves and boundary-layer friction. It is assumed that the calorically perfect gas constants of the combustion products have been properly chosen to reflect their behavior for the *actual* expansion conditions.

This development will closely parallel that of Sec. 5.4 in both purpose and methodology. The principal reason for examining the definition of each expansion component performance measure is, as before, to provide insight into its physical meaning and behavior. The principal reason for deriving the relationships between various expansion component performance measures is, again, to allow the conversion of values into any system we prefer. Since the analyses employ the same assumptions, share the same caveats, and use the same general line of attack as those for compression components, they will be reduced to their bare essentials in what follows.

Several generalizations, nevertheless, need to be stated at the outset. First, the expansion component one-dimensional performance measures also represent a mixture of engineering intuition and tradition, and are predominantly based upon the experiences of the designers of nozzles for aircraft and rocket engines. You will find that they are different from those used to describe compression components. Second, the main difference between compression component and expansion component performance analysis is that the static temperature ratio T_3/T_0 of the former is replaced by the static pressure ratio p_4/p_0 of the latter as a principal independent variable. By virtue of the earlier discussion, the static pressure ratio p_4/p_0 will be used in the determination of *both* ideal and real flow quantities. Third, in both compression and expansion components, the upstream or entry conditions are known in their entirety. Fourth, the ratio of specific heats γ_e will continue to be taken to be 1.240 in the examples that follow, although the equations will allow the use of any value you wish.

7.5.1 Total Pressure Ratio

The *total pressure ratio* across the expansion component is defined as the total pressure at the exit divided by the total pressure at the entry, and is denoted by the symbol π_e. It follows immediately from

Eqs. (2-50) and (2-54) that

$$\pi_e \doteq \frac{p_0}{p_4} \left\{ \frac{1 + \dfrac{\gamma_e - 1}{2} M_{10}^2}{1 + \dfrac{\gamma_e - 1}{2} M_4^2} \right\}^{\gamma_e/(\gamma_e - 1)} \qquad (7\text{-}14)$$

and

$$\frac{T_4}{T_{10}} = \left\{ \frac{1 + \dfrac{\gamma_e - 1}{2} M_{10}^2}{1 + \dfrac{\gamma_e - 1}{2} M_4^2} \right\} \qquad (7\text{-}15)$$

so that

$$\pi_e = \frac{p_0}{p_4} \left(\frac{T_4}{T_{10}} \right)^{\gamma_e/(\gamma_e - 1)} \qquad (7\text{-}16)$$

Combining and rearranging Eqs. (4-19), (4-27), and (7-16) produce the desired results, namely

$$\pi_e = \left\{ \frac{1}{\eta_e + (1 - \eta_e) \left(\dfrac{p_4}{p_0} \right)^{(\gamma_e - 1)/\gamma_e}} \right\}^{\gamma_e/(\gamma_e - 1)} \quad \leq 1 \qquad (7\text{-}17)$$

or

$$\eta_e = \frac{1 - \left(\dfrac{1}{\pi_e} \cdot \dfrac{p_0}{p_4} \right)^{(\gamma_e - 1)/\gamma_e}}{1 - \left(\dfrac{p_0}{p_4} \right)^{(\gamma_e - 1)/\gamma_e}} \quad \leq 1 \qquad (7\text{-}18)$$

The behavior of these equations is portrayed for typical pressure ratios in Fig. 7.11, where it is evident that even modest reductions of adiabatic expansion efficiency are associated with enormous decreases of total pressure ratio. The violently nonlinear nature of this relationship, as well as the absolute magnitude of the total pressure ratios (approximately 0.1–0.4 versus the 0.980–0.995 experienced in conventional nozzles), also help explain why π_e is only occasionally employed by the hypersonic airbreathing propulsion community.

7.5.2 Velocity Coefficient

The expansion component *velocity coefficient* is defined as the actual average total velocity at the exit of the expansion system divided by the ideal total velocity, and is denoted by the symbol C_{ev}. It follows

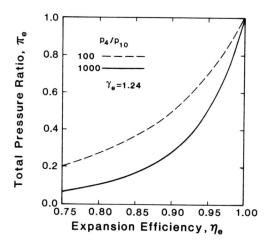

Fig. 7.11 Expansion component total pressure ratio as a function of adiabatic expansion efficiency and static pressure ratio.

immediately that

$$C_{ev} \doteq \frac{V_{10}}{V_Y} \qquad (7\text{-}19)$$

Please note that this is a close relative of the compression system kinetic energy efficiency, particularly so because the ideal exit velocity is based on isentropic expansion to freestream static pressure in both cases.

Combining Eqs. (2-48), (4-19), and (7-19) produces the desired results, namely

$$C_{ev} = \sqrt{\eta_e + (1 - \eta_e)\left(\frac{V_4}{V_Y}\right)^2} \le 1 \qquad (7\text{-}20)$$

or

$$\eta_e = \frac{C_{ev}^2 - \left(\frac{V_4}{V_Y}\right)^2}{1 - \left(\frac{V_4}{V_Y}\right)^2} \le 1 \qquad (7\text{-}21)$$

It is most convenient to plot these relationships as functions of V_4/V_Y, a quantity that has already been observed to increase with freestream Mach number as in Table 7.1. The behavior of these equations is portrayed in Fig. 7.12 for values of V_4/V_Y that correspond to the three example cases of Table 7.1. The expansion component velocity coefficient is both sensitive to the flight conditions and close to 1, which makes the accurate conversion to adiabatic expansion efficiency demanding. The velocity coefficient is, nevertheless, the most

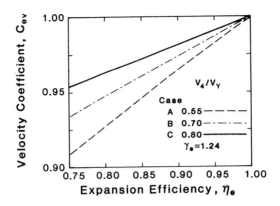

Fig. 7.12 Expansion component velocity coefficient as a function of adiabatic expansion efficiency and ideal velocity ratio.

widely used expansion component performance measure, largely because its effect on engine performance is so readily apparent.

7.5.3 Dimensionless Entropy Increase

The expansion system *dimensionless entropy increase* is defined as the entropy increase divided by the specific heat at constant pressure of the combustion products. Combining Eqs. (2-55) and (7-17) produces the desired results, namely

$$\frac{s_{10} - s_4}{C_{pe}} \doteq \ell n \left\{ \eta_e + (1 - \eta_e) \left(\frac{p_4}{p_0}\right)^{(\gamma_e - 1)/\gamma_e} \right\} \geq 0 \qquad (7\text{-}22)$$

or

$$\eta_e = \frac{1 - \left(\dfrac{p_0}{p_4}\right)^{(\gamma_e - 1)/\gamma_e} e^{(s_{10} - s_4)/C_{pe}}}{1 - \left(\dfrac{p_0}{p_4}\right)^{(\gamma_e - 1)/\gamma_e}} \leq 1 \qquad (7\text{-}23)$$

The behavior of these equations is shown in Fig. 7.13 for typical values of p_4/p_0. The dimensionless entropy increase and adiabatic expansion efficiency are again well-behaved functions of one another, but the former is virtually never used in practice.

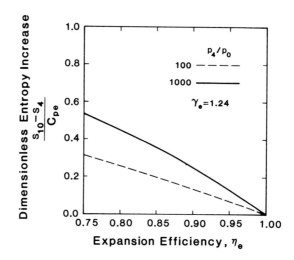

Fig. 7.13 Expansion component dimensionless entropy increase as a function of adiabatic expansion efficiency and static pressure ratio.

7.5.4 Expansion Component Performance Measure Summary

In order to facilitate future expansion component analysis, the missing relationships between the four performance measures examined above were derived, and the complete set is presented in matrix form in Table 7.2.

As anticipated by Eq. (2-55), π_e decays exponentially with $s_{10} - s_4$. This strong dependence makes π_e less practical as an indicator of the performance of flows having significant entropy increases.

7.5.5 Experiential Information

Reliable experimental data for the performance of scramjet exhaust systems are extremely scarce, primarily because of the difficulty of properly simulating either the internal exhaust flow or the external environmental conditions. Most of what is known is drawn from CFD, buttressed at the low Mach number end by test results previously obtained for turbojets, ramjets, and rockets.

The CFD results, largely reported in terms of the expansion system velocity coefficient C_{ev}, are encouraging.[7.3] In terms of the example cases of Table 7.1, they suggest that a fixed geometry exhaust component designed for case B can have a C_{ev} in excess of 0.99 at the design point, and still maintain C_{ev} at approximately 0.95 for cases A and C. Some of the loss at the case C high Mach number end is due to a complex internal shock wave that can be traced back

Table 7.2 Interrelationships between the various expansion component performance measures. For example, in order to express the adiabatic expansion efficiency η_e in terms of the velocity coefficient C_{ev}, read across the η_e row to the C_{ev} column to find that

$$\eta_e = \frac{C_{ev}^2 - \left(\dfrac{V_4}{V_Y}\right)^2}{1 - \left(\dfrac{V_4}{V_Y}\right)^2}$$

Also note that, within the limits of our assumptions, following Example Case 2.5

$$\left(\frac{V_4}{V_Y}\right)^2 = \left\{ \frac{1 - \left(\dfrac{p_4}{p_{t4}}\right)^{(\gamma_e-1)/\gamma_e}}{1 - \left(\dfrac{p_0}{p_{t4}}\right)^{(\gamma_e-1)/\gamma_e}} \right\} \le 1$$

	η_e
$\eta_e =$	
$\pi_e =$	$\left\{ \dfrac{1}{\eta_e + (1 - \eta_e)\left(\dfrac{p_4}{p_0}\right)^{(\gamma_e-1)/\gamma_e}} \right\}^{\gamma_e/(\gamma_e-1)}$
$C_{ev} =$	$\sqrt{\eta_e + (1 - \eta_e)\left(\dfrac{V_4}{V_Y}\right)^2}$
$\dfrac{s_{10} - s_4}{C_{pe}} =$	$\ell n \left\{ \eta_e + (1 - \eta_e)\left(\dfrac{p_4}{p_0}\right)^{(\gamma_e-1)/\gamma_e} \right\}$
$\eta_e =$	$\dfrac{1 - \left(\dfrac{1}{\pi_e} \cdot \dfrac{p_0}{p_4}\right)^{(\gamma_e-1)/\gamma_e}}{1 - \left(\dfrac{p_0}{p_4}\right)^{(\gamma_e-1)/\gamma_e}}$
$\pi_e =$	π_e
$C_{ev} =$	$\sqrt{\dfrac{1 - \left(\dfrac{1}{\pi_e} \cdot \dfrac{p_0}{p_4}\right)^{(\gamma_e-1)/\gamma_e}}{1 - \left(\dfrac{p_0}{p_4}\right)^{(\gamma_e-1)/\gamma_e}} \cdot \left[1 - \left(\dfrac{V_4}{V_Y}\right)^2\right] + \left(\dfrac{V_4}{V_Y}\right)^2}$
$\dfrac{s_{10} - s_4}{C_{pe}} =$	$\dfrac{\gamma_e - 1}{\gamma_e} \ell n \dfrac{1}{\pi_e}$

(continued on next page)

Table 7.2 (continued)

$$\eta_e = \frac{C_{ev}^2 - \left(\dfrac{V_4}{V_Y}\right)^2}{1 - \left(\dfrac{V_4}{V_Y}\right)^2}$$

$$\pi_e = \left\{ \frac{1}{\dfrac{C_{ev}^2 - \left(\frac{V_4}{V_Y}\right)^2}{1 - \left(\frac{V_4}{V_Y}\right)^2} \cdot \left[1 - \left(\dfrac{p_4}{p_0}\right)^{(\gamma_e-1)/\gamma_e}\right] + \left(\dfrac{p_4}{p_0}\right)^{(\gamma_e-1)/\gamma_e}} \right\}^{\gamma_e/(\gamma_e-1)}$$

$$C_{ev} = C_{ev}$$

$$\frac{s_{10} - s_4}{C_{pe}} = \ell n \left\{ \frac{C_{ev}^2 - \left(\frac{V_4}{V_Y}\right)^2}{1 - \left(\frac{V_4}{V_Y}\right)^2} \cdot \left[1 - \left(\dfrac{p_4}{p_0}\right)^{(\gamma_e-1)/\gamma_e}\right] + \left(\dfrac{p_4}{p_0}\right)^{(\gamma_e-1)/\gamma_e} \right\}$$

$$\eta_e = \frac{1 - \left(\dfrac{p_0}{p_4}\right)^{(\gamma_e-1)/\gamma_e} e^{(s_{10}-s_4)/C_{pe}}}{1 - \left(\dfrac{p_0}{p_4}\right)^{(\gamma_e-1)/\gamma_e}}$$

$$\pi_e = e^{-\frac{\gamma_e}{\gamma_e-1} \cdot \frac{s_{10}-s_4}{C_{pe}}}$$

$$C_{ev} = \sqrt{\frac{1 - \left(\dfrac{p_0}{p_4}\right)^{\frac{\gamma_e-1}{\gamma_e}} e^{(s_{10}-s_4)/C_{pe}}}{1 - \left(\dfrac{p_0}{p_4}\right)^{\frac{\gamma_e-1}{\gamma_e}}} \cdot \left[1 - \left(\dfrac{V_4}{V_Y}\right)^2\right] + \left(\dfrac{V_4}{V_Y}\right)^2}$$

$$\frac{s_{10} - s_4}{C_{pe}} = \frac{s_{10} - s_4}{C_{pe}}$$

to an excessively large expansion fan turning angle (see Fig. 7.6). Translating these into adiabatic expansion efficiency η_e via Fig. 7.12 leads to the conclusion that cases A, B, and C would have η_e approximately equal to 0.86, 0.96, and 0.73, respectively. Our previous cycle calculations have shown that, integrated over this operating range, these values are in the right ballpark. Presumably, this performance could only be improved by variable geometry. The downside to this information is that the engine may run out of net thrust at too low a freestream Mach number.

The physics of the boundary-layer friction situation are also promising because of the rapid and vigorous expansion that takes place within the entry corner expansion fan. The importance of friction in these analyses is usually expressed by a group such as $C_f \cdot A_w / A_{10}$ [e.g., Eq. (5-24) or (5-26)]. On the one hand, the expansion fan reduces the magnitude of A_w / A_{10} for the entire expansion process. On the other, the rapid acceleration of the flow at the corner can cause "laminarization" or "relaminarization" of the boundary layer to occur, greatly reducing C_f in the neighborhood of the entrance and for whatever downstream distance is required for the turbulent boundary layer to be restored.[7.4]

7.5.6 Exit Plane Angularity Coefficient

The sketch in Fig. 7.14 shows that the exit flow can suffer from at least two types of angularity, namely the *inclination* of the average streamline to the desired direction α, and the dispersion or *splay* about the average streamline $\pm\theta$. The two-dimensional, uniform velocity, radial exhaust flow model pictured in Fig. 7.14 can be integrated to obtain an estimate for the *expansion angularity coefficient*,

$$C_{ea} = \frac{u_{10}}{V_{10}} = \cos \alpha \cdot \frac{\sin \theta}{\theta} \qquad (7\text{-}24)$$

The results of calculations based on Eq. (7-24) for the expected ranges of α and θ are also found in Fig. 7.14. They reveal that the gross thrust loss due to exit flow angularity is not likely to exceed 1 or 2 percent, an amount that, while relatively small, must still be taken into account. Additional losses would accrue at the sides of the finite two-dimensional expansion component, where the flow is free to expand outward.

7.5.7 Gross Thrust Coefficient

We can now include the effects on gross thrust of flow angularity at the exit plane. The expansion component *gross thrust coefficient* is defined as the actual value of F_{eg} as given by Eq. (7-13) divided

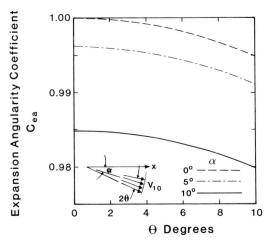

Fig. 7.14 Expansion system angularity coefficient as a function of angle of dispersion and angle of inclination.

by the ideal value, and is denoted by the symbol C_{eg}. It follows immediately that

$$C_{eg} \doteq \frac{\dot{m}_4 u_{10} + p_0 A_9 - \dot{m}_4 S a_4}{\dot{m}_4 V_Y + p_0 A_9 - \dot{m}_4 S a_4} \leq 1 \qquad (7\text{-}25)$$

Since $V_Y \approx S a_Y$ and $p_0 A_9$ makes a minor contribution to both the numerator and denominator, Eq. (7-25) may be approximated with sufficient accuracy for the purposes of exploration by

$$C_{eg} \cong \frac{C_{ea} C_{ev} S a_Y - S a_4}{S a_Y - S a_4} = \frac{C_{ea} C_{ev} - \dfrac{S a_4}{S a_Y}}{1 - \dfrac{S a_4}{S a_Y}} \leq 1 \qquad (7\text{-}26)$$

where the definitions of C_{ea} and C_{ev} have been used.

The behavior of the expansion system gross thrust coefficient according to Eq. (7-26) is shown in Fig. 7.15 for the three example cases of Table 7.1. The two most important features of this information are that the adiabatic expansion efficiency must exceed *at least* 0.85 in order to capitalize on the available gross thrust, and that C_{eg} is *quite* sensitive to flow angularity.

7.5.8 Net Thrust Coefficient

The definition of the expansion component gross thrust coefficient implies that the ideal exit momentum flux is a resource that must be

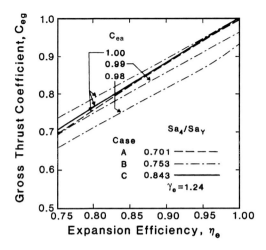

Fig. 7.15 Expansion component gross thrust coefficient as a function of adiabatic expansion efficiency for the three example cases of Table 7.1. The upper three curves correspond to $C_{ea} = 1.00$ and the lower two curves correspond to case B with $C_{ea} = 0.99$ and 0.98.

conserved in some absolute sense. This is certainly a rational viewpoint for those charged only with the performance of the expansion system. A different interpretation arises when the entire airbreathing engine is considered, for which the difference between the expansion system exit and compression system entry momentum fluxes is the primary determinant of the uninstalled thrust (i.e., Eq. (4-37) with $p_{10} \cong p_0$).

Assuming that the freestream flow is aligned with the desired thrust direction, and accounting for the momentum loss due to exit flow angularity, the system oriented performance viewpoint leads to the definition of the *net thrust coefficient* C_{en}, namely the difference between the actual engine exit and entry momentum fluxes divided by the difference between the ideal engine exit and entry momentum fluxes, or

$$C_{en} \doteq \frac{\dot{m}_0(u_{10} - V_0)}{\dot{m}_0(V_Y - V_0)} = \frac{C_{ea}C_{ev}\dfrac{V_Y}{V_0} - 1}{\dfrac{V_Y}{V_0} - 1} \leq 1 \qquad (7\text{-}27)$$

where the engine entry and exit mass flow rates have been taken to be equal in this exercise, a convenient but easily remedied assumption.

The main value of the net thrust coefficient is to emphasize the extreme importance of preserving the thrust potential available at the entry of the expansion system via high expansion efficiency and small flow angularity. This can be seen in Fig. 7.16, which shows how C_{en} varies with η_e for the three example cases of Table 7.1. A remarkable feature of Fig. 7.16 is the strong effect that C_{ea} has on C_{en}. Even small amounts of flow angularity can significantly reduce the net thrust of the engine.

The expansion system net thrust coefficient is occasionally used in the open literature, and can always be swapped with the gross thrust coefficient if V_Y/V_0 and Sa_4/Sa_Y are known. An enduring message, however, is that the adiabatic expansion efficiency must be *at least* in the range of 0.85 – 0.90 in order to capitalize on the thrust potential of the expansion system.

7.5.9 Accounting for Nonequilibrium Chemistry

Thus far in Chap. 7, *ideal performance* has meant the most that can be isentropically obtained from the *actual* chemical state of the flow, which, depending upon reaction rates and residence times, can be anywhere between frozen and equilibrium. It is always interesting and sometimes useful to base performance instead on the *equilibrium* chemical state because this maximizes the enthalpy available for exit kinetic energy. Furthermore, when exhaust system performance is referenced to isentropic, equilibrium chemistry, it provides a more universal and clearly defined measure, and is more consistent with the equilibrium thermodynamic cycle discussions of Chap. 4.

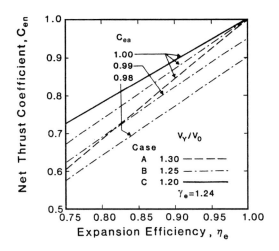

Fig. 7.16 Expansion component net thrust coefficient as a function of adiabatic expansion efficiency for the three cases of Table 7.1. The upper three curves correspond to C_{ea} = 100 and the lower two curves correspond to case B with C_{ea} = 0.99 and 0.98.

The exhaust system performance measures are easily converted from *actual* to *equilibrium* versions via the steps described below. Please note that the equilibrium parameters are all designated in this development by an asterisk above, so that, for example, the isentropic, equilibrium chemistry exit velocity is V_Y^*.

Defining the *equilibrium adiabatic expansion efficiency* as

$$\eta_e^* \doteq \frac{\dfrac{V_{10}^2}{2} - \dfrac{V_4^2}{2}}{h_4 - h_Y^*} \leq 1 \tag{7-28}$$

and observing that the *actual adiabatic expansion efficiency* can be written

$$\eta_e \doteq \frac{\dfrac{V_{10}^2}{2} - \dfrac{V_4^2}{2}}{h_4 - h_Y} \leq 1 \tag{7-29}$$

it follows immediately that

$$\eta_e^* = \left(\frac{h_4 - h_Y}{h_4 - h_Y^*}\right) \eta_e \leq \eta_e \tag{7-30}$$

the implications of Eq. (7-30) can be examined using the information generated by HAP(Equilibrium) that is gathered in Table 7.3 for the two extreme conditions that produce the largest possible values of $(h_4 - h_Y^*)/(h_4 - h_Y)$, namely isentropic equilibrium and frozen composition expansion after combustion. Table 7.3 discloses that this ratio averages approximately 1.17 for stoichiometric hydrogen and methane combustion, so that η_e^* might be expected to be as little as $0.85\eta_e$. Fortunately, the truth lies somewhere between equilibrium and frozen flow, and a more likely estimate for η_e^* is about $0.93\eta_e$. Nevertheless, these results show that a major shortfall due to nonequilibrium expansion system processes is possible for scramjet engines.

The overall or "effective" value of γ_e for both the isentropic equilibrium and frozen expansion processes was calculated from the integrated Gibbs equation in the form

$$\gamma_e = \frac{1}{1 - \dfrac{\ell n \dfrac{T_4}{T_Y}}{\ell n \dfrac{p_4}{p_0}}} \tag{7-31}$$

Table 7.3 contains other information that is helpful to the understanding and analysis of expansion system flows. On one hand, the

Table 7.3 Summary of computational results for the stoichiometric, constant-pressure combustion of hydrogen and methane fuels with air, and their subsequent isentropic, *equilibrium* and *frozen* chemical composition expansion to freestream static pressure. The computations include a base case and systematic variations of the principal variables bracketing their expected operating ranges for each fuel. Both fuels were assumed to be in the gaseous state at a temperature of 1000 °R (556 K).

Hydrogen Fuel, H_2

	T_3 °R	$p_3 = p_4$ atm	$\dfrac{p_4}{p_0}$	$\dfrac{h_4 - h_Y^*}{h_4 - h_Y}$	γ_e^*	γ_e Frozen
Base:	2800	1.00	250	1.167	1.192	1.285
	2400	1.00	250	1.147	1.203	1.284
	3200	1.00	250	1.186	1.180	1.286
	2800	0.10	250	1.230	1.167	1.296
	2800	10.0	250	1.112	1.213	1.276
	2800	1.00	100	1.151	1.175	1.278
	2800	1.00	1000	1.185	1.213	1.297

Methane Fuel, CH_4

	T_3 °R	$p_3 = p_4$ atm	$\dfrac{p_4}{p_0}$	$\dfrac{h_4 - h_Y^*}{h_4 - h_Y}$	γ_e^*	γ_e Frozen
Base:	2800	1.00	250	1.179	1.187	1.287
	2400	1.00	250	1.158	1.199	1.286
	3200	1.00	250	1.199	1.175	1.289
	2800	0.10	250	1.237	1.164	1.296
	2800	10.0	250	1.128	1.207	1.279
	2800	1.00	100	1.161	1.171	1.280
	2800	1.00	1000	1.200	1.208	1.298

trends of the observed $(h_4 - h_Y^*)/(h_4 - h_Y)$ agree with our intuitive expectations. For example, this parameter decreases with increasing combustion pressure and decreasing expansion ratio because higher static pressures always drive the chemistry toward equilibrium, and decreases with decreasing burner entry temperature because that reduces the amount of dissociation present at the start of the expansion process. On the other hand, overall or "effective" values of γ_e calculated for the equilibrium and frozen expansion processes reflect the additional enthalpy release of the former. Finally, the average of the equilibrium and frozen γ_e for the two base cases of Table 7.3, as well as the average of *all* the γ_e shown there, is very nearly 1.240, which explains *at last* why this value was used in the example calculations of this chapter.

The equilibrium expansion system velocity coefficient is defined as

$$C_{ev}^* \doteq \frac{V_{10}}{V_Y^*} = C_{ev} \frac{V_Y}{V_Y^*} \tag{7-32}$$

Applying the first law as expressed by Eq. (2-46) separately to the equilibrium chemistry and actual flows and rearranging lead to the expression

$$\frac{V_Y}{V_Y^*} = \sqrt{\frac{\left(\dfrac{h_4 - h_Y}{h_4 - h_Y^*}\right)}{1 - \left\{1 - \left(\dfrac{h_4 - h_Y}{h_4 - h_Y^*}\right)\right\}\left(\dfrac{V_4}{V_Y}\right)^2}} \leq 1 \tag{7-33}$$

Equations (7-32) and (7-33) are used in combination in order to evaluate C_{ev}^*. Since the first approximation for V_Y/V_Y^* from Eq. (7-33) is

$$\sqrt{\frac{h_4 - h_Y}{h_4 - h_Y^*}}$$

it follows that C_{ev}^* is substantially less than C_{ev}.

The equilibrium expansion system net thrust coefficient is defined as

$$C_{en}^* = \frac{\dot{m}_0 (u_{10} - V_0)}{\dot{m}_0 (V_Y^* - V_0)} = \frac{C_{ea} C_{ev} \dfrac{V_Y}{V_0} - 1}{\dfrac{V_Y^*}{V_Y} \cdot \dfrac{V_Y}{V_0} - 1} \leq C_{en} \tag{7-34}$$

which must also be substantially less than C_{en} .

7.6 EXPANSION COMPONENT CFD EXAMPLE

Thus far, uniform, supersonic, burner-exit or expansion-component entry flow has been assumed. This assumption is extremely useful for conducting conceptual design of expansion components and for understanding their performance. But in reality, the burner-exit flow is nonuniform. As the freestream is processed by the compression component and the burner, it is highly distorted. For example, the processing of the flow in crossflow planes of the burner by localized, air and fuel interactions (mixing and burning) and by protuberances (such as fuel injectors), if any, further distorts the nonuniform flow exiting the compression component. Before the core region of the burner flow becomes uniform, this flow enters the expansion component. The performance of this component based on this nonuniform flow may significantly differ from that computed assuming uniform flow.

The real, rather than the idealized, performance may be determined, if the flow quantities at the burner-exit plane are known, either by measurement or computation. Another alternative is to obtain a trend or a qualitative nature of the effect of the nonuniformity in the burner-exit flow by conducting a sensitivity analysis. This approach is suitable in conceptual design activities, and a procedure for conducting it is described below.

The processing of the freestream by the compression component of a hypersonic propulsion system is analyzed using computational fluid dynamics (CFD) tools to provide two-dimensional or axisymmetric or three-dimensional, viscous or inviscid characterization of the distorted flow entering the burner. To minimize the complexity of analyzing the flow in the combustor, this entry flow is "averaged" to provide uniform flow conditions. Next, one-dimensional analysis using combustion efficiency and mixing correlations provides the one-dimensional flow characteristics at the burner exit. This uniform flow is systematically varied, and this variation may consider the nature of nonuniformity at the exit of the compression component. This variation has to be done such that the resultant entry profiles satisfy the flow governing equations at the entrance of the expansion component, and the averaged values of these profiles and the ideal gross thrust are the same as those obtained by the one-dimensional flow profiles. For instance, static pressure, temperature, and streamwise velocity are adjusted such that mass, momentum, and energy of the uniform flow are kept the same. Furthermore, species are equilibrated to these new flow profiles. Utilizing the nonuniform entry flow, the governing equations are solved with CFD tools to determine the flow in the expansion component. From this flow, the integrated performance quantities, such as thrust and thrust-vector angle, are obtained. These quantities are then compared with those obtained for uniform entry flow to determine the performance sensitivity to the entry flow nonuniformity.

Following a procedure similar to that outlined above, Goel et al.[7.5] provide an example. The expansion component studied in this example is shown in Fig. 7.17. The one-dimensional burner-exit conditions are the following: $p_4 = 3.39 \times 10^3$ lbf/ft^2 (1.6 × 10^5 N/m^2), $\rho_4 = 1.11 \times 10^{-2}$ lbm/ft^3 (1.77 × 10^{-1} kg/m^3), and $M_4 = 2.45$. The ambient freestream flow conditions are the following: $p_0 = 17.3$ lbf/ft^2 (8.30 × 10^2 N/m^2), $\rho_0 = 7.87 \times 10^{-4}$ lbm/ft^3 (1.26 × 10^{-2} kg/m^3), and $M_0 = 10.0$.

The nonuniform profiles are developed from these burner-exit conditions such that the mass flow rate $\dot{m}_4$, the stream (axial) thrust, the stagnation pressure p_{t4}, and the stagnation density ρ_{t4} are kept constant. Different profiles for the ratio of static pressure to stagnation pressure are assumed (Fig. 7.18). Corresponding Mach number profiles are obtained by considering the flow at each point of the pressure profile to be one-dimensional and isentropic. Subsequently, these Mach profiles are used to determine profiles for the ratio of density to stagnation density, again using a locally one-dimensional, isentropic flow assumption. The streamwise velocity profiles are computed from these profiles for pressure, Mach number, and density. The ratio of the specific heats γ is taken to be 1.25. The nonuniformity of the pressure profiles is characterized by the pressure distortion parameter D_p which is defined as the nondimensionalized center of pressure measured from a point located at mid-height of the nozzle at its entrance. Please note that the center of pressure is simply the length of the moment arm owing to the force exerted by the pressure distribution. The uniform pressure profile corresponds to $D_p = 0.0$.

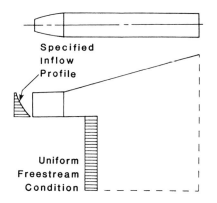
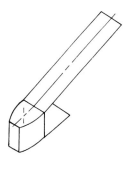

Fig. 7.17 An expansion component with a constant-area section, an entry flap, and an expansion surface, and with a finite lateral dimension.

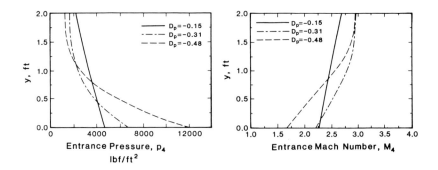

Fig. 7.18   A set of distorted pressure and Mach number profiles at the entry of the expansion component.

Figure 7.18 shows three profiles of pressure and Mach numbers, corresponding to $D_p = -0.15$, -0.31, and -0.48, at the entrance of the expansion component. Mirror images of these profiles have positive distortion parameters. The same profiles are used at all lateral positions in the entry plane.

All profiles have the same mass flow rate ($\dot{m}_4 = 1.90 \times 10^2$ lbm/s or 86.2 kg/s) and the same stream thrust (5.75×10^4 lbf or 2.56×10^5 N).[7.6] The area-averaged, entry static pressure at $D_p = \pm 0.15$ is the same as that for $D_p = 0.0$, whereas those at $D_p = \pm 0.31$ and ± 0.48 are, respectively, 11 percent lower and 19.5 percent higher than that at $D_p = 0.0$. Likewise, the entry energy levels at $D_p = 0.0$ and ± 0.15 are identical. But those at $D_p = \pm 0.31$ and ± 0.48 are, respectively, 5 percent lower and 9 percent higher than the level at $D_p = 0.0$. Please note that the creation of equivalent, nonuniform profiles from uniform profiles or of equivalent, uniform profiles from nonuniform profiles can be a nontrivial task.

The three-dimensional Euler equations are considered for the conduct of the flow in the expansion component and of the interaction of this flow with the freestream. Effects of viscosity and chemical kinetics are neglected, and the air is assumed to be a calorically perfect gas.

To determine trends or relative changes, modeled numerics and physics and CFD input parameters, except the trend causing parameters (entry profiles), are kept the same, when no appreciable change in physics is expected. The determination of the absolute values of performance parameters or that of relative changes was not the objective of this study. Since accurate values were not sought, no effort was made to establish the computational accuracy, for example, by conducting a grid-size sensitivity analysis. As long as the bias and precision of computations are kept the same, the assumptions regarding the physics (including chemistry) are acceptable, and the computed global features are consistent with this physics, trends determined without establishing the level of computational accuracy are meaningful. The CFD tool used for this study has been well developed and tested, and this tool was properly applied to the problem at hand in order to achieve the objective of the study.

Plate 5 (at end of book) shows static pressure contours. When pressure is higher along the lower half of the entry plane than along the upper half, for example, $D_p = -0.31$, the entry flow is diverted toward the expansion surface to force the high-pressure portion of this flow to interact with this surface and to impart energy to it. When the entry flow has pressures higher next to the expansion surface than those next to the surface with the flap, for example, $D_p = +0.31$, the entry flow bends away from the expansion surface. On the other hand, Mach numbers are higher along the expansion surface when $D_p = +0.31$ than when $D_p = -0.31$. Obviously, high pressures along this surface are essential for utilizing it efficiently. This is verified by a higher gross thrust coefficient resulting from higher axial thrust at $D_p = -0.31$ than those at $D_p = +0.31$ (Fig. 7.19). This coefficient is directly related to this thrust. As the center of pressure of the entry static pressure moves toward the cowl surface, the axial thrust generally increases. There is a limit to how close this center can be to this surface before the axial thrust is adversely affected.

Both pitching moment and thrust-vector angle are of great significance to the overall design and performance of flight vehicles utilizing hypersonic airbreathing propulsion systems. In the present study, the pitching moment generally increases as D_p is reduced (Fig. 7.19). This is consistent with the fact that high pressures on the expansion surface increase the pitching moment. The variation of thrust-vector angle shows the opposite trend. While the thrust force increases, the lift force decreases, as pressures on the cowl side increase. Consequently, the thrust-vector angle decreases (Fig. 7.19).

As the distortion parameter is changed from $+0.15$ (or -0.15) to $+0.48$ (or -0.48), trends in the normalized pitch moment and the thrust-vector angle are not consistent. This discrepancy is probably a result of the fact that within this range of the distortion parameter the area-averaged static pressure values and, consequently, the energy levels at the expansion component entry plane are not the same. As stated previously, these values are lower at $D_p = \pm 0.31$ and higher at $D_p = \pm 0.48$ than those at $D_p = \pm 0.15$. Nevertheless, the trends in these parameters indicated by each pair of D_p's having the same magnitude but opposite signs are consistent.

The message of this study is that the performance parameters are significantly sensitive to the nonuniformity in the entry static pres-

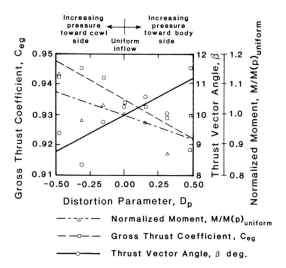

Fig. 7.19   Gross thrust coefficient, pitching moment, and thrust-vector angle as functions of the distortion parameter, adapted from Ref. 7.5. (In this reference, "gross thrust coefficient" is called *nozzle efficiency*.) These integrated performance parameters are determined from the computed three-dimensional flowfields.

sure profile. Other types of nonuniformities at the entry plane of an expansion component may also significantly change the performance of this component. For example, a change in the thickness of the boundary layer at the burner exit can affect the performance of the expansion component. These parameters, in turn, affect the specification of a fluid dynamics system. For instance, a 1-percent change in the gross thrust coefficient makes an order-of-magnitude percentage-change in the takeoff gross weight of a flight vehicle. Therefore, the accurate definition of entry flow is essential for designing this component during preliminary and final design of hypersonic propulsion systems. Please note that $C_{eg} \cong 0.93$ and $\beta \cong 10$ deg. are consistent with prior discussions.

As demonstrated in previous chapters, this parametric study provides further evidence that CFD tools are essential for designing and analyzing hypersonic propulsion systems. An attempt to conduct this study at a ground-based test facility instead of through the utilization of CFD tools at a computational facility would have been excessively expensive. Often, a controlled parametric investigation may not be feasible in a test facility. Under such circumstances, CFD is the only tool that can then be used. Moreover, this tool computes complete flowfields that can provide a better understanding about fluid dynamics phenomena and about cause-and-effect relationships than test facilities can, because those facilities cannot supply complete flowfields.

7.7 CONCLUSION

Our journey through the individual systems or components of the airbreathing engine is now complete. With the information in hand it will be possible to make estimates of overall engine performance that are considerably more realistic and accurate than heretofore, and the results will be quite easily interpreted and understood. Moreover, the framework has been put in place that will allow us to deal in a straightforward way with other important aspects of ramjet and scramjet technology, such as alternate cycles, methods for enhancing thrust, and rigorously accounting for installation effects.

In short, we will now reassemble the pieces that we have been examining, and the whole will exceed the sum of the parts.

REFERENCES

[7.1] Shapiro, A. H., *The Dynamics and Thermodynamics of Compressible Fluid Flow,* Ronald Press, New York, 1953.

[7.2] Emanuel, G., *Gasdynamics: Theory and Applications,* AIAA Education Series, New York, 1986.

[7.3] Harloff, G. J., Lai, H. T., and Nelson, E. S., "Two-Dimensional Viscous Flow Computations of Hypersonic Scramjet Nozzle Flowfields at Design and Off Design Conditions," NASA CR-182150, June 1988.

[7.4] Kays, W. M. and Crawford, M. E., *Convective Heat and Mass Transfer,* McGraw-Hill, New York, 1980.

[7.5] Goel, P., Barson, S. L., and Halloran, S. D., "The Effect of Combustor Flow Nonuniformity on the Performance of Hypersonic Nozzles," *Hypersonic Combined Cycle Propulsion,* AGARD CP-479, 1990.

[7.6] Goel, P., private communication, 1992.

PROBLEMS

7.1 In order to develop your intuition for the unusual properties of hypersonic exhaust systems, use the methodology of Sec. 7.3 to design and draw ideal, minimum-length, two-dimensional expansion systems for these two cases:

(a) $M_4 = 3.0$, $M_Y = 6.5$, and $\gamma_e = 1.24$.
(b) $M_4 = 3.0$, $M_Y = 6.5$, and $\gamma_e = 1.20$.

Sketch the wave crossing Zone IV as best you can, while complying with the constraints imposed by the flow in adjacent regions.

Do your findings corroborate the information given in Figs. 7.4 through 7.7? What is the static pressure ratio for each of these designs? What is the approximate relative height and length of the entrance region for each of these designs? How long a flap would be needed?

7.2 Calculate and graph the relationship between M_4 and M_Y for constant static pressure ratios p_4/p_0 of 100, 250, and 1000 and $\gamma_e = 1.24$. Locate cases A, B, and C on this graph. How sensitive is M_Y to p_4/p_0 for a given M_4?

7.3 How would the analysis of Sec. 7.3 change if p_4/p_0 were specified instead of M_Y?

7.4 Carry out the stream thrust analysis for cases A and C of Sec. 7.3 and compare the results with those of Fig. 7.8. Do you agree with the assertion that they are virtually identical to those of case B?

7.5 A designer wishes to use flap deflection to make the slip line parallel to the freestream flow at the same time that the static pressure at the end of the flap matches the freestream static pressure.

Demonstrate that the designer's efforts are doomed to futility if nothing new happens between the hinge line and the end of the flap.

Do this by using oblique shock wave and simple expansion fan calculations to show that the cumulative effect of two consecutive turns of equal magnitude but opposite sign of a supersonic flow (regardless of which sign comes first) on the static pressure is virtually nil.

What can happen between the hinge line and the end of the flap to bring about the desired result? What will the static pressure be on the slip line relative to the freestream static pressure when the desired result is achieved?

7.6 Derive the following expansion component performance measure interrelationships of Table 7.2, and plot the results in a format similar to Fig. 7.11. You will find that this is most easily accomplished by starting with the interrelationships already developed in Sec. 7.5.

(a) π_e as a function of C_{eg}.
(b) π_e as a function of $(s_{10} - s_4)/C_{pe}$.

Which of these, if any, appear to hold some promise as a simple means for portraying the physical behavior of expansion component performance?

7.7 Reproduce the "polytropic efficiency" derivation of Example 5.4 for the case of expansion component performance. In particular, using the same reasoning and assumptions, prove that

$$\eta_e = \frac{1 - \left(\dfrac{p_0}{p_4}\right)^{\left(\frac{\gamma_e-1}{\gamma_e}\cdot\varepsilon_e\right)}}{1 - \left(\dfrac{p_0}{p_4}\right)^{(\gamma_e-1)/\gamma_e}} \leq 1$$

Plot η_e as a function of p_4/p_0 for several constant values of ε_e. Why does η_e increase with p_4/p_0 for a constant value of ε_e? Why is this the *same* as the case of turbines for turbojet engines?

7.8 Employ Eq. (7-21) to calculate and graph how C_{ev} varies with V_4/V_Y for several constant values of η_e. Why do these curves look this way?

Locate the three cases of Table 7.1 on this figure and reconcile these points with the experiential information discussion of Sec. 7.5.5.

7.9 One rule of thumb for the "laminarization" of accelerating flows is that it happens whenever [7.4]

$$\frac{\mu}{\rho V^2} \cdot \frac{\Delta V}{\Delta L} > 10^{-5}$$

Assuming that $\Delta V/V$ is approximately 0.20, and finding the viscosity from Eq. (2-4), estimate the maximum allowable acceleration length ΔL that will lead to laminar boundary-layer flow downstream of the expansion system entry corner expansion fan. Is this length likely to be achieved in a practical vehicle?

7.10 Reproduce the boundary-layer friction development of Sec. 5.5.3 for the case of expansion component performance. In particular, if $(C_f/2)/(A_w/A_{10})$ is of the order of 0.005, estimate the expansion efficiency η_e for $\gamma_e = 1.24$.

7.11 Can the expansion system efficiency be determined directly from global measurements, and, if so, under what conditions?

Apply the approach and assumptions of Sec. 5.5.5 in order to establish the system of equations that relate exhaust system entry conditions, exit conditions, net axial force or thrust, net heat transfer, and expansion efficiency together. Assume that the exit static pressure is equal to the freestream static pressure and that the exit velocity is uniform and parallel to the entry flow.

Calculate the force that would be exerted on the exhaust system by the isentropic flow of Fig. 7.1 in the absence of heat transfer. What would the expansion efficiency be if the measured force is 10 percent less than the isentropic value? How would η_e and V_{10} be affected if there were no friction, but net heat transfer of 2 percent the exhaust system entry total enthalpy flux (i.e., $\varepsilon = -0.02$)?

7.12 Show that the expansion system transverse angularity or lift coefficient (i.e., normal to the intended thrust direction) for a two-dimensional, uniform velocity, radial exhaust flow having an inclination angle of α and a divergence angle of $\pm\theta$ is given by the expression:

$$\frac{v_{10}}{V_{10}} = \sin \alpha \cdot \frac{\sin \theta}{\theta}$$

7.13 Show that the expansion system angularity loss coefficient for a *conical*, uniform velocity, radial exhaust flow having a divergence angle of $\pm\theta$ but no inclination angle is given by the expression:

$$C_{ea} = \frac{u_{10}}{V_{10}} = \frac{1 + \cos \theta}{2}$$

Compare this with Eq. (7-24) in order to determine how much greater the divergence angle penalty is for conical flow than for two-dimensional flow.

7.14 Combine Eqs. (7-26) and (7-27) to show that

$$C_{en} = \frac{C_{eg} + \dfrac{Sa_4}{Sa_Y}(1 - C_{eg}) - \dfrac{V_0}{V_Y}}{1 - \dfrac{V_0}{V_Y}}$$

Plot C_{en} as a function of C_{eg} for the three cases of Table 7.1. What conclusions do you draw from this information?

7.15 Modify Eq. (7-27) to include fuel mass flow, all else being equal. Redraw Fig. 7.16 for the fuel/air ratios of several typical fuels. How large an impact does the fuel mass flow make? Why?

7.16 Starting from Eq. (7-28), show that

$$C_{ev} = \sqrt{\eta_e^* \left\{ \left(\frac{V_Y^*}{V_Y}\right)^2 - \left(\frac{V_4}{V_Y}\right)^2 \right\} + \left(\frac{V_4}{V_Y}\right)^2}$$

Combine this result with Eq. (7-34) to plot C_{en}^* as a function of η_e^* and several typical combinations of C_{ea}, V_Y^*/V_Y, V_4/V_Y, and V_0/V_Y. To which of these parameters is C_{en}^* most sensitive? Why?

7.17 Repeat the computations of Table 7.3 for octane fuel, C_8H_{18}. Does the presence of the additional carbon significantly change the results from those previously obtained for hydrogen and methane fuels? Why, or why not? Are these trends consistent with the differences between the behavior of hydrogen and methane fuels?

7.18 Repeat the computations of Table 7.3 for fuel/air ratios that are 0.80 and 1.20 times the respective stoichiometric values for hydrogen and methane. Does the stoichiometry significantly change the previous results? Why, or why not?

8
AIRBREATHING
PROPULSION SYSTEMS

8.1 INTRODUCTION

The purpose of the material presented in this chapter is to bring together a number of important and closely interrelated, yet intellectually separable, subjects that deal with the hypersonic airbreathing propulsion system as a *whole*. These topics will include the total performance of the engine from takeoff to orbit, with special emphasis on its integration into the vehicle.

You will see that several distinct types of operation are considered here. In addition to the usual attention to scramjet behavior, we will also consider propulsion devices intended to provide the necessary thrust in the so-called *low speed regime* (from takeoff to scramjet takeover) and at the very high flight speeds and altitudes where scramjet performance flags.

The discussions that follow will draw heavily upon the concepts and their consequences as developed in earlier chapters, but will take for granted your working knowledge of and facility with the methods and their applications.

8.2 REAL SCRAMJET PERFORMANCE

We are now in the position to more realistically estimate the uninstalled performance of scramjet engines over their intended range of operation. This will be done by substituting our best estimates of the behavior of the individual components into the *stream thrust analysis* of Chap. 4 as embodied by HAP(Performance) and reviewing the results.

8.2.1 Stream Thrust Analysis Modifications

More precisely, we will attempt to home in on the potential of the scramjet by computing performance based upon three sets of component estimates that are optimistic or *high*, pessimistic or *low*, and roughly midway between or *average*. The analysis proceeds exactly as detailed in Sec. 4.4.2, except that the losses in the components are

429

analytically modeled as follows:

Compression System (See Chap. 5)

$$\eta_c = \eta_1 \left(\frac{1 - \dfrac{1}{\psi_1}}{1 - \dfrac{1}{\psi}} \right) \tag{5-20}$$

where

$$\psi_1 = 1 + \frac{\gamma_c - 1}{2} M_0^2 \left\{ 1 - \left[1 + \frac{C_f}{2} \cdot \frac{A_w}{A_3} \right]^2 \left[1 + \frac{2(1 - \psi)}{(\gamma_c - 1)M_0^2} \right] \right\} \tag{5-26}$$

Combustion System (See Chap. 6)

Unaltered, but the example is only done for *constant pressure* combustion. You may, of course, use some of the methods of Chap. 6 to estimate η_b separately.

Expansion System (See Chap. 7)

$$\eta_e = \frac{C_{ev}^2 - \left(\dfrac{V_4}{V_Y} \right)^2}{1 - \left(\dfrac{V_4}{V_Y} \right)^2} \tag{7-21}$$

where

$$T_Y = T_4 \left(\frac{p_0}{p_4} \right)^{R/C_{pe}} = T_4 \left(\frac{p_0}{p_3} \right)^{R/C_{pe}} \tag{8-1}$$

and

$$\left(\frac{V_Y}{V_4} \right)^2 = 1 + \frac{2C_{pe}T_4}{V_4^2} \left(1 - \frac{T_Y}{T_4} \right) \tag{8-2}$$

Note that nonequilibrium chemistry is accounted for via different estimates of C_{pe}. Also, $u_{10} = C_{ea}V_{10}$ replaces V_{10} in every computation *after* Eq. (4-53). For example,

$$Sa_{10} = C_{ea}V_{10} \left\{ 1 + \frac{RT_{10}}{(C_{ea}V_{10})^2} \right\} \tag{4-54}$$

$$\frac{A_{10}}{A_0} = (1 + f) \cdot \frac{p_0}{p_{10}} \cdot \frac{T_{10}}{T_0} \cdot \frac{V_0}{C_{ea}V_{10}} \tag{4-55}$$

are to be used in

$$\frac{F}{\dot{m}_0} = (1+f)Sa_{10} - Sa_0 - \frac{RT_0}{V_0}\left(\frac{A_{10}}{A_0} - 1\right) \qquad (4\text{-}38)$$

8.2.2 Stream Thrust Analysis Results

The stream thrust analysis was carried out for a standard atmosphere using HAP(Performance) and HAP(Trajectory) along a constant q_0 trajectory for $5 < M_0 < 25$. The required input information is compiled below.

$$\psi = 7.0$$

$$q_0 = 1000 \text{ lbf/ft}^2 \quad (47.88 \text{ kN/m}^2)$$

$$f = 0.0291$$

$$fh_{\text{PR}} = 1510 \text{ BTU/lbm} \quad (3510 \text{ kJ/kg})$$

$$h_f = 0.0$$

$$T^\circ = 400 \text{ }^\circ\text{R} \quad (222 \text{ K})$$

$$R = 1730 \text{ (ft/s)}^2/^\circ\text{R} \quad [289.3 \text{ (m/s)}^2/\text{ K}]$$

$$C_{pc} = 0.260 \text{ BTU/lbm} \cdot {}^\circ\text{R} \quad (1.09 \text{ kJ/kg} \cdot \text{K})$$

$$\gamma_c = 1.362$$

$$C_{pb} = 0.360 \text{ BTU/lbm} \cdot {}^\circ\text{R} \quad (1.51 \text{ kJ/kg} \cdot \text{K})$$

$$\gamma_b = 1.238$$

$$\frac{V_{fx}}{V_3} = 0.50$$

$$\frac{V_f}{V_3} = 0.50$$

$$\frac{p_{10}}{p_0} = 1.0$$

	Component Estimates		
	High	Average	Low
η_1	0.95	0.90	0.85
$\left(\dfrac{C_f}{2}\cdot\dfrac{A_\omega}{A_3}\right)_c$	0.01	0.02	0.04
η_b	0.90	0.85	0.80
$\left(C_f\cdot\dfrac{A_\omega}{A_3}\right)_b$	0.10	0.20	0.40
C_{ev}	0.99	0.98	0.97
C_{pe}	0.38	0.36	0.34	BTU/lbm $\cdot$ °R
	(1.59	1.51	1.42	kJ/kg $\cdot$ K)
γ_e	1.222	1.238	1.255
C_{ea}	1.00	0.99	0.98

The results of the computations are summarized in Fig. 8.1. The most important message is that the overall scramjet performance, as expressed by $F/\dot{m}_0$, I_{sp}, or η_0, is extremely sensitive to the performance levels of the individual components. If the *high* estimates are achieved, the aerospace plane becomes almost irresistible. If the *low* estimates pertain, it will be difficult to find even a reasonable hypersonic cruise application. The *average* or most probable perfor-

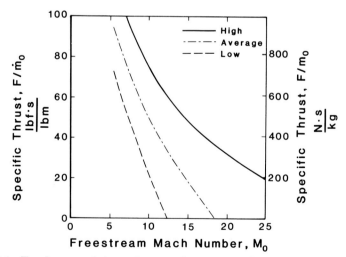

Fig. 8.1 Real scramjet engine performance obtained from stream thrust analysis as a function of freestream Mach number and type of component estimate.

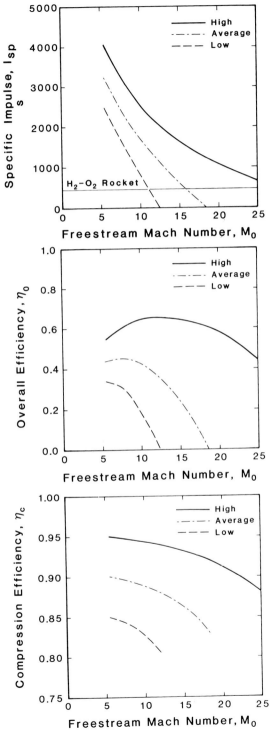

Fig. 8.1 (continued) Real scramjet engine performance obtained from stream thrust analysis as a function of freestream Mach number and type of component estimate.

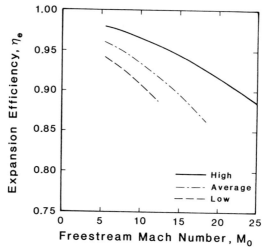

Fig. 8.1 (continued) Real scramjet engine performance obtained from stream thrust analysis as a function of freestream Mach number and type of component estimate.

mance suggests that practical airbreathing hypersonic cruise vehicles up to freestream Mach numbers of 15 and beyond are possible, but that the transatmospheric option will require superior components and/or some ingenuity.

The compression and expansion efficiencies are also given in Fig. 8.1 as examples of the internal performance of the engine. This information shows that these efficiencies behave just as we expect, and forms the basis of more accurate estimates of their values for use in simpler analyses.

8.3 PERFORMANCE ESTIMATION VIA KINETIC ENERGY EFFICIENCY

There is always a desire to identify simpler, less laborious methods for estimating uninstalled airbreathing engine performance. An example of this type is the first law performance analysis of C. H. Builder as presented in Sec. 4.3. Another intriguing approach is based on extending the notion of *kinetic energy efficiency* so popular for compression systems to the less familiar territories of combustion and expansion systems.[8.1] A shortcoming of this method is that the working medium must be assumed to behave as a calorically perfect gas with *the same* constants *throughout* the engine in order to avoid unwieldy algebra. The favorable benefits of this tradeoff will be quite evident from the results.

The heart of this matter is to generalize the definition of kinetic energy efficiency for the three major engine systems as follows:

Compression: $\qquad \eta_{KE,c} = \dfrac{V_X^2}{V_0^2} \leq 1$ $\qquad$ (8-3)

Combustion: $\qquad \eta_{KE,b} = \dfrac{V_Y^2}{V_X^2} \cdot \dfrac{T_{t3}}{T_{t4}} = \dfrac{V_Y^2}{V_X^2} \cdot \dfrac{1}{\tau_b} \leq 1$ $\qquad$ (8-4)

Expansion: $\qquad \eta_{KE,e} = \dfrac{V_{10}^2}{V_Y^2} \leq 1$ $\qquad$ (8-5)

where the nomenclature of Fig. 4.2 has been used. These definitions may be combined to yield

$$V_{10} = V_0\sqrt{\tau_b \eta_{KEO}}$$ (8-6)

where

$$\eta_{KEO} = \eta_{KE,c} \cdot \eta_{KE,b} \cdot \eta_{KE,e}$$ (8-7)

is the *overall kinetic energy efficiency.* Equation (8-6) may now be substituted into Eq. (3-11) in order to derive an expression for uninstalled specific thrust, namely

$$\frac{F}{\dot{m}_0} = V_0 \left\{ (1+f)\sqrt{\tau_b \eta_{KEO}} - 1 \right\}$$ (8-8)

This relationship is especially transparent because it reveals exactly how the mass flow rate and chemical energy release τ_b of the fuel compete with overall kinetic energy efficiency of the engine in the production of thrust. The interrelationships of Table 3.2 may, as ever, be used to convert specific thrust to any other desired performance measure.

A closer investigation of airbreathing engine performance is made possible by replacing the burner total temperature ratio with the expression resulting from the first law analysis of the adiabatic combustion process:

$$\tau_b = \frac{1}{1+f}\left(1 + \frac{\eta_b f h_{PR}}{C_p T_{t0}}\right)$$ (8-9)

where the kinetic energy of the fuel has either been neglected or included in h_{PR}. Combining Eqs. (8-8) and (8-9) yields

$$\frac{F}{\dot{m}_0} = V_0 \left\{ \sqrt{\eta_{KEO}(1+f)\left(1 + \frac{\eta_b f h_{PR}}{C_p T_{t0}}\right)} - 1 \right\}$$ (8-10)

or

$$\frac{F}{\dot{m}_0} = a_0 M_0 \left\{ \sqrt{\eta_{KEO}\,(1+f)\left[1 + \frac{\eta_b f h_{\mathrm{PR}}}{C_p T_0 \left(1 + \frac{\gamma-1}{2}M_0^2\right)}\right]} - 1 \right\}$$

(8-11)

and

$$I_{sp} = \frac{1}{g_0 f} \cdot \frac{F}{\dot{m}_0}$$

(8-12)

Note that the correct interpretation of f in the above requires that it not exceed the stoichiometric value in the chemical energy release term, while it can take on any value in the mass addition term.

The behavior of Eq. (8-12) for the stoichiometric combustion of hydrogen fuel in air is portrayed in Fig. 8.2, where $\gamma = 1.28$ has been chosen to represent the entire process. Among the features worth noting are that this analysis places ramjets and scramjets on a continuum, the maximum performance for any given η_{KEO} is in the general vicinity of a freestream Mach number of 3, and for any given η_{KEO} there is a freestream Mach number above which the uninstalled thrust is less than 0. Moreover, the qualitative and quantitative agreement with the average real scramjet performance taken from Fig. 8.1 is an encouraging sign for this approach. Finally, this presentation allows for the direct comparison of airbreathing and rocket engine performance, the horizontal line marking typical hydrogen-oxygen rocket specific impulse of 460 s. This comparison reinforces

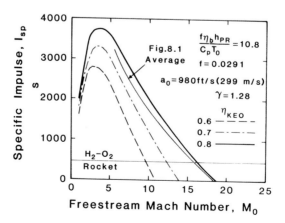

Fig. 8.2 Specific impulse as a function of freestream Mach number and overall airbreathing engine kinetic energy efficiency for the stoichiometric combustion of hydrogen in air.

our previous observations that the airbreathing engine is vastly superior to rocket propulsion up to $10 < M_0 < 15$, depending upon η_{KEO}, after which the reverse is true.

The acid test of the utility of the kinetic energy approach is comparison with experimental data, as reproduced from Ref. 8.1 in Figs. 8.3 and 8.4. The information assembled there suggests that this is indeed a sensible approach for the *ballpark* estimation of performance, and that the overall kinetic energy efficiency has been in the range $0.65 < \eta_{KEO} < 0.75$. Since $\eta_{KE,c}$ is liable to be about 0.90 (see Chap. 5) and $\eta_{KE,e}$, which is the square of C_{ev}, is liable to be about 0.98 (see Chap. 7) for this range of M_0, we see that thermodynamic cycle efficiency and losses due to friction, shock waves, and mixing reduced $\eta_{KE,b}$ to about 0.80. Please note that $\eta_{KE,b}$ does not reflect the incomplete release of any available chemical energy, although overall performance is penalized starting with Eq. (8-9). We must therefore conclude that substantially improving overall engine performance very much depends on understanding and increasing the kinetic energy efficiency of the combustor.

This method can also be used to trace the relationship between overall kinetic energy efficiency and the freestream Mach number at which the specific impulse or thrust reaches 0. This may also be interpreted as the minimum η_{KEO} required to produce thrust at the given M_0 or the maximum possible M_0 for a given η_{KEO}, and is

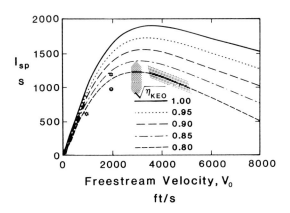

Fig. 8.3 Specific impulse as a function of freestream velocity for airbreathing ramjet engines burning hydrocarbon fuels, taken from Ref. 8.1. The curves are lines of constant overall kinetic energy efficiency.

obtained from the expression

$$\eta_{KEO} = \cfrac{1}{(1+f)\left\{1 + \cfrac{\eta_b f h_{PR}}{C_p T_0 \left(1 + \cfrac{\gamma - 1}{2} M_0^2\right)}\right\}} \qquad (8\text{-}13)$$

Figure 8.5 contains the boundaries established by Eq. (8-13) for the stoichiometric combustion of both hydrogen and a typical hydrocarbon fuel in air. This presentation makes it painfully clear that an η_{KEO} of 0.75 can at most power the vehicle to an M_0 of about 15, and that an η_{KEO} in excess of 0.85 would be required to propel the vehicle to orbital speed.

The kinetic energy efficiency approach is both handy and revealing, and therefore clearly meets the requirements for analyses of this type. It is particularly appealing to the intuition because it tracks the management and preservation of the available kinetic energy, easily the most important physical property of these "kinetic energy machines."

8.4 INSTALLED AIRBREATHING ENGINE THRUST

Up to this point we have dealt primarily with the so-called uninstalled airbreathing engine thrust, as described in detail in Sec. 4.4.1. Even though this concept is somewhat artificial, it is a creative fiction

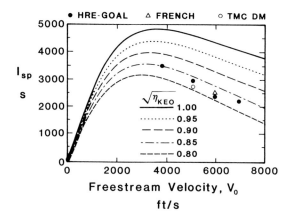

Fig. 8.4 Specific impulse as a function of freestream velocity for airbreathing scramjet engines burning hydrogen, taken from Ref. 8.1. The curves are lines of constant overall kinetic energy efficiency.

that has very effectively served the important purpose of allowing the behavior of the engine to be considered independent of any particular application. This, in turn, has allowed propulsion engineers to concentrate their efforts on improving the *internal* performance as well as to quote their measures of performance unambiguously.

Eventually, of course, the engine must be incorporated or *integrated* into a specific vehicle, and means must be provided to convert the uninstalled thrust to the net propulsive force that the vehicle will *feel*. The latter is widely known as the *installed thrust* and, as described in Sec. 4.4.1, is found by adjusting or *correcting* the uninstalled thrust in order to account for the *installation drag* or *penalty*. It must be emphasized that the conventions that describe and define installed thrust are somewhere between arbitrary and mythological, and are largely chosen to conform to the experience and convenience of the users, namely the vehicle and engine developers. This latter fact leads to the two significant corollaries described below.

First, there are many different possible methodologies for defining (and then bookkeeping) installed thrust, each of which has its own peculiar strengths and weaknesses. Any methodology that is rigorously based on fundamentals and is applied consistently will provide the desired information, so the final choice is at least partially subjective. Our approach will be quite similar to that used by the aircraft turbine engine community, allowing us to build upon

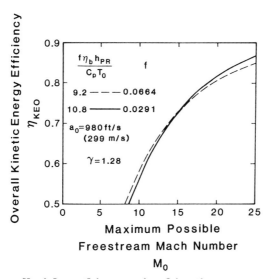

Fig. 8.5 Overall airbreathing engine kinetic energy efficiency as a function of the maximum achievable freestream Mach number for the stoichiometric combustion of hydrogen and a typical hydrocarbon fuel in air.

engine integration methods that have stood the test of time and have been carefully molded to present the essential ingredients in easily recognized packages.[8.2]

Second, the determination of installed thrust involves the *interface* between vehicle and engine, one of the most complex areas of any aerospace system development. Engine integration has itself become a major technology in modern aerospace system design and development, and can have beneficial and satisfying results when considered from the outset and done properly. When done poorly (or not at all), development crises are invited that can lead to reduced system performance, increased program time and cost, and/or bitter recriminations.

8.4.1 Isolated or Podded Airbreathing Engines

Figure 8.6 contains the control volume and terminology for the analysis of the installed performance of airbreathing engines that are isolated in the sense that they are separated from the primary vehicle and supported by means of a pod or its equivalent. For the purposes of this discussion, the engine may be regarded as axisymmetric, although the conclusions apply equally well to other geometries. Figure 8.6 is the logical successor to the upper portion of Fig. 4.8, the major difference being that the leading oblique shock wave is not *swallowed* by the physical opening at the face of the engine.

When the control volume form of the conservation of momentum is applied to the configuration of Fig. 8.6, the net *axial force* or *installed thrust* is found to be

$$T = \dot{m}_{10}u_{10} - \dot{m}_0 u_0 + \int_b^c (p - p_0)\, dA_x$$

$$+ \int_c^d (p - p_0)dA_x + \int_{10} (p - p_0)\, dA_x - F_{fx} \qquad (8\text{-}14)$$

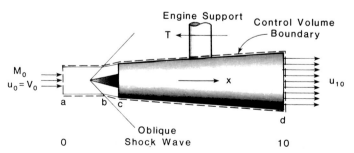

Fig. 8.6   Control volume and terminology used in the evaluation of installed airbreathing engine thrust for isolated or podded engines.

where u_{10} is understood to be the mass-average value. The term dA_x denotes the axial projection of the differential area and is counted as positive when the x component of the outwardly directed normal is in the positive x direction. The local static pressure has been reduced to the gauge pressure by subtracting the freestream static pressure in order to make the bookkeeping consistent with that of the vehicle. This has no effect on the installed thrust because the area integral of a constant pressure over a closed volume is zero force. The axial frictional force F_{fx} includes the contribution from the entire control volume surface, where the shear stresses at the gas-gas interfaces are usually (but not always) neglected. Next we interpret and redefine the quantities appearing in Eq. (8-14).

The *uninstalled* thrust is given by the familiar collection of terms of Eq. (4-37a)

$$F = \dot{m}_{10}u_{10} - \dot{m}_0 u_0 + (p_{10} - p_0)A_{10} \qquad (8\text{-}15)$$

where the exit plane pressure integral has been simplified by using the area-average static pressure p_{10}.

The *additive drag* is given by the integral

$$D_{\text{add}} = -\int_b^c (p - p_0)\, dA_x \qquad (8\text{-}16)$$

which will be 0 if the compression system is at its design point (i.e., shock-on-lip) freestream Mach number *or* beyond (i.e., the shock is swallowed). When the freestream Mach number is below the design point (as it is in Fig. 8.6), there will generally be a positive additive drag for supersonic flight because the pressure on the dividing streamline is above freestream static pressure and the outward normal points in the negative x direction. As explained in Sec. 4.4.1, and even though the details of the compression system flowfield are accounted for in the determination of the exhaust conditions, the additive drag must be included in the installed thrust.

The *external drag* is given by the integral

$$D_{\text{ext}} = -\int_c^d (p - p_0)\, dA_x \qquad (8\text{-}17)$$

which is also generally positive for supersonic flight for the same reasons as the additive drag. One has the option of including the external drag in the vehicle drag and setting it equal to 0 when reckoning installed thrust. However, it is traditionally left in place in Eq. (8-14) as a strong inheritance from subsonic flight, where the external drag can actually be a thrust (originating in leading-edge suction) that as much as cancels the additive drag for well-designed nacelles.[8.2] Thus, the additive drag and the external drag are usually grouped together.

Finally, the axial frictional force F_{fx} is usually included in the vehicle drag and deleted from Eq. (8-14).

When these considerations are combined in Eq. (8-14) we obtain the working form for evaluating the installed thrust for isolated or podded airbreathing engines, namely

$$T = F - (D_{\text{add}} + D_{\text{ext}}) \tag{8-18}$$

8.4.2 Integral Airbreathing Engines

As we have seen, hypersonic vehicles are most likely to have airbreathing engines that cannot be isolated from the vehicle and can exert an overwhelming influence on the entire configuration. Figure 8.7 contains the control volume and nomenclature for the analysis of the installed performance of *integral* airbreathing engines that are intimately joined with the primary vehicle. For the purposes of this discussion, the engine may be regarded as two-dimensional, although the conclusions apply equally well to other geometries. Figure 8.7 is the logical successor to the lower portion of Fig. 4.8, with the same caveats as applied to Fig. 8.6. In fact, this development is greatly condensed because the arguments and terminology for integral engines are virtually the same as for isolated engines. Only the exceptions will be discussed in what follows.

When the control volume form of the conservation of momentum is applied to the configuration of Fig. 8.7, the net axial force or *installed thrust* is found to be

$$T = \dot{m}_{10}u_{10} - \dot{m}_0 u_0 + \int_b^c (p - p_0)\,dA_x + \int_c^d (p - p_0)\,dA_x$$

$$+ \int_d^e (p - p_0)\,dA_x + \int_e^f (p - p_0)\,dA_x + \int_f^g (p - p_0)\,dA_x - F_{fx} \tag{8-19}$$

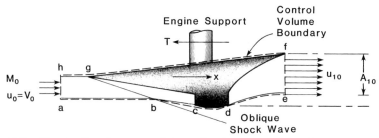

Fig. 8.7 Control volume and terminology used in the evaluation of installed airbreathing engine thrust for integral engines.

The uninstalled thrust is again given by the collection of terms,

$$F = \dot{m}_{10}u_{10} - \dot{m}_0 u_0 + (p_{10} - p_0) A_{10} \qquad (8\text{-}20)$$

where p_{10} is the area-average static pressure. The *additive drag* is given by the sum of the integrals,

$$D_{\text{add}} = -\int_b^c (p - p_0) \, dA_x - \int_d^e (p - p_0) \, dA_x \qquad (8\text{-}21)$$

where A_{10} is formed from the axial projection of point d upon the exit plane.

The *external drag* is now given by the sum of the integrals,

$$D_{\text{ext}} = -\int_c^d (p - p_0) \, dA_x - \int_f^g (p - p_0) \, dA_x \qquad (8\text{-}22)$$

where the latter is more imaginary than most, and is therefore almost always removed from the installed thrust and included instead in the forces acting on the vehicle.

With the foregoing in mind, the expression for the installed thrust of integral engines is the same as Eq. (8-18), and is therefore formally *identical* to that of isolated engines.

Please also note that we are now able to more precisely write the installation loss coefficient ϕ_e of Chap. 3 as

$$\phi_e = 1 - \frac{T}{F} = \frac{D_{\text{add}} + D_{\text{ext}}}{F} \qquad (8\text{-}23)$$

and that an equivalent set of *installed* performance measures (e.g., Table 3.2) can be developed for a specific engine integrated into a specific vehicle by using T in place of F.

8.4.3 Additive Drag and External Drag

In stark contrast to the subsonic flight situation, where the external drag can almost cancel the additive drag, the sum of the additive and external drags on hypersonic airbreathing engines can be formidable. This can best be illustrated by means of the example case sketched in Fig. 8.8, which continues the two oblique shock wave case begun in the upper portion of Fig. 5.12. The static pressure ratio increase across the oblique shock waves and the static pressure ratio decrease across the ensuing simple isentropic expansion waves were calculated with HAP(Gas Tables) using $\gamma = 1.36$.

Before examining the example case of Fig. 8.8, it will be worthwhile to distance the additive and external drag results as much as

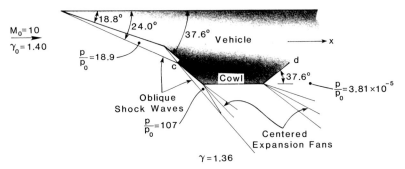

Fig. 8.8 Example case for the calculation of airbreathing engine cowl or nacelle external drag.

possible from the specific geometrical details by reducing them to drag coefficients in accordance with

$$C_D = \frac{D}{q_0 A_x} = \frac{(p - p_0)A_x}{\dfrac{\gamma_0 p_0 M_0^2}{2} A_x} = \frac{2}{\gamma_0 M_0^2}\left(\frac{p}{p_0} - 1\right) \qquad (8\text{-}24)$$

Substituting the values of Fig. 8.8 into Eq. (8-24) shows that the *external drag coefficients* of the cowl are 1.518 on the leading surface and 0.0143 on the trailing surface. Thus, even though there is close to a vacuum on the trailing surface, the external drag contribution of the leading surface per unit area far outweighs that of the trailing surface for hypersonic freestream Mach numbers. It should also be noted that the contribution to the *additive drag coefficient* of the oblique shock waves for this example case is 0.256 after the first deflection of the dividing streamline and 1.518 after the second deflection, where the conditions are identical to those along the cowl leading surface.

In order to predict whether such drag coefficients would significantly reduce overall performance, we rearrange Eqs. (8-18) and (8-24) into the form

$$\frac{T}{\dot{m}_0 V_0} = \frac{F}{\dot{m}_0 V_0} - \frac{(C_D A_x)_{\text{add}}}{2A_0} - \frac{(C_D A_x)_{\text{ext}}}{2A_0} \qquad (8\text{-}25)$$

Since a typical value of the dimensionless quantity $F/\dot{m}_0 V_0$ for these conditions is only about 0.20 (see Sec. 4.4.4), the numbers of this example show that it is extremely urgent for the designer to avert as much additive and external drag as possible. This can be done by minimizing the deflection of the dividing streamline and reducing the exposed leading surface area and drag of the cowl by keeping it parallel to the freestream flow. Here we can observe the art of compromise at work, because *both* the internal and external performance of the airbreathing engine are influenced by the shape of the cowl, particularly when a variable-geometry compression system is contemplated.

To summarize, although uninstalled thrust is the first concern of the engine designer, it is the installed thrust that matters to vehicle performance. The conversion from uninstalled thrust to installed thrust can be accomplished in a fundamentally correct and consistent manner, as demonstrated by the methodology presented here. One must remain aware that many other legitimate accounting procedures exist, and are likely to be encountered in practice. Since installation penalties have the potential to greatly diminish the performance of scramjet engines, they must be minimized which means, in turn, that they must be taken seriously from the outset.

8.5 THRUST AUGMENTATION

A variety of schemes for generating thrust beyond that of the basic ramjet or scramjet engine are available for application at critical points of the mission. These *thrust augmentation* methods loom very large when the net thrust (or the effective overall efficiency or specific impulse) of the vehicle approaches 0 for whatever reason. The purpose of this section is to highlight the importance of such techniques, and to describe enough of them for you to see that this field of study is the beneficiary of considerable imaginative attention.

The most desirable thrust augmentation devices are those that naturally integrate themselves geometrically and mechanically into the existing ramjet or scramjet engine flowpath. This, in general, minimizes the additional volume, frontal area, surface area, drag, complexity, weight, and cost required to bring them onboard. Thus, although separate turbojet engines for takeoff thrust or separate rocket engines for extra thrust anywhere in the mission certainly appear in the thrust augmentation catalog, they are not necessarily the first or best choices.

Please note that the emphasis of thrust augmentation is primarily upon the *magnitude* of the thrust, rather than upon the uninstalled engine specific performance measures, because the amount of fuel consumed during the mission depends upon the *effective* or installed performance (see Chap. 3). Moreover, it will be impossible to perform the intended mission if the net thrust of the vehicle prematurely reaches 0.

The techniques are described in the order of the flight Mach number of their most likely application, from low to high.

8.5.1 The Ejector Ramjet

The basic property of ordinary ejectors is that they multiply the original or *primary* mass flow by drawing in a supplemental or *secondary* mass flow from the surrounding atmosphere, as indicated in Fig. 8.9. In the same process, the total pressure of the secondary flow is raised to a value between that of the ambient and that of the primary flow. Ejectors are mechanically simple, requiring only an enclosing passage or *shroud* around the primary flow long enough to enable complete mixing with the secondary flow. The ejector is therefore crudely analogous to the mixed exhaust flow bypass turbofan engine, although the energy transfer efficiency is low because it is accomplished by viscous shear forces rather than rotating turbomachinery.[8.3]

Since ramjets produce little or no thrust during takeoff, a rocket engine or its equivalent must be part of the vehicle. This device could either operate independently or act as the primary of an ejector for which the existing ramjet or scramjet passage could serve as a shroud. In the latter case, the pressurized flow leaving the ejector can be decelerated, mixed with fuel, and burned in a combustor, and then accelerated through a nozzle to produce thrust. The net effect of the ejector is to supply the burner with a flow of pressurized air that would be roughly equivalent to the *ram* conditions of a much higher forward speed. Thus, the ejector-burner-nozzle combination is aptly referred to as the *ejector ramjet*.

Ejector ramjets are attractive low speed propulsion candidates because of their mechanical simplicity and because they can easily be integrated into the existing flowpath. You will soon see that they can increase the thrust substantially beyond that of the primary flow acting alone. Thus, they meet all the stated criteria of desirable thrust augmentation devices.

8.5.1.1 Ideal ejector ramjet analysis.   Referring to Fig. 8.9 for geometry and nomenclature, a one-dimensional analysis of the behavior of ideal (i.e., no friction or heat transfer at the wall and no shock waves) ejector ramjets will be performed. A number of assumptions will be made that simplify the ensuing algebra, but they exert a relatively minor influence on the outcomes because the fundamental mechanism at work is the distribution of a fixed pool of momentum and energy between two flows.

The ejector portion of the device will have constant area and fixed geometry, and the flows will all be treated as steady and compressible and having the same values of the appropriate calorically perfect gas constants. The inlet primary flow will be supersonic, the inlet secondary flow will be subsonic, and the ejector exit plane flow will be choked or sonic. These assumptions may be checked during the

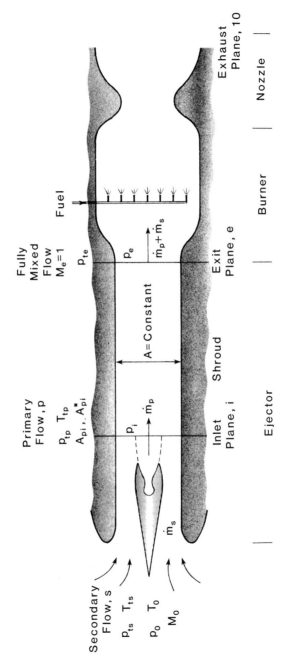

Fig. 8.9   Schematic diagram for ideal ejector ramjet analysis.

course of the computations, particularly the latter, which requires that the ejector exit plane static pressure exceed the ambient static pressure. The combustor will have no total pressure losses (i.e., the energy is added at 0 Mach number and therefore experiences no Rayleigh heating loss), and the added energy will bring the total temperature of the mixture at the burner exit up to some fraction of the initial total temperature of the primary flow. The flow is then expanded isentropically to ambient static pressure. The material for the primary flow is stored on the vehicle, and the mass flow of the fuel consumed in the burner will be neglected.

The reader is cautioned that compressible flow ejectors are capable of many other *modes* of operation (e.g., choked secondary inlet flow and/or unchoked exit plane flow) that would necessitate different analyses.[8.4] The general performance of ejectors is, however, not fundamentally altered by these subtleties, and the results of the present analysis are a fair representation. The equation set is arranged to be solved by trial-and-error using the inlet plane static pressure as the iteration variable, but the solution can also be found automatically with root- or zero-seeking computer software.

The method of solution will be explained and illustrated by means of the following typical example, for which the correct, converged inlet static pressure is furnished from the beginning so that the results will be meaningful. The input parameters are:

Primary flow:

$$\frac{p_{tp}}{p_0} = 15.0 \qquad \frac{T_{tp}}{T_0} = 10.0 \qquad \frac{A}{A_p^*} = 12.0 \qquad \gamma = 1.35$$

Secondary flow:

$$M_0 = 0.50 \qquad \frac{p_{ts}}{p_0} = 1.180 \qquad \frac{T_{ts}}{T_0} = 1.044 \qquad \gamma = 1.35$$

Exhaust Flow:

$$\frac{p_{10}}{p_0} = 1.0 \qquad \frac{T_{t10}}{T_{tp}} = 1.0 \qquad \gamma = 1.35$$

The total temperature and total pressure ratios of the secondary flow were chosen to be greater than 1 in order to simulate forward motion. The inlet plane static pressure is:

$$\frac{p_i}{p_0} = 0.912 \qquad \left\{ \text{Note: } \frac{p_{ts}}{p_0} > \frac{p_i}{p_0} > \left(\frac{2}{\gamma+1} \right)^{\frac{\gamma}{\gamma-1}} = 0.537 \right\}$$

First, the inlet plane flow properties are computed based on isentropic flow. Note that the inlet plane is the axial location at which the primary and secondary flows are at static pressure *equilibrium*.

Thus, this analysis ignores any mixing that might occur upstream of this location, an effect that is minimized by the natural attempts of the designer to match the exit pressure of the primary flow nozzle to the local secondary flow static pressure.

$$M_{pi} = \left\{ \frac{2}{\gamma - 1} \left[\left(\frac{p_{tp}}{p_0} \cdot \frac{p_0}{p_i} \right)^{\frac{\gamma-1}{\gamma}} - 1 \right] \right\}^{\frac{1}{2}} \qquad M_{pi} = 2.47 \qquad (8\text{-}26)$$

$$\frac{A_{pi}}{A_p^*} = \frac{1}{M_{pi}} \left\{ \frac{2}{\gamma + 1} \left(1 + \frac{\gamma - 1}{2} M_{pi}^2 \right) \right\}^{\frac{\gamma+1}{2(\gamma-1)}} \qquad \frac{A_{pi}}{A_p^*} = 2.70 \qquad (8\text{-}27)$$

$$\frac{A_{pi}}{A} = \frac{A_{pi}}{A_p^*} \cdot \frac{A_p^*}{A} \qquad \frac{A_{pi}}{A} = 0.225 \qquad (8\text{-}28)$$

$$\frac{A_{si}}{A} = 1 - \frac{A_{pi}}{A} \qquad \frac{A_{si}}{A} = 0.775 \qquad (8\text{-}29)$$

$$M_{si} = \left\{ \frac{2}{\gamma - 1} \left[\left(\frac{p_{ts}}{p_0} \cdot \frac{p_0}{p_i} \right)^{\frac{\gamma-1}{\gamma}} - 1 \right] \right\}^{\frac{1}{2}} \qquad M_{si} = 0.628 \qquad (8\text{-}30)$$

$$\alpha = \frac{\dot{m}_s}{\dot{m}_p} = \text{bypass ratio}$$

$$= \frac{p_{ts}}{p_0} \cdot \frac{p_0}{p_{tp}} \cdot \frac{A_{si}}{A} \cdot \frac{A}{A_{pi}} \cdot \frac{M_{si}}{M_{pi}} \sqrt{\frac{T_{tp}}{T_0} \cdot \frac{T_0}{T_{ts}}} \left\{ \frac{1 + \frac{\gamma - 1}{2} M_{pi}^2}{1 + \frac{\gamma - 1}{2} M_{si}^2} \right\}^{\frac{\gamma+1}{2(\gamma-1)}} \qquad (8\text{-}31)$$

$$\alpha = 1.954$$

Next, the equations of conservation of energy and mass are employed to find the exit plane properties.

$$\frac{T_e}{T_{tp}} = \frac{2}{\gamma + 1} \left(\frac{1 + \alpha \dfrac{T_{ts}}{T_0} \cdot \dfrac{T_0}{T_{tp}}}{1 + \alpha} \right) \qquad \frac{T_e}{T_{tp}} = 0.347 \qquad (8\text{-}32)$$

$$\frac{p_e}{p_0} = (1 + \alpha) \frac{A_p^*}{A} \cdot \frac{p_{tp}}{p_0} \cdot \sqrt{\frac{T_e}{T_{tp}}} \left(\frac{2}{\gamma + 1} \right)^{\frac{\gamma+1}{2(\gamma-1)}} \qquad \frac{p_e}{p_0} = 1.266 \qquad (8\text{-}33)$$

Next, the equations of conservation of momentum and mass are employed to test the selected p_i/p_0. This ratio will be unity for the correct solution.

$$\frac{\frac{p_e}{p_0}(1+\gamma)}{\frac{p_i}{p_0}\left\{\frac{A_{pi}}{A}\left(1+\gamma M_{pi}^2\right)+\frac{A_{si}}{A}\left(1+\gamma M_{si}^2\right)\right\}} = 1.0 \qquad (8\text{-}34)$$

Finally, the traditional performance measures of the ejector ramjet are computed. They include the *ejector exit total pressure ratio*:

$$\frac{p_{te}}{p_0} = \frac{p_e}{p_0}\left(\frac{\gamma+1}{2}\right)^{\frac{\gamma}{\gamma-1}} \qquad\qquad \frac{p_{te}}{p_0} = 2.36 \qquad (8\text{-}35)$$

and the *thrust augmentation ratio*, which is defined as

$$\phi_p = \frac{\dot{m}_e V_{10} - \dot{m}_s V_0}{\dot{m}_p V_{p0}} = (1+\alpha)\frac{V_{10}}{V_{p0}} - \alpha\frac{V_0}{V_{p0}} \qquad (8\text{-}36)$$

where

$$M_{p0} = \left\{\frac{2}{\gamma-1}\left[\left(\frac{p_{tp}}{p_0}\right)^{\frac{\gamma-1}{\gamma}}-1\right]\right\}^{\frac{1}{2}} \qquad M_{p0} = 2.41 \qquad (8\text{-}37)$$

$$\frac{V_0}{V_{p0}} = \frac{M_0}{M_{p0}}\left\{\frac{T_{ts}}{T_0}\cdot\frac{T_0}{T_{tp}}\left(\frac{1+\frac{\gamma-1}{2}M_{p0}^2}{1+\frac{\gamma-1}{2}M_0^2}\right)\right\}^{\frac{1}{2}}\frac{V_0}{V_{p0}} = 0.0931 \qquad (8\text{-}38)$$

$$M_{10} = \left\{\frac{2}{\gamma-1}\left[\left(\frac{p_{te}}{p_0}\right)^{\frac{\gamma-1}{\gamma}}-1\right]\right\}^{\frac{1}{2}} \qquad M_{10} = 1.193 \qquad (8\text{-}39)$$

$$\frac{V_{10}}{V_{p0}} = \frac{M_{10}}{M_{p0}}\left\{\frac{T_{t10}}{T_{tp}}\left(\frac{1+\frac{\gamma-1}{2}M_{p0}^2}{1+\frac{\gamma-1}{2}M_{10}^2}\right)\right\}^{\frac{1}{2}}\frac{V_{10}}{V_{p0}} = 0.629 \qquad (8\text{-}40)$$

whence

$$\phi_p = 1.675$$

Before moving on to the results, please note that at very low forward speeds the ejector ramjet thrust is partially due to suction forces exerted on the contoured portion of the secondary flow inlet. Consequently, the latter must be carefully shaped in order to avoid separation losses and achieve the potential thrust. Also, the effects of various possible losses could easily be included in the above equation set if desired.

8.5.1.2 Ideal ejector ramjet performance.
The results of computations based on the ideal ejector ramjet analysis are presented for the typical case as a function of freestream Mach number in Fig. 8.10. The principal result is that the ejector ramjet has the potential to significantly increase the thrust above that of the primary flow alone, as much as by a thrust augmentation ratio of about 1.6–2.2 in the Mach number range for which a ramjet could produce little or no thrust. The escalation of thrust with freestream Mach number should also prove helpful in overcoming the transonic drag rise of the vehicle. Although additional fuel is required to raise the temperature of the flow in the burner, the additional thrust is regarded as *free* in the sense that it relies only on existing hardware.

The remaining information of Fig. 8.10 is both interesting and comforting. Both the bypass ratio and the ejector exit total pressure ratio increase rapidly with freestream Mach number once Mach 1 is reached because of the ram pressure of the oncoming air. The ejector exit total pressure ratio is always large enough to ensure that $p_e > p_0$ [see Eq. (8-35)]. The secondary inlet flow is always subsonic, in conformance with the assumptions. The primary inlet flow, not shown in Fig. 8.10, is always supersonic.

At a freestream Mach number of about 2.0, the primary flow may be turned off and the engine allowed to continue on as an ordinary ramjet. The computations summarized in Fig. 8.10 correctly suggest that the contribution of the ejector has become small by that point, and that the device has undergone a smooth transition to ramjet operation anyway. Altogether, the ejector ramjet is a very promising concept for hypersonic applications.

8.5.2 External Burning

The static pressure on the rearward-facing surface of the engine cowl and/or flap is often less than the freestream static pressure, resulting in an external drag on the vehicle and an installation penalty for the ramjet or scramjet. The two primary causes for the low pressure or *suction* on the aft surfaces are sketched in the upper portion of Fig. 8.11, where it is assumed for simplicity that the cowl is aligned with the freestream flow and therefore that the conditions adjacent to the cowl are those of the freestream. In supersonic flight the corner expansion fan can, as we have seen in Chap. 7, reduce the

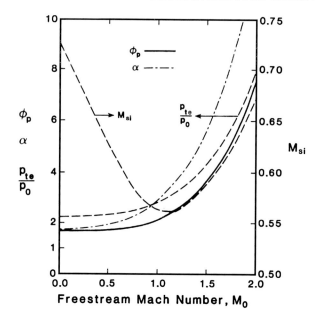

Fig. 8.10 Ideal ejector ramjet performance for the typical case as a function of freestream Mach number.

downstream static pressure virtually to 0. In subsonic or transonic flight, the adverse pressure gradient downstream of the corner can separate the boundary layer, resulting in what is commonly called wake or form drag. The latter is especially critical because transonic flight is often a *pinch point* for the vehicle, where high drag and low thrust combine to minimize the desired acceleration.

These problems can be partially solved by releasing chemical energy in the vicinity of the suction surface for the purpose of raising the local static pressure, a process known as *external burning*. Since no closed thermodynamic cycle is involved, external burning is perhaps more precisely thought of as a *drag reduction* scheme. A useful mental image then is that the higher temperature air expands to *fill* the void and presents the external flow with an imaginary streamlined vehicle boundary.

External burning is closely related to *base burning,* a similar technique for reducing supersonic drag on the blunt aft ends of missiles and projectiles, and for which an abundance of material exists in the open literature.[8.5] The base burning of solid pyrotechnics intended to make *tracer* bullets luminous and easily visible also causes them to take a different trajectory from the majority of undesignated bullets, partly defeating their purpose.

A first-order one-dimensional analysis for exploring the potential of external burning is easily devised with the help of Fig. 8.11 and the results of the calorically perfect gas constant pressure heating Example Case 2.8. In this model, external burning encloses the surface in question with a region having a constant static pressure equal to ambient. Other levels of the constant static pressure could be chosen and analyzed, but this one has the special virtue of not deflecting or otherwise disturbing the adjacent freestream flow.

Applying Eq. (2-120) of Example Case 2.8 to Fig. 8.11, we find immediately that

$$\frac{A_e}{A_i} = 1 + \frac{A_b}{A_i} = \tau_e \left(1 + \frac{\gamma_0 - 1}{2} M_0^2\right) - \frac{\gamma_0 - 1}{2} M_0^2 \qquad (8\text{-}41)$$

where A_i is, for the moment, the arbitrary throughflow area of freestream flow involved in the external burning, and

$$\tau_e = \frac{T_{te}}{T_{t0}} = 1 + \frac{\eta_b f h_{PR}}{C_{p0} T_0 \left(1 + \dfrac{\gamma_0 - 1}{2} M_0^2\right)} \qquad (8\text{-}42)$$

Since the drag reduction is

$$\Delta D = (p_0 - p_b) A_b \qquad (8\text{-}43)$$

where p_b is the axial projected area average of the base surface static pressure *prior* to external burning, a reasonable measure of specific impulse performance is

$$I_{sp} = \frac{\Delta D}{g_0 \dot{m}_f} \qquad (8\text{-}44)$$

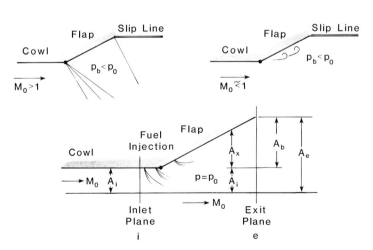

Fig. 8.11   Schematic diagrams for the one-dimensional analysis of external burning.

which, when combined with Eqs. (8-41) through (8-43), yields the desired relationship :

$$\frac{I_{sp}}{\left(1 - \frac{p_b}{p_0}\right)} = \frac{\gamma_0 - 1}{\gamma_0} \cdot \frac{\eta_b h_{PR}}{g_0 V_0} \tag{8-45}$$

This amazingly simple expression shows that external burning performance is independent of the magnitude of *both* f and A_i, varying directly with the heating value of the fuel and inversely as the freestream velocity. The numerical results for typical fuels shown in Fig. 8.12 reveal specific impulse levels in the transonic regime comparable to that of the ramjet, providing, of course, that p_b is only a small fraction of p_0.

The greatest challenge for external burning is to *distribute* the chemical energy release so that the region of constant static pressure is achieved. One important aspect of this can be seen by combining Eqs. (8-41) and (8-42) to show that

$$\frac{A_i}{A_b} = \frac{C_{p0} T_0}{\eta_b f h_{PR}} \tag{8-46}$$

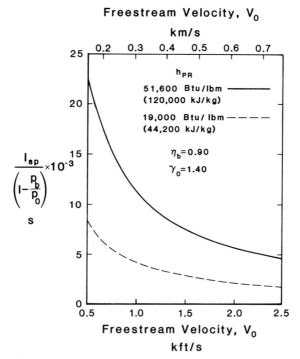

Fig. 8.12 Specific impulse performance of external burning as a function of freestream velocity and fuel heating value.

This leads to the calculation of A_i/A_b, which is a measure of the distance the injected fuel must penetrate the external flow. The injection distance can be minimized by stoichiometric burning (i.e., the largest possible f). If you substitute typical quantities into Eq. (8-46), you will find that the minimum value of A_i/A_b is in the vicinity of 0.07, which can mean significant penetration distances for large vehicles. The other important aspect is the need to tailor the location of the fuel injection and the subsequent mixing and combustion to conform to the schedule required by constant pressure heating over a range of freestream Mach numbers. That schedule can be found by writing Eq. (8-41) for an arbitrary axial location and rearranging to obtain

$$\tau_x = \frac{T_{tx}}{T_{t0}} = 1 + \frac{1}{\left(1 + \frac{\gamma_0 - 1}{2}M_0^2\right)} \cdot \frac{A_x}{A_i} \tag{8-47}$$

Given the discussions of Chap. 6 regarding mixing and chemical reaction, and recognizing that burning at relatively low static temperature and pressure in an unconfined space makes things harder, we can appreciate what a daunting task this is for the designer.

The available experimental data[8.6, 8.7] indicate that this simple model is correct and that most of the benefits of external burning can be achieved in practice, provided primarily that there is sufficient residence time for the desired chemical energy release schedule to be realized. The data and more complex analyses also show that further increasing the chemical energy addition will result in higher surface static pressure and greater thrust, although not necessarily higher I_{sp}.

External burning is an attractive candidate for thrust augmentation because, like the ejector ramjet, it requires little additional hardware. It is especially appealing in the transonic flight regime where the external drag can dominate performance, one main reason being that the flap will be in the closed position, thus reducing p_b while increasing A_b. Finally, there would be additional thrust benefits, not accounted for here, because the static pressure on the ramjet or scramjet engine expansion surface must increase when the exhaust flow is exposed to higher static pressures along the slip line.

8.5.3 Fuel and Oxidizer Enrichment

As the vehicle moves along a constant q_0 trajectory to the higher hypersonic velocities and altitudes, the specific thrust declines (see Fig. 8.1) and the increasingly rarefied atmosphere causes the airflow rate captured by the vehicle to diminish (see Fig. 2.5). Consequently, the *total* airbreathing engine thrust can decrease to the point at which the desired acceleration can no longer be maintained. Several solutions have been proposed to provide the required thrust, all of which fall in the general category of *fuel and oxidizer enrichment.*

To begin with, we have already seen that the *thrust* (rather than specific fuel consumption or specific impulse) is improved by supplying a fuel flow rate above the stoichiometric level at any flight Mach number. An extension of this idea is to carry oxidizer onboard as well, and to burn the excess fuel and simultaneously increase the mass flow rate. This approach is superior to having a separate rocket engine because it uses the existing scramjet flowpath and fuel system, and can have better specific fuel consumption or specific impulse because the combustion energy release is spread out over more fluid. Other combinations and permutations are clearly feasible, such as burning only some of the excess fuel with onboard oxidizer.

Obviously, the correct design choice must be based on cycle analyses of the most promising possibilities. These analyses, however, can be accomplished within the performance framework already established in this textbook because the enrichment methods primarily alter the analysis within the combustion system as the reactants and products change.

A fascinating feature of enrichment is that a flow of *pure* hydrogen, owing to its extraordinarily low molecular weight, will have a high specific impulse when heated to relatively modest total temperatures. The thermal energy can even be transferred from the hot surfaces of the vehicle or engine. Since hydrogen enrichment will only occur at high altitudes (and low static pressures), the vacuum exhaust velocity of Example Case 2.5 is relevant. Taking $C_p = 3.42$ BTU/lbm· °R (14.3 kJ/kg·K) and $T_t = 2000$ °R (1111 K), we find that

$$V_e = \sqrt{2C_p T_t} = 18,500 \text{ ft/s (5640 m/s)}$$

or

$$I_{sp} = \frac{V_e}{g_0} = 575 \text{ s}$$

This is considerably more than the vacuum specific impulse of about 460 s for hydrogen-oxygen rocket engines, and does not require the availability of oxygen or involve either chemical reactions or extremely high temperature gases.

8.6 COMBINED CYCLE AIRBREATHING ENGINES

The favored solution to the hypersonic airbreathing propulsion problem of maintaining acceptable thrust and fuel consumption over the entire flight spectrum is to unite several different propulsion concepts within the same internal flowpath. Airbreathing engines of this type are known as *combined cycle* or *composite* engines. Even though most combined cycle engines are aimed at the low speed regime, you will find that this is not exclusively true.

As a general rule, the concepts incorporated in combined cycle engines are based upon familiar turbojet, ramjet, scramjet, and rocket propulsion, but it is fair to include anything that will help, such as ejectors and air liquefiers. Happily, combined cycle engines are an inventor's paradise, their imaginations having brought forth an amazing diversity of promising ideas. The *turbo ramjet* power plant of the Griffon II is an early example of a successful design.

Our goal will be to capture the significance of combined cycle engines by means of a few carefully selected examples, rather than to carry out an exhaustive survey of the many possible combinations and permutations. The essential point to be made is that combined cycles are likely to be key players in the cast of hypersonic propulsion engines.

References 8.8 and 8.9 are good starting points for additional material on this rather vast subject. The reader will soon discover that the same concepts or devices are called by a variety of different names in the open literature, a situation that we will not attempt to reconcile.

8.6.1 The Turbo Ramjet

The most widely used method of generating thrust at very low flight speeds rests upon the mechanical compression of turbojet engines, which both induces a flow of air and leads to a positive thermal cycle efficiency for any ensuing energy release. The increased temperatures and pressures resulting from the mechanical compression are also conducive to the efficient and stable combustion of fuel. Figure 8.13 contains a diagram of the *turbo ramjet,* frequently called the *air turbo ramjet,* a combined cycle engine configuration that takes advantage of these principles.

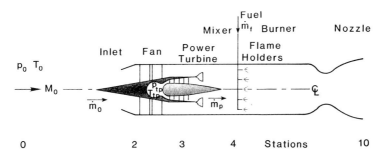

Fig. 8.13 Schematic diagram for ideal turbo ramjet analysis.

The main components of the turbo ramjet, moving in the streamwise direction from left to right, are:

1. A modest pressure ratio, high throughflow axial flow compressor, more commonly known as a *fan* because it resembles one
   in form and function. At high freestream Mach numbers, say
   > 5.0 or so, the compression ratio of the fan is no longer needed
   for thermal efficiency, and its survivability is threatened by the
   high stagnation temperature of the captured air. Consequently,
   if the engine is intended for use at even higher freestream Mach
   numbers, provision must be made to remove or retract the fan
   from the main flowpath.

2. A turbine that is driven by high pressure, high temperature
   gases generated in a separate combustion chamber. This turbine provides the power required by the fan, and enjoys the
   special advantage that its inlet conditions are isolated from and
   therefore independent of the flight conditions. The turbine flow
   will be referred to here as the *primary* stream. The presence
   of the fan and turbine require that at least the forward portion
   of this engine be axisymmetric, rather than two-dimensional,
   which can complicate integration with the vehicle.

3. A mixer that blends the airflow with the primary flow. This
   increases the total temperature of the airflow, and distributes
   any unreacted fuel throughout the air, where it can burn. The
   primary flow can be deliberately designed to be overly rich in
   fuel, and thus to provide hot, gaseous fuel for easy combustion
   with the air.

4. Fuel injectors to provide additional unreacted fuel to the flow,
   as desired.

5. A burner, including flameholders, with sufficient residence time
   to bring the chemical reactions to completion.

6. A nozzle that properly expands the exhaust flow to the ambient
   static pressure.

One might expect from its superficial appearance that the turbo
ramjet behaves like a turbojet at very low flight speeds and like a
ramjet or rocket engine at higher flight speeds. This conjecture will
be tested by the forthcoming analysis.

8.6.1.1 Ideal turbo ramjet analysis.   With reference to Fig. 8.13 for geometry and nomenclature, a one-dimensional analysis of ideal (i.e., no
friction or heat transfer at the wall and no shock waves) turbo ramjets

will be performed. This will be our final opportunity to demonstrate the application of this approach to the analysis of a complex engine configuration. Any doubts that you may have about the versatility and power of this methodology can be put to rest by reading Refs. 8.2 and 8.10, in which extraordinarily complicated configurations are faithfully modeled, *including* the effects of losses, variable gas constants, mixing, bleed air, mechanical power extraction, and control systems. In some important cases the algebra is transparent enough that the precise conditions of optimal performance can be identified.

As ever, the flows will be treated as steady and compressible and having the same values of the appropriate calorically perfect gas constants. The compression process to the fan inlet and the rotating machinery processes will be treated as isentropic, the power generated by the turbine being equal to that absorbed by the fan. The influence of the fan on the airflow will be characterized by its total pressure ratio $\pi_F = p_{t3}/p_{t2}$, a common practice in the turbojet community and an honest representation of reality. The turbine exit flow will contain no unreacted fuel, and mixing with the airflow will be completed before any additional fuel is injected and burned. The mixing and combustion processes will take place at essentially zero Mach number, so that there are no total pressure losses due to kinetic energy mixing or Rayleigh heating, and the *static* pressures of the primary flows and airflows are equal to each other *and* to their respective *total* pressures when they meet. The station designations leave room for the possibility of burning fuel from the primary flow by equating the mixer with the main burner. The burning of the injected fuel in this analysis is therefore analogous to afterburning in ordinary turbojets. The burning will bring the total temperature of the mixture up to some fraction of the initial total temperature of the primary flow. The flow is then expanded isentropically to ambient static pressure. The materials for the primary flow and injected fuel flow are stored on the vehicle, and their respective mass flow rates will both be accounted for in the analysis.

The method of solution is direct and will be explained and illustrated by means of the following typical example. Please note that the primary flow stagnation conditions and the freestream Mach number are identical to those of the ejector ramjet example, thus enabling fair comparison of their performance. The input parameters are:

Primary flow:

$$\frac{p_{tp}}{p_0} = 15.0 \qquad \frac{T_{tp}}{T_0} = 10.0 \qquad \gamma = 1.35$$

Airflow:

$$M_0 = 0.50 \qquad \frac{p_{t0}}{p_0} = 1.180 \qquad \frac{T_{t0}}{T_0} = 1.044$$

$$\pi_F = 1.50 \qquad a_0 = 1116 \text{ ft/s } (340 \text{ m/s}) \qquad \gamma = 1.35$$

Injected fuel flow:

$$\frac{\eta_b h_{PR}}{C_p T_0} = 485$$

Exhaust flow:

$$\frac{p_{10}}{p_0} = 1.00 \qquad \frac{T_{t10}}{T_{tp}} = 1.00 \qquad \gamma = 1.35$$

First, the exhaust plane Mach number is computed based on isentropic flow.

$$M_{10} = \left\{ \frac{2}{\gamma - 1} \left[\left(\pi_F \cdot \frac{p_{t0}}{p_0} \right)^{\frac{\gamma-1}{\gamma}} \right] - 1 \right\}^{\frac{1}{2}} \qquad M_{10} = 0.955 \quad (8\text{-}48)$$

Next, the *bypass ratio* is computed based on the work balance between the fan and turbine. This relationship reflects the important physical truth that when the freestream Mach number becomes large enough that $\pi_F \left(p_{t0}/p_0 \right) = p_{tp}/p_0$, then the turbine can no longer provide power for the fan, and the fan airflow and the bypass ratio vanish together. At this point the propulsive thrust is due entirely to the primary flow, and the device reverts to behaving like a rocket engine.

$$\alpha = \frac{\dot{m}_0}{\dot{m}_p} = \frac{T_{tp}}{T_0} \cdot \frac{T_0}{T_{t0}} \left\{ \frac{1 - \left(\pi_F \cdot \frac{p_{t0}}{p_0} \cdot \frac{p_0}{p_{tp}} \right)^{\frac{\gamma-1}{\gamma}}}{\pi_F^{\frac{\gamma-1}{\gamma}} - 1} \right\} \qquad \alpha = 36.8$$

$$(8\text{-}49)$$

Next, the mixed total temperature is computed based on conservation of energy and mass of the primary flow plus airflow. This quantity is useful for determining the thrust in the absence of downstream fuel injection or afterburning, where $T_{t10} = T_{t4}$, or for determining the flow rate of injected fuel necessary to achieve a specified exhaust total temperature.

$$\frac{T_{t4}}{T_0} = \frac{\dfrac{T_{t0}}{T_0} \cdot \pi_F^{\frac{\gamma-1}{\gamma}} \left\{ \alpha + \dfrac{T_{tp}}{T_0} \left(\pi_F \cdot \dfrac{p_0}{p_{tp}} \right)^{\frac{\gamma-1}{\gamma}} \right\}}{\alpha + 1} \qquad \frac{T_{t4}}{T_0} = 1.28 \quad (8\text{-}50)$$

The latter case is of more interest here, and leads to the computation of the fuel/air ratio in accordance with

$$f = \frac{\dot{m}_f}{\dot{m}_0} = \frac{\dfrac{T_{t10}}{T_{tp}} \cdot \dfrac{T_{tp}}{T_0} - \dfrac{T_{t4}}{T_0}}{\dfrac{\eta_b h_{PR}}{C_p T_0}} \qquad f = 0.0180 \quad (8\text{-}51)$$

Finally, the *thrust augmentation ratio* is computed from

$$\phi_p = \frac{F}{\dot{m}_p V_{p0}} = \{1 + \alpha(1 + f)\} \frac{V_{10}}{V_{p0}} - \alpha \frac{V_0}{V_{p0}} \qquad (8\text{-}52)$$

where

$$M_{p0} = \left\{ \frac{2}{\gamma - 1} \left[\left(\frac{p_{tp}}{p_0} \right)^{\frac{\gamma-1}{\gamma}} - 1 \right] \right\}^{\frac{1}{2}} \qquad M_{p0} = 2.41 \quad (8\text{-}53)$$

$$\frac{V_0}{V_{p0}} = \frac{M_0}{M_{p0}} \left\{ \frac{T_{t0}}{T_0} \cdot \frac{T_0}{T_{tp}} \left(\frac{1 + \dfrac{\gamma - 1}{2} M_{p0}^2}{1 + \dfrac{\gamma - 1}{2} M_0^2} \right) \right\}^{\frac{1}{2}} \qquad \frac{V_0}{V_{p0}} = 0.0931 \quad (8\text{-}54)$$

$$\frac{V_{10}}{V_{p0}} = \frac{M_{10}}{M_{p0}} \left\{ \frac{T_{t10}}{T_{tp}} \left(\frac{1 + \dfrac{\gamma - 1}{2} M_{p0}^2}{1 + \dfrac{\gamma - 1}{2} M_{10}^2} \right) \right\}^{\frac{1}{2}} \qquad \frac{V_{10}}{V_{p0}} = 0.522 \quad (8\text{-}55)$$

whence

$$\phi_p = 16.7$$

Since the turbo ramjet is a total propulsion system, it is also worthwhile to compute the traditional performance measures, including

$$\frac{F}{\dot{m}_0} = a_0 M_0 \left\{ \left(1 + f + \frac{1}{\alpha} \right) \frac{V_{10}}{V_{p0}} \cdot \frac{V_{p0}}{V_0} - 1 \right\} \qquad (8\text{-}56)$$

$$\frac{F}{\dot{m}_0} = 84.4 \text{ lbf·s/lbm} \quad (827 \text{ N·s/kg})$$

$$S = \frac{\dot{m}_f + \dot{m}_p}{F} = \left(f + \frac{1}{\alpha}\right) \cdot \frac{\dot{m}_0}{F} \tag{8-57}$$

$$S = 1.95 \text{ lbm}/(\text{lbf·h}) \quad (55.2 \text{ g}/(\text{kN·s})$$

$$I_{sp} = \frac{F}{g_0\,(\dot{m}_f + \dot{m}_p)} = \frac{1}{g_0\left(f + \dfrac{1}{\alpha}\right)} \cdot \frac{F}{\dot{m}_0} \qquad I_{sp} = 1850 \text{ s} \tag{8-58}$$

8.6.1.2 Ideal turbo ramjet performance. The results of computations based on the ideal turbo ramjet are presented for the typical case as a function of freestream Mach number in Fig. 8.14. They concur with our general expectations, and reveal that this is a promising combined cycle engine configuration.

The static and low freestream Mach number levels of $F/\dot{m}_0$ and I_{sp} are comparable to those of nonafterburning turbojet engines, which places them somewhat below those of afterburning turbojet engines and far above those of ramjets or high bypass ratio turbofan engines. In short, the turbo ramjet should do well during takeoff conditions.

In the transonic range, the increased ram effect enhances the thermal cycle efficiency, compensating for the falling α and causing $F/\dot{m}_0$, I_{sp}, and ϕ_p to remain more or less constant throughout. This matches the requirements of the vehicle, which needs high thrust from takeoff through transonic flight speeds.

Finally, as $\pi_F(p_{t0}/p_0)$ approaches p_{tp}/p_0, the bypass ratio diminishes rapidly and the turbo ramjet changes character to that of a rocket engine. In fact, when $M_0 = 2.16$ and the bypass ratio is 0, the I_{sp} is *identical* to that of the primary flow acting alone. This event can be postponed to higher freestream Mach numbers by a number of factors. To begin with, since p_0 decreases with M_0 along a constant q_0 trajectory, it follows that p_{tp}/p_0 must increase with M_0. Furthermore, the freestream compression process is less likely to be isentropic at $M_0 > 1.0$, thus reducing $\pi_F(p_{tp}/p_0)$. Finally, the design value of p_{tp}/p_0 could be made larger. Any or all of these effects can be included in the ideal turbo ramjet analysis.

It is especially instructive to compare the performance of the turbo ramjet with that of the ejector ramjet (or ERJ when fashioned into a complete engine), as summarized for identical design parameters in Figs. 8.10 and 8.14. Since ideal turbo ramjet replaces the inefficient viscous energy transfer mechanism of the ideal ejector ramjet with the perfectly efficient energy transfer mechanism of rotating machinery, one would expect it to have superior performance, and would not be disappointed. In particular, both ϕ_p and α are approximately an order of magnitude larger for the ideal turbo ramjet.

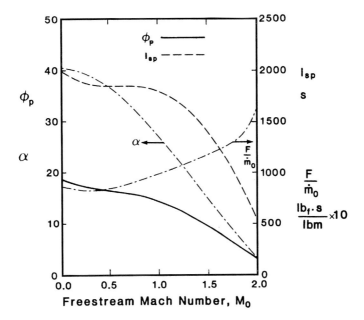

Fig. 8.14 Ideal turbo ramjet performance for the typical case as a function of freestream Mach number.

Moreover, for the ejector ramjet at $M_0 = 0.5$,

$$(I_{sp})_p = \frac{F}{g_0 \dot{m}_p} = \frac{a_0 M_0}{g_0} \cdot \phi_p \cdot \frac{V_{p0}}{V_0} = 312 \text{ s}$$

which is a small fraction of the corresponding value for the turbo ramjet, not even accounting for the mass flow rate of *fuel* consumed in the burner of the ejector ramjet.

Thus, we are presented with the classical contest between a simple machine of modest performance and a complex machine of high performance. Of such things are systems studies made and designers' salaries earned.

8.6.1.3 The turbo ramjet rocket. An interesting variation on the present theme is the *turbo ramjet rocket,* depicted in Fig. 8.15. The primary reason for adding the internal rocket engine is to further supplement the thrust available at any forward speed, particularly at the lower and higher Mach numbers for which the ramjet and scramjet may not be adequate. As Fig. 8.15 indicates, the extra rocket integrates nicely into the overall configuration, and the existing exhaust nozzle helps to provide the very large area ratio required for proper expansion at the highest Mach numbers and altitudes.

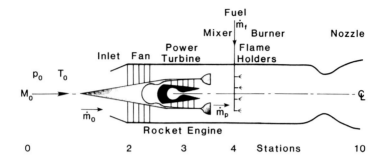

Fig. 8.15 Schematic diagram of the turbo ramjet rocket.

Another member of the turbo ramjet rocket family is the *supercharged ejector ramjet,* or SERJ, depicted in Fig. 8.16. In this device the fan power is extracted from a separate gas generator exhaust by means of a turbine located at the tips of the fan blades.

The higher thrust of the turbo ramjet rockets would, of course, be paid for by a lower specific impulse and more mechanical complexity, a tradeoff that in some circumstances could be favorable. Any analysis of these airbreathing engines would have to include the ejector action of the rocket exit flow on the adjacent flow.

8.6.2 The Liquid Air Cycle Engine

An entirely separate class of hypersonic airbreathing engines is made possible by the availability of very low temperature or *cryogenic* liquid hydrogen fuel. Even though liquid hydrogen is generally thought to be destined for hypersonic flight because of its high specific energy release (i.e., heat of combustion per unit mass) and vehicle cooling capacity, its low boiling point may well become the leading attraction.

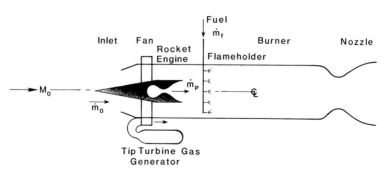

Fig. 8.16 Schematic diagram of the supercharged ejector ramjet (SERJ).

The basic principle of the *liquid air cycle engine,* or LACE, is illustrated in Fig. 8.17. The cooling capacity of the cryogenic hydrogen is used to produce liquid air (LAIR) from the atmosphere so that it can be mechanically compressed easily and injected together with the now gaseous hydrogen into a rocket engine, where they chemically react to provide thrust. This is a direct way of obtaining oxygen for combustion from the surrounding atmosphere rather than carrying it onboard. The process relies on the fact that the temperature of liquid hydrogen [36.7 °R (20.4 K) at 1 atm] is considerably less than that of liquid air [142 °R (78.9 K) at 1 atm]. Since only the fuel must be transported, and since the air contains nitrogen that adds to the exhaust mass flow rate, the performance of the liquid air cycle engine will in general be superior to that of a pure hydrogen-oxygen rocket engine.

Your ability to visualize the operation of the liquid air cycle engine may be improved by recognizing that, even under static conditions, the rapid condensation of the air on the surfaces of the heat exchangers lowers the local static pressure so that the surrounding air is literally sucked inside.

8.6.2.1 The heat exchange process. The fundamental determinant of liquid air cycle engine performance is the ratio of the mass of air liquefied per unit mass of hydrogen expended, a quantity referred to as the *condensation ratio* and denoted by the symbol CR. The condensation ratio is, in fact, the inverse of the fuel/air ratio f, and it would indeed be a pleasing outcome if the result of the heat exchange process were that the latter is at or near the stoichiometric value $f_{st} = 0.0291$. Our reasoning process is therefore simplified by employing the *equivalence ratio*

$$\phi = \frac{f}{f_{st}} = \frac{1}{f_{st}CR} = \frac{34.4}{CR} \tag{8-59}$$

as the principal indicator.

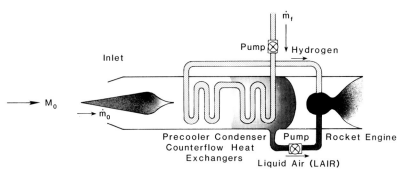

Fig. 8.17 **Schematic diagram of the liquid air cycle engine (LACE).**

Under standard sea level static conditions, the ratio of the enthalpy that can be *absorbed* by hydrogen going from the liquid state to atmospheric temperature to the enthalpy that must be *removed* from the air to bring it to the liquid state at 1 atm is very nearly 10. Thus, the largest imaginable condensation ratio is also 10, and the corresponding equivalence ratio is 3.44, which means that the rocket engine reactant mixture will at best be very rich in fuel.

Unfortunately, it is not possible to approach this upper limit in practice because of the underlying nature of the heat exchange process, which will be explained with the help of Fig. 8.18.[8.8,8.11] This diagram contains two typical *temperature-enthalpy trajectories* for equilibrium air being cooled toward the liquid state at different static pressures, and four typical temperature-enthalpy trajectories for equilibrium hydrogen being heated away from the liquid state at constant static pressure and different values of ϕ. Since this is a

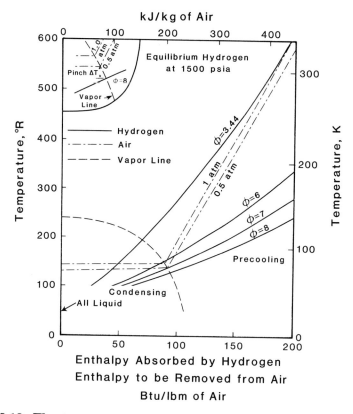

Fig. 8.18 The temperature-enthalpy diagram for a typical liquid air cycle engine heat counterflow exchanger.

counterflow heat exchange process (i.e., the lowest temperature air is being cooled by the lowest temperature hydrogen), the enthalpy scale of this diagram may be interpreted either as the enthalpy already absorbed by the hydrogen or the enthalpy yet to be removed from the air at the *same physical location* in the heat exchanger. In other words, the *local* hydrogen and air temperatures are found by looking up their values at the *same* position on the enthalpy scale.

The heat exchange process can only proceed in the proper direction if the *local* air temperature is everywhere *greater* than the local hydrogen temperature, which means that the hydrogen trajectory must be everywhere *below* the air trajectory. Using the $CR = 10$, $\phi = 3.44$ example, Fig. 8.18 shows that this rule is definitely violated, even though the total amount of cooling would be enough to completely liquefy the air. The diagram also reveals that obeying this rule will require that ϕ exceed 6.0, and be perhaps as large as 7.0 or 8.0. The critical point in the trajectories is found at the enthalpy for which the smallest temperature difference exists between the hydrogen and the air. This minimum temperature "clearance" is called the *pinch* ΔT, as illustrated by the sketch in the upper left-hand corner of Fig. 8.18. In order to maintain reasonably sized heat exchangers, the pinch ΔT must be about 10–30 °R (5–15 K), which suggests in turn that ϕ and CR must be approximately 8.0 and 4.3, respectively. The required equivalence ratio would be even larger if the static pressure were lowered due, for example, to pressure losses in the inlet and heat exchanger passages. Please note that once the pinch point has been passed, there is an abundance of cooling capacity to complete the process.

This situation can be improved by several techniques aimed at increasing the pinch ΔT. First, an expander turbine could be used to reduce the hydrogen temperature upstream of the pinch location. Second, some hydrogen could be recycled to the fuel tank so that the *apparent* ϕ at the pinch location would be greater than the final value. This method would, however, deplete the heat sink capacity remaining in the fuel tank for the rest of the journey. Third, the hydrogen could be subcooled to the triple point and beyond to the partially solidified or *slush* state. Depending upon the percent of slush, this can gain an additional 25–60 BTU/lbm (58–140 kJ/kg) of hydrogen, but the designer must contend with a two-phase fuel in the tanks. Fourth, a catalyst can be used to obtain endothermic (i.e., heat-absorbing) conversion of the hydrogen from its tanked parahydrogen state (i.e., opposing proton spins and lower internal energy) to the equilibrium balance of parahydrogen and orthohydrogen (i.e., parallel proton spins and higher internal energy). The endothermic reaction equates to additional cooling capacity for the hydrogen fuel. Each of these methods inevitably adds weight and complexity, so they must pay for themselves in performance in order to be adopted.

Finally, it should be noted that, as indicated in Figs. 8.17 and 8.18, the heat exchanger is divided into a *precooler* for the gaseous state and a *condenser* for the saturated state in order to satisfy their differing needs. A special requirement for the precooler is the prevention of ice formed from the freezing of the water contained in the air at low altitudes. This can be done by continuously spraying antifreeze or humectant compounds such as ethanol, glycerol, methanol, ethylene glycol, or propylene glycol into the precooler airflow and providing a collecting and removal system for the mixture of liquids. A special requirement for the condenser is to collect and deliver the liquid air to the pressurizing pumps under all atmospheric and flight maneuvering conditions.

8.6.2.2 Ideal liquid air cycle engine performance. The performance of the liquid air cycle engine will be based on the ideal exhaust velocity of the rocket engine alone. This analysis specifically includes the impact of the most important variable, the condensation ratio (or its counterpart, the equivalence ratio). The performance obtained is equivalent to sea level static behavior. Although it would be an easy matter to deduct the inherited momentum of the freestream airflow from the thrust, the effect of flight on the condensation ratio is difficult to determine because it alters both the stagnation pressure and enthalpy of the captured airflow. For this reason, a wide range of condensation ratios was included in the computations. The result, as you shall see, is a reasonable basis of fair comparison with other airbreathing propulsion concepts.

Assuming typical stagnation enthalpies and a typical stagnation pressure for the fuel and air delivered to the rocket engine combustion chamber, HAP(Equilibrium) was used to compute the state of the combustion products and then isentropically expand them to their equilibrium condition at atmospheric pressure. The enthalpy difference between the combustion chamber and exhaust conditions was used to find the ideal exhaust velocity according to

$$V_{10} = \sqrt{2\left(h_{t10} - h_{10}\right)} \tag{8-60}$$

The remaining performance measures are defined in the usual way, and are written as

$$I_{sp} = \frac{F}{g_0 \dot{m}_f} = \frac{(\dot{m}_0 + \dot{m}_f)\,V_{10}}{g_0 \dot{m}_f} = (CR+1)\frac{V_{10}}{g_0} \tag{8-61}$$

and

$$\frac{F}{\dot{m}_0} = \frac{(\dot{m}_0 + \dot{m}_f)V_{10}}{\dot{m}_0} = \left(1 + \frac{1}{CR}\right)V_{10} \tag{8-62}$$

where it should be emphasized that the airflow rate is *free* as far as fuel consumption is concerned.

The results of this analysis are shown in Fig. 8.19 for typical liquid air cycle engine parameters. As one might expect, the effect of operating at low condensation ratios (or very fuel-rich) is to dramatically reduce the I_{sp} because the available chemical energy of the excess hydrogen is wasted. When ϕ is near its probable value of 8.0 and CR is about 4.3, the I_{sp} is slightly greater than 1000 s, which is much better than the ejector ramjet but about half of that provided by the turbo ramjet. This outcome underscores the depressing effect of low condensation ratios on performance, and emphasizes the benefits that can be obtained if condensation ratios are increased.

It is enlightening to observe the peculiar influences of the very low molecular weight hydrogen on the other quantities of Fig. 8.19. The first influence is that the very high specific heat of the hydrogen greatly reduces the combustion chamber temperature T_{t10} as ϕ increases or CR decreases. At their probable values, T_{t10} will be less than 1800 °R (1000 K). The second influence is that the very low molecular weight compensates for the reduced T_{t10} to produce

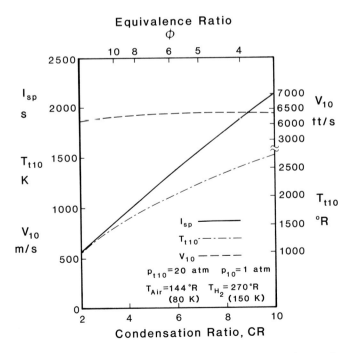

Fig. 8.19 Ideal liquid air cycle engine sea level static performance as a function of condensation ratio or equivalence ratio.

an almost constant exhaust velocity of about 6300 ft/s (1920 m/s). The latter guarantees that $F/\dot{m}_0$, as given by Eq. (8-62), will also remain almost constant until the contribution of the fuel mass flow rate as represented by the term $1/CR$ becomes important. For the range of CR shown in Fig. 8.19, the range of $F/\dot{m}_0$ is only 218–286 lbf · s/lbm (2140–2810 N · s/kg), a quantity much greater than that of the turbo ramjet.

The time has come to invent the *liquid air cycle ejector ramjet engine,* or *LACERJ* for short. This device would simply use the exhaust of the LACE as the primary flow for an ERJ, which would augment the original thrust both by spreading the energy among more mass flow and by burning some of the excess hydrogen fuel with the oxygen of the ejector secondary flow. One good combined cycle engine concept evidently leads to another.

8.6.3 The Inverse Cycle Engine

As a final testimony to the fertile imaginations of engine cycle designers, witness the *inverse cycle engine,* or ICE, a concept that also depends on the heat sink capacity of the cryogenic hydrogen fuel.[8.8] Whereas the total pressure ratio across the ideal turbojet engine created by the compressor-burner-turbine *gas generator* set is[8.2]

$$\left(\frac{p_{t10}}{p_{t0}}\right)_{TJ} = \pi_c \left\{1 - \frac{T_{t0}}{T_{t4}}\left(\pi_c^{\frac{\gamma-1}{\gamma}} - 1\right)\right\}^{\frac{\gamma}{\gamma-1}} \quad (8\text{-}63)$$

where π_c is the compressor total pressure ratio p_{t3}/p_{t0}, the total pressure ratio created across the ideal ICE of Fig. 8.20 by the turbine-

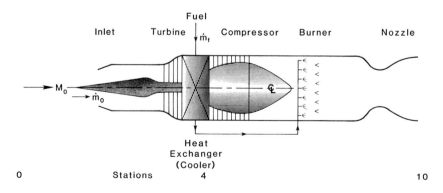

Fig. 8.20 Schematic diagram of the inverse cycle engine (ICE).

heat exchanger (cooler)-compressor set is

$$\left(\frac{p_{t10}}{p_{t0}}\right)_{\text{ICE}} = \pi_c \left\{ 1 - \frac{T_{t4}}{T_{t0}} \left(\pi_c^{\frac{\gamma-1}{\gamma}} - 1 \right) \right\}^{\frac{\gamma}{\gamma-1}} \tag{8-64}$$

The total pressure ratio serves, as ever, to increase the ideal exhaust velocity in accordance with Eq. (2-81). Since T_{t4}/T_{t0} for a turbojet is about 3000 °R/500 °R = 6.0 and T_{t0}/T_{t4} for the inverse cycle engine is about 500 °R/125 °R = 4.0 (the air must remain gaseous), then, for $\pi_c = 10.0$ and $\gamma = 1.40$, the respective total pressure ratios are 5.54 and 3.96. These are very much greater, of course, than those of an ejector ramjet.

As advertised, one of the benefits of this type of cycle analysis is that optimal solutions may be found analytically. In the case of the inverse cycle engine, it can be shown that,[8.2] all other things being equal, the maximum p_{t10}/p_{t0} is produced when

$$(\pi_c)_{\text{ICE}} = \left\{ \frac{1 + \dfrac{T_{t0}}{T_{t4}}}{2} \right\}^{\frac{\gamma}{\gamma-1}} \tag{8-65}$$

For $T_{t0}/T_{t4} = 4.0$ and $\gamma = 1.40$, the *best* $\pi_c = 24.7$, which is probably more compression than one would prefer to carry around, but well within the state-of-the-art. The p_{t10}/p_{t0} for this optimal compression ratio obtained from Eq. (8-64) is 4.77.

The basic manifestation of the inverse cycle engine concept is shown schematically in Fig. 8.20, although other variations with potentially superior performance have been proposed.[8.8] The main advantage of this configuration is that the most severe environmental conditions are found at the turbine entry, downstream of which they become progressively less demanding because energy is being removed from the flow to drive the compressor. This is opposite to the normal turbojet case, in which the environmental conditions within the compressor become progressively more demanding as energy is added to the flow. Consequently, turbojet design limitations are often set by the properties of the compressor exit flow. Thus, the ICE should be able to operate at higher freestream Mach number than the ordinary turbojet engine. The energy removed in the heat exchanger is not lost, but is returned with the hydrogen fuel injected into the burner.

The inverse cycle engine also has some drawbacks. For one thing, the initial turbine expansion increases the throughflow area, thus increasing the frontal area, volume, and difficulty of integration. For another, the rotating machinery makes at least the core of the engine axisymmetric, also increasing the difficulty of integration. Finally, the water present in the air at low altitudes must be safely removed somewhere in the turbine expansion and cooling process. Once again, cost/benefit tradeoff studies will be required.

Before closing, it should be noted that the ICE concept gave birth to the idea of the *precooled turbojet,* in which a heat exchanger (cooler) is placed ahead of the compressor in order to reduce the power required for a given compression ratio, the throughflow area required for a given air mass flow rate, and the severity of the environment (i.e., stagnation temperature) as seen by the compressor. This can extend the freestream Mach number range over which the precooled turbojet can operate and reduce the size and weight of the rotating turbomachinery, but requires a heat exchanger that can tolerate harsh environments. There is, in fact, a long history of precooling turbojet engines with water injection and even piston engines for similar reasons.

8.7 SUMMARY

Anyone entering the hypersonic airbreathing propulsion community will find that a surprisingly large fraction of its human energy is consumed on the systems topics of this chapter. This is largely due to the fact that the solutions to all the individual technical problems must fit gracefully into a single propulsion system that integrates harmoniously into the vehicle. Moreover, that propulsion system must have satisfactory performance over the entire flight spectrum, including the low and high speed regimes.

The overpowering need for good component performance has also been amply demonstrated, as well as the need to capitalize on almost any variety of propulsion in order to provide thrust when necessary. These special needs will continue to be at the center of gravity of the hypersonic vehicle community, so stay tuned for further creativity and progress.

REFERENCES

[8.1] Curran, E. T., Leingang, J. L., Carreiro, L. R., and Petters, D. P., "Further Studies of Kinetic Energy Methods in High Speed Ramjet Cycle Analysis," AIAA Paper 92-3805, July 1992.

[8.2] Oates, G. C., *The Aerothermodynamics of Gas Turbine and Rocket Propulsion,* Revised and Enlarged Edition, AIAA Education Series, Washington, DC, 1988.

[8.3] Heiser, W. H., "Thrust Augmentation," Transactions of the ASME, *Journal of Engineering for Power,* Vol. 89, Series A, No. 1, Jan. 1967.

8.4 Dutton, J. C., and Carroll, B. F., "Limitation of Ejector Performance Due to Exit Choking," *ASME Journal of Fluids Engineering*, Vol. 110, No. 1, Mar. 1988.

8.5 Murthy, S. N. B., (ed.), *Aerodynamics of Base Combustion*, AIAA Progress in Astronautics and Aeronautics Series, Vol. 40, New York, 1976.

8.6 Trefny, C., "On the Use of External Burning to Reduce Aerospace Vehicle Transonic Drag," AIAA Paper 90-1935, July 1990.

8.7 Billig, F. S., "External Burning in Supersonic Streams," Johns Hopkins University Applied Physics Laboratory Report TG-912, May 1967.

8.8 Murthy, S. N. B., and Curran, E. T., (eds.), *High-Speed Flight Propulsion Systems*, AIAA Progress in Astronautics and Aeronautics Series, Vol. 137, Washington, DC, 1991.

8.9 Kors, D. L., "Combined Cycle Propulsion for Hypersonic Flight," Thirty-Eighth I.A.F. Congress, Paper IAF-87-263, Brighton, UK, Oct. 1987.

8.10 Mattingly, J. D., Heiser, W. H., and Daley, D. H., *Aircraft Engine Design*, AIAA Education Series, New York, 1987.

8.11 Brewer, G. D., *Hydrogen Aircraft Technology*, CRC Press, Boca Raton, FL, 1991.

PROBLEMS

8.1 Continue the exploration of the baseline hydrogen-fueled real scramjet performance analysis of Sec. 8.2 using HAP(Performance) and HAP(Trajectory). Vary each of the individual component loss parameters and the cycle static temperature ratio in order to determine the sensitivity of performance to each over the usual freestream Mach number range. Also, compare constant pressure combustion to constant area.

What conclusions do you draw from this study? In particular, which loss parameters would you concentrate most on reducing, and what cycle static temperature ratio would you prefer?

8.2 Do you think that real scramjet performance is sensitive to the dynamic pressure q_0 of the trajectory? Find out by repeating the hydrogen-fueled scramjet performance analysis of Sec. 8.2 for $q_0 = 2000$ lbf/ft^2 (95.8 kN/m^2) using HAP(Performance) and HAP(Trajectory).

Explain these results in terms of the governing equations.

8.3   Even though Eq. (8-11) shows that airbreathing engine specific thrust steadily increases with fuel/air ratio, this is not true for specific impulse. In order to visualize this, compute and plot I_{sp} from Eq. (8-12) as a function of M_0 for a several different values of f. Note that there is an envelope that encloses the entire family of curves. You can win great respect by proving that the optimum f at a given M_0 is given by

$$f_{opt} = \frac{2\sqrt{1 - \eta_{KEO}}}{(1 - \sqrt{1 - \eta_{KEO}})\left\{\dfrac{\eta_b h_{PR}}{C_p T_0 \left(1 + \frac{\gamma - 1}{2} M_0^2\right)}\right\} - (1 + \sqrt{1 - \eta_{KEO}})}$$

provided that $f_{opt} < f_{st}$ and therefore that the envelope is described by the expression

$$(I_{sp})_{env} = \frac{a_0 M_0}{y_0}\left\{\frac{(1 - \sqrt{1 - \eta_{KEO}})}{2}\left[\frac{\eta_b h_{PR}}{C_p T_0 \left(1 + \frac{\gamma - 1}{2} M_0^2\right)} - 1\right] + 1\right\}$$

Superimpose this expression on your computed results in order to prove that it works. What use can designers make of this result?

8.4   The airbreathing engine may be turned off at the rocket *takeover Mach number,* namely the freestream Mach number at which the specific impulse equals that of a rocket engine using the same fuel. Use Eqs. (8-11) and (8-12) to prepare a plot for hydrogen fuel similar to that of Fig. 8.5, but for the takeover Mach number at which the airbreathing $I_{sp} = 460$ s.
Are these results significantly different from those of Fig. 8.5? Why (or why not)?

8.5   Use Eq. (8.11) to examine the influence of η_b on specific thrust, analytically and/or computationally. Do these results mirror the sensitivity of $F/\dot{m}_0$ to η_b that we have seen before?

8.6   Repeat the example of Fig. 8.8 for $5 < M_0 < 25$ in order to determine the variation of $C_{D_{add}}$ and $C_{D_{ext}}$ with freestream Mach number. What generalizations can you make about these results?

8.7   Repeat Problem 8.6 for the two oblique shock-wave compression system of the lower portion of Fig. 5.12. What generalizations can you make about these results?

8.8 Combine the results of Eqs. (8-11) and (8-23) and Problem 8.6 to examine and estimate the behavior of $F/\dot{m}_0 V_0$ and $T/\dot{m}_0 V_0$ with freestream Mach number. Assume that $\eta_{KEO} = 0.85$ and that $A_{x\,ext}/A_0 = 0.20$ and $A_{x\,add}/A_0 = 0$ for all M_0. What generalizations can you make about these results?

8.9 Write a computer program for *ideal ejector ramjet analysis.* How does static (i.e., $M_0 = 0$) performance vary with p_{tp}/p_0, T_{tp}/T_0, A/A_p^*, and T_{t10}/T_{tp}?
Select the best (i.e., highest ϕ_p) combination you can find and compare its performance for $0 < M_0 < 2.0$ with that of Fig. 8.10. Based on this experience, what advice would you give to ejector ramjet designers?
How sensitive is ejector ramjet performance to γ?

8.10 Modify the ideal ejector ramjet computer program of Problem 8.9 to include the most obvious losses. First, add correction factors to p_{tp}/p_0, p_{ts}/p_0, and p_{te}/p_0 to reflect their individual irreversibilities. These factors should be less than, but close to, 1. Then modify the conservation of momentum equation [i.e., Eq. (8-34)] to include the effects of wall friction of the constant area mixer.
How do these effects influence the performance of the ejector ramjet of Fig. 8.10? Which losses have the greatest influence on performance? Why?

8.11 Since the performance of the ejector ramjet approaches that of an ordinary ramjet as the freestream Mach number increases, it should be enlightening to examine the behavior of a thrust augmentation measure based on *ideal ramjet thrust* (rather than ideal primary thrust), or

$$\phi_s = \frac{\dot{m}_e V_{10} - \dot{m}_s V_0}{\dot{m}_s V_{RJ} - \dot{m}_s V_0} = \frac{\dfrac{1+\alpha}{\alpha} \cdot \dfrac{V_{10}}{V_0} - 1}{\dfrac{V_{RJ}}{V_0} - 1} = \frac{\dfrac{1+\alpha}{\alpha} \cdot \dfrac{V_{10}}{V_0} - 1}{\sqrt{\dfrac{T_{t10}}{T_{tp}} \cdot \dfrac{T_{tp}}{T_0} \cdot \dfrac{T_0}{T_{t0}}} - 1}$$

Use the computer program of Problem 8.9 to reproduce Fig. 8.10 for ϕ_s. At what M_0 would you turn off the primary flow? Why?

8.12 Prove by means of their fundamental definitions and/or the interrelationships of Table 3.2 that the other standard measures of performance for *external burning* may be written:

$$\frac{F}{\left(1 - \frac{p_b}{p_0}\right)\dot{m}_0} = \frac{\gamma_0 - 1}{\gamma_0} \cdot \frac{\eta_b f h_{PR}}{V_0}$$

$$\left(1 - \frac{p_b}{p_0}\right) S = \frac{\gamma_0}{\gamma_0 - 1} \cdot \frac{V_0}{\eta_b h_{PR}}$$

$$\frac{\eta_0}{\left(1 - \frac{p_b}{p_0}\right)} = \frac{\gamma_0 - 1}{\gamma_0} \cdot \eta_b$$

Evaluate these measures of performance for typical values of the physical quantities involved. How does external burning stack up against other types of airbreathing propulsion?

8.13 Write a computer program for *ideal turbo ramjet* analysis. How does static (i.e., $M_0 = 0$) performance vary with p_{tp}/p_0, T_{tp}/T_0, π_F, and T_{t10}/T_{tp}?

Select the best (i.e., highest ϕ_p) combination you can find and compare its performance for $0 < M_0 < 2.0$ with that of Fig. 8.14. Based on this experience, what advice would you give to turbo ramjet designers?

How sensitive is turbo ramjet performance to γ?

8.14 A designer would like to determine whether turbo ramjets can work effectively up to a freestream Mach number of 5.0. Use the computer program of Problem 8.13 to examine the influence of promising variables on performance.

(a) What is the effect of increasing p_{tp}/p_0?

(b) What is the effect of the diminishing p_0 that accompanies increasing M_0 at constant q_0 and γ?

(c) What is the effect of the total pressure loss on the airflow for supersonic M_0? Use the following military specification MIL-E-5008E correlation for the expected inviscid performance of high speed aircraft compression components:

$$\frac{p_{t2}}{p_{t0}} = 1 - 0.075\,(M_0 - 1)^{1.35} \qquad 1 \le M_0 \le 5$$

(d) What would happen if you scheduled π_F to decrease with increasing M_0? What schedule would you recommend?

(e) Do you have any other ideas? How well do they work? What advice would you give the designer?

8.15 Determine the static performance of an *ideal liquid air cycle engine* using HAP(Equilibrium). Assume that a condensation ratio of 5.0 can be achieved in practice, and use the same quantities as Fig. 8.19 except:

(a) $p_{t10} = 15.0$ atm

(b) $p_{t10} = 30.0$ atm

How sensitive is performance to p_{t10}?

8.16 Combine the results of Problems 8.9 and 8.15 to evaluate the static performance of an *ideal liquid air cycle ejector ramjet engine*. Install the LACE of Problem 8.15 in the ideal ejector ramjet engine example of Sec. 8.5, being careful to recognize that T_{t10}/T_{tp} is not given, but must instead be calculated on the basis of the combustible mixture that exists after the ejector mixing is complete.

How do the LACERJ and the ERJ compare in static performance? Remember that the ERJ must carry both fuel *and* oxidizer onboard.

8.17 The *inverse cycle engine* may be regarded as an afterburning turbojet with an inverse compressor-burner-turbine arrangement. Make the usual ideal cycle analysis assumptions, namely isentropic flow in the inlet, rotating machinery and nozzle, no total pressure losses in the heat exchanger or burner, f negligible compared to 1, and the calorically perfect gas constants fixed.

(a) Show that the performance of the *ideal* ICE is given by the set of expressions:

$$\frac{F}{\dot{m}_0} = V_{10} - V_0 = a_0 M_0 \left\{ \left[\frac{2}{(\gamma - 1)M_0^2} \cdot \frac{\frac{T_{t10}}{T_{t0}}}{\tau_c \tau_t} \left(\tau_c \tau_t \frac{T_{t0}}{T_0} - 1 \right) \right]^{\frac{1}{2}} - 1 \right\}$$

where

$$\tau_c = \pi_c^{\frac{\gamma-1}{\gamma}}, \quad \tau_t = \pi_t^{\frac{\gamma-1}{\gamma}} \qquad \text{(Isentropic machinery)}$$

$$\tau_t = 1 - \frac{T_{t4}}{T_{t0}}(\tau_c - 1)$$ (Machine work balance)

$$f = \frac{C_p T_{t0}}{h_{PR}}\left(\frac{T_{t10}}{T_{t0}} - 1\right)$$ (Burner energy balance)

$$S = \frac{f}{F/\dot{m}_0}$$

$$I_{sp} = \frac{1}{g_0 S}$$

(b) Earn special praise by using the set of equations above to derive Eq. (8-65).

(c) Noting that the ideal ICE is identical to the ideal ramjet when there is no mechanical compression, or $\tau_c = 1.0$, show that the performance of the ideal RJ is given by the *same* set of expressions as the ideal ICE except that:

$$\frac{F}{\dot{m}_0} = a_0 M_0 \left\{\sqrt{\frac{T_{t10}}{T_{t0}}} - 1\right\}$$

(d) Compare the performance of the ideal ICE and ideal RJ over the freestream Mach number range $0 < M_0 < 5$ for

$$\pi_c = 10 \qquad \frac{T_{t10}}{T_{t0}} = 5.0 \qquad \gamma = 1.40$$

You may find Ref. 8.2 very helpful for this exercise.

SPECIAL HYPERSONIC
AIRBREATHING PROPULSION TOPICS

9.1 INTRODUCTION

There is an array of topics surrounding the performance issues of hypersonic airbreathing propulsion about which serious questions always seem to arise. The questions are serious because they deal with the success or failure, or possibly even the survival, of the aerospace system. The purpose of this chapter is to answer, or at least provide you with the tools to begin to answer, the most important of these questions. Since each of these topics has occupied the careers of a great many expert investigators, we will only be able to scratch the surface of these intriguing fields of study.

9.2 ENGINE STRUCTURES

A complex but inevitable connection exists between the durability (or life) and the weight and cost of the engine. Increasing the former generally requires increasing the latter. As we learned in Chap. 3, the structural or empty mass fraction *must* be reduced in order to achieve a reasonable initial mass ratio for hypersonic aerospace systems. The primary goal of the structural designer is, therefore, to achieve the *required* life with *minimum* weight and cost.

In order to perform this delicate balancing act, the designer must have an accurate knowledge both of the forces that act to *consume* the life of engine parts and of the ability of available materials to *withstand* deterioration. We are therefore flirting with the most impenetrable and "proprietary" domains of the engine companies, because of their heavy investment in these technologies, as well as the enormous competitive advantages that accrue to proven superiority. Even easily obtained, the enormous capabilities of their sophisticated computer programs, material properties data, and seasoned technical judgment cannot be reproduced in a textbook or the classroom.

Consequently, we will focus on some of the leading phenomena, and place strong emphasis on cause and effect. Despite this limited horizon, anyone following this development and applying some imagination will be awed by the challenges and accomplishments of the "real world."

9.2.1 Convective Heat Transfer

Although the static pressures acting on the engine surfaces cause forces and stresses that must be properly accounted for in durability analyses, the special feature of hypersonic flight is that the life of the airbreathing engine is severely threatened by the very large freestream stagnation enthalpies, otherwise known as the "thermal environment." These stagnation enthalpies can heat known materials beyond their allowable working temperature ranges for strength and/or to conditions under which destructive chemical reactions with the adjacent gas (e.g., oxidation or hydrogen embrittlement) occur too rapidly, leading to the need for wall cooling. The resulting temperature gradients within the cooled walls, in turn, generate thermal stresses that can exceed the strength of the material. We therefore begin by investigating the magnitude of the local wall heat transfer rate, by which we precisely mean the heat flux to or from the wall at the surface.

Wall heat transfer will take place because of *radiation* to or from the line-of-sight surroundings and by *conduction* to or from the nearby boundary layer gas, a process loosely referred to as *convective heat transfer* because it is strongly coupled to the overall motion of the fluid. Convective heat transfer is sometimes referred to as *aerodynamic heating.* We will neglect the radiative heat transfer because it tends to cool the exposed surfaces of the compression and the expansion systems, and because it usually has a relatively minor impact on the combustion system, although it makes good sense to check this assumption in practice. The convective heat transfer from the "hot side" of the wall is eventually absorbed by the coolant that flows on the "cold side." Since this coolant is usually fuel that is later injected into the engine throughflow, the energy involved remains available for the thermodynamic cycle.

Despite the complex and interwoven influences of freestream pressure gradient, streamline curvature and turbulence, crossflows, wall roughness, protuberances, cavities, boundary layer transition, and surface catalycity, preliminary estimates of convective heat transfer for *thermal boundary layers* with variable properties are commonly based on an equation of the form

$$\dot{q}_w = St \cdot \rho_e V_e (h_{aw} - h_w) \qquad (9\text{-}1)$$

so that the heat flux at the surface may be visualized as the product of a conductance and an enthalpy difference.[9.1,9.2,9.3] Put simply, perhaps the most important attribute of this relationship is that people use it because it works. Note that a positive value of $\dot{q}_w$ indicates that heat is being transferred to the wall, and a negative value of $\dot{q}_w$ indicates that heat is being transferred from the wall.

In Eq. (9-1) St is the *Stanton number,* the dimensionless heat transfer coefficient quantity that must be most carefully selected on the basis of data and/or experience to give valid results for the convective heat transfer situation under consideration, ρ_e is the density

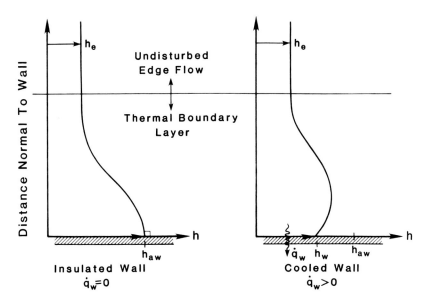

Fig. 9.1   Schematic illustration of the behavior of the static enthalpy in boundary layers.

of the gas in the adjacent inviscid flow at the outer *edge* of the boundary layer, and V_e is the velocity of the gas at the edge.

Referring to Fig. 9.1, the term h_{aw} represents the enthalpy the bounding- or zero-velocity streamline would attain if the wall were perfectly insulated, and is therefore known as the *adiabatic wall enthalpy* or the *recovery enthalpy*. Similarly, the prevailing temperature under this condition is known as the *adiabatic wall temperature* or the *recovery temperature*. If the fluid had no thermal conductivity, applying the first law to this situation would show that h_{aw} must be the same as the stagnation enthalpy of the adjacent flow, so that the net effect of viscosity would be to convert all the original freestream kinetic energy to internal or thermal energy. Since the real fluid has finite thermal conductivity, some of the thermal energy is conducted away from the higher temperature region of the wall to the cooler fluid, thus lowering h_{aw} in accordance with the relationship

$$r = \frac{h_{aw} - h_e}{\dfrac{V_e^2}{2}} \tag{9-2}$$

The term r is known as the *recovery factor,* and must be less than 1 by virtue of the foregoing reasoning.

The term h_w represents the *selected* design value of the static enthalpy at the surface, and must obviously correspond to a temperature lower than the allowable working temperature of the wall material.

Before continuing the analysis, one may use Eq. (9-1) in its raw form to reach some interesting conclusions. First, all other things being equal, Eq. (9-1) shows that $\dot{q}_w$ is directly proportional to $\rho_e V_e$, which means that $\dot{q}_w$ increases in inverse proportion to the through-flow area. In other words, we should expect the convective heat transfer to be greatest in the combustion system where the through-flow area is least. Second, since h_{aw} is the highest enthalpy the surface can reach, it is intuitively satisfying to find it in the driving enthalpy difference term of Eq. (9-1). Third, since there are many good reasons to limit the convective heat transfer, Eq. (9-1) suggests that materials with the highest possible T_w should be sought. This, however, can be somewhat illusive because even the best materials have maximum working temperatures that are far lower than the adiabatic wall temperatures encountered in hypersonic flight. For example, titanium alloys have lost most of their load-carrying capability by 1600 °R (890 K), superalloys by 1800 °R (1000 K), metal matrix composites by 4500 °R (2500 K), and coated carbon-carbon composites by 5400 °R (3000 K). Since adiabatic wall temperatures can easily reach 5,000–15,000 °R (2778–8333 K), the ability of high material working temperatures to reduce convective heat transfer is tempered by the law of diminishing returns of Eq. (9-1).

Some idea of the magnitude of the convective heat transfer rates to be expected in practice will now be obtained by applying Eq. (9-1) to the classic and well-characterized cases of zero-pressure gradient, constant wall temperature, flat plate, laminar and turbulent flows. You should bear in mind that it has been successfully applied to a great variety of much more complex flows. Note should be taken of the fact, however, that this simple type of flow is frequently found on the compression, combustion, and expansion system surfaces of ramjet and scramjet engines.

9.2.1.1 The Eckert reference enthalpy method.

9.2.1.1 The Eckert reference enthalpy method. The technique for estimating the heat transfer from high speed compressible flows most compatible with the spirit of this textbook is the *Eckert reference enthalpy method*.[9.1,9.2,9.3,9.4] It is based upon the hypothesis that the overall effects of a variable property boundary layer can be replaced by those of a constant property boundary layer whose properties correspond to a suitable *average* or *reference* enthalpy (and temperature) *that can be found a priori*. Reference quantities are designated by an asterisk. Thus, for example, Eq. (9-1) becomes

$$\dot{q}_w = St^* \cdot \rho^* V_e (h_{aw} - h_w) \tag{9-3}$$

This remarkable leap of intuition not only provides us with a valuable design tool, but, even more amazingly, has been proven analytically to be exact for several familiar cases.[9.5]

An important restriction on this method is that it cannot be applied when *dissociation* occurs. The fortunate corollary to this restriction is that the molecular weight of the gas must therefore be constant, so that density is easily eliminated from any expressions by means of the perfect gas law and the usual boundary layer assumption that the static pressure is impressed by and therefore equal to that of the inviscid external flow.

The Eckert reference enthalpy method proceeds as follows. First, the *reference enthalpy* is found from the simultaneous solution of

$$h^* = \frac{h_e + h_w}{2} + 0.22r\frac{V_e^2}{2} \tag{9-4}$$

and

$$\text{Laminar flow} \qquad r = \sqrt{Pr^*} = \sqrt{\frac{\mu^* C_p^*}{k^*}} \tag{9-5a}$$

$$\text{Turbulent flow} \qquad r = \sqrt[3]{Pr^*} = \sqrt[3]{\frac{\mu^* C_p^*}{k^*}} \tag{9-5b}$$

where Pr is the Prandtl number. This solution requires iteration because the gas properties are functions of h^* (or T^*), and vice versa. For example, when dealing with air, HAP(Air) may be used to determine h_e, h_w, and the C_p^* that corresponds to the h^* (or T^*). Moreover, μ^* may be determined from Eq. (2-4) and k^* from Eq. (2-5) and T^*. Some authors recommend that the best results for turbulent flow are obtained when the reference Prandtl number Pr^* is set at a fixed value close to but less that 1, such as 0.9. We have chosen instead to employ Eq. (9-5b) both for uniformity and because it always yields a Pr^* quite close to 1.

For the classical case of zero-pressure gradient, constant wall temperature, flat plate flows, the reference Stanton number is

$$\text{Laminar flow} \qquad St^* = \frac{0.332}{Pr^{*\frac{2}{3}} Re_x^{*\frac{1}{2}}} \tag{9-6a}$$

$$\text{Turbulent flow} \qquad St^* = \frac{0.0287}{Pr^{*\frac{2}{5}} Re_x^{*\frac{1}{5}}} \tag{9-6b}$$

so that the convective wall heat transfer is

Laminar flow
$$\dot{q}_w = \frac{0.332\rho^* V_e (h_{aw} - h_w)}{Pr^{*\frac{2}{3}} Re_x^{*\frac{1}{2}}} \tag{9-7a}$$

$$= \frac{0.332 p_e V_e (h_{aw} - h_w)}{RT^* Pr^{*\frac{2}{3}} Re_x^{*\frac{1}{2}}} \tag{9-7b}$$

Turbulent flow
$$\dot{q}_w = \frac{0.0287\rho^* V_e (h_{aw} - h_w)}{Pr^{*\frac{2}{5}} Re_x^{*\frac{1}{5}}} \tag{9-7c}$$

$$= \frac{0.0287 p_e V_e (h_{aw} - h_w)}{RT^* Pr^{*\frac{2}{5}} Re_x^{*\frac{1}{5}}} \tag{9-7d}$$

where

$$Re_x^* = \frac{\rho^* V_e x}{\mu^*} = \frac{p_e V_e x}{RT^* \mu^*} \tag{9-8}$$

An interesting and useful corollary to this analysis is that the skin friction coefficient can be obtained from

Laminar flow
$$C_f = \frac{\tau_w}{\frac{\rho_e V_e^2}{2}} = C_f^* \cdot \frac{\rho^*}{\rho_e} = \frac{0.664}{Re_x^{*\frac{1}{2}}} \cdot \frac{T_e}{T^*} \tag{9-9a}$$

Turbulent flow
$$C_f = \frac{0.0574}{Re_x^{*\frac{1}{5}}} \cdot \frac{T_e}{T^*} \tag{9-9b}$$

When dealing with *air*, the form of Eqs. (2-4) and (2-5) are such that a sufficiently accurate approximation for the reference Prandtl number is

$$Pr^* = \frac{\mu^* C_p^*}{k^*} = \frac{\mu_{SL}}{k_{SL}} \cdot C_{p_{SL}} \cdot \frac{C_p^*}{C_{p_{SL}}} = Pr_{SL} \cdot \frac{C_p^*}{C_{p_{SL}}} \tag{9-10}$$

Before moving on to example calculations, two general comments are in order. First, available experimental and analytical evidence supports the conclusion that convective heat transfer estimates based on the Eckert method are very likely to be within 10–20 percent of the correct value, which is more than adequate for our purposes. Second, as demonstrated in Ref. 9.2, the main influence of the viscous dissipation in the boundary layer is to cause the boundary layer thickness to grow more rapidly because the higher local temperatures both reduce the density and increase the viscosity of the gas. The thicker boundary layer, in turn, has less convective heat transfer and skin friction, as reflected by the reference quantities appearing in Eqs. (9-7) and (9-9).

Example 9.1

Consider the flow of *air* in the vicinity of the leading edge of the compression surface, where the external flow is essentially undisturbed, and the edge conditions are the same as the freestream conditions (i.e., the subscript *e* is replaced with the subscript *o* everywhere). Assuming that the boundary layer flow is *laminar*, the Eckert reference enthalpy method will be used to calculate the convective heat transfer and the skin friction coefficient for the following conditions:

$$M_0 = 10.0$$
$$q_0 = 1000 \text{ lbf/ft}^2 \quad (47.88 \text{ kN/m}^2)$$
$$\gamma_0 = 1.40$$
$$T_w = 2000 \text{ °R} \quad (1111 \text{ K})$$
$$x = 1 \text{ ft} \quad (0.3048 \text{ m})$$

Step 1. Equation (2-12).

$$\frac{p_0}{p_{SL}} = \frac{2q_0}{\gamma_0 M_0^2 p_{SL}} \qquad \frac{p_0}{p_{SL}} = 6.75 \times 10^{-3}$$

Step 2. Appendix B, linear interpolation, or HAP(Trajectory).

$$\frac{T_o}{T_{SL}} = 0.809$$

Step 3. Appendix B, linear interpolation, or HAP(Trajectory).

$$\frac{a_0}{a_{SL}} = 0.900$$

Step 4.

$$V_0 = \frac{a_0}{a_{SL}} \cdot a_{SL} \cdot M_0 \qquad V_0 = \begin{array}{l} 10,040 \text{ ft/s} \\ (3060 \text{ m/s}) \end{array}$$

Step 5.

$$\frac{V_0^2}{2} = \begin{array}{l} 2010 \text{ BTU/lbm} \\ (4680 \text{ kJ/kg}) \end{array}$$

Step 6.   HAP(Air).

$$h_0 = -28.3 \text{ BTU/lbm}$$
$$(-65.8 \text{ kJ/kg})$$

Step 7.   HAP(Air).

$$h_w = 380 \text{ BTU/lbm}$$
$$(884 \text{ kJ/kg})$$

Step 8.   Equations (9-4), (9-5a), and (9-10), and HAP(Air), solved by iteration.

$h^* = 600 \text{ BTU/lbm}$ $T^* = 2760 \text{ °R}$ $C_p^* = 0.299 \text{ BTU/lbm·°R}$
    (1400 kJ/kg) (1530 K) (1.25 kJ/kg · K)

Step 9.   Equations (2-4), (2-5), and (9-10).

$Pr^* = 0.743 \dfrac{C_p^*}{C_{p_{SL}}}$ $C_{p_{SL}} = 0.242 \text{ BTU/lbm·°R}$ $Pr^* = 0.917$
                              $(1.015 \text{ kJ/kg · K})$

Step 10.   Equations (9-2) and (9-5a).

$$h_{aw} = h_e + r\frac{V_e^2}{2}$$ $h_{aw} = 1900 \text{ BTU/lbm}$
                                             (4420 kJ/kg)

Step 11.   HAP(Air).

$$T_{aw} = 4910 \text{ °R}$$
$$(2730 \text{ K})$$

Step 12.   Equation (9-8).

$$Re_x^* = 2.72 \times 10^4$$

Step 13.   Equation (9-7b).

$$\dot{q}_w = 0.0219 \text{ BTU/in}^2 \cdot \text{s}$$
$$(3.58 \text{ W/cm}^2)$$

Step 14.   Equation (9-9a).

$$C_f = 0.612 \times 10^{-3}$$

Example 9.2

Consider the flow of *air* in the vicinity of the entry of the combustion system, where the edge conditions are essentially those of the usual engine Station 3 (i.e., the subscript e is replaced with the subscript 3 everywhere). Assuming that the boundary layer flow is *turbulent*, the Eckert reference enthalpy method will be used to calculate the convective heat transfer and the skin friction coefficient for the same freestream conditions as Example 9.1, and:

$$\psi = 7.0$$
$$\eta_c = 0.90$$
$$\gamma_c = 1.36$$
$$x = 10 \text{ ft} \quad (3.048 \text{ m})$$

Step 1.   Equation (5-4).

$$\frac{p_3}{p_{SL}} = \frac{p_0}{p_{SL}} \cdot \frac{p_3}{p_0} = 1.78$$

Step 2.

$$\frac{T_3}{T_{SL}} = \psi \cdot \frac{T_o}{T_{SL}} = 5.67 \qquad \frac{a_3}{a_{SL}} = \sqrt{\frac{T_3}{T_{SL}}} = 2.38$$

Step 3.   Equation (4-9).

$$M_3 = 3.09$$

Step 4.

$$V_3 = \frac{a_3}{a_{SL}} \cdot a_{SL} \cdot M_3 \qquad\qquad V_3 = \begin{array}{l} 8200 \text{ ft/s} \\ (2500 \text{ m/s}) \end{array}$$

Step 5.

$$\frac{V_3^2}{2} = \begin{array}{l} 1340 \text{ BTU/lbm} \\ (3120 \text{ kJ/kg}) \end{array}$$

Step 6.   HAP(Air).

$$h_3 = \begin{array}{l} 654 \text{ BTU/lbm} \\ (1520 \text{ kJ/kg}) \end{array}$$

Step 7.   HAP(Air).

$$h_w = \begin{array}{l} 380 \text{ BTU/lbm} \\ (884 \text{ kJ/kg}) \end{array}$$

Step 8.   Equations (9-4), (9-5b), and (9-10), and HAP(Air), solved by iteration.

$$h^* = \begin{array}{l} 807 \text{ BTU/lbm} \\ (1880 \text{ kJ/kg}) \end{array} \quad T^* = \begin{array}{l} 3440 \text{ °R} \\ (1910 \text{ K}) \end{array} \quad C_p^* = \begin{array}{l} 0.315 \text{ BTU/lbm·°R} \\ (1.32 \text{ kJ/kg · K}) \end{array}$$

Step 9.   Equations (2-4), (2-5), and (9-10).

$$Pr^* = 0.743 \frac{C_p^*}{C_{p_{SL}}} \qquad C_{p_{SL}} = \begin{array}{l} 0.242 \text{ BTU/lbm·°R} \\ (1.015 \text{ kJ/kg} \cdot \text{K}) \end{array} \qquad Pr^* = 0.967$$

Step 10.   Equations (9-2) and (9-5b).

$$h_{aw} = h_3 + r\frac{V_3^2}{2} \qquad\qquad \begin{array}{l} h_{aw} = 1980 \text{ BTU/lbm} \\ (4610 \text{ kJ/kg}) \end{array}$$

Step 11.   HAP(Air).

$$\begin{array}{l} T_{aw} = 6000 \text{ °R} \\ (3330 \text{ K}) \end{array}$$

Step 12.   Equation (9-8).

$$Re_x^* = 4.16 \times 10^7$$

Step 13.   Equation (9-7d).

$$\begin{array}{l} \dot{q}_w = 1.63 \text{ BTU/in}^2 \cdot \text{s} \\ (266 \text{ W/cm}^2) \end{array}$$

Step 14.   Equation (9-9b).

$$C_f = 1.47 \times 10^{-3}$$

One fairly obvious lesson of these exercises is that even this easy method requires quite a bit of work, although the results are rewarding. Imagine what the next higher level of computational difficulty must entail.

9.2.1.2 Summary of results.   The Eckert reference enthalpy method was used repeatedly to generate the results for a constant q_0 trajectory summarized in Table 9.1. The compression system calculations were done in the manner of Example 9.1, and the combustion system in the manner of Example 9.2. The $M_0 = 12.0$ compression surface conditions violate the restriction against dissociation and are therefore not reported. Please keep in mind that the numerical values presented in Table 9.1 are rounded off after calculations to at least four significant figures. Of the many possible conclusions to be drawn from this information, the following five seem to be the most relevant and interesting.

First, as anticipated, the combustor surface convective heat transfer exceeds that of the compression surface by a large factor, and all

Table 9.1 Compression surface and combustion surface boundary layer behavior of air as predicted by the Eckert reference enthalpy method for a hypersonic vehicle trajectory of $q_o = 1000\,\text{lbf/ft}^2$ (47.88 kN/m²). Also:

$$T_w = 2000\ °R \qquad \psi = 7.0$$
$$(1111\ K) \qquad \eta_c = 0.90$$
$$\gamma_o = 1.40 \qquad \gamma_c = 1.36$$

		Freestream Mach Number M_0					
		6		10		12	
		Laminar	Turbulent	Laminar	Turbulent	Laminar	Turbulent
Compression Surface ($x = 1.0$ ft/0.3048 m)							
$\dot{q}w$	$\dfrac{\text{BTU}}{\text{in}^2 \cdot \text{s}}$	0.00480	0.0131	0.0219	0.0401	—	—
	$\dfrac{\text{W}}{\text{cm}^2}$	0.785	2.14	3.58	6.56	—	—
$C_f\ \times 10^3$		0.504	1.33	0.612	1.13	—	—
$Re_x^*\ \times 10^{-4}$		9.04	8.92	2.72	2.69	—	—
T^*	°R	1760	1780	2760	2780	—	—
	K	979	987	1530	1550	—	—
T_{aw}	°R	2770	2840	4910	4940	—	—
	K	1540	1580	2730	2740	—	—
Combustor Surface ($x = 10$ ft/3.048 m)							
$\dot{q}w$	$\dfrac{\text{BTU}}{\text{in}^2 \cdot \text{s}}$	0.0147	0.233	0.0981	1.63	0.140	2.10
	$\dfrac{\text{W}}{\text{cm}^2}$	2.40	38.1	16.0	266	22.8	343
$C_f\ \times 10^3$		0.123	2.01	0.0880	1.47	0.0909	1.35
$Re_x^*\ \times 10^{-7}$		3.84	3.84	4.17	4.16	2.78	2.81
T^*	°R	2450	2450	3440	3440	4190	4170
	K	1360	1360	1910	1910	2330	2320
T_{aw}	°R	2980	2980	5990	6000	7230	7180
	K	1650	1660	3330	3330	4020	3990

the more so as the flight speed increases. This can be traced back as far as Eq. (9-1) as being fundamentally caused by the reduced throughflow area.

Second, the magnitude of the turbulent combustor convective heat transfer is simply enormous by almost any standard of comparison. Even the dreaded piloted vehicle re-entry heating is unlikely to approach 1.0 BTU/(in$^2 \cdot$ s) (160 W/cm^2) (largely because the nosetip boundary layer is laminar due to the favorable pressure gradient).[9.1,9.2] Turbulent combustor convective heating easily exceeds this amount at $M_0 = 10.0$ and is increasing rapidly with flight speed. A more picturesque and easily remembered measure is that, since solar insolation (the amount of solar energy flux reaching the surface of the Earth) is about 1 kW/m^2 = 0.1 W/cm^2, combustor heating is equivalent to many *thousands* of "suns."

Third, all other things being equal, the difference between laminar and turbulent convective heat transfer is substantial. If we use Re_x^* as the indicator of transition, it is apparent that the flow within the combustor is turbulent and that transition takes place somewhere along the compression surface. This puts the compression system designer under great pressure to predict the location of boundary layer transition correctly. If transition occurs later than predicted, some portion of the structure will be colder than expected and some cooling potential will be wasted. If transition occurs earlier than predicted, some portion of the structure will be hotter than expected, possibly leading to a serious loss of durability. Small wonder, then, that closed-loop control of coolant flow to provide the desired surface temperature distribution is an attractive concept.

Fourth, keeping in mind the fact that the calculated skin friction coefficient C_f represents the lower, smooth-wall, two-dimensional boundary layer limit, the values reported in Table 9.1 support the earlier estimates of this textbook that it is quite reasonable to expect that $0.001 < C_f < 0.005$.

Fifth, the calculated adiabatic wall temperatures support the earlier assertion that they exceed the working temperature of any known structural material, making cooling an absolute necessity.

Without putting too fine a point on it, many other phenomena are at work that *increase* the convective heat transfer. Briefly, then, the estimates quoted above should be interpreted to be the *lower* limits of heat fluxes to be expected. Next we examine some of the other mechanisms and their possible impacts.

9.2.1.3 Convective heat transfer enhancement. The phenomena that can significantly increase $\dot{q}_w$ are listed below, along with descriptions of the primary causes and estimates of the extent of their influence.

The order of appearance is by increasing flowfield complexity, and the effects can be, but are not necessarily, additive.

Freestream or Flight Conditions

The influence of the dynamic pressure q_0 and freestream velocity V_0 on convective heat transfer can be *very crudely* estimated for hypersonic flight from Eq. (9-7) as

$$\text{Laminar flow} \qquad \dot{q}_w \propto \frac{\rho_o V_0 \cdot V_0^2}{(\rho_0 V_0)^{\frac{1}{2}}} \propto q_0^{\frac{1}{2}} V_0^{\frac{3}{2}} \qquad (9\text{-}11a)$$

$$\text{Turbulent flow} \qquad \dot{q}_w \propto \frac{\rho_o V_0 \cdot V_0^2}{(\rho_o V_0)^{1/5}} \propto q_0^{\frac{4}{5}} V_0^{\frac{6}{5}} \qquad (9\text{-}11b)$$

where the reference quantities have been replaced by freestream quantities, the variation of transport properties with temperature has been ignored, and $(h_{aw} - h_w)$ has been set equal to $V_0^2/2$. The information presented in Table 9.1 gives credence to this method of estimation for the combustor, as follows. The ratio of the freestream velocity at $M_0 = 12$ to that at $M_0 = 10$ is $(12,230/10,040) = 1.218$. For laminar flow, $(1.218)^{3/2} = 1.344$ compares reasonably well with $(0.140/0.0981) = 1.427$. For the more important case of turbulent flow, $(1.218)^{6/5} = 1.267$ compares remarkably well with $(2.10/1.63) = 1.288$. In both cases, the rough estimate underpredicts the influence of increasing freestream velocity.

With Eq. (9-11) in hand, it is evident that convective heat transfer increases with both q_0 and V_0. In fact, for the critical case of turbulent flow in the combustor, the combined effect of doubling both q_0 and V_0 is to increase $\dot{q}_w$ by a factor of 4. Equation (9-11) also demonstrates the value of moving to a lower q_0, higher altitude trajectory in order to reduce convective heat transfer.

Surface Roughness

For flows having high Reynolds numbers, even small departures from the perfect smoothness of walls can substantially increase the turbulent skin friction and convective heat transfer.[9.3] In the combustor, for example, where Re_x^* is of the order of 10^7, the experimental evidence suggests that relative wall roughnesses of only 0.0001 and 0.001 would increase C_f and $\dot{q}_w$ by factors of about 1.5 and more than 2.0, respectively.

Three-Dimensional Flows

There are a host of three-dimensional flows that carry fresh fluid from the inviscid region to the vicinity of the wall, thus increasing the "scrubbing" and, with it, the skin friction and convective heat transfer. To mention a few, these include the streamwise vorticity that can be naturally generated in corners and on curved surfaces, and the streamwise and transverse vorticity deliberately generated in order to force fuel and air to mix and/or to prevent boundary layer separation. It is difficult to place a value on these effects because they depend strongly on the details of the configuration, but they are always substantial and must be accounted for by the designer in order to avoid structural problems.

The three phenomena treated so far have in common the fact that they impact *broad expanses* of the engine surface, rather than being confined to local regions. They can also act collectively, making it possible to understand why the designer must be prepared to face maximum convective heat transfer rates that are at least a factor of 10 above those of Table 9.1. Thus, turbulent combustor convective heat transfer could easily be 20 $BTU/in^2 \cdot s$ (3300 W/cm^2 or 33,000 suns). Not surprisingly, the degree of difficulty is about the same as the daunting Space Shuttle Main Engine (SSME). Using brochure values of SSME flow quantities, the maximum convective heat transfer (at the sonic throat) is estimated to be in the range of 20–40 $BTU/in^2 \cdot s$ (3300–8300 W/cm^2).

We will now look at three phenomena that primarily impact *local regions* of the flow. Although they have a minor effect on the overall heat balance of the cooling system, they can easily cause catastrophic damage due to hot spots and *burnthrough* if not anticipated, and they often call for sophisticated, concentrated, expensive cooling techniques.

Shock—Boundary Layer Separation

When an impinging or originating shock wave causes boundary layer separation, the additional scrubbing of the surface within the separated region or "bubble" increases the convective heat transfer beyond what otherwise would have been expected there. One correlation proposed for these circumstances is[9.6]

$$\text{Laminar flow} \qquad \frac{\dot{q}_{w,pk}}{\dot{q}_w} = \left(\frac{p_d}{p_u}\right)^{1.30} \qquad (9\text{-}12a)$$

$$\text{Turbulent flow} \qquad \frac{\dot{q}_{w,pk}}{\dot{q}_w} = \left(\frac{p_d}{p_u}\right)^{0.85} \qquad (9\text{-}12b)$$

where $\dot{q}_{w,pk}/\dot{q}_w$ is the ratio of the local *peak* heat flux to the unseparated value, and p_d/p_u is the static pressure ratio applied by the shock wave that separates the flow. The *minimum* value of the latter for a given M_u is obtained by substituting the criteria of Eqs. (5-35) and (5-36) into Eq. (5-37). The *minimum* value of the former is then obtained by substituting those results into Eq. (9-12).

Table 9.2   Minimum ratio of wall peak heat flux to unseparated heat flux due to shock—boundary layer separation (γ_c = 1.36).

| | Upstream Mach Number, M_u | | | |
| | 3.0 | | 5.0 | |
	$\dfrac{p_d}{p_u}$	$\dfrac{\dot{q}_{w,pk}}{\dot{q}_w}$	$\dfrac{p_d}{p_u}$	$\dfrac{\dot{q}_{w,pk}}{\dot{q}_w}$
Laminar flow	1.60	1.84	1.87	2.25
Turbulent flow	2.78	2.38	3.82	3.13

The results for several representative cases are tabulated in Table 9.2. The clear message is that the *minimum* increase in local heat flux caused by shock—boundary layer separation is a factor of 2–3. Larger factors are, of course, possible.

Leading-Edge Shock Wave Interference

Any blunt body immersed in a supersonic flow generates its own bow shock wave that begins the process of bringing the approaching flow to rest at its stagnation point or along its stagnation line. Because of the importance of this type of flowfield to re-entry vehicles, it has received a lot of attention, and quite satisfactory analysis methods have been developed.[9.1, 9.2] When, as sketched in Fig. 9.2, even a relatively weak shock wave impinges on or *interferes* with the bow shock wave or the "shock layer," some extraordinarily complex aerodynamics and very high convective heat transfer rates can result.[9.7] It was, in fact, pylon leading-edge ramjet cowl shock wave interference that burned through the ventral fin of the X-15A-2.

Six types of interference patterns have been identified, of which Type IV, shown in Fig. 9.3, is one of the worst from the heat transfer standpoint. Type IV interference occurs when the impinging shock wave strikes the subsonic region of the shock layer and the angle of inclination between the flow in region ABC and the body is relatively large. A sizable embedded supersonic "freejet," similar to that of an open nozzle exhaust, develops downstream of the line BC that stagnates underneath the line DF and scrubs the surface beneath the lines DE and FG.

Experimental evidence indicates that the Type IV interference increases the heat flux above the undisturbed, maximum stagnation point value by a factor of 5–10 when $M_u = 4.6$, with even larger multiples being possible at higher approach Mach numbers.[9.7]

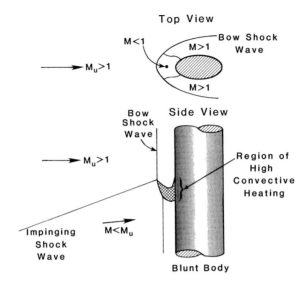

Fig. 9.2 Schematic diagram of leading-edge shock wave impingement leading to interference heating. The blunt body can be, for example, a fin, pylon, strut, fuel injector, or protuberance.

Intersections or Junctions

When the wall boundary layer encounters a blunt body a *horseshoe vortex* is formed within the viscous layer because the stagnation pressure at the outer edge is greater than that near the wall, as illustrated in Fig. 9.4. This familiar rotational flow pattern at the intersection or junction of the blunt body with the wall scrubs the surface with fresh fluid that increases the local heat flux. The horseshoe vortex proceeds to wrap itself around the base of the blunt body, thus increasing the heat flux over a larger surface area.

One correlation proposed for *turbulent* flow in these circumstances is[9.8]

$$\dot{q}_{w,pk} = 0.7155 \times 10^{-4} M_u \cdot p_{td}^{1.176}(T_{aw} - T_w) \quad \text{BTU/in}^2 \cdot \text{s} \quad (9\text{-}13)$$

$$\left[210.6 \times 10^{-4} M_u \cdot p_{td}^{1.176}(T_{aw} - T_w) \text{ W/cm}^2 \right]$$

where M_u is the supersonic Mach number upstream of the blunt body and p_{td} is the stagnation pressure downstream of the normal shock wave in atmospheres.

Applying this method to the $M_u = 10$, turbulent flow cases of Table 9.1, we find that $\dot{q}_{w,pk} = 1.99$ BTU/in$^2 \cdot$ s (325 W/cm^2) for the compression surface and 37.3 BTU/in$^2 \cdot$ s (6100 W/cm^2) for the combustor surface. If we divide these by the corresponding values found in Table 9.1, we find that an intersection or junction causes a peak heat flux that is $(1.99/0.0401) = 49.5$ times that of the undisturbed compression surface and $(37.3/1.63) = 22.9$ times that of the undisturbed combustor surface. These are both very large multiples and absolute heat flux magnitudes.

9.2.1.4 Design considerations.   The preceding list is anything but exhaustive. It does not include the influence of such geometrical variations as gaps, rearward-facing steps, cavities, and wakes, or the fact that the wall heat transfer rate of even constant static pressure conical flow exceeds that of wedge flow by the factor $\sqrt{3}$.[9.1] It does not include the influence of the injection of fuel and/or coolants or the increased throughflow enthalpy after combustion has taken place. It does not include the influence of unsteadiness, which can be caused either by large-scale fluctuations of the main flow, by vibrations of the engine surfaces, or by acoustic fatigue in the 50–300 Hz range, the level of the latter having been estimated to be as much as 180 db in the combustor and 190 db on the exhaust system surfaces.

The basic message is, nonetheless, quite clear. The designer must be prepared to cope with very high global and extremely high local heat transfer rates. As a result, many clever structural designs have been devised to provide the necessary thermal protection.

Figure 9.5 shows a portion of a typical *convective cooling* panel intended for use on the broad expanses of engine flowpath surfaces. The hot side or "front side" convective heat flux is removed by the coolant that flows in the cold side or "back side" passages directly

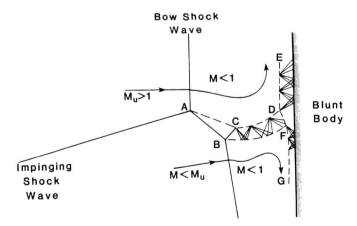

Fig. 9.3   Schematic representation of a Type IV interference pattern.

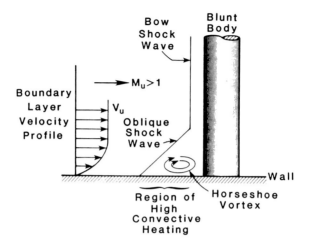

Fig. 9.4   Schematic diagram of the flow in the vicinity of an intersection or junction of a blunt body and a wall that leads to a horseshoe vortex and increased heating.

behind the engine wall. The coolant passages can be made narrow or more complex in order to increase the convective heat transfer rate on the cold side.

Figure 9.6 shows a typical configuration for use at *stagnation points or lines* that can be applied to any local region of intense convective heating. The internal convective heat transfer is raised in several different ways. First, the impingement creates a "mirror image" internal stagnation point flow that intensely cools the region. Second, the narrow passages provide high coolant velocity and exposed surface area, both of which increase heat transfer. Third, a variety of labyrinthine obstacles built into the passages stimulate the

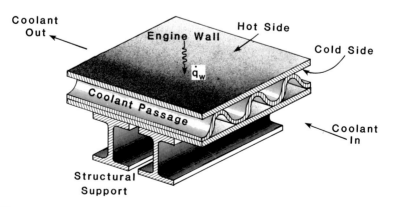

Fig. 9.5   A portion of a typical engine flowpath cooling panel.

turbulence of and wall scrubbing by the coolant, as well as further increase local velocities and surface areas.

When convective cooling is not sufficient, external *film cooling* may be considered, the coolant being laid directly down on the hot side in order to provide an additional physical barrier to heat transfer. The effectiveness of the film is gradually lost as it moves downstream and mixes with the external flow, so it must be periodically replenished. Figure 9.7 shows the configuration of Fig. 9.6 augmented by film cooling. The costs of film cooling include more sophisticated manufacturing and sometimes losing the coolant and its energy to the surroundings, yet this is often the best or only available solution to a severe heating problem. Film cooling could provide additional thermal protection for the convective cooling panel of Fig. 9.5, if necessary, and the coolant and its energy would become part of the engine throughflow.

9.3 THERMAL STRESS

Keeping temperatures within the working ranges of the structural materials is only part of the story. The other part is that the heat transfer is accompanied by temperature gradients that, because they cause differential thermal expansion or growth that must be countered by internal strains, also generate internal stresses known as *thermal stresses.* As you will see, these thermal stresses can be so large that the material may rupture the first time it is heated, or in thermal fatigue after only a few thermal or hot-cold cycles. Similarly, the *thermal strains* can be equally destructive, especially in modern, high-temperature materials that tend to be strong but "brittle" (i.e., they can tolerate only relatively small amounts of plastic strain).

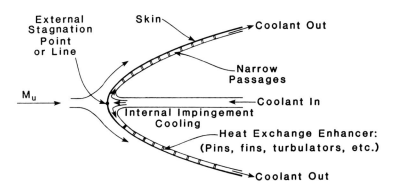

Fig. 9.6 A typical stagnation point or line convective cooling configuration.

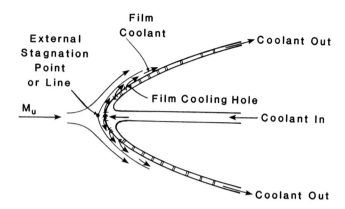

Fig. 9.7   A typical film cooling arrangement.

In order to better understand and put some numbers on these coupled phenomena, we will examine the classical case of a long circular cylinder having free ends that is pressurized and heated from within, as shown in Fig. 9.8. This is certainly a reasonable mental model for the structure of the combustor, and has the usual advantage that the closed form mathematical solution is available in the literature.[9.9]

Referring to Fig. 9.8 for configuration and symbology, the radial distribution of *radial* normal stress within the material of the cylinder is given by

$$
\sigma_r = \left\{ \frac{1 - \left(\dfrac{r_o}{r}\right)^2}{\left(\dfrac{r_o}{r_i}\right)^2 - 1} \right\} \left\{ p - \frac{E\alpha \dot{q}_w r_i}{2k(1-\nu)} \cdot \ln \frac{r_o}{r_i} \right\} - \frac{E\alpha \dot{q}_w r_i}{2k(1-\nu)} \cdot \ln \frac{r_o}{r}
$$

$$(9\text{-}14)$$

and radial distribution of the *hoop* or circumferential normal stress is given by

$$
\sigma_h = \left\{ \frac{\left(\dfrac{r_o}{r}\right)^2 + 1}{\left(\dfrac{r_o}{r_i}\right)^2 - 1} \right\} \left\{ p - \frac{E\alpha \dot{q}_w r_i}{2k(1-\nu)} \cdot \ln \frac{r_o}{r_i} \right\} + \frac{E\alpha \dot{q}_w r_i}{2k(1-\nu)} \left(1 - \ln \frac{r_o}{r}\right)
$$

$$(9\text{-}15)$$

where E is the modulus of elasticity or Young's modulus, α is the coefficient of thermal or linear expansion, k is the thermal conductivity, and ν is Poisson's ratio of the cylinder material. The quantity p is the uniform absolute pressure *inside* the cylinder (the absolute *external* pressure is 0) and $\dot{q}_w$ is the uniform convective heat transfer rate *into* the inside surface.

These radial and hoop normal stresses are not the whole story. The axial normal stress is comparable to the hoop normal stress (see Problem 9.11), and the latter can be 25 percent higher near the free ends. Moreover, the *transient* thermal stresses can be even greater because of the large, temporary temperature gradients that occur.

We will now explore the general nature of σ_r and σ_h by means of the following typical round number example (these material properties are similar to those of a high carbon steel):

$p = 10$ atm

$\dot{q}_w = 1.0$ BTU/in$^2\cdot$ s   (163.5 W/cm^2)

$r_o = 1.0$ ft   (0.3048 m)

$r_i = 0.98$ ft   (0.2987 m)

$E = 3.0 \times 10^4$ klbf/in^2 (20.7×10^7 kN/m^2)

$\alpha = 7.0 \times 10^{-6}$ 1/°R   (12.6×10^{-6} 1/K)

$k = 5.0 \times 10^{-4}$ BTU/in·s · °R   (0.374 W/cm · K)

$\nu = 0.30$

Using these values, we find that

$$\frac{E\alpha\dot{q}_w r_i}{2k(1-\nu)} \cdot \ell n \frac{r_o}{r_i} = 4850 \text{ atm}$$

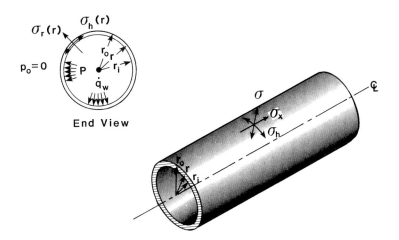

Fig. 9.8 Schematic diagram of the pressurized and heated circular cylinder used in the stress analysis.

which, as previously asserted, is enormously larger than any anticipated p, so that the latter makes an insignificant contribution to the
radial and hoop normal stresses, and therefore will be neglected hereinafter. The behavior of the remaining terms calculated according to
Eqs. (9-14) and (9-15) is presented in Fig. 9.9. The information there
shows that σ_r is relatively small, entirely compressive, and maximum
near the center of the material.

In sharp contrast, σ_h is alarmingly large [the yield stress for most
materials being less than 100,000 lbf/in^2 (689,000 kN/m^2)] compressive on the inside, tensile or expansive on the outside, and minimum
near the center of the material. This reflects the fact that the attempt of the warmer inside material to expand more than the cooler
outside material is resisted by hoop stresses that require them to
move in unison.

We will therefore focus on σ_h rather than σ_r in the discussions
that follow.

9.3.1 Material Properties

Engineers eventually find that current material properties data are
essential for progress. This information is jealously guarded as pro-

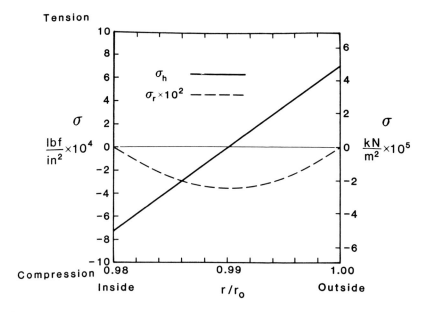

**Fig. 9.9 Calculated radial distribution of hoop and radial normal
stresses in a circular cylinder with free ends due to internal pressurization and heating. Note that the radial normal stresses have
been multiplied by a factor of 100 in this presentation.**

prietary by the owners because of the competitive advantage it provides. Nevertheless, good estimates can usually be made with the help of recent material properties handbooks, technical papers, and textbooks.[9.10]

One of the most profitable features of Eqs. (9-14) and (9-15) is that the magnitude of the thermal stresses are directly proportional to the combination of material properties contained in the *thermal stress factor* or group

$$\frac{E\alpha}{k(1-\nu)} \qquad (9\text{-}16)$$

In words, the thermal stresses are determined by the collective actions of several vastly different phenomena as represented by their material properties. The form of the result is intuitively satisfying and explains why designers strive to obtain materials that are not only strong, but also have low thermal expansion and high thermal conductivity. This can lead to the selection of some surprising materials.

The form of Eq. (9-15) compels one to present the generalized version of the behavior of the maximum thermal hoop normal stress (in this case the compressive stress at the inside radius, which is almost the same as the tensile stress at the outside radius) as in Fig. 9.10. There we see that σ_h is exactly proportional to the thermal stress factor of Eq. (9-16) and approximately proportional to $(1 - r_i/r_o)$. The latter suggests that the wall should be made as thin as possible, until some other consideration such as pressure forces, manufacturability, or susceptibility to foreign object damage takes over. The value of the thermal stress group of the familiar aerospace materials of Table 9.3 is also displayed in Fig. 9.10, exposing the general (but, fortunately, imperfect) tendency of stronger materials to experience greater thermal stresses.

The principal message of Fig. 9.10, however, is that the thermal stresses in available materials are likely to be very large, even at the relatively modest convective heat flux of 1.0 BTU/in²·s (164 W/cm²). Recalling that thermal stresses are directly proportional to heat flux, and that heat flux values at least ten times that of Fig. 9.10 are possible, it is obvious that thermal stress is a fundamental issue of engine structural design. We can also see in a more absolute sense why it may be necessary to move to a lower dynamic pressure trajectory at some point in the flight in order to keep the convective heat transfer and thermal stresses within safe limits.

The biggest surprise of Fig. 9.10 is that the best materials are not necessarily what we might have expected. In fact, a good index of the inherent *thermal durability* of a given material is the ratio of *yield stress* σ_y to thermal stress, or, equivalently, to the thermal stress

Table 9.3 Properties of several aerospace materials related to thermal stress and thermal durability. A constant value for Poisson's ratio of 0.30 was used in these calculations.

Material Properties

Materials:	E $\frac{\text{klbf}}{\text{in}^2}$ $\left(\frac{\text{kN}}{\text{m}^2}\right)$	α $\frac{1}{°\text{R}}$ $\left(\frac{1}{\text{K}}\right)$	k $\frac{\text{BTU}}{\text{in}\cdot\text{s}\cdot°\text{R}}$ $\left(\frac{\text{W}}{\text{cm}\cdot\text{K}}\right)$	$\frac{E\alpha}{k(1-\nu)}$ $\frac{\text{lbf}/\text{in}^2}{\text{BTU}/\text{in}\cdot\text{s}}$ $\left(\frac{\text{kN}/\text{m}^2}{\text{W}/\text{cm}}\right)$	σ_y $\frac{\text{klbf}}{\text{in}^2}$ $\left(\frac{\text{kN}}{\text{m}^2}\right)$	$\frac{\sigma_y k(1-\nu)}{E\alpha}$ $\frac{\text{BTU}\cdot\text{s}}{\text{in}}$ $\left(\frac{\text{W}}{\text{cm}}\right)$
Molybdenum TZ (TZM)	3.75×10^4 (25.9×10^7)	3.30×10^{-6} (5.94×10^{-6})	16.7×10^{-4} (1.25)	10.6×10^4 (1.76×10^3)	106 (73.1×10^4)	1.00 (415)
Copper (Cu)	1.70×10^4 (11.7×10^7)	9.30×10^{-6} (16.7×10^{-6})	49.0×10^{-4} (3.66)	4.61×10^4 (0.763×10^3)	35.0 (24.1×10^4)	0.759 (316)
Aluminum (Al 6061)	1.00×10^4 (6.90×10^7)	13.0×10^{-6} (23.4×10^{-6})	33.0×10^{-4} (2.47)	5.63×10^4 (0.934×10^3)	35.0 (24.1×10^4)	0.622 (258)
Incoloy 909	2.30×10^4 (15.9×10^7)	5.10×10^{-6} (9.18×10^{-6})	3.01×10^{-4} (0.225)	55.7×10^4 (9.27×10^3)	105.0 (72.4×10^4)	0.189 (78.1)
Lockalloy (Al-Be alloy)	2.80×10^4 (19.3×10^7)	10.0×10^{-6} (18.0×10^{-6})	27.8×10^{-4} (2.08)	14.4×10^4 (2.39×10^3)	26.0 (17.9×10^4)	0.181 (74.9)
Titanium (Ti 6-4)	1.17×10^4 (8.07×10^7)	5.60×10^{-6} (10.1×10^{-6})	2.15×10^{-4} (0.161)	43.5×10^4 (7.23×10^3)	76.0 (52.4×10^4)	0.175 (72.5)
Steel (high carbon)	3.00×10^4 (20.7×10^7)	7.00×10^{-6} (12.6×10^{-6})	4.40×10^{-4} (0.329)	68.2×10^4 (11.3×10^3)	65.0 (44.8×10^4)	0.095 (39.6)

group of Eq. (9-16). Returning to Table 9.3, we see that Molybdenum TZ promises the best thermal durability because it combines comparatively low thermal expansion with high thermal conductivity. Copper and Al 6061 are also interesting contenders for the same reason. In contrast, the relatively high-strength materials Incoloy 909 and Titanium 6-4 are among the least attractive materials because they have high thermal expansion and low thermal conductivity.

Materials properties can be sensitive to temperature.[9.10] For example, for Lockalloy, although α is nearly constant from 200–1000 °F, σ_y drops from about 45 to 25 klbf/in^2 from 0–800 °F, and E drops from 30×10^6 to 16×10^6 from 200–800 °F. Similarly, they cannot be used near their melting points, which for Lockalloy is 1193 °F. Finally, σ_y is affected by time at temperature due to creep and a variety of mechanisms such as grain sliding.

9.3.2 Thermal Strain

The designer may always elect to allow the material to yield into plastic flow in order to relieve the elastic stresses and avoid an immediate failure, but at a considerable cost in reduced cyclic life and increased complexity of structural analysis. This may be of no benefit to modern, high-strength, high-temperature materials because they are brittle and cannot tolerate total strains (elastic plus plastic) in excess of about 1 percent without breaking.

The maximum total thermal hoop normal strain for the circular cylinder is very nearly[9.9]

$$\epsilon_h = \frac{\sigma_h}{E} \tag{9-17}$$

Thus, referring to Fig. 9.10 and Table 9.3, the total strain for Molybdenum TZ for those conditions and $r_i/r_o = 0.98$ is merely 0.32 percent, while for the relatively brittle Incoloy 909 it is a completely unacceptable 2.9 percent. This may help you understand why designers of high-temperature engine parts must consider (and must have the material characterization data available to analyze) many aspects of structural behavior before they make their decisions.

9.3.3 Wall Temperature Drop

From the cooling system standpoint, it is desirable to have the smallest possible temperature difference or *wall temperature drop* between the hot and the cold sides of a cooling panel because that makes the best use of each unit of coolant by raising it to its highest possible temperature. As you will soon see, the more thermal energy each unit of fuel used as coolant can absorb, the less the likelihood of having to "waste" fuel because the equivalence ratio of the engine cycle

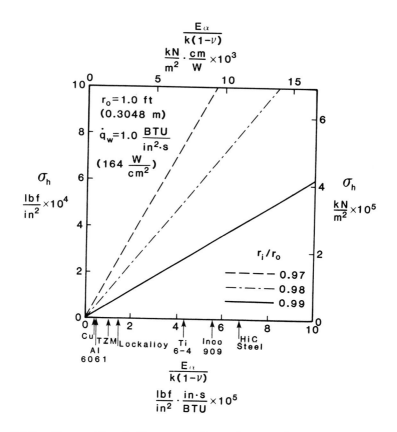

Fig. 9.10　Generalized diagram of maximum thermal hoop normal stress of the circular cylinder of Fig. 9.8 as a function of the thermal stress group of Eq. (9-16) and radius ratio.

exceeds 1 (i.e., there is not enough oxygen to burn all the fuel acting as a coolant).

Since the temperature drop is given by

$$T_i - T_o = \frac{\dot{q}_w r_i}{k} \cdot \ell n \frac{r_o}{r_i} \cong \frac{\dot{q}_w (r_o - r_i)}{k} \qquad (9\text{-}18)$$

the conditions of Fig. 9.10 and Table 9.3 can be used to show that the temperature drop for Molybdenum TZ when $r_i/r_o = 0.98$ is 144 °R (80 K), while for Incoloy 909 it is 800 °R (440 K). This is a significant difference that can compensate for a lower material working temperature and/or raise the coolant to a higher temperature. We see again in this example the importance of evaluating the impact of all the properties of the materials under consideration.

9.4 System Cooling Requirements

As in other things, the local heat flux is only one part of the thermal protection story. Another important part is whether the existing fuel flow rate is sufficient to satisfy the total cooling requirements of the entire system. If not, performance suffers either because extra coolants or fuel that cannot be burned (i.e., $f > f_{st}$) will have to be carried aloft.

The leading tendencies in this arena are easily explored for the engine flowpath in hypersonic flight by capitalizing on the general appreciation of heat flux magnitudes gained in Sec. 9.2 and the rough estimates for the behavior with flight conditions of Eq. (9-11). To begin with, the total rate at which the fuel *can* remove the thermal energy as heat is

$$\dot{Q}_f = \dot{m}_f h_{fc} \tag{9-19}$$

where h_{fc} is the amount of thermal energy that can be absorbed by each unit mass of fuel. This is obviously a fundamental and important property of the fuel. To continue, following Eq. (9-11) and assuming that the majority of the heating is due to turbulent flow, the total rate at which heat *must be* removed from the engine surfaces is

$$\dot{Q}_r = \bar{\dot{q}}_w A_w \propto \frac{\rho_o V_0 V_0^2 A_w}{(\rho_o V_0)^{1/5}} \propto \dot{m}_0 \frac{V_0^{\frac{11}{5}}}{q_0^{\frac{1}{5}}} \cdot \frac{A_w}{A_0} \tag{9-20}$$

where A_w is the area to be cooled by the fuel and $\bar{\dot{q}}_w$ is the surface area-averaged heat flux. Thus, the ratio of the *available* fuel cooling to the *required* fuel cooling is

$$\frac{\dot{Q}_f}{\dot{Q}_r} = \frac{f h_{fc}}{K} \cdot \frac{q_0^{\frac{1}{5}}}{V_0^{\frac{11}{5}}} \cdot \frac{A_0}{A_w} \tag{9-21}$$

which can be rearranged to reveal the value of $f h_{fc}$ necessary to exactly balance the heat load or break even (i.e., $\dot{Q}_f = \dot{Q}_r$), namely

$$f h_{fc} = K \cdot \frac{V_0^{\frac{11}{5}}}{q_0^{\frac{1}{5}}} \cdot \frac{A_w}{A_0} \tag{9-22}$$

The missing constant K can be obtained for the specific engine configuration under scrutiny from a consistent set of design information at any reference point by the appropriate combination of Eqs. (9-19) through (9-21), depending on the type of information given.

As a representative example, consider a typical airbreathing aerospace vehicle flying at

$$M_0 = 10.0$$
$$q_0 = 1000 \text{ lbf/ft}^2 \quad (47.88 \text{ kN/m}^2)$$

with

$$A_0 = 300 \text{ ft}^2 \quad (27.9 \text{ m}^2)$$
$$A_w = 1000 \text{ ft}^2 \quad (92.9 \text{ m}^2)$$
$$\overline{\dot{q}_w} = 1 \text{ BTU/in}^2 \cdot \text{s} \quad (163.5 \text{ W/cm}^2)$$

so that

$$K = \frac{\overline{\dot{q}_w}}{\rho_o V_0} \cdot \frac{q_0^{\frac{1}{5}}}{V_0^{\frac{11}{5}}}$$

$$= 1.41 \times 10^{-7} \frac{\text{BTU}}{\text{lbm}} \frac{(\text{lbf/ft}^2)^{\frac{1}{5}}}{(\text{ft/s})^{\frac{11}{5}}} \left(9.67 \times 10^{-6} \frac{\text{kJ}}{\text{kg}} \frac{(\text{N/m}^2)^{\frac{1}{5}}}{(\text{m/s})^{\frac{11}{5}}} \right)$$

The results for this example, compiled using HAP(Trajectory) to evaluate trajectory properties, are presented in Fig. 9.11 for the usual

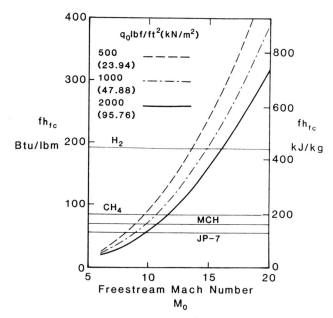

Fig. 9.11   The quantity fh_{fc} required for fuel cooling to balance total engine surface heat flux as a function of freestream Mach number and dynamic pressure.

range of flight conditions. It can be seen there, in accordance with Eq. (9-22), that the break even fh_{fc} increases rapidly with V_0 and decreases slowly with q_0, the latter being due to the fact that the fuel flow rate increases slightly more rapidly than the heat flux with q_0 at a given V_0. This effect would be even more pronounced for laminar flow.

9.4.1 Fuel Properties

A realistic dimension is now easily added to this example by considering the fh_{fc} that can be provided by available fuels. Table 9.4 is a collection of such information,[9.11] and is logically based on $f_{st}h_{fc}$ because that is the largest usable value. Table 9.4 presents latent plus sensible or *physical heat* absorbed by four standard aerospace fuels in going from tank conditions to 1800 °R (1000 K), consistent with the cooling system back side temperatures discussed earlier. In all cases, the concept is that the fuel will be stored in the liquid state and heated to the vapor state or "boiled" before being introduced into the flowpath as a gas.

When the information of Table 9.4 is superimposed on Fig. 9.11, several very important conclusions emerge. We can see immediately that hydrogen is without peer as an engine coolant, and that it offers our typical engine the potential of flight to freestream Mach numbers of 15 and beyond without assistance. It is an interesting coincidence that, as we found in Chap. 8, the performance of the airbreathing engine tapers off at approximately the same M_0, so that additional propellant, coolants, or alternative means of propulsion may relieve or avoid the heating problems for higher flight Mach numbers.

However, the costs and complexities associated with the logistic support of cryogenic hydrogen, as well as the very large vehicles that directly result from the exceedingly low density of liquid hydrogen, continually force reconsideration of conventional hydrocarbons. Unfortunately, as Fig. 9.11 shows for CH_4, MCH (methylcyclohexane), and JP-7, their use severely limits the maximum possible M_0. These results confirm and support the conventional wisdom that methane-fueled vehicles can cruise to about $M_0 = 10$ and JP-7 vehicles to about $M_0 = 8$.

These results could be considered somewhat optimistic because they ignore the cooling requirements of the internal vehicle equipment and assume that the entire cooling capacity of the fuel is utilized. They could also be considered pessimistic because they ignore radiative cooling and the increasing likelihood of laminar boundary layer flow at higher freestream Mach numbers and altitudes (recall Fig. 2.6). On balance, then, it should not be surprising that they agree reasonably well with more sophisticated analyses.

Table 9.4 Physical heat absorption capacities of several typical hydrocarbon aerospace fuels. Since JP-7 (the fuel developed for the SR-71) is a blend of hydrocarbon compounds designed to provide the desired operational characteristics, only the average molecular ratio of carbon to hydrogen is given. The h_{fc} of MCH is from its freezing point of 265 °R (147 K), while that of JP-7 is from standard ambient conditions.

		Fuel Properties		
Fuel	Liquid Density lbm/ft^3 (kg/m^3)	f_{st}	Physical h_{fc} BTU/lbm (kJ/kg)	$f_{st}h_{fc}$ BTU/lbm (kJ/kg)
Hydrogen H$_2$	4.74 (75.9)	0.0291	6500 (15,100)	189 (440)
Methane CH$_4$	18.7 (300)	0.0583	1460 (3400)	85.1 (198)
Methylcyclohexane CH$_3$C$_6$H$_{11}$	48.3 (774)	0.0680	1020 (2360)	69.1 (161)
JP-7 C$_{12.5}$H$_{26}$	49.5 (793)	0.0675	860 (2000)	58.0 (135)

This leads to perhaps the most important conclusion, namely that the *entire* cooling capacity of the fuel will probably be needed for the engine *alone*. Thus, the remaining acreage of the vehicle surfaces (of the order of 10,000 ft^2 or 1000 m^2) must be made of materials that can survive with only radiative cooling (with the exception of local regions of intense heating, such as leading edges and sharply concave corners). Since these external surfaces are a substantial fraction of the empty mass of the aerospacecraft, one can easily see why there is such an intense interest in increasing the specific strength (i.e., yield strength divided by density, the basic figure of merit for airborne materials) and the working temperature of promising candidates.

9.4.2 Endothermic Fuels

Much thought and effort have gone into developing ways to increase the cooling or heat sink capacity of hydrocarbons fuels in order to provide greater freedom for the designer. Although the predominant motivation has been to enable the partial or complete substitution of conventional liquid hydrocarbon compounds for liquid hydrogen, this issue also receives attention because high-performance aircraft wish to fly faster, carry more heat generating electronic equipment, and employ higher temperature engine cycles, and therefore demand greater total cooling capacity from the fuel initially in their tanks.

It is interesting and relevant, then, that hydrocarbon fuels are capable of undergoing heat-absorbing or *endothermic* chemical reactions related to those of traditional petroleum "cracking" or "pyrolysis" processes.[9.11] The decomposition of hydrocarbons, in fact, requires additional energy. The increased cooling capacity or heat sink is referred to as *chemical heat* absorption. Other benefits of the chemical reactions are that they occur at almost constant temperature, and that they convert the high energy density hydrocarbon fuels to smaller or weakly bonded molecules that can burn rapidly, such as hydrogen, ethylene, and propylene. Conversely, H_2 and CH_4 cannot provide chemical heat absorption.

The most promising approaches for the proposed hydrocarbon fuel endothermic reactions are *thermal cracking* and *catalytic dehydrogenation*. Virtually any nonaromatic type of fuel can be thermally cracked, so the main considerations in fuel selection are the amount of heat sink available and the combustion reactivity and environmental impact of the natural products of the pyrolysis. The technology of catalysts has advanced rapidly in recent times, and they can serve to increase the rate of particular chemical reactions (especially at lower temperatures), increase the selectivity of desired endothermic reactions, and tailor the final products of the chemical reactions. The increased control over the endothermic processes offered by catalysts is offset by their increased complexity, weight, maintenance, and cost, so the designers have another tradeoff on their hands.

From a purely thermodynamic standpoint, the chemical heat absorption of a given fuel depends only upon the final products of decomposition and not upon the path taken to manufacture them. Thus, the potential chemical heat absorption for a given fuel is the *same* for thermal cracking and catalytic dehydrogenation, provided paths can be found to the same final chemical products. The theoretical potential for conventional liquid hydrocarbon fuels is in the range of an additional 1000 BTU/lbm (2326 kJ/kg), and this has been demonstrated for MCH using Pt/Al_2O_3 as a catalyst.[9.11] The physical *plus* chemical heat absorption for MCH and JP-7 would then place their $f_{st}h_{fc}$ at about 137 BTU/lbm (319 kJ/kg) and 125 BTU/lbm (292 kJ/kg), respectively.

Figure 9.12 reproduces Fig. 9.11 with the substitution of these enhanced endothermic values for MCH and JP-7. This presentation makes it clear that this would be an impressive advance for conventional hydrocarbon fuels, moving them well beyond methane to a freestream Mach number of about 13.

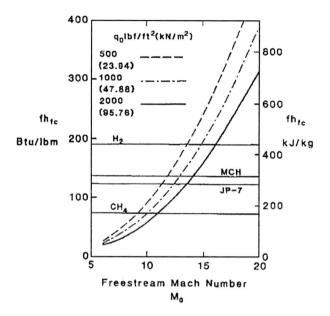

Fig. 9.12 The quantity fh_{fc} **required for fuel cooling to balance total engine surface heat flux as a function of freestream Mach number and dynamic pressure.**

The concept of *multi-fuel* or *dual-fuel* orbital vehicles is that high energy density liquid hydrocarbons are burned during takeoff and initial acceleration in order to reduce the vehicle size and drag and the amount of cryogenic fuel required, high specific energy liquid hydrogen being used to complete the mission. The "optimum" transition point between fuels depends on the details of the mission to be accomplished and the goals of the program. In any case, however, endothermic fuels are advantageous to the designer because they relax the maximum M_0 constraint imposed by heat transfer on hydrocarbon fuels.

We should expect to hear a great deal more about endothermic fuels in the future.

9.5 ENVIRONMENTAL IMPACT

One measure of the closeness to reality of an aerospace system is the amount of attention being paid to assessing and controlling its impact on the environment. Although hypersonic vehicles will not be flown routinely for some time to come, their advent has been anticipated, and Environmental Impact Studies have already begun. One reason for this is that it was the supersonic transport (SST) programs of the 1970's that first heightened public awareness and galvanized public opinion about several aircraft environmental issues.

Today, any major library contains a surprisingly large number of books, journals, reports, and articles related to the technical, climatological, biological, agricultural, medical, financial, legal, social, political, historical, and emotional aspects of the effects of human activities on the environment. These writings make it clear that this field of study is energetically fermenting, and that the last word is long from being written. It is therefore too early to know precisely what the most challenging problems for hypersonic airbreathing propulsion will be, but one can easily identify several environmental concerns that are sure to receive heavy scrutiny.

9.5.1 Low Altitude Operations

The federal government currently exerts its most stringent control over the atmospheric pollution associated with low altitude aircraft operations in the immediate vicinity of the airport. The *acoustic emissions* of subsonic jet aircraft are controlled by the U.S. Federal Aviation Administration (FAA) FAR Part 36 Stage 3 or the equivalent International Civil Aviation Organization (ICAO) Annex 16 Chapter 3 certification procedures, in order to maintain acceptable noise exposure levels in the adjacent communities.[9.12] The engine *exhaust emissions* (the chemical products of combustion) are regulated by U.S. Environmental Protection Agency (EPA) standards in order to avoid direct (e.g., such "noxious effluents" as soot/smoke/particulates and CO) and indirect [e.g., formation of the photochemical ozone (O_3) component of smog from unburned hydrocarbons (HC) and oxides of nitrogen (NO_x)] threats to the health of the inhabitants.[9.13] Attention is also paid to the storage, handling, and disposal of hazardous materials, such as the fluids used for propulsion, lubrication, cooling, de-icing, fire suppression, and cleaning.

It is a tribute to the aircraft engine community that all of these additional demanding requirements have been met since the 1970's without perceptibly diminishing the rate of advance of the usual measures of performance. One important lesson of these times was that it was much harder to retrofit an existing engine to meet new requirements than it was to start from scratch. Another was that the need to understand and predict the behavior of the acoustic field and of the minuscule or "trace" quantities of the combustion products that are pollutants led to enormously more sophisticated and capable experimental and analytical tools than were previously available (or would have been without the environmental impetus). Any future hypersonic vehicle program will benefit from this history.

Although any hypersonic airbreathing vehicle will be held to similar standards during low altitude operations, the exact nature of the problems to be solved will escape definition until the type of low speed propulsion is chosen. Moreover, until operations become frequent and/or widespread, hypersonic airbreathing vehicles will probably enjoy the luxury of flying from remote, sparsely populated sites that offer minimal environmental concerns.

9.5.2 High Altitude Operations

A legacy of the SST debate is public anxiety about sonic booms and stratospheric ozone depletion. Both of these take place at higher altitudes and over large expanses far from the airport. They definitely pertain to hypersonic airbreathing vehicles, and will now be dealt with in turn

9.5.2.1 Sonic booms.

A *sonic boom* is felt on the ground as the far-field pattern of compression and expansion waves attached to and trailing downward from a vehicle are rapidly swept across a stationary observer. The pattern is ultimately created by the lift needed to support the vehicle, and sounds like a sharp crack or boom because it begins with the coalesced compression waves emanating from the extreme leading edges of the vehicle. Please note that the strength of the arriving shock wave is immediately doubled upon reflection from the ground. The result can be particularly disturbing because there is no warning and because the initial increase of static pressure can be so large as to cause emotional or physical damage.[9.14] The United States and many other countries forbid the Concorde from flying supersonically over their territories in order to avoid these consequences.[9.12]

Although the strength of the sonic boom is determined by the total geometrical configuration of the vehicle and its flight condition, it is strongly influenced by the propulsion system, and especially by the oblique shock waves generated on the external compression surface. One difference connected with airbreathing propulsion is that any compression system shock waves that are "swallowed" inside the cowl cannot participate in the sonic boom.

As the vehicle accelerates and climbs along its flight path, the severity of the initial static pressure rise tends to diminish, partly because the pattern is spreading laterally, partly because the compression and expansion waves have more room to intersect and cancel, and partly because the weight of the vehicle is diminishing as fuel is consumed. These effects are reflected in the ballpark estimate for the *maximum far-field overpressure* (along the flight path) or the static pressure rise felt by the stationary observer immediately after the ground reflection of the initial shock wave [9.14]

$$\Delta p = \frac{K W^{\frac{1}{2}} (M_0^2 - 1)^{\frac{3}{8}}}{M_0 H^{\frac{3}{4}}} \qquad (9\text{-}23)$$

where

$$K = 28 \text{ lbf/ft}^2 \ (\text{ft}^{\frac{3}{4}}/\text{lbf}^{\frac{1}{2}})$$

$$\left[260 \text{ N/m}^2 \ (\text{m}^{3/4}/\text{N}^{\frac{1}{2}})\right]$$

and W is the instantaneous weight of the vehicle. The constant K was deduced from data from the Concorde and the Space Shuttle,[9.11] which you will find to be fairly consistent with an overpressure of 2.0 lbf/ft^2 (95.8 N/m^2) at a freestream Mach number of 2.0, an altitude of 50,000 ft (15,240 m), and a weight of 100,000 lbf (444,800 N). Note that this corresponds to a dynamic pressure of 682 lbf/ft^2 (32.6 kN/m^2).

The behavior of the maximum far-field overpressure of Eq. (9-23) with flight Mach number for $q_0 = 1000$ lbf/ft^2 (47.88 kN/m^2) is depicted in Fig. 9.13. The human reaction to sonic boom is extremely subjective and difficult to quantify, but ranges from awareness and "acceptable" annoyance below overpressures of about 1 lbf/ft^2 or 50 N/m^2 to "unacceptable" annoyance, anxiety, fear, and, finally, terror, as the overpressures increase to about 3 lbf/ft^2 or 150 N/m^2.[9.12] The annoyance of sonic booms is amplified because they come without warning, and surprise or startle human beings and animals, causing emotional distress and unpredictable involuntary reactions. Sonic booms can also do physical damage to buildings, dislodging loose objects from shelves, breaking windows, cracking plaster walls and ceilings, and, ultimately, harming structures and equipment, particularly after repeated or cumulative exposure. These effects are also difficult to quantify because of the variability of construction quality and the paucity of data, but the threshold at which the probability of structural damage becomes unacceptable (say one broken window pane per 10 million panes per boom), is about 2 lbf/ft^2 or 100 N/m^2.[9.14] Overlaying these results on Fig. 9.13 reveals that sonic boom will probably have unacceptable consequences during the supersonic phase of flight, but this is much less likely during the hypersonic phase.

The hazards of sonic boom can be minimized by tailoring the shape of the vehicle to weaken compression waves and by first climbing out subsonically and then accelerating to supersonic speeds in designated corridors having little (e.g., deserts or mountains) or essentially no (e.g., oceans) population. However, there are a number of

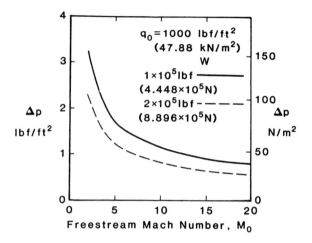

Fig. 9.13   Estimate of maximum sonic boom overpressure (along flight path) as a function of freestream Mach number and vehicle instantaneous weight for a constant dynamic pressure trajectory.

factors that are known to increase the amount of overpressure, some of which are not under our control. Flight maneuvers, including accelerating, turning, climbing, or diving, tend to change the shock pattern to produce either "superbooms" (that double the overpressure because of focusing effects) or multiple booms. Atmospheric effects, including standard day refraction of the wave pattern along and across the flight path, nonstandard day statistical variations, wind, and turbulence, and the shape of the local terrain and buildings, make prediction virtually impossible, and can easily double the overpressure at some times and places.[9.14]

With regard to deliberately selecting a benign flight path, note that the minimum turning radius in the horizontal plane is

$$r_c = \frac{V^2}{ng} \qquad (9\text{-}24)$$

where n is the maximum allowable number of g's of horizontal acceleration. Thus, a hypersonic vehicle flying at "merely" 5000 ft/s (1524 m/s) will require a turning radius of 74 mi (118 km) for a 2 g turn (a reasonable estimate for the structural limit). At 10,000 ft/s (3048 m/s) this becomes 294 mi (474 km). Defining a safe corridor under this constraint could be difficult, and, if the pilot must depart from the original flight plan, it could be nearly impossible.

9.5.2.2 Stratospheric ozone depletion.   Ozone (O_3) is a pariah at ground level because it is a highly oxidizing agent that harms living things as well as the material of inanimate objects. Ironically, the reverse

is true in the stratosphere, where relatively small concentrations of
O_3 ($\leq$ 10 parts per million) are responsible for absorbing a substantial fraction of the incident, cancer-causing ultraviolet radiation of the sun. The energy absorbed by the O_3 is also responsible for the increase of temperature with altitude that characterizes
the stratosphere.[9.13] The temperature gradient, in turn, makes the
stratosphere hydrostatically stable and inhibits vertical mixing. It
should not be surprising, therefore, that the "ozone layer" and the
stratosphere are virtually synonymous.

Anything that threatens to deplete the protective shield of stratospheric ozone is regarded with appropriate dread by the public below.
Interestingly, much of this awareness can be traced back to the SST,
where the initial observation that stratospheric engine exhaust emissions of oxides of nitrogen (NO_x) could reduce ozone led to great
interest in and study of the chemistry of the stratosphere.

HAP(Equilibrium) may be used to obtain realistic estimates of
the types and amounts of the chemical constituents found in airbreathing engine exhaust flows. Figure 9.14 displays the *emission
index* (i.e., grams of constituent per kilogram of fuel, a standard
yardstick employed by the EPA as a fair gauge of the environmental impact per unit mass of fuel) for hydrogen and methane fuels of
several of the incomplete products of combustion that are referred
to here as "candidate pollutants." In these computations, the entry
conditions to the burner were determined using HAP(Trajectory),
and the combustion was taken to be stoichiometric, equilibrium, and
constant pressure (i.e., constant static enthalpy). The results equate
to frozen flow in the expansion system and therefore represent what
may be presumed to be the upper limit for the candidate pollutants,
although this disregards the complex reactions that can take place
in the combustor boundary layers.

The information of Fig. 9.14 reveals that the emission index of
NO is almost independent of fuel type and freestream Mach number, and has a value of about 24. Since all other oxides of nitrogen
are entirely negligible under these conditions, NO and NO_x are the
same. The emission indices of OH and O increase steadily but gradually with freestream Mach number, and are dependent on fuel type.
The reduced values for OH and O that go with methane fuel are
accompanied by the appearance of an emissions index of about 90
for CO.

The challenge of computing emission indices of candidate pollutants pales in comparison to determining their eventual fate in the
stratosphere because the chemical processes there are so perplexing.
The following examples will amply illustrate the problem.[9.13]

First, consider the direct interaction of NO with the O_3 of the
stratosphere. On the one hand, in the mid-to-upper stratosphere,

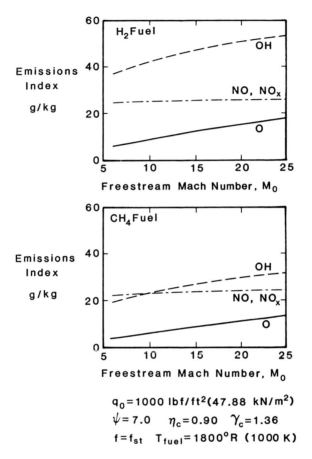

$q_0 = 1000$ lbf/ft^2(47.88 kN/m^2)

$\psi = 7.0$ $\eta_c = 0.90$ $\gamma_c = 1.36$

$f = f_{st}$ $T_{fuel} = 1800°R$ (1000 K)

Fig. 9.14 The emission indices of three candidate pollutants as functions of freestream Mach number and fuel type.

the NO participates in a catalytic cycle that eliminates O_3 without affecting the NO, to wit,

$$
\begin{array}{r}
NO + O_3 \rightarrow NO_2 + O_2 \\
NO_2 + O \rightarrow NO + O_2 \\
\hline
\text{Net}: \quad O_3 + O \rightarrow 2O_2
\end{array}
\qquad (9\text{-}25)
$$

This is the ozone depleting mechanism that attracted attention to the impact of the SST exhaust emissions on the stratosphere in the first place. On the other hand, in the lower stratosphere, where O atoms are less abundant, the photolysis of NO_2 to form O, which produces O_3, becomes a competitive reaction, in which

$$NO + O_3 \rightarrow NO_2 + O_2$$
$$NO_2 + h\nu \rightarrow NO + O$$
$$\underline{O + O_2 \rightarrow O_3}$$

Net : $\underline{\hphantom{xxxxxxxxx}}$ (9-26)

Second, consider the interactions of the candidate pollutants with $C\ell$, which is a leading destroyer of stratospheric O_3 because of the catalytic chain reaction [similar to that of Eq. (9-25)]

$$C\ell + O_3 \rightarrow C\ell O + O_2$$
$$\underline{C\ell O + O \rightarrow C\ell + O_2}$$

Net : $\underline{O_3 + O \rightarrow 2O_2}$ (9-27)

On the one hand, any unburned CH_4 can eliminate $C\ell$ and terminate its destruction of O_3 by producing $HC\ell$ in the reaction

$$CH_4 + C\ell \rightarrow CH_3 + HC\ell \tag{9-28}$$

On the other, OH can restore the $C\ell$ by recycling it from $HC\ell$ in the reaction

$$HC\ell + OH \rightarrow H_2O + C\ell \tag{9-29}$$

The impact of exhaust emissions on stratospheric ozone is therefore dependent upon the altitude at which it is emitted. Moreover, the stability of the stratosphere can support pollutant lifetimes of 100 years or more. Consequently, it is not unusual to find studies that project that over the next century increased N_2O (due to the use of fertilizers) will deplete stratospheric ozone by about 4 percent, an effect that will persist for 100 years, while the increase of NO_2 (due to subsonic aircraft) will replenish stratospheric ozone by about 1 percent, an effect that will last for less than one year.[9.13] It is also virtually impossible to predict what the public will regard as an "acceptable" impact. The most that can be done at this time, therefore, is to heighten your awareness of the importance and complexity of this topic. As noted earlier, the last word on stratospheric ozone depletion is far from being written.

9.6 ENGINE SUBSYSTEMS

Behind the engine wall, literally, there is a metropolis of collections of equipment known as *engine subsystems*. This simple label fails to betray the challenge of their technology and the vital importance of their success to the aerospace system. Every propulsion engineer needs to be aware of their presence.

Even though engine subsystems only constitute a few percent of the dry or empty weight of the vehicle, they provide a wide variety of services for the propulsion system, as well as for the parent aerospacecraft. Figures 9.15 and 9.16 underscore the dimensions of the job by presenting schematic and cutaway diagrams for the subsystems of a typical airbreathing hypersonic engine. Although the particular subsystems must differ from one application to another, and a standard lexicon of terminology does not exist, a reasonable collection would include at least the following (not prioritized):

1. Fuel and oxidizer tanks (both cryogenic and ordinary temperatures).

2. Fuel and oxidizer delivery (including turbopumps, piping, and valves).

3. Fuel flow and engine geometry controls (including the computational brains, the sensors, and the activating brawn).

4. Rotating shafts, bearings, lubrication, and supports (especially for combined cycle engines).

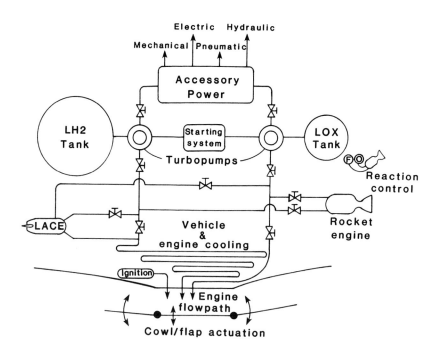

Fig. 9.15 Schematic diagram showing the subsystems of a typical hypersonic airbreathing engine.

5. Starting, restarting, and ignition.

6. Auxiliary propulsion (such as reaction control for position and attitude control in space, and emergency propulsion for landing).

7. Heat exchangers (for the engine proper, such as the wall cooling panels, the LACE, and bearing lubricants, as well as air conditioning for the electronics and crew).

8. Mechanical, electric, pneumatic, and hydraulic power (for all accessories).

9. Vehicle pressurization, sealing, purging, venting, and draining (including the use of inert gases to reduce leakage hazards and prevent cryopumping).

10. Fire prevention, detection, and suppression.

Accomplishing all these functions repeatedly and reliably in a harsh and demanding environment, while minimizing weight and cost, is obviously the dedicated work of many experts. Another textbook would be required to do justice to their efforts. Instead, we can only attempt to provide an introduction to the field by means of a few examples. Success in the design and development of advanced subsystem components is usually tied to the details, more often than not to improved materials and/or structural concepts, and hypersonic airbreathing propulsion is no exception.

Fuel piping is a pervasive challenge because there can be miles of passages that must handle hydrogen at temperatures up to at least 1800 °R (1000 K) and pressures of the order of 100 atm. The piping must be lightweight because it is not a structural element, and the material must not be subject to hydrogen embrittlement (on the inside) or oxidation (on the outside), or reliable protective coatings must be provided. Available materials do not offer adequate specific strength (i.e., yield strength divided by density) at these temperatures, and altogether new materials, such as ceramic matrix composites and carbon/carbon composites must be considered.

Heat exchanger tubing for the LACE, for example, must be extremely fine (i.e., long and narrow, with very thin walls) because of heat transfer considerations. It must also be joined at each end (i.e., welded or brazed) to the heat exchanger headers, and be able to withstand thermal stress and foreign object damage. Like fuel piping, there is a lot of it, and it is not part of the structure. Finding a relatively light material that can be reliably formed and joined into this configuration is an important matter.

Fuel valves for hydrogen are subject to self-welding because the hydrogen removes the oxide layers and exposes clean metal surfaces. When the valve is closed, the gate welds to the seat, and it cannot

Fig. 9.16 Cutaway drawing showing the subsystems of a typical hypersonic airbreathing vehicle. Courtesy of the National Aerospace Laboratory of Japan.

be reopened. Since the fuel valves could be opened and closed many times during a mission, this is a situation that must be avoided. An obvious but untested and uncertain solution would be to use non-metallic materials, such as ceramics.

Fuel turbopumps for hydrogen would be similar to those of most liquid rocket engines in the sense that they "bootstrap" the process by burning a small fraction of fuel and oxidizer in a gas turbine to provide pumping power for the remainder. Since the maximum hydrogen flow rates for hypersonic airbreathing vehicles are expected to be similar to those of a Space Shuttle Main Engine (SSME), about 150 lbm/s (68 kg/s), this might appear to be an off-the-shelf technology. However, the SSME turbopumps were designed for only 7.5 hours of operation over a lifespan of 100 missions and are considered to be too heavy for airbreathing applications. Thus, the designer is confronted with the daunting task of making a complex machine both lighter and more durable. New turbomachinery concepts, new materials, and even new rotating shaft bearing concepts will have to be considered.

Cryogenic fuel tanks for liquid hydrogen are a dominant feature of hypersonic airbreathing vehicles. They are very large because of the low density of hydrogen. For example, 100,000 lbm (45,360 kg) of liquid hydrogen occupies a volume of about 21,000 ft^3 (600 m^3), or a sphere with a diameter of about 34 ft (10 m). This is probably larger than the room you are in right now. The tank must provide the structural strength to support the fuel at rest and during accelerations and the insulation to maintain the cryogenic liquid state, while being as lightweight as possible. Advanced materials, such as aluminum-lithium and carbon-carbon composites, must be given strong consideration, as well as such advanced concepts as "integral" tanks that provide primary structural strength for the entire vehicle.[9.15] Figure 9.17 is a photograph of a NASP prototype 900-gallon, insulation-wrapped, multi-bubble, graphite-epoxy cryogenic fuel tank that was successfully tested for mechanical and thermal loads inside a fuselage structure. The "multiple-lobe" fuel tank cross section is a structurally efficient design that puts the strength of cylindrical vessels into a configuration compatible with the aerodynamics of the vehicle.

9.7 VEHICLE STABILITY AND CONTROL

Directing the flight of hypersonic airbreathing vehicles from take-off to cruise or orbit and back will be a great achievement for stability and control. The airbreathing engine will be an important player in this drama.

Fig. 9.17 Prototype National Aero-Space Plane liquid hydrogen fuel tank. Although it is a one-third linear scale model, it is still about 10 ft or 3 m in diameter.

An interesting characteristic of the jet age is that the influence of the engine on the airframe has grown steadily with time. In the beginning, the engine was relatively small compared to the airframe, and was merely plugged in or hung on at some convenient location. As airplanes flew higher and faster, they required more and/or larger engines and more airflow, and it became essential to carefully integrate them together in order to avoid harmful engine-airframe installation effects (e.g., increased airframe drag or stall-inducing engine inlet distortion). The enormous propulsive power onboard enabled previously unheard of opportunities, such as vertical flight and maneuvering enhanced by deflected thrust.

The ramjet and scramjet represent the culmination of this trend for, as you have seen, they are equal partners with the airframe in terms of external territory, occupying almost the entire lower half of the vehicle, and their propellants and subsystems account for much of the weight and volume. Indeed, it has become impossible to clearly separate propulsion from airframe. Spirited members of the propulsion community now call the vehicle an "engine frame." These trends have serious implications for the stability and control of the hypersonic vehicle that must be faced during the design process.

Stability and control analysis is based on the moments acting about the instantaneous center of gravity of the vehicle, rather than on the forces acting on the vehicle. Excellent textbooks are available for anyone wishing to brush up on or dig more deeply into this important subject.[9.16, 9.17] Figure 9.18 summarizes the customary terminology and nomenclature needed for a basic discussion. Rotation about the x axis is called *roll* or *lateral motion*, and is primarily controlled by differential or opposite deflection of the ailerons. Rotation about the y axis is called *pitch* or *longitudinal motion*, and is primarily controlled by coordinated deflection of the elevators. Rotation about the z axis is called *yaw* or *directional motion*, and is primarily controlled by deflection of the rudder.

A vehicle that initially returns toward its *equilibrium* or *trimmed* position after being disturbed is said to be *statically stable*. A vehicle that initially moves away from its equilibrium position after being disturbed is said to be *statically unstable*. A vehicle that eventually returns to and remains at its equilibrium position after being disturbed is said to be *dynamically stable*. A vehicle that does not return to its equilibrium position and regain trimmed flight after being disturbed is said to be *dynamically unstable*. The foregoing definitions all refer to the unaltered configuration of the vehicle. The control surfaces may be moved in order to compel the vehicle to return to the original, or any other, equilibrium position, as well as perform nonequilibrium maneuvers.

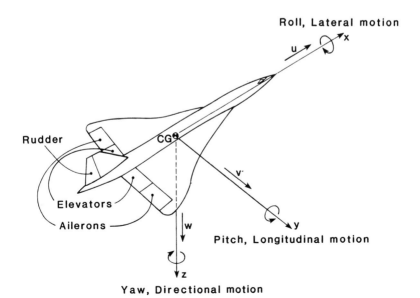

Fig. 9.18 Stability and control terminology and nomenclature.

The criterion for static longitudinal stability is that the center of gravity must be forward of the *neutral point,* a geometric location that depends upon the configuration of the vehicle, particularly the wing and horizontal stabilizer, and the Mach number. The location of the neutral point succinctly quantifies the combined influence of the wing and the horizontal stabilizer on the longitudinal motion of the vehicle. An arrow, which is very stable, has a neutral point located close to the tail feathers. The *static margin* of a vehicle is the distance that the neutral point lies behind the center of gravity divided by the wing chord, and is positive when the vehicle has static longitudinal stability. Similarly, when the neutral point lies ahead of the center of gravity, the vehicle will be statically unstable for longitudinal motion, and when they coincide, the vehicle has neutral static longitudinal stability.

Many modern civil and military high-performance aircraft operate routinely with negative static margins because they are able to maneuver more rapidly (being essentially unstable), and can have less trim drag and horizontal stabilizer area. The increased pilot workload that would otherwise be required to overcome the longitudinal instability is shouldered by a redundant collection of computers that constantly monitor flight conditions and position control surfaces to maintain stability, a system known as *fly-by-wire.* Fly-by-wire can only effectively compensate for a certain amount of negative static margin, of the order of 0.20.

Hypersonic aerospacecraft, for similar reasons, will also have negative static margins, at least over some portions of their flight trajectories. The magnitude of the negative static margin depends directly upon the instantaneous location of the center of gravity which, in turn, is determined by the distribution of the fuel and oxidizer remaining onboard. Since the fuel and oxidizer constitute such an unusually large fraction of the takeoff weight of a typical hypersonic vehicle, it is clear that the longitudinal distribution of the remaining fuel and oxidizer must be carefully managed in order to guarantee adequate static margin under all foreseeable conditions over the entire flight spectrum.

Moving up the ladder of complexity, we next observe that the airbreathing engine itself produces moments about the instantaneous center of gravity of the vehicle. The magnitude of the moments can be rigorously obtained by applying control volume conservation of moment of momentum to the flow entering and leaving the engine (see Fig. 8.7). For our present purposes, however, it will suffice to observe that moments are created at least because the axial thrust may not be applied through the center of gravity (and is probably offset below, causing the vehicle to pitch up) and because the exhaust flow may have a nonaxial component of velocity and momentum (see

Fig. 7.9) upward in the overexpanded condition, causing the vehicle to pitch up, and downward in the underexpanded condition, causing the vehicle to pitch down. These moments must be accounted for in determining the necessary control authority of the vehicle. Moreover, since the thrust-to-weight ratio is unusually large for hypersonic vehicles, and since the thrust and its moments can be varied by throttle setting independent of flight conditions, the flight spectrum must be thoroughly explored in order to avoid unpleasant surprises. Since longitudinal moment of inertia of the vehicle varies considerably as the fuel is burned, the pitch rate resulting from a given moment is also variable, and a wide range of responsiveness must be considered. Similar remarks apply to roll rate and yaw rate responsiveness.

An optimistic view of the engine moment situation is that combinations of throttle setting and engine geometry can be found that contribute to the positive longitudinal stability of the vehicle. This would, of course, be the ultimate integration of propulsion and aerodynamics. A pessimistic view is that a set of baroque conditions will slip through that creates havoc during the flight-test program. An example of this would be ground effects on the exhaust stream during takeoff that increase the pressure at the rear and cause the vehicle to pitch over onto its nose.

The next level of complexity is that engine performance and flight conditions are closely coupled, even when the throttle is fixed. That is, a change in Mach number, altitude, or angle of attack alters the entry conditions to the combustor and the resulting engine thrust vector, which alters the flight conditions, and so on. Analyzing this phenomenon ideally requires a detailed knowledge of the temporal response of the engine to changing flight conditions, which may not be available before the first flight, thus necessitating a conservative approach to the control system's design (and possibly excess margin and control power).

Published material is beginning to appear on this subject. One paper[9.18] reports the investigation of the dynamic longitudinal stability of lifting hypersonic vehicles, employing a relatively simple model of variation of thrust with freestream Mach number and altitude. The authors examined two long-period longitudinal modes, namely the conventional oscillatory phugoid mode (constant angle of attack) and the aperiodic height mode (caused by the variation of atmospheric density with altitude). The authors conclude that the propulsion system is a dominant influence on the stability of these modes. The propulsion system significantly destabilized both modes in regions where thrust increased with Mach number (e.g., as ramjet thermal and propulsive efficiency are improving near $M_0 = 3.0$). Increasing the thrust-to-weight ratio from 1 to 4 further destabilized the phugoid mode, but stabilized the aperiodic height mode. Nevertheless, both modes were dynamically *unstable* for both thrust-to-weight ratios for their vehicle.

Several other stability and control issues peculiar to hypersonic airbreathing vehicles have also emerged. One is the impact on the maneuver envelope (i.e., turn rate, climb rate, and descent rate) of the stringent constraints imposed by the propulsion system on allowable vehicle angle of attack, sideslip and dynamic pressure variations, and throttle modulation. The need to execute precise, coordinated maneuvers during excursions from a hypersonic cruise condition can sharply restrict the maneuver envelope.[9.19] Another is the impact of the physical deformations of the vehicle caused by the low frequency oscillations of the slender, flexible, cooled structure on the static longitudinal stability of the vehicle. These so-called aerothermoelastic effects can further reduce the negative static margin.[9.20]

Finally, there is the impact of losing the thrust of an engine, or engine-out behavior, on the yaw stability of the vehicle. For highly integrated configurations, like the National Aero-Space Plane, which has a number of engine modules laid side by side within the cowl, module-out is equivalent to engine-out. The loss of an engine off the centerline produces yawing moments, and the yawing motion can induce coupled rolling or pitching moments that must be quickly compensated for by the control system. The importance of this problem is confirmed by one of the truly heroic ground test programs of the Concorde development. The concern was that the rapid rotation of the aircraft in supersonic flight due to the loss of one engine would cause others to stall because their inlet distortion would grow faster than could be compensated for by their control systems. In order to prove that the Rolls Royce SST engines could tolerate this situation without stalling, the National Gas Turbine Establishment tested them in a facility with a pivoting supersonic nozzle that could simulate the effects of changing yaw and pitch by swinging the upstream flow across the engine inlet at several degrees per second.

These, and other, hypersonic airbreathing vehicle stability and control problems can be solved by providing suitable control strategies and sufficient control authority. This discussion is primarily intended to verify that the problems are serious and real, that they are best discovered and solved during the design phase rather than during the flight-test program, and that the airbreathing propulsion system is more influential than ever before. In addition, we hope you see that stability and control is a truly fascinating field of study that is critical to the success of hypersonic flight.

9.8 HYPERSONIC AIRBREATHING PROPULSION TESTING

The successful aircraft and space propulsion systems in service today are visible evidence of a disciplined development process that is founded on decades of experience and designed to ensure that the products meet customer needs. One of the most important steps in this development process is captive *ground testing,* the aim of which is to permit highly instrumented articles to be precisely evaluated under carefully controlled and realistic environmental conditions until they are proven to be ready for flight. This allows performance, reliability, and durability to be compared with the prescribed requirements and, where necessary, improvements to be made or unexpected problems solved. Also important, where possible, is *flight testing,* either on the parent vehicle or, occasionally, on a slave vehicle that provides a suitable platform for the propulsion device. Flight testing can offer more realistic, sometimes otherwise unobtainable, environmental conditions, but usually at a sacrifice of control over those conditions, of the quantity and quality of the instrumentation and data, of safety, and of economy relative to ground testing.

Much has been written through the years about the status of testing capabilities for hypersonic airbreathing propulsion.[9.21−9.24] A powerful reason for this interest is that major ground test facilities require at least 10 years to design, build, shake down, and calibrate. This period can be even longer if the technology necessary to design the facility is not in hand, as is the case for $M_0 > 10$ facilities, and must first be developed through research or pilot plants. When you add this to the 10 years of ground testing required by major aerospace systems before they are ready for production, you find that test facility planning must precede initial operational capability by more than 20 years. Thus, one must start early in order to arrive on time.

The material to follow echoes the major themes and conclusions of the most recent reports on hypersonic airbreathing propulsion testing, which have much in common with each other. Reference 9.21 has special historical significance because it describes the outstanding pioneering work and progress that was made during the hypersonic revolution prior to 1967, as well as the fate of the facilities involved. Reading these reports will also convince any reasonable person that propulsion testing is a massive, expensive, sophisticated, dynamic, and essential undertaking that demands the skills, energies, and imaginations of thousands of very talented and fiercely devoted people. We hope that this brief summary of their endeavors also conveys this message.

One of the most important findings of these investigations is that the classical method of reducing test facility demands by means of dimensionless parameters offers only limited refuge in the case of hypersonic airbreathing propulsion, mainly because it is necessary to correctly reproduce the combustion chemistry until we know more than we do now. Thus, the combustor entry conditions *and* the scale of the test article must be as close as possible to those expected in flight in order to provide reliable experimental results.

A useful classification to bear in mind when considering ground-test facilities is their division into R&D (research and development) or T&E (test and evaluation) facilities. R&D facilities are generally highly instrumented and generate high resolution data to help understand phenomena, to develop and validate computational methods, and to carry out parametric studies on components and subcomponents. In short, they provide the detailed database for design. T&E facilities are used to evaluate the resulting vehicle design in terms of overall system performance, operability, and durability, and are generally larger scale, emphasize measurement of global quantities near flight conditions, cost more to build and operate, and have a very long lead time for acquisition.

9.8.1 Continuous Flow Ground Testing

The bread-and-butter method of testing propulsion devices on the ground is to simulate the flight conditions in a full-scale *continuous flow facility* conceptually similar to that depicted in the so-called *freejet* facility of Fig. 9.19. The air is drawn in from the nearby atmosphere and compressed, dried, and heated or cooled, as necessary, to duplicate the actual freestream stagnation conditions, introduced to the test cell from the upstream plenum by a facility nozzle that brings the flow to the freestream Mach number, static pressure, and direction, and removed along with any combustion products by an exhauster system containing vacuum pumps and water spray cooling that maintains freestream static pressure downstream of the test article. The total airflow required for freejet testing must exceed that being swallowed by the engine, usually by several multiples, in order to assure adequate replication of the overall flowfield.

Where possible, continuous flow, full-scale testing is the most desirable ground test alternative because it provides the most precise control over the environmental simulation and propulsion system variables, and allows the actual hardware (rather than a scaled-down or otherwise modified version) to be used. The long duration of testing also allows the propulsion system to come to mechanical and thermal equilibrium, which is one way of precisely defining the configuration of the hardware, and is the condition most likely to be found in flight. Continuous flow facilities are also used to test the response of the integrated propulsion system to transient throttle motions to demonstrate its operability (i.e., its resistance to such disruptive events as unstart and flame-out and the authority of its control system). For the same purpose, some modern, sophisticated, continuous flow facilities can rapidly vary the test conditions and

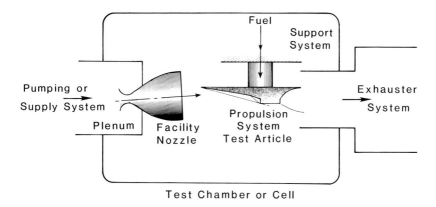

Fig. 9.19   Schematic diagram of a typical airbreathing propulsion system freejet test facility.

angle of attack in order to simulate the effect of transient maneuvers on the propulsion system.

The Aeropropulsion Systems Test Facility (ASTF) located at the U.S. Air Force Arnold Engineering Development Center (AEDC) in Tennessee, pictured in Fig. 9.20, is a very advanced version of a continuous flow ground test facility. The ASTF can carry out full-scale testing of turbine engines up to 100,000 lbf (444.8 kN) of thrust for $500 < q_0 < 2000$ lbf/ft^2 and $M_0 < 3.8$. Thus, even at a construction cost of \$625 million, it is useful for takeoff and ramjet, but not scramjet, testing. You will now see why continuous flow, full-scale testing is a distinct luxury, and is not always possible for hypersonic airbreathing propulsion.

Since the flow in the upstream plenum has a low, subsonic Mach number, and the flow in the facility freejet nozzle is nearly reversible, the plenum conditions approximate those that would result from bringing the desired flight conditions isentropically to their rest or stagnation conditions. The almost incomprehensible magnitude of the stagnation conditions of hypersonic flight along a constant dynamic pressure trajectory are quite easily obtained by first using HAP(Trajectory) to establish the freestream conditions, and then using HAP(Air) to find the stagnation quantities of interest (e.g., temperature and pressure).

The staggering results of this analysis for hypersonic flight are presented in Fig. 9.21, one of the few graphs truly worthy of a logarithmic scale in this entire textbook. This information makes it clear that the stagnation temperature is primarily dependent upon freestream Mach number, that stagnation pressure increases somewhat with dynamic pressure, and, above all, that stagnation pressure

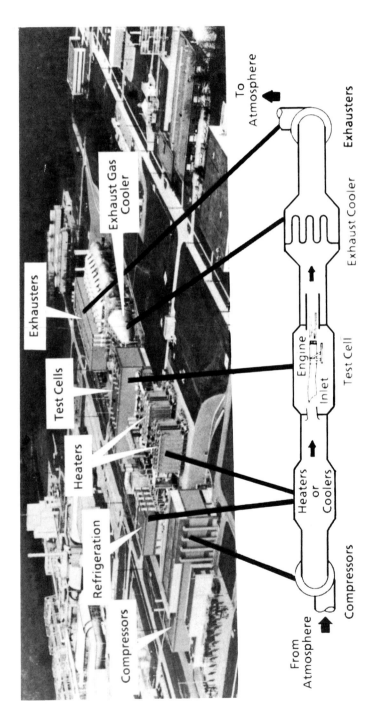

Fig. 9.20 The Aeropropulsion Systems Test Facility at the U.S. Air Force Arnold Engineering Development Center in Tennessee, an advanced continuous flow ground test facility.

increases dramatically with freestream Mach number despite the fact that the freestream static pressure is decreasing as its inverse square.

If you think that making upstream equipment that can withstand many thousands of atmospheres and tens of thousands of degrees is an overwhelming engineering task, you are right. One way to see this is to recognize that it is hard to build a practical pressure vessel that can contain more than about 1000 atm (the pressure on a submarine more than 6 mi or 10 km down), and even mighty carbon melts by about 7000 °R or 3900 K. Both of these are exceeded for $M_0 > 12$. Another way is to calculate the minimum compressor

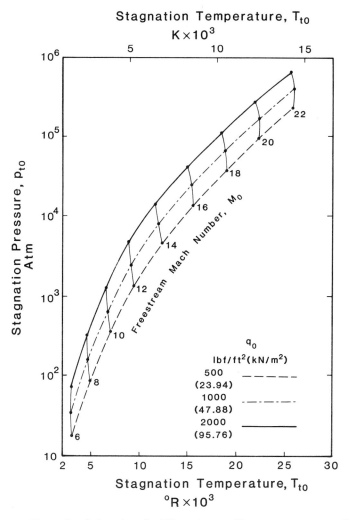

Fig. 9.21 Required freejet facility stagnation pressure and stagnation temperature as functions of freestream Mach number and trajectory dynamic pressure.

power and total power required for some nominal propulsion-system air throughflow, say 100 lbm/s (45.36 kg/s). An estimate for the lower limit of required compressor power is obtained from availability considerations as

$$\dot{W} \geq \dot{m}\left\{(h_{to} - T_{SL}s_0) - (h_{SL} - T_{SL}s_{SL})\right\} \qquad (9\text{-}30)$$

and the total power is obtained from first law considerations as

$$\dot{W} + \dot{Q} = \dot{m}\,(h_{to} - h_{SL}) \qquad (9\text{-}31)$$

Recall from Chap. 2 that $\dot{W}$ and $\dot{Q}$ are positive when added to the flow.

The results of calculations based on Eqs. (9-30) and (9-31) and the previously computed stagnation enthalpies are summarized in Table 9.5. Regardless of the units used, these quantities are huge, especially for such a modest amount of airflow. For example, it takes about ten high bypass ratio turbofan compression systems to produce 1000 kHP, and 1000 MW is about the electrical power generated by a major nuclear power plant. The estimates of Table 9.5 also ignore the power required to return the exhaust flow to the atmosphere, an amount that equals or exceeds the upstream value.

An interesting feature of the information of Table 9.5 is that the minimum compressor power approaches the total power as the freestream Mach number increases. This is merely a reflection of the fact that the vast majority of the freestream enthalpy is in kinetic energy or mechanical form for truly hypersonic flight, and therefore that most of the upstream enthalpy must be generated by the compressor. The inequality in Eq. (9-30) reminds us that the real compression process is imperfect, and that the actual compressor power will be larger. Indeed, if the compressor efficiency is too low, the compressor power will exceed the total power required, and the excess will have to be removed by cooling (unavoidable or deliberate) in order to provide correct simulation.

Some relief from these enormous demands can be obtained by means of *direct connect* or *semi-direct* testing, as depicted in Fig. 9.22. The main advantages of this approach are that a lower stagnation pressure (corresponding to the flow conditions downstream of all or some of the compression system losses) is required and the same hardware can be tested with less upstream flow. The main disadvantages are that the same stagnation enthalpy must be provided and the combustor entry and external flowfields do not reproduce those of true flight.

The stagnation pressure relief afforded by direct connect testing may also be estimated using HAP(Trajectory) and HAP(Air), but

Table 9.5   Estimated power requirements for a continuous flow facility providing 100 lbm/s (45.36 kg/s) at a dynamic pressure of 1000 lbf/ft² (47.88 kN/m²).

	Minimum Compressor Power	Total Power
Freestream Mach Number M_0	kHP (MW)	kHP (MW)
6	78.2 (58.3)	93.7 (69.9)
10	259 (193)	281 (210)
20	1240 (928)	1270 (951)

somewhat differently than before. First, HAP(Trajectory) is used to estimate the conditions entering the combustor, and then HAP(Air) is used as before to compute the isentropic stagnation conditions.

The results of this analysis for $q_0 = 1000$ lbf/ft² (47.88 kN/m²) are plotted in Fig. 9.23, alongside those of Fig. 9.21 for comparison. This information reveals that the required stagnation pressure can be reduced by a factor of about 6, even for $\eta_c = 0.90$, at almost any hypersonic freestream Mach number. Although this is a considerable reduction, the required stagnation pressures remain, in an absolute sense, enormous. We can also see the influence of equilibrium chem-

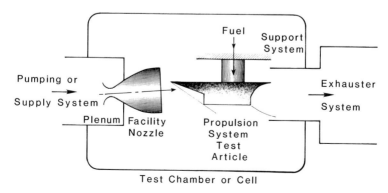

Fig. 9.22   Schematic diagram of a typical airbreathing propulsion system semi-direct connect test facility.

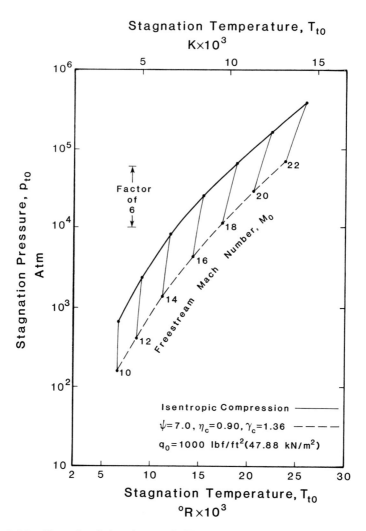

Fig. 9.23 Required freejet and direct connect facility stagnation pressure and stagnation temperature as functions of freestream Mach number for a constant q_0 trajectory.

istry on these results in that the stagnation temperature is reduced for a given freestream Mach number (and stagnation enthalpy) because the lower stagnation pressure permits the energy to be invested in dissociation rather than random motion.

9.8.2 Short Duration Ground Testing

A close relative of continuous flow ground testing is provided by *short duration* or *blowdown facilities,* in which the air is slowly compressed and stored in high-pressure vessels until ready, and heated by various means while making its way to the test article. Although they are not, strictly speaking, continuous, the run times are of the order of a few seconds to a few minutes, depending on the test conditions, and therefore long enough to approach aerothermodynamic equilibrium. Two of the preferred methods for heating the air are to pass it through a lattice of heated bricks or pebbles, known as a *thermal storage heater* or *pebble bed heater,* or to burn it in a combustor that replenishes the consumed oxygen in order to supply the engine with suitable chemical constituents, known as a *vitiated heater.* The latter method produces somewhat higher temperatures than the former, but raises questions about the fidelity of the propulsion-system combustion process, particularly when operating away from the design point where molecular weight differences and trace constituents can significantly alter the mixing and chemical reactions.

Figure 9.24 contains a schematic illustration of an advanced blow-down facility, the Aerodynamic and Propulsion Test Unit (APTU) located at the AEDC. The diameter of the upstream nozzle is about 3 ft or 1 m. The APTU pressure vessels are capable of providing as much as 135 atm, and stagnation temperatures of about 3600 °R (1800 K) with alumina pebble bed heating or 4500 °R (2500 K) with vitiated heating. Thus, referring to Fig. 9.20, we conclude that the APTU can provide faithful simulation for fairly large test articles for a dynamic pressure of 1000 lbf/ft^2 (47.88 kN/m^2) and freestream Mach numbers up to about 6 with pebble bed heating and about 8 with vitiated heating. This represents a truly hypersonic airbreathing propulsion testing capability.

The stagnation temperature of the air can be further increased in both continuous and short duration ground test facilities by means of the electrical resistance heating of an *arc heater,* perhaps as high as 16,000 °R (8889 K), but true aerothermodynamic simulation would require that the stagnation pressure also be greatly increased beyond the current limit of about 100 atm. Arc heaters can contaminate the flow with the molecular debris of sputtered electrode materials and become less stable as mass flow rate increases. Nonetheless, arc heaters can provide the stagnation temperatures necessary for the testing of materials and structures up to freestream Mach numbers of about 15, and are therefore an important weapon in the ground test arsenal.

A final variant of short duration ground testing is to replace the upstream equipment with a liquid rocket engine whose exhaust products are similar to air in terms of oxygen concentration and molecular weight. The NASP program funded the development of one such large-scale ground test facility by the Aerojet TechSystems Company, which provides true stagnation pressure and temperature simulation for a dynamic pressure of 1000 lbf/ft^2 (47.88 kN/m^2) up to freestream Mach numbers of about 8. One may therefore safely conclude that

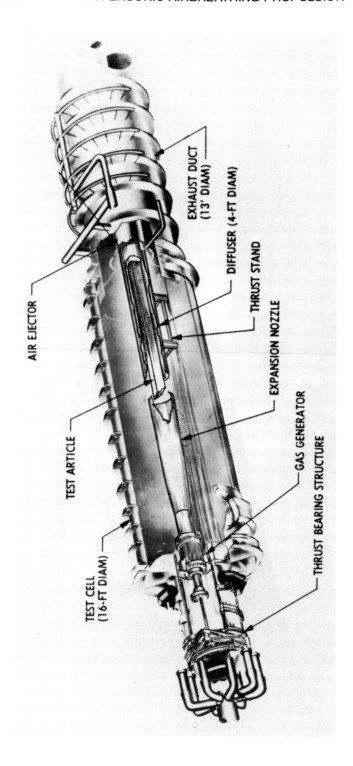

Fig. 9.24   The Aerodynamic and Propulsion Test Unit at the U.S. Air Force Arnold Engineering Test Center in Tennessee, an advanced blowdown ground test facility.

flying at Mach 8 is roughly equivalent to living in the exhaust of a high-performance rocket engine. No one said it was going to be easy.

9.8.3 Pulsed Flow Ground Testing

An important benchmark in ground testing is the ability to simulate flows that have high enough stagnation temperatures to produce dissociation of the air or "real gas effects," at least in the regions where the flow is brought completely or nearly to rest, such as stagnation points, boundary layers, and separation zones. Again, this is primarily because the real gas effects will change the character of the chemistry in complex ways that must be discovered or confirmed through credible testing. Since the real gas effects turn on gradually, rather than abruptly, with increasing stagnation temperature (see Figs. 2.7 through 2.10) the boundary is blurred, but generally begins to take place in the vicinity of stagnation temperatures of about 4500 °R (2500 K) or freestream Mach numbers of about 8 for typical constant dynamic pressure trajectories (see Fig. 9.21).

Unfortunately, *none* of the continuous flow or short duration facilities described above can test under conditions for which real gas effects would be strong or dominant. Instead, an ingenious assortment of *pulsed flow* ground test facilities has been devised that can provide the desired conditions, three of which are depicted in Fig. 9.25. Pulsed flow facilities share the common operating principle of "leveraging" the energy of a larger mass of gas by transferring it rapidly to a smaller mass, and then "leveraging" the energy of the smaller mass by concentrating its release in space and time. They also share the principal drawback of very short duration times, usually in the range of a fraction of a millisecond to several milliseconds. A rule of thumb is that a flow comes to equilibrium in about the time it takes the average particle to traverse the region of interest, an interval known as the *passage time* or *fill time*. Since the passage time for scramjet combustors is of the order of 1 ms, it would appear that pulse facilities provide adequate test time. However, the rule of thumb is more appropriate to "well-behaved" flows, and is likely to be optimistic where slowly developing flow structures (such as regions of separation) are involved or heat transfer is important. Hence, considerable effort is applied toward increasing their duration times.

The three pulsed flow facility concepts of Fig. 9.25 will now be very briefly described. Since facility operating parameters are subject to many qualifications, those of typical state-of-the-art examples are included only to illustrate their comparative merits. The devices are called tunnels by analogy to wind tunnels because they were originally intended to test the flying characteristics of hypersonic aerospace vehicles.

In the *shock tunnel*, the high-pressure gas in the driver tube is released by bursting the primary (p) diaphragm, causing a strong normal shock wave to propagate through the working gas down the driven or shock tube. When the normal shock wave reflects off the secondary (s) diaphragm, it bursts, allowing the highly compressed air behind the shock wave to escape through the facility nozzle into the test chamber or article, and then into a vacuum dump tank. The Calspan 96-in (4.0 ft/1.22 m nozzle diameter) hypersonic shock tunnel has replicated $M_0 = 10$ conditions, and is capable of creating stagnation temperatures to about $M_0 = 15$, although the stagnation pressure is somewhat less than that required for freejet testing. The test duration is a few milliseconds.

In the *free-piston shock tunnel*, the piston is driven forward by the force of the compressed gas accumulated in the reservoir. The piston compresses a light gas in the compression tube, which then plays the role of the driver gas of the ordinary shock tunnel. The University of Queensland free-piston shock tunnel generates stagnation temperatures corresponding to $M_0 = 22$, but the stagnation pressures are about 1 percent of the desired value. The test duration is 0.05–0.10 ms.

In the *expansion tube*, the high-pressure gas in the driver tube is released by bursting the primary diaphragm, causing a strong normal shock wave to propagate through the working gas down the driven tube. When the normal shock wave reaches the secondary diaphragm it bursts, sending a normal shock wave through the low pressure acceleration gas but allowing the working gas to expand into the emptying acceleration tube and eventually into the facility. The resulting Mach number of the working gas is high enough that no facility nozzle is required. The test time begins after the slug of compressed acceleration gas has passed through the test section and ends with the arrival of the first waves that disrupt the uniform test flow, for example, the downstream edge of the expansion wave also generated by the rupturing of the secondary diaphragm. The fundamental advantages of the expansion tube are that the working gas flow never stagnates, thus reducing the extent of dissociation, and several conditions can be obtained by altering the initial filling pressures. The disadvantage is that the test durations are relatively short. The NASA Langley Research Center Hypersonic Pulse (HY-PULSE) expansion tube facility generates stagnation temperatures corresponding to $M_0 = 18$, but the stagnation pressure roughly corresponds to direct connect conditions. The test duration is about 0.2 ms.

Although these pulsed flow facilities are primarily based on one-dimensional flow principles, their simulation capability is degraded by nonuniform flows due to the presence of boundary layers and

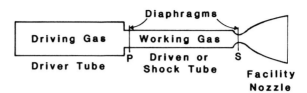

A: Shock Tunnel

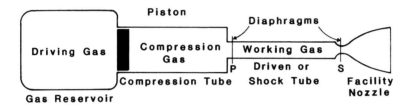

B: Free-piston Shock Tunnel

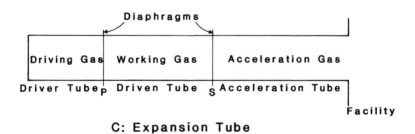

C: Expansion Tube

Fig. 9.25   Schematic diagrams of three pulsed flow ground test facility concepts.

diaphragms, as well as by the axial diffusion of material at the contact surfaces or interfaces between different gases and debris resulting from the rupture of diaphragms.

Finally, arc heating can be superimposed even on a pulsed flow in order to increase its enthalpy, and devices that combine the two are picturesquely titled *hotshot tunnels*. The AEDC Hotshot 2 can supply correct freestream stagnation conditions up to Mach numbers of around 12 with test durations of about 100 ms. The flow is, however, contaminated by the arc striking process.

9.8.4 Magnetohydrodynamic Accelerators

The ground test facilities described so far (except for the expansion tube) all bring the flow to rest or nearly to rest, however briefly, before it is introduced into the facility nozzle, raising two concerns for the operator. First, the facility must be designed to withstand the brunt of the stagnation temperature and stagnation pressure whenever and wherever they occur. Second, paradoxically, real gas effects also occur in these stagnation regions, and can be frozen into the flow by the rapid nozzle expansion, degrading the simulation of the freestream chemical constituents.

The *magnetohydrodynamic (MHD) accelerator,* which increases the stagnation temperature and stagnation pressure of the *moving* air by the application of electromagnetic $(j \times B)$ body forces, provides a novel means for circumventing these problems for both continuous flow and short duration test facilities. Some of the energy supplied by the electromagnetic field is converted directly into kinetic energy, easing the job of the facility nozzle. The remainder is dissipated as joule heating that reduces the stagnation pressure of the flow. Since the electromagnetic force results from the cross product of electrical current and magnetic field, the action of an MHD accelerator takes place in a duct placed between the poles or jaws of superconducting magnets capable of efficiently generating fields of about 5–10 Tesla.

An MHD accelerator supplied by an arc heater is capable of producing stagnation temperatures corresponding to orbital speeds, but the stagnation pressures tend to be somewhat less than those needed for direct connect testing at typical trajectory dynamic pressures. The test duration can easily be of the order of seconds. Unfortunately, this device also combines several sources of test gas contamination, including electric arc debris, a residue of elements deliberately introduced to improve the conductivity of air, and equilibrium products of dissociation created in the MHD accelerator during the energy addition process. The energy addition is not completely uniform, causing a nonuniform flow profile.

9.8.5 Flight Testing

At some point in every aerospace vehicle program, flight testing is required in order to expose the complete system to the real environment and thereby either demonstrate its readiness for routine operation and/or serial production, or identify critical problems. It is apparent that flight testing will occupy a special niche for hypersonic airbreathing vehicles because, as we now know, only partial simulation can be achieved in existing or foreseeable ground test facilities. One is driven to the conclusion that the main purpose of the first hypersonic airbreathing vehicles will be to provide the platform required to carry their propulsion systems to the "real world."

This process will entail more risk than usual because many critical propulsion-system operating features will be incompletely characterized before flight, and because the real environment could arouse those sinister "unknown-unknowns." Thus, attention often turns to two familiar options, one being *piloted experimental* or "X" vehicles, like the X-15 of the past or the proposed X-30, and the other being *unpiloted subscale* flight testing, such as models carried on or launched from aircraft or missiles. Piloted vehicles offer the comfort of direct human control and observation, as well as enough flexibility and margin that the flight envelope can be gradually and deliberately explored and expanded. Unpiloted vehicles, such as the CIAM scale model test of Sec. 1.2.5, offer relatively lower risk and investment, but do not duplicate the scale of the ultimate propulsion hardware and can explore only relatively small portions of the flight envelope. In either case, it is difficult to maintain precise test conditions, and to separate propulsion-system performance from that of the parent vehicle. Both approaches are also very costly per unit of data obtained when compared to ground testing. Nevertheless, one or both will probably be an essential part of the development process.

9.8.6 Instrumentation

Instrumentation is an unsung, but critical, participant in every form of testing. The quality of the experimental information from ground and flight tests is only as good as the instrumentation. Devising instruments for hypersonic airbreathing propulsion applications will be especially difficult because of the extremely harsh environment and because of the bewildering amount and types of data to be taken. In order to drive this point home, here is a partial list of the measurements known to be required.

Local Surface Measurements
   * Average wall static pressure
   * Fluctuating wall static pressure
   * Acoustic intensity
   * Surface temperature and heat transfer
   * Skin friction
   * Boundary layer separation and transition
   * Flow direction
   * Static and dynamic stress and strain
   * Wall deformation
   * Catalycity
   * Vibration and acceleration

Flowfield Measurements
   * Static pressure
   * Static temperature and enthalpy

* Static density
* Velocity magnitude and direction
* Chemical constituents
* Turbulence, vorticity, and mixing
* Prominent features, such as shock waves, organized vortices, separation and recirculation zones, and injection regions

Integrated and Derived Quantities
* Air throughflow
* Fuel flow
* Component forces and moments
* Fuel/air mixing effectiveness
* Combustion efficiency
* Overall net thrust and moments
* Operability
* Combustion stability
* Starting and restarting characteristics
* Environmental impact
* Variable, geometry positioning
* Fuel status

This list must be compounded by the variety of techniques and devices potentially available, each of which has its own strengths and weaknesses, in order to appreciate the magnitude of the task at hand. Moreover, the situation is more difficult for flight testing than ground testing because space and weight are more precious and the environment is more severe. Fortunately, the advent of such modern technologies as lasers and diffusion bonded gauges, coupled with the data processing power of modern computers, has created many new intrusive and nonintrusive measurement techniques. Nevertheless, the historical record shows that sustained effort and investment are required to develop the necessary tools in a timely fashion.

Taking all things together, we see that the *singular* challenge for hypersonic airbreathing propulsion instrumentation will be to gather the critical data from the experimental vehicles, because their primary purpose will be to allow the propulsion systems to be tested in the correct environment.

9.8.7 CFD Example: The Free-Piston Shock Tunnel

Many aspects of ground testing benefit from CFD, including the planning and design of facilities, models, and experiments, the monitoring and control of tests, and the interpretation and interpolation/extrapolation of data. This is especially true for hypersonic

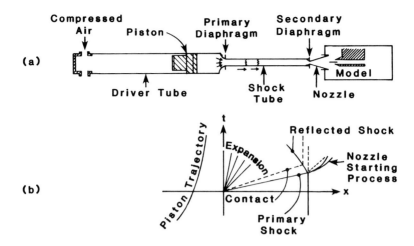

Fig. 9.26   Schematic diagram of the free-piston shock tunnel and starting process, taken from Ref. 9.25.

propulsion because the existing database is meager and the stakes are high. Modern ground test facilities must have a powerful, indigenous, complementary CFD capability or face extinction.

An area of special CFD focus is the understanding and design of complex, high-performance facilities, perhaps exemplified by the study of the behavior of pulsed flow tunnels and tubes.[9.25, 9.26] As an example, Fig. 9.27 presents some results of the CFD analysis of the shock-wave reflection and facility nozzle starting process of the axisymmetric free-piston shock tunnel shown in Fig. 9.26. This study emphasizes such multi-dimensional effects as the facility boundary layers and shock-wave interactions in the nozzle throat region, and is intended to gain a better understanding of the processes that delay the establishment of flow in the facility nozzle. The stagnation conditions are approximately 11,400 °R (6330 K) and 500 atm, and the nozzle Mach number is 8.

Figure 9.27 shows the contours of constant density behind the reflected shock wave and within the nozzle at a series of times during the starting process beginning at $t = 0$ with the shock wave propagating toward the nozzle throat. The computed starting time of the nozzle agrees with experimental data. The detail of the CFD results is sufficiently fine to identify a novel candidate mechanism for the premature contamination of the test gas, namely the establishment and propagation of an axisymmetric or ring vortex near the centerline of the shock tube and just behind the reflected shock wave. This phenomenon is discernible in the lower portion of Fig. 9.27.

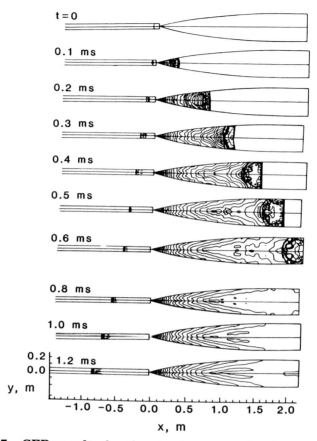

Fig. 9.27   CFD results for the shock wave reflection and facility nozzle starting process of an axisymmetric, Mach 8, free-piston shock tunnel, taken from Ref. 9.25. The contours are of constant density, and the times span the nozzle starting process from just before the arrival of the driving shock wave to the establishment of full flow in the nozzle.

9.8.8 Quo Vadis?

This discussion of hypersonic airbreathing propulsion testing reinforces the conclusions reached elsewhere in this textbook, especially those in Chap. 2 regarding CFD. The customary luxury of full simulation of all conditions in ground test facilities will not be possible. Consequently, a *combination* of carefully designed partial-simulation ground tests, CFD, and flight testing will be needed to provide the design tools and the development and certification procedures for hypersonic propulsion systems. Defining and perfecting this combination will be among the major goals and accomplishments of the hypersonic era.

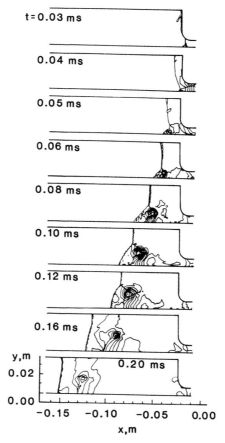

Fig. 9.27 (continued) CFD results for the shock wave reflection and facility nozzle starting process of an axisymmetric, Mach 8, free-piston shock tunnel, taken from Ref. 9.25. The contours are of constant density, and the times span the nozzle starting process from just before the arrival of the driving shock wave to the establishment of full flow in the nozzle.

9.9 OBLIQUE DETONATION WAVE PROPULSION

Just when you think it's over, it isn't. Our motive for presenting this subject last is to show that new ideas in hypersonic airbreathing propulsion are still coming.

In order to avoid the burdens and uncertainties associated with conventional mixing and diffusive burning, consider the possibility of injecting and mixing the fuel with the air upstream of the "combustor," and fashioning the oblique shock-wave system so that the chemical energy is instantly released by the terminal oblique shock wave because it brings the mixture to spontaneous ignition conditions, as illustrated in Fig. 9.28. The terminal combination of oblique shock wave and rapid chemical reaction is known as an *oblique detonation wave (ODW)*. The immediate advantages of the ODW process are that the drag, convection heating, length, weight, cost, and maintenance of the combustor are almost entirely eliminated. The propulsion system of Fig. 9.28 is known as an *oblique detonation wave engine (ODWE)*.

9.9.1 ODWE Theory

The ODW and the ODWE have been amply treated in the open literature,[9.27, 9.28] and our goal here is merely to summarize some of the key results.

Stationary or standing ODW's certainly occur in nature, so the most important question to be answered is whether they can be stabilized under conditions that are suitable for hypersonic airbreathing propulsion. One-dimensional flow analysis will for one last time prove equal to the task.

The schematic diagram of a one-dimensional ODW is shown in Fig. 9.29. The supersonic flow is deflected or turned through an angle δ and an ODW is formed at an angle θ. The upstream Mach number M_u is, in general, less than the freestream Mach number M_0 owing to previous deceleration in the upstream compression system. The release of chemical energy (modeled, as usual, as an addition of heat but no mass) is what differentiates the ODW from an oblique shock wave. *Detonation* is said to occur when a shock wave–induced combustion wave follows so closely behind the igniting shock wave that the two waves are pressure-coupled. Detonation waves are classified as *overdriven* or *Chapman-Jouguet,* depending on whether the normal component of the downstream Mach number M_{dn} is *subsonic* or *sonic*. In contrast, a *shock wave–induced combustion wave* results when the shock wave is followed by a distinct, spatially resolvable combustion wave. Shock wave–induced combustion waves

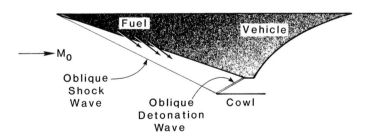

Fig. 9.28 Schematic diagram of an oblique detonation wave engine (ODWE).

occur when the normal component of the downstream flow velocity is *supersonic*.

The steady, one-dimensional conservation equations for the ODW of Fig. 9.29 are

Mass
$$\rho_u u_u = \rho_d u_d \tag{9-32}$$

Normal momentum
$$p_u + \rho_u u_u^2 = p_d + \rho_d u_d^2 \tag{9-33}$$

Tangential momentum
$$(\rho_u u_u)v_u = (\rho_d u_d)v_d \tag{9-34}$$

using Eq. (9-32) $\quad v_u = v_d$

Energy
$$h_u + q + \frac{u_u^2 + v_u^2}{2} = h_d + \frac{u_d^2 + v_d^2}{2} \tag{9-35}$$

using Eq. (9-34) $\quad h_u + q + \dfrac{u_u^2}{2} = h_d + \dfrac{u_d^2}{2}$

where q is the heat added per unit mass of fluid to represent the chemical energy release due to combustion. Whereas most property changes are best expressed as ratios, the energy addition is incremental. Further, the fluid will be assumed to be a calorically perfect gas with constant properties.

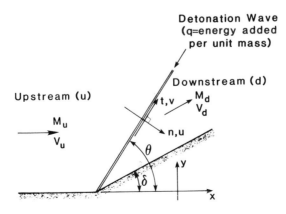

Fig. 9.29 Schematic diagram of an oblique detonation wave (ODW). Note that the natural coordinates for analysis are normal and tangential to the ODW.

Equations (9-32) through (9-35) can be combined to yield

$$\tilde{q} = -\frac{\gamma+1}{2}\chi^2 M_u^2 \sin^2\theta + \left(1 + \gamma M_u^2 \sin^2\theta\right)\chi$$
$$- \left(1 + \frac{\gamma-1}{2}M_u^2 \sin^2\theta\right) \tag{9-36}$$

where

$$M = \frac{V}{\sqrt{\gamma RT}} \tag{9-37}$$

$$\tilde{q} = \frac{q}{C_p T_u} \tag{9-38}$$

$$\chi = \frac{\rho_u}{\rho_d} = \frac{u_d}{u_u} = \frac{\tan(\theta-\delta)}{\tan\theta} \tag{9-39}$$

A generalized diagram of the solution of Eq. (9-36) for several values of $\tilde{q}$ and constant upstream conditions, showing the classifications of the different possible downstream flowfields, is given in Fig. 9.30. This diagram should not be confused with the familiar oblique shock wave diagram showing oblique shock wave angle as a

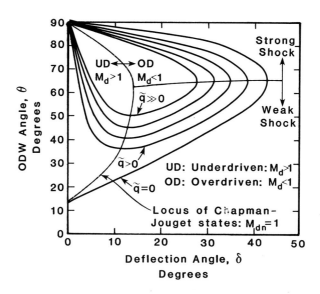

Fig. 9.30 Generalized form of oblique detonation wave angle as a function of deflection angle and heat addition, for constant upstream conditions.

function of deflection angle and upstream Mach number, which it resembles. Instead, upstream Mach number is fixed in Fig. 9.30, and only the chemical energy release is varied. The outermost contour, corresponding to zero chemical energy release is, nonetheless, common to both diagrams. In particular, note that the ODW angle for zero deflection and chemical energy release, and the least ODW angle appearing on the diagram, is the local Mach angle

$$\theta = \sin^{-1}\left(\frac{1}{M_u}\right) \tag{9-40}$$

There are two values of θ for each value of δ, the larger belonging to a *strong shock wave* and the smaller belonging to a *weak shock wave*. For all contours of constant $\tilde{q}$, the minimum value of θ corresponds to the Chapman-Jouguet detonation wave with $M_{dn} = 1$ (see Fig. 9.30). Points to the left of the minimum are underdriven detonation waves with $M_{dn} > 1$. Points to the right of the minimum (including all strong shock waves) are overdriven detonation waves with $M_{dn} < 1$. Just like ordinary strong oblique shock waves, strong overdriven oblique detonation waves are associated with static pressure increases so large that they detach and are therefore not found naturally. Consequently, the region of interest for ODW's is that of the *weak overdriven detonation waves*, bounded by the Chapman-Jouguet angle and the detachment angle, as indicated in Fig. 9.30.

Figure 9.31 contains two specific solution diagrams for identical operating parameters except for the upstream Mach numbers. The contour for a given M_u and $\tilde{q}$ is most easily computed by assuming a value for θ, solving the quadratic Eq. (9.36) for its two real roots, and using Eq. (9.39) in the form

$$\delta = \tan^{-1}\left\{\frac{(1-\chi)\tan\theta}{1+\chi\tan^2\theta}\right\} \tag{9-41}$$

to find the two values of δ. This process is then repeated for a series of θ, until the entire contour is established.

These diagrams show that, for these typical conditions, there are many possible ODW configurations that will work. The range of acceptable deflection angles is an important outcome, especially if a fixed geometry compression system is being sought. Thus, it is important that Fig. 9.31 shows that the "solution space" becomes wider and shallower as M_u increases, which means both that the range of acceptable deflection angles increases and that the variation of ODW angle with $\tilde{q}$ decreases. Thus, the ODW "prefers" higher upstream Mach numbers, and exhibits an asymptotic behavior analogous to that of the previously described Mach number invariance principle.

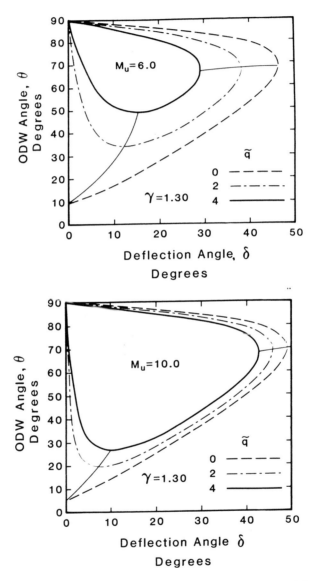

Fig. 9.31 Oblique detonation wave angle as a function of deflection angle and heat addition, for constant upstream conditions, with M_u = 6.0 and 10.0.

The overall efficiency η_o of an ODWE will differ little from its conventional combustion counterpart. Discussions of the practicality of the ODWE center on uniformly premixing the fuel and air prior to the ODW without wasting any in the external stream, and avoiding combustion upstream of the intended location of the ODW. This bring us to our final twist.

9.9.2 The Ram Accelerator

One way of avoiding the problems just mentioned is to "fly" a fixed geometry ODWE projectile/centerbody down a circular tube containing a uniform mixture of combustible gases, as depicted in Fig. 9.32. Since the tube wall plays the role of the cowl, no fuel is wasted. This apparatus is best known as a *ram accelerator,* and is frequently described as a ramjet or scramjet in a tube.[9.29, 9.30]

Like its aircraft counterparts, the ODWE inside the ram accelerator tube cannot set itself into motion, and must be fired or launched into the tube by a gun or other device at a sufficient speed that a stable ODW will arise. The minimum required injection speed is known as the *takeover velocity,* and is found from diagrams such as Fig. 9.31. That is, for any selected geometry of the projectile and combustible mixture, these diagrams can be used in conjunction with ordinary oblique shock wave calculations to determine what freestream Mach number M_u and velocity V_u is required to locate the initial operating point within the weak overdriven detonation wave solution space. Figure 9.31 makes it obvious that this becomes easier as the projectile speed increases. Typical takeover velocities are about a mile or a few kilometers per second, or about 20 percent greater than the Chapman-Jouguet velocity, which can be provided by present-day guns.

Because the ram accelerator projectile has no vehicle attached, and, being relatively small (say about an inch or a few centimeters in diameter and having a mass of about a tenth of a pound or kilogram),

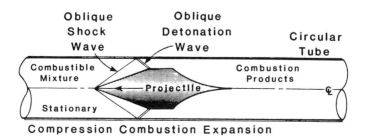

Fig. 9.32 Schematic diagram of a ram accelerator.

has a frontal area/volume ratio very favorable to propulsion, initial acceleration in hydrogen/air mixtures is of the order of *10,000 g's*. This is an usually impressive capability. A tube of less than about 1000 ft or **300 m** in length would produce *orbital velocities* if that acceleration rate were to be sustained. It is also interesting to note that the usual overall performance measure of ram accelerators is the *ballistic efficiency,* which is defined as the instantaneous ratio of the increase of the projectile kinetic energy to the energy released through the combustion process, or

$$\eta_{\text{ballistic}} = \frac{F \cdot V}{\dot{m}_f \cdot h_{PR}} \qquad (9\text{-}42)$$

As you can see, the ballistic efficiency is identical to our overall propulsive efficiency η_o, and should be similar in magnitude.

Although the ram accelerator would not be attractive to human passengers or for delicate cargo, it does have many potential applications. These include delivering "bricks and mortar" to orbit from the Earth or moon, powering military weapons, and providing a new ground test technique for simulating high speeds, ironically even of use for developing hypersonic airbreathing propulsion systems.

The technical problems of ram accelerators are classical, and include performance, stability (including aerodynamic "centering" during flight inside the tube), durability, and repeatability. At the moment, none of these seem to be insuperable, and there is a general feeling that nature is on the side of the ram accelerator.

9.10 JOURNEY'S END

This chapter could have been called *Surprising Topics in Hypersonic Airbreathing Propulsion.*

Human beings, whose fingertips tell them that water is *unbearably* "hot" or "cold" at less than 1 W/cm^2, may well wonder what a thousand W/cm^2 would feel like. Shock interactions can enormously increase local convective heat transfer rates. Steel is not always more durable than copper. Hypersonic vehicles using endothermic fuels may resemble flying petroleum refineries. Hypersonic airbreathing flight inevitably will lead to new concerns, real or imagined, about our impact on local and distant environments. The "subsystems" that support the propulsion system will make or break this enterprise. The propulsion system dominates every aspect of the vehicle, including stability and control, and will have to participate actively in providing stable flight. Ground test capabilities will be stretched to their limits and beyond in order to ensure safe, reliable, and high-performance hypersonic airbreathing propulsion, but the truth will not be known until we fly into new "waters." Oblique detonation wave engines offer inventors a host of new possibilities.

This is, of course, the essence of the beauty and excitement of exploring new frontiers. Once again, we have the chance to discover that our familiar scientific tools, based on fundamental principles, yield an understanding of phenomena unimaginably remote from everyday experience. In the end, we gain renewed confidence in our methods, new insight into their application to any situation, new faith in the ingenuity and creativity of our fellow travelers, and, in this case, the methodology required for the next generation of air-breathing propulsion. Despite the initial appearance of extraordinary complexity and difficulty, sustained hypersonic flight now seems not only possible, but inevitable.

This portion of your hypersonic voyage is over. We hope you have found it enjoyable, enlightening, and rewarding, as we have, and wish you success in the future, whatever your final destination may be.

REFERENCES

[9.1] Bertin, J. J., *Hypersonic Aerothermodynamics*, AIAA Education Series, Washington, DC, 1994.

[9.2] Anderson, J. D., Jr., *Hypersonic and High Temperature Gas Dynamics*, McGraw-Hill, New York, 1989.

[9.3] Kays, W. M., and Crawford, M. E., *Convective Heat and Mass Transfer*, McGraw-Hill, New York, 1980.

[9.4] Eckert, E. R. G., "Engineering Relations for Heat Transfer and Friction in High-Velocity Laminar and Turbulent Boundary Layer Flow Over Surfaces With Constant Pressure and Temperature," Transactions of the ASME, Aug. 1956.

[9.5] Dorrance, W. H., *Viscous Hypersonic Flow*, McGraw-Hill, New York, 1962.

[9.6] Markarian, C. F., "Heat Transfer in Shock Wave–Boundary Layer Interaction Regions," Naval Weapons Center Rept. NWC TP-4485, China Lake, CA, Nov. 1968.

[9.7] Emanuel, G., *Gasdynamics: Theory and Applications*, AIAA Education Series, Washington, DC, 1986.

[9.8] Neumann, R. D., and Hayes, J. R., "Protuberance Heating at High Mach Numbers—A Critical Review and Extension of the Data Base," AIAA Paper 81-0420, Jan. 1981.

[9.9] Hearn, E. J., *Mechanics of Materials*, Pergamon Press, Elmsford, NY, 1985.

[9.10] Smith, W. F., *Structure and Properties of Engineering Alloys*, McGraw-Hill, New York, 1993.

[9.11] Lander, H., and Nixon, A. C., "Endothermic Fuels for Hypersonic Vehicles," *Journal of Aircraft*, Vol. 8, No. 4, Apr. 1971.

[9.12] Smith, M. J. T., *Aircraft Noise,* Cambridge University Press, Cambridge, 1989.

[9.13] Finlayson-Pitts, B. J., and Pitts, J. N., Jr., *Atmospheric Chemistry: Fundamentals and Experimental Techniques,* John Wiley, New York, 1986.

[9.14] Wiggins, J. H., Jr., *Effects of Sonic Boom,* J. H. Wiggins, Palos Verdes Estates, CA, 1969.

[9.15] Shih, P. K., Prunty, J., and Mueller, R. N., "Thermostructural Concepts for Hypervelocity Vehicles," *Journal of Aircraft,* Vol. 28, No. 5, May 1991.

[9.16] Anderson, J. D., Jr., *Introduction to Flight,* McGraw-Hill, New York, 1989.

[9.17] Roskam, J., *Airplane Flight Dynamics and Automatic Flight Controls—Parts I and II,* Roskam Aviation and Engineering Corp., Ottawa, KS, 1979.

[9.18] Berry, D. T., "National Aerospace Plane Longitudinal Long-Period Dynamics," *Journal of Guidance, Control, and Dynamics,* Vol. 14, No. 1, Jan. 1991.

[9.19] Raney, D. L., and Lallman, F. J., "Control Concept for Maneuvering in Hypersonic Flight," AIAA Paper 91-5055, Dec. 1991.

[9.20] Cheng, P. Y., Chan, S. Y., Myers, T., Klyde, D. H., and McRuer, D. T., "Aeroservoelastic Stabilization Techniques for Hypersonic Flight Vehicles," AIAA Paper 91-5056, Dec. 1991.

[9.21] Hallion, R. P., *The Hypersonic Revolution,* Vol. I: From Max Valier to Project Prime, 1924–1967, Aeronautical Systems Division, Wright-Patterson AFB, OH, 1987.

[9.22] National Research Council, "Future Aerospace Ground Test Facility Requirements for the Arnold Engineering Development Center," National Academy Press, Washington, DC, 1992.

[9.23] United States General Accounting Office, "AeroSpace Plane Technology—Research and Development Efforts in Europe," GAO/NSIAD-91-194, Washington, DC, July 1991.

[9.24] Murthy, S. N. B., and Curran, E. T., (eds.), *High-Speed Flight Propulsion Systems,* AIAA Progress in Astronautics and Aeronautics Series, Vol. 137, Washington, DC, 1991.

[9.25] Jacobs, P. A., "Simulation of Transient Flow in a Shock Tunnel and a High Mach Number Nozzle," NASA CR-187606, July 1991.

[9.26] Jacobs, P. A., "Numerical Simulation of Transient Hypervelocity Flow in an Expansion Tube," NASA CR-189615, Mar. 1992.

[9.27] Pratt, D. T., Humphrey, J. W., and Glenn, D. E., "Morphology of Standing Oblique Detonation Waves," *Journal of Propulsion and Power,* Vol. 7, No. 5, Sep. 1991.

[9.28] Glenn, D. E., and Pratt, D. T., "Numerical Modeling of Standing Oblique Detonation Waves," AIAA Paper 88-0440, Jan. 1988.

[9.29] Hertzberg, A., Bruckner, A. P., and Bogdanoff, D. W., "Ram Accelerator: A New Chemical Method for Accelerating Projectiles to Ultrahigh Velocities," *AIAA Journal*, Vol. 26, No. 2, Feb. 1988.

[9.30] Bogdanoff, D. W., and Brackett, D. C., "A Computational Fluid Dynamics Code for the Investigation of Ramjet-in-Tube Concepts," AIAA Paper 87-1978, June 1987.

PROBLEMS

9.1 Employ the Eckert reference enthalpy method to determine the variation of convective heat transfer $\dot{q}_w$ with dynamic pressure q_0 for Examples 9.1 and 9.2. Repeat those calculations with q_0 equal to 500 lbf/ft^2 (23.94 kN/m^2) and 2000 lbf/ft^2 (95.76 kN/m^2).

Do the trends follow the simple relationships of Eq. (9-11)? What are the consequences of this conclusion?

How sensitive is the skin friction coefficient C_f to the dynamic pressure?

9.2 Employ the Eckert reference enthalpy method to determine the variation of convective heat transfer $\dot{q}_w$ with wall temperature T_w for Examples 9.1 and 9.2. Repeat those calculations with T_w equal to 1000 °R (555.6 K) and 3000 °R (1667 K).

How sensitive is the convective heat transfer to the allowable wall temperature? What other virtues might there be to increasing allowable wall temperature?

How sensitive is the skin friction coefficient C_f to the wall temperature?

9.3 Employ the Eckert reference enthalpy method to determine the variation of convective heat transfer $\dot{q}_w$ with adiabatic compression efficiency η_c for Examples 9.1 and 9.2. Repeat those calculations with η_c equal to 0.85 and 0.95.

How sensitive is the convective heat transfer to the adiabatic compression efficiency? Why is this so?

How sensitive is the skin friction coefficient C_f to the adiabatic compression efficiency?

9.4 The rate at which energy is radiated from an exposed surface is given by the Stefan-Boltzmann equation.

$$\dot{q}_r = \epsilon \sigma T_w^4$$

where the total emissivity $0 < \epsilon < 1$ is a physical property of the exposed surface, and the Stefan-Boltzmann constant is

$$3.30 \times 10^{-15} \text{ BTU/in}^2 \cdot \text{s} \cdot {}^\circ R^4 \quad (5.57 \times 10^{-12} \text{ W/cm}^2 \cdot \text{K}^4)$$

(a) Use the Stefan-Boltzmann equation and the information of Table 9.1 to calculate the maximum possible energy radiation rate from the compression and combustor surfaces (i.e., $\epsilon = 1$) for all conditions shown there. How does the rate of radiative energy transfer compare with the corresponding convective heat transfer? What general conclusions do you draw from these results?

(b) For a section of uncooled insulated outer skin having a convective heat transfer rate of $0.1 \text{ BTU/in}^2 \cdot \text{s}$ (16.4 W/cm^2) and a total emissivity of 0.8, use the Stefan-Boltzmann equation to calculate the local equilibrium temperature. What materials can be used at this temperature? How sensitive is this result to any of the input quantities? What general conclusion do you draw from this result?

9.5 Equation (9-11) provides a means for crudely estimating the ratio of laminar to turbulent convective heat transfer rate at any point of flight. Calculate and sketch this ratio as a function of freestream Mach number for several typical values of trajectory dynamic pressure.

Devise a similar means for estimating the ratio of laminar to turbulent skin friction at any point of flight. Calculate and sketch this ratio as above.

How do these results compare with those of Table 9.1? What general conclusions do you draw from these results?

9.6 A fragment of experimental data reveals that $\dot{q}_{w,pk}/\dot{q}_w$ for a shock—boundary layer separation is 4.0. What can you say about the Mach number just upstream of the separation region?

9.7 Fragments of experimental data for the convective heat transfer rate at a strut junction exposed to freestream flight conditions reveal that $\dot{q}_{w,pk} = 4.0 \text{ BTU/in}^2 \cdot \text{s}$ and $T_{aw} - T_w = 8100 \, {}^\circ R$ (4500 K). What can you say about the possible combinations of q_0 and M_0 that existed during the test? Assume turbulent flow.

9.8 The *effectiveness* of film cooling is customarily defined as

$$\eta_{fc} = \frac{T_{aw} - T_w}{T_{aw} - T_c} \qquad 0 \leq \eta_{fc} \leq 1$$

where T_c is the coolant temperature just prior to injection and η_{fc} is a function at least of the amount of coolant flow, the geometry of the cooling configuration, and the condition of the approaching flow. Use this definition and the information of Table 9.1 to estimate the film cooling effectiveness *required* for the conditions given there. Assume that $T_c = 900\ °R\ (500\ K)$.

Referring to the open literature, determine whether these levels of cooling effectiveness will be easy or difficult to achieve.

9.9 Starting from Eqs. (9-14) and (9-15), prove analytically that, as the wall thickness t approaches 0 and r_o/r_i approaches 1, then

$$|\sigma_r|_{\max} \rightarrow \frac{E\alpha r_i \dot{q}_w}{8k(1-\nu)} \cdot \epsilon^2$$

and

$$|\sigma_h|_{\max} \rightarrow \frac{E\alpha r_i \dot{q}_w}{2k(1-\nu)} \cdot \epsilon$$

where

$$\epsilon = \frac{r_o}{r_i} - 1 \rightarrow 0$$

Do these results qualitatively agree with your intuition? Do they quantitatively agree with those displayed in Fig. 9.10? Would you be willing to use them as design approximations?

Use these results to prove also that the wall temperature drop approaches

$$T_i - T_o \rightarrow \frac{(1-\nu)|\sigma_h|_{\max}}{E\alpha}$$

What use can you make of this unexpected relationship?

9.10 A cylinder such as that of Fig. 9.8 has $\dot{q}_w = 2.0\ \text{BTU/in}^2 \cdot \text{s}$ $(33\ \text{W/cm}^2)$, $r_o/r_i = 1.05$, and $p = 1$ atm. The maximum hoop stress is experimentally found to be $20{,}000\ \text{lbf/in}^2\ (960{,}000\ \text{kN/m}^2)$. What can you say about the properties of the cylinder material?

9.11 The axial stress σ_x far from the free ends in the cylinder of Fig. 9.8 is given by the expression[9.9]

$$\sigma_x = \frac{E\alpha \dot{q}_w r_i}{2k(1-\nu)} \left\{ -\frac{2\left(\dfrac{r_i}{r_o}\right)^2}{1-\left(\dfrac{r_i}{r_o}\right)^2} \cdot \ell n\,\frac{r_o}{r_i} + 1 - 2\ell n\,\frac{r_o}{r} \right\}$$

(a) Show that σ_x equals σ_h on the inside and outside surfaces of the cylinder.

(b) Sketch the variation of σ_x and σ_h with radius from the inside to the outside surface. How would you say they compare overall?

9.12 Make a list of four other materials you believe would be good candidates for scramjet combustors. (Some possibilities not covered in this textbook are carbon, red brass, Rene 41, silicon carbide reinforced titanium metal matrix composite, and 316 stainless steel.) Assemble the information of Table 9.3 for your materials from the open literature, locate them on Fig. 9.10, and calculate their indices of thermal durability, thermal strains, and temperature drops.

Did you find any surprises or new, outstanding candidates? What did you learn about the availability of material property characterization in the open literature?

9.13 Recreate Fig. 9.11 for one-half and twice the surface area averaged heat flux $\bar{q}_w$ of the original example. How sensitive is the maximum possible M_0 to $\bar{q}_w$ for the representative fuels of Table 9.4?

9.14 Make a list of four other fuels you believe would be good candidates for scramjet cooling. (Some possibilities not covered in this textbook are ethane, hexane, octane, cetane, and decalin.) Assemble the information of Table 9.4 for your materials from the open literature and locate them on Fig. 9.11.

Did you find any surprises or new, outstanding candidates? What did you learn about the availablility of fuel property characterization in the open literature?

9.15 Suppose that a known operating point for a hypersonic vehicle is

$$M_0 = 8.0$$

$$q_0 = 2000 \text{ lbf/ft}^2 \qquad (95.8 \text{ kN/m}^2)$$

$$\bar{q}_w = 2.0 \text{ BTU/in}^2 \cdot s \qquad (327 \text{ W/cm}^2)$$

$$\frac{A_w}{A_0} = 4.0$$

What is the maximum M_0 for this vehicle if the fuel is endothermic MCH?

9.16 A vehicle is observed to be flying at an altitude of approximately 92 kft (28 km) and a velocity of 5900 ft/s (1800 m/s). The measured maximum far-field overpressure is about 1.9 lbf/ft^2

(91 N/m^2). Estimate the dynamic pressure and weight of the vehicle. Assume standard atmospheric conditions.

9.17 Reproduce the emissions index information of Fig. 9.14 for octane. How do these results differ from those of H_2 and CH_4? Why?

9.18 Using the mental model and nomenclature of Fig. 7.14 for exit plane two-dimensional dispersion or splay, show that the mass average downward component of velocity that contributes to the pitching or longitudinal motion is given by the expression

$$\frac{\overline{V}_y}{\overline{V}_{10}} = \frac{\sin \alpha_0 \sin \theta_0}{\theta_0}$$

9.19 The information of Fig. 9.21 is often presented in the open literature as freestream stagnation pressure versus freestream total enthalpy for lines of constant M_0 and constant q_0. Convert Fig. 9.21 to that format, with the help of HAP(Air).

Does this method of presentation alter our original conclusions about freejet facility requirements?

9.20 A popular method of presenting required hypersonic facility freejet stagnation conditions in flight envelope coordinates is sketched below. Using the information of Fig. 9.21 and HAP(Trajectory), construct this graph for typical hypersonic airbreathing propulsion flight conditions.

What new conclusions do you draw from this presentation?

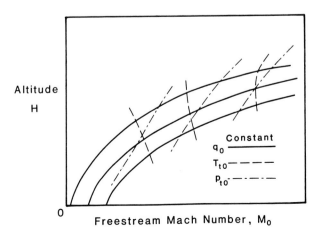

9.21 Compute and plot the minimum $\dot{W}$ [from Eq. (9-30)] and the total $\dot{W} + \dot{Q}$ [from Eq. (9-31)] for freejet facility operating conditions corresponding to $q_0 = 1000$ lbf/ft^2 (47.88 kN/m^2) and $A_0 = 100$ ft^2 (30.48 m^2) as a function of $6 < M_0 < 25$. Note that A_0 is expected to be a weak function of M_0 for hypersonic flight.

Do these quantities increase continuously with M_0, or do they have local maxima? Why, or why not? What are the implications of these results for facility designers?

9.22 Reproduce the direct connect results of Fig. 9.23 for q_0 one-half and twice the q_0 of the original example. Is the factor of 6 sensitive to the dynamic pressure of the trajectory?

Overlay these results on those of Problem 9.19 in order to produce another encapsulated view of direct connect versus freejet facility requirements.

9.23 Confirm to your own satisfaction, with the help of HAP(Air), that "real gas" or dissociation effects commence at a stagnation temperature of about 4500 °R (2500 K) for $q_0 = 1000$ lbf/ft^2 (47.88 kN/m^2). Use the rapid increase of C_p above the thermally perfect value as the primary indicator of the onset of "real gas" effects. How much dissociation actually occurs before the onset is observed?

9.24 Calculate and plot graphs equivalent to Fig. 9.31, but for higher values of M_u, say 15 and 20. Toward what limit are the contours of constant $\tilde{q}$ converging? Is this behavior inherent in the original algebraic equations? What is the physical meaning of this trend? What are the implications of this for the ODWE designer?

Finally, earn our highest esteem and enduring respect by completing the following two coupled problems that shed a considerable amount of light on the behavior of ram accelerators and ODWE's.

9.25 A ram accelerator tube of 1-in diameter (0.0254 m) is filled with a combustible mixture into which a projectile having a mass of 0.50 lbm (0.227 kg) is moving at a Mach number of 10.0. The static pressure of the combustible mixture in the tube is 10 atm. The projectile has a cone semi-vertex angle of 15 deg. and is configured so that the oblique shock wave reflected from the wall is absorbed at the cone-cylinder "shoulder," as shown in Fig. 9.32, terminating the oblique shock wave system.

The flow may be treated as frictionless and steady in the frame of reference of the projectile, the gas may be taken as calorically perfect with a constant value of $\gamma = 1.30$ throughout, and the chemical energy released in the combustion process doubles the stagnation temperature of the gas (i.e., $\tau_b = 2.0$) without increasing the mass flow.

Estimate the performance of this ram accelerator by means of the following series of one-dimensional models found in HAP(Gas Tables):

(a) Compute the conditions downstream of the conical shock wave and upstream of the reflected shock wave. Use the values on the conical surface rather than those immediately downstream of the conical shock wave.

(b) Assume that the reflected shock wave is without energy release and two-dimensional or planar, and compute the downstream properties. When calculating the ratio of the annulus area to the freestream or tube area, remember that the geometry is axisymmetric.

(c) Calculate the "combustor" exit conditions in accordance with simple frictionless, constant area heat release or Rayleigh line theory. This is best accomplished using conventional Rayleigh line analysis techniques.

(d) Calculate the "nozzle" exit conditions based upon a one-dimensional isentropic expansion process that returns the flow to the original area.

Use these results to determine the overall performance of the ram accelerator.

(i) Calculate

$$\frac{F}{p_0 A_0} = \frac{p_{10}}{p_0}\left(1 + \gamma M_{10}^2\right) - \left(1 + \gamma M_0^2\right)$$

based on control volume momentum considerations [c.f., Eq. (2-58)].

(ii) Calculate the net thrust on the projectile.

(iii) Calculate the overall efficiency or ballistic coefficient of the projectile. Do this by first showing that Eqs. (2-73) and (9-42) may be manipulated to yield

$$\eta_o = \eta_{\text{ballistic}} = \frac{F \cdot V_0}{\dot{m}_0 f h_{PR}} = \frac{F}{p_0 A_0} \cdot \frac{\gamma - 1}{\gamma} \cdot \frac{1}{(\tau_b - 1)\left(1 + \dfrac{\gamma - 1}{2} M_0^2\right)}$$

Can you buttress this result by means of η_{th} and η_p arguments?

(iv) Calculate the acceleration in g's of the projectile.

(v) Calculate the total pressure ratio across the projectile.

9.26 Repeat the analysis of Problem 9.25 except that the reflected shock wave and combustion processes are combined into a single ODW calculation by means of Eqs. (9-32) through (9-41). Note that the geometry of the projectile will not be identical to that above because the reflected oblique shock wave angle is different.

The ODW calculation is facilitated by first finding the ODW angle θ by trial and error from Eq. (9-36), and then modifying the usual oblique shock wave algebraic jump condition relations to include $\tilde{q}$ in order to solve for the remaining quantities. If your approach yields two solutions, you must choose the one corresponding to the overdriven condition and therefore the lower value of M_{dn}. You will find it revealing to locate the ODW operating point on Fig. 9.31, for which M_u is similar to that of this problem.

How do these results compare with those of Problem 9.25? What causes the greatest differences? What advice would you give a designer about estimating the performance of ram accelerators and ODWE's?

NOMENCLATURE

A	air axial vortex mixing empirical parameter kinetic reaction rate empirical parameter throughflow area
A_i	chemical formula of the i-th species
AFT	adiabatic flame temperature
a	acceleration acoustic speed or speed of sound
B	Breguet range factor kinetic reaction rate empirical parameter
b	number of moles of an element passage height
C	carbon
°C	degrees in the Celsius temperature scale
C	coefficient molecular concentration
C_a	axial force coefficient
C_D	coefficient of drag
C_f	skin friction coefficient
C_L	coefficient of lift
C_p	specific heat at constant pressure
C_v	specific heat at constant volume
C_{pt}	stagnation point pressure coefficient
CI	contact index
D	diameter drag, vehicle drag molecular diffusivity
D_i	gross destruction rate of the i-th species
D_p	profile pressure distortion parameter
E	kinetic reaction rate empirical parameter modulus of elasticity system internal energy

E_m	mass-basis entrainment ratio
E_n	mole-basis entrainment ratio
E_v	volumetric-basis entrainment ratio
e	specific internal energy
$°F$	degrees in the Fahrenheit temperature scale
F	force fuel uninstalled engine thrust
$F/\dot{m}_0$	specific thrust (air mass flow rate specific thrust)
f	mass-basis fuel/air ratio
f_i	net production rate of the i-th species
G	system Gibbs potential function
g	acceleration of gravity mass-specific Gibbs potential function variance (second moment) of a distribution
g_k	partial molal Gibbs function of the k-th species
g_o	acceleration of gravity at the surface of the Earth
H	hydrogen
H	dimensionless static enthalpy, C_pT/C_pT_t geometric altitude height of duct or passage
h	specific enthalpy system enthalpy
h_k	molal enthalpy of the k-th species
h_{PR}	heat of reaction or heating value
Δh_f^0	enthalpy of formation
I	impulse function
I_s	intensity of segregation segregation index unmixedness
I_{sp}	specific impulse (fuel weight flow rate specific thrust)
i	intermittency
JJ	number of reactions

j	molal diffusive flux
K	Kelvins in the SI temperature scale
K	dimensionless kinetic energy, $V^2/2C_pT_t$
K_D	process efficiency
K_{p_j}	equilibrium constant for the j-th reaction
k	thermal conductivity
k_j	chemical kinetic or reaction rate constant of the j-th reaction
L	Legendre transformation length of duct or passage lift, vehicle lift vehicle length scale
l	large-structure spacing
M	Mach number third body catalyst
MFP	mass flow parameter
m	mass vehicle mass
$\dot{m}$	mass flow rate
N	nitrogen
N	number of molecules number of moles
NLM	number of elements
NS	number of species
n	number of degrees of freedom number of g's number of molecules number of moles polytropic process exponent
O	oxygen
oe	mass specific orbital energy
P	probability density function (PDF)
Pr	Prandtl number, $\mu C_p/k$
p	absolute pressure

Q_i gross production rate of the i-th species

$\dot{Q}$ total heat flux or interaction rate

q dynamic pressure

$\dot{q}$ local heat flux or interaction rate

$\tilde{q}$ reduced or dimensionless ODW heat flux

°R   degrees in the Rankine temperature scale

R cruise range
     perfect or universal gas constant

R_j forward rate of the j-th chemical reaction

R_{-j} reverse rate of the j-th chemical reaction

Re Reynolds number, $\rho V L/\mu$

r radius
     recovery factor
     turning radius
     velocity ratio

S specific fuel consumption (thrust specific fuel mass flow rate)
     system entropy

Sa stream thrust function

Sc Schmidt number, $\mu/\rho D$

St Stanton number

s density ratio
     specific entropy

T absolute temperature
     installed engine thrust

t time

u magnitude of the axial velocity

u_c magnitude of the convective velocity

V magnitude of the total velocity
     system volume

v magnitude of the transverse velocity
     specific volume

W vehicle weight

$\dot{W}$ total shaft power or work interaction rate

x	axial direction axial location number of carbon atoms per fuel molecule
Y	mole fraction
y	number of hydrogen atoms per fuel molecule transverse direction transverse location
Z	compressibility factor
z_e	energy height
α	angle of attack bypass ratio chemical mass fraction coefficient of thermal expansion dispersion or splay angle energy split normal injection empirical mixing parameter
α'_{ij}	stoichiometric coefficient of reactant species i in reaction j
α''_{ij}	stoichiometric coefficient of product species i in reaction j
Γ	initial mass ratio
γ	ratio of specific heats
Δ	finite difference
δ	boundary layer thickness flow deflection or turning angle (compression) shear layer width
δ_D	Dirac delta function
δ_m	mixing layer thickness
ϵ	polytropic efficiency small fraction strain
η	efficiency
η_M	mixing efficiency
η_{pr}	static pressure recovery
θ	angle between flow and oblique detonation wave angle between flow and oblique shock wave boundary layer momentum thickness flow or streamline angle fuel-air mixing empirical parameter vehicle flight path angle

Λ system performance parameter

μ absolute viscosity
Mach wave angle

ν kinematic viscosity
Poisson's ratio

Π mass fraction

π total pressure ratio

ρ density

σ measurement uncertainty
normal component of material stress

τ shear stress
total temperature ratio

Φ dimensionless stream thrust function

ϕ engine external loss coefficient
fuel/air equivalence ratio
thrust augmentation ratio

χ dimensionless axial position
ODW density or velocity ratio

ψ cycle static temperature ratio

ω flow turning angle (expansion)
magnitude of the vorticity

Subscripts

A air

AFT   adiabatic flame temperature

a angularity
location on H-K diagram

add   additive

aw adiabatic wall

b base
control volume boundary
heat addition process
location on the H-K diagram

c compression process
compressor
critical or sonic point
flight path curvature
isolator exit core flow

E	Earth
e	empty entropy limit exit or exhaust expansion process external outer edge of boundary layer
eq	equilibrium
ext	external
F	fan fuel
f	friction fuel
fc	fuel cooling
g	gross
h	hoop or circumferential direction
ICE	inverse cycle engine
i	initial injection conditions inlet or entry
ind	ignition delay time induction time
KE	kinetic energy
KEO	overall kinetic energy
L	based on vehicle length scale
m	location on H-K diagram mixing mixing transition point within the mixing layer
n	net normal or perpendicular
o	measured from the center of the Earth overall
P	product
p	location on H-K diagram payload primary propulsive

pk   peak value

R   reactant

r   heat rejection process
           radial direction

SL   sea level reference

s   due to normal shock
           isentropic process
           reattachment point (end of separation)
           secondary

st   stoichiometric

TJ   turbojet

t   tangential
           throat
           total or stagnation condition
           turbine

tc   thermodynamic cycle

th   thermal

u   separation point (beginning of separation)
           upstream of normal or oblique shock wave

v   velocity

w   wall

X   location on *T*-*s* diagram

x   axial direction
           axial location

Y   location on *T*-*s* diagram

y   downstream of normal shock wave
           transverse direction
           yield stress

0   undisturbed freestream conditions as seen from the refer-
           ence frame of the airbreathing engine or vehicle standard
           atmospheric pressure

1–10   airbreathing engine reference station numbers (see Fig. 4.1)

σ total mass flow

*   choking or critical point

Superscripts

$\sim$ mass average quantity

0   standard atmospheric pressure

$*$ based on equilibrium chemistry
based on reference enthalpy or temperature

o reference temperature for absolute enthalpy

APPENDIX A

1. Basic Definitions and Constants

Time	$1 \text{ h} = 3600 \text{ s}$
Length	$1 \text{ in} = 2.5400 \text{ cm}^*$
	$1 \text{ ft} = 12 \text{ in}$
	$1 \text{ mi} = 5280 \text{ ft}$
Mass	$1 \text{ lbm} = 0.45359 \text{ kg}^*$
Force	$1 \text{ lbf} = 32.174 \text{ lbm·ft/s}^2$
	$1 \text{ N} = 1 \text{ kg·m/s}^2$
Energy	$1 \text{ BTU} = 778.16 \text{ ft·lbf}^*$
	$1 \text{ J} = 1 \text{ N·m}$
Power	$1 \text{ hp} = 550 \text{ ft·lbf/s}$
	$1 \text{ W} = 1 \text{ J/s}$
Pressure	$1 \text{ atm} = 14.696 \text{ lbf/in}^{2*} = 2116.2 \text{ lbf/ft}^2$
	$1 \text{ Pa} = 1 \text{ N/m}^2$
Temperature	The Fahrenheit scale is $T(°F) = 1.8 \ T(°C) + 32$
	where $T(°C)$ is the International Celsius scale.
	The Rankine scale is
	$T(°R) = T(°F) + 459.69$
	$T(°R) = 1.8 \ \{T(°C) + 273.16\}$
	$T(°R) = 1.8 \ T(K)$
	where $T(K)$ is the Kelvin scale.*
Acceleration of standard gravity	$g_0 = 9.8067 \text{ m/s}^{2*} = 32.174 \text{ ft/s}^2$
Newton constant	$c = F/ma = 1$

Scale Factors**

Number	Prefix	Symbol	Example
10^6	mega	M	megawatt (MW)
10^3	kilo	k	kilometer (km)
10^{-2}	centi	c	centimeter (cm)
10^{-3}	milli	m	milliwatt (mW)

*Table 63, Keenan, J. H., and Kaye, J., *Gas Tables*, John Wiley and Sons, New York, 1945.

**Table 1.3, Reynolds, W. C., and Perkins, H. C., *Engineering Thermodynamics*, McGraw-Hill, New York, 1977.

2. Unit Conversion Factors

Quantity	BE units	SI units	Conversion Factor (Number of SI units per one BE unit)
Length	ft	m	0.3048
	mile	km	1.609
Mass	lbm	kg	0.4536
Force	lbf	N	4.448
Density	lbm/ft^3	kg/m^3	16.02
Energy	BTU	kJ	1.055
Enthalpy	ft · lbf	J	1.356
Power	hp	W	745.7
	BTU/h	W	0.2931
Pressure	lbf/ft^2	Pa	47.88
	lbf/in^2	kPa	6.895
Temperature	°R	K	$(1.8)^{-1}$
Fuel heating value Specific enthalpy	BTU/lbm	kJ/kg	2.326
Heat transfer	$BTU/(in^2 \cdot s)$	W/cm^2	163.5
Specific entropy	$BTU/(lbm \cdot °R)$	$kJ/(kg \cdot K)$	4.187
Specific heat	$ft^2/(s^2 \cdot °R)$	$m^2/(s^2 \cdot K)$	0.1672
Specific impulse	s	s	1
Specific thrust	$lbf \cdot s/lbm$	$N \cdot s/kg$	9.807
Thermal conductivity	$BTU/(s \cdot ft \cdot °R)$	$kJ/(s \cdot m \cdot K)$	6.231
Thrust specific fuel consumption	$lbm/(lbf \cdot h)$	$g/(kN \cdot s)$	28.33
Viscosity	$lbf \cdot s/ft^2$	$N \cdot s/m^2$	47.88

APPENDIX B

U.S. STANDARD ATMOSPHERE, 1976

1. British Engineering (BE) Units

Geometric Altitude (kft)	Pressure Ratio (p/p_{SL})	Temperature Ratio (T/T_{SL})	Density Ratio (ρ/ρ_{SL})	Speed of Sound Ratio (a/a_{SL})
0	1.0000	1.0000	1.0000	1.0000
2	0.9298	0.9863	0.9428	0.9931
4	0.8637	0.9725	0.8881	0.9862
6	0.8014	0.9588	0.8359	0.9792
8	0.7429	0.9450	0.7861	0.9721
10	0.6878	0.9313	0.7386	0.9650
12	0.6362	0.9175	0.6933	0.9579
14	0.5877	0.9038	0.6502	0.9507
16	0.5422	0.8901	0.6092	0.9434
18	0.4997	0.8763	0.5702	0.9361
20	0.4599	0.8626	0.5332	0.9288
22	0.4227	0.8489	0.4980	0.9214
24	0.3880	0.8352	0.4646	0.9139
26	0.3557	0.8215	0.4330	0.9063
28	0.3256	0.8077	0.4031	0.8987
30	0.2975	0.7940	0.3747	0.8911
32	0.2715	0.7803	0.3480	0.8834
34	0.2474	0.7666	0.3227	0.8756
36	0.2250	0.7529	0.2988	0.8677
38	0.2044	0.7519	0.2719	0.8671
40	0.1858	0.7519	0.2471	0.8671
42	0.1688	0.7519	0.2245	0.8671
44	0.1534	0.7519	0.2040	0.8671
46	0.1394	0.7519	0.1854	0.8671
48	0.1267	0.7519	0.1685	0.8671
50	0.1151	0.7519	0.1531	0.8671
52	0.1046	0.7519	0.1391	0.8671
54	0.9507 -1	0.7519	0.1265	0.8671
56	0.8640	0.7519	0.1149	0.8671
58	0.7852	0.7519	0.1044	0.8671
60	0.7137	0.7519	0.9492 -1	0.8671

(continued on next page)

575

1. British Engineering (BE) Units (continued)

Geometric Altitude (kft)	Pressure Ratio (p/p_{sL})		Temperature Ratio (T/T_{sL})	Density Ratio (ρ/ρ_{sL})		Speed of Sound Ratio (a/a_{sL})
62	0.6486	−1	0.7519	0.8627	−1	0.8671
64	0.5895		0.7519	0.7841		0.8671
66	0.5358		0.7520	0.7125		0.8672
68	0.4871		0.7542	0.6459		0.8684
70	0.4429		0.7563	0.5857		0.8696
72	0.4028	−1	0.7584	0.5312	−1	0.8708
74	0.3665		0.7605	0.4820		0.8720
76	0.3336		0.7626	0.4374		0.8732
78	0.3036		0.7647	0.3971		0.8744
80	0.2765		0.7668	0.3606		0.8756
82	0.2518	−1	0.7689	0.3276	−1	0.8768
84	0.2294		0.7710	0.2976		0.8780
86	0.2091		0.7731	0.2705		0.8792
88	0.1906		0.7752	0.2459		0.8804
90	0.1738		0.7772	0.2236		0.8816
92	0.1585	−1	0.7793	0.2034	−1	0.8828
94	0.1446		0.7814	0.1851		0.8840
96	0.1320		0.7835	0.1684		0.8852
98	0.1204		0.7856	0.1533		0.8864
100	0.1100		0.7877	0.1396		0.8875
105	0.8769	−2	0.7930	0.1106	−1	0.8905
110	0.7011		0.8066	0.8692	−2	0.8981
115	0.5629		0.8213	0.6854		0.9063
120	0.4537		0.8359	0.5428		0.9143
125	0.3671		0.8506	0.4316		0.9223
130	0.2982	−2	0.8652	0.3446	−2	0.9302
135	0.2430		0.8798	0.2762		0.9380
140	0.1988		0.8944	0.2222		0.9458
145	0.1631		0.9091	0.1794		0.9534
150	0.1343		0.9237	0.1454		0.9611
155	0.1109	−2	0.9383	0.1182	−2	0.9686
160	0.9176	−3	0.9393	0.9770	−3	0.9692
165	0.7593		0.9393	0.8084		0.9692
170	0.6283		0.9354	0.6717		0.9672
175	0.5188		0.9208	0.5634		0.9596

(continued on next page)

1. British Engineering (BE) Units (continued)

Geometric Altitude (kft)	Pressure Ratio (p/p_{SL})		Temperature Ratio (T/T_{SL})	Density Ratio (ρ/ρ_{SL})		Speed of Sound Ratio (a/a_{SL})
180	0.4271	−3	0.9063	0.4713	−3	0.9520
185	0.3505		0.8917	0.3931		0.9443
190	0.2868		0.8772	0.3270		0.9366
195	0.2339		0.8626	0.2711		0.9288
200	0.1901		0.8481	0.2242		0.9209
205	0.1540	−3	0.8336	0.1847	−3	0.9130
210	0.1243		0.8191	0.1517		0.9050
215	0.9992	−4	0.8046	0.1242		0.8970
220	0.8003		0.7901	0.1013		0.8888
225	0.6384		0.7756	0.8232	−4	0.8807
230	0.5072	−4	0.7611	0.6664	−4	0.8724
235	0.4012		0.7466	0.5374		0.8640
240	0.3161		0.7358	0.4296		0.8578
245	0.2482		0.7254	0.3422		0.8517
250	0.1943		0.7151	0.2717		0.8456

Reference values: $p_{SL} = 2116 \text{ lbf/ft}^2$
$T_{SL} = 518.7 \text{ °R}$
$\rho_{SL} = 0.07647 \text{ lbm/ft}^3$
$a_{SL} = 1116 \text{ ft/s}$

Notation: Single digit preceded by a minus sign indicates power of 10 by which the associated and following tabulated values should be multiplied, e.g., $0.2468 \, -2 = 0.002468$.

U.S. STANDARD ATMOSPHERE, 1976
2. Système International (SI) Units

Geometric Altitude (km)	Pressure Ratio (p/p_{SL})		Temperature Ratio (T/T_{SL})	Density Ratio (ρ/ρ_{SL})		Speed of Sound Ratio (a/a_{SL})
0	1.0000		1.00000	1.00000		1.0000
1	0.8870		0.9774	0.9075		0.9887
2	0.7846		0.9549	0.8217		0.9772
3	0.6920		0.9324	0.7423		0.9656
4	0.6085		0.9098	0.6689		0.9538
5	0.5334		0.8873	0.6012		0.9420
6	0.4660		0.8648	0.5389		0.9299
7	0.4057		0.8423	0.4817		0.9178
8	0.3519		0.8198	0.4292		0.9054
9	0.3040		0.7973	0.3813		0.8929
10	0.2615		0.7748	0.3376		0.8802
11	0.2240		0.7523	0.2978		0.8673
12	0.1915		0.7519	0.2546		0.8671
13	0.1636		0.7519	0.2176		0.8671
14	0.1399		0.7519	0.1860		0.8671
15	0.1195		0.7519	0.1590		0.8671
16	0.1022		0.7519	0.1359		0.8671
17	0.8734	-1	0.7519	0.1162		0.8671
18	0.7466		0.7519	0.9930	-1	0.8671
19	0.6383		0.7519	0.8489		0.8671
20	0.5457		0.7519	0.7258		0.8671
21	0.4667	-1	0.7551	0.6181	-1	0.8690
22	0.3995		0.7585	0.5266		0.8709
23	0.3422		0.7620	0.4490		0.8729
24	0.2933		0.7654	0.3832		0.87491
25	0.2516		0.7689	0.3272		0.8769
26	0.2160	-1	0.7723	0.2797	-1	0.8788
27	0.1855		0.7758	0.2392		0.8808
28	0.1595		0.7792	0.2047		0.8827
29	0.1372		0.7826	0.1753		0.8847
30	0.1181		0.7861	0.1503		0.8866

(continued on next page)

2. Système International (SI) Units (continued)

Geometric Altitude (km)	Pressure Ratio (p/p_{SL})		Temperature Ratio (T/T_{SL})	Density Ratio (ρ/ρ_{SL})		Speed of Sound Ratio (a/a_{SL})
31	0.1018	−1	0.7895	0.1289	−1	0.8885
32	0.8774	−2	0.7930	0.1107		0.8905
33	0.7573		0.8016	0.9447	−2	0.8953
34	0.6547		0.8112	0.8071		0.9007
35	0.5671		0.8208	0.6909		0.9060
36	0.4920	−2	0.8304	0.5925	−2	0.9113
37	0.4276		0.8400	0.5090		0.9165
38	0.3722		0.8496	0.4381		0.9217
39	0.3245		0.8591	0.3777		0.9269
40	0.2834		0.8688	0.3262		0.9321
41	0.2478	−2	0.8784	0.2822	−2	0.9372
42	0.2171		0.8880	0.2445		0.9423
43	0.1904		0.8976	0.2122		0.9474
44	0.1673		0.9072	0.1844		0.9525
45	0.1472		0.9168	0.1605		0.9575
46	0.1296	−2	0.9263	0.1399	−2	0.9625
47	0.1143		0.9359	0.1222		0.9674
48	0.1010		0.9393	0.1075		0.9692
49	0.8916	−3	0.9393	0.9492	−3	0.9692
50	0.7874		0.9393	0.8383		0.9692
55	0.4197	−3	0.9050	0.4638	−3	0.9513
60	0.2167		0.8573	0.2528		0.9259
65	0.1079		0.8096	0.1332		0.8998
70	0.5153	−4	0.7621	0.6762	−4	0.8730
75	0.2357		0.7232	0.3259		0.8504
80	0.1039		0.6894	0.1507		0.8303

Reference values: $p_{SL} = 1.013 \times 10^5 \text{ N/m}^2 \text{ (Pa)}$
$T_{SL} = 288.2 \text{ °K}$
$\rho_{SL} = 1.225 \text{ kg/m}^3$
$a_{SL} = 340.3 \text{ m/s}$

Notation: Single digit preceded by a minus sign indicates power of 10 by which the associated and following tabulated values should be multiplied, e.g., $0.2468 - 2 = 0.002468$.

INDEX

COLOR PLATES

Table of Contents for Colored Plates

Plate Number	Chapter
1	1
2	1
3	5
4	6
5	7

Plate 1   X-15A-2 attached to its parent B-52 aircraft prior to launch; the dummy ramjet and large drop tanks are clearly visible. Courtesy of the U.S. Air Force Flight Test Center. (From Chapter 1.)

Plate 2   Artist's conception of the National Aero-Space Plane X-30 single-stage-to-orbit experimental vehicle in flight. Courtesy of the NASP Joint Program Office. (From Chapter 1.)

(a)

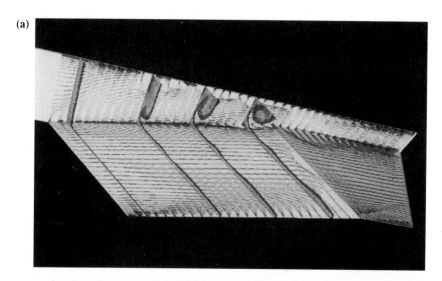

(b)

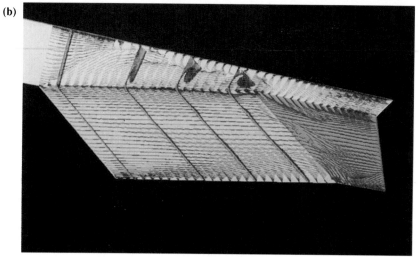

Plate 3   Contours of u-velocity at three axial stations in the vicinity of forebody plane and the sidewall, and simulated surface oil flow patterns on these surfaces for (a) Re=0.55 x 10^6/ft (1.80 x·10^6/m), and (b) Re=2.15 x 10^6/ft (7.05 x 10^6/m) (see Ref. 5.24). The streamwise corner vortex is larger at lower Reynolds number than that at higher Reynolds number. (From Chapter 5.)

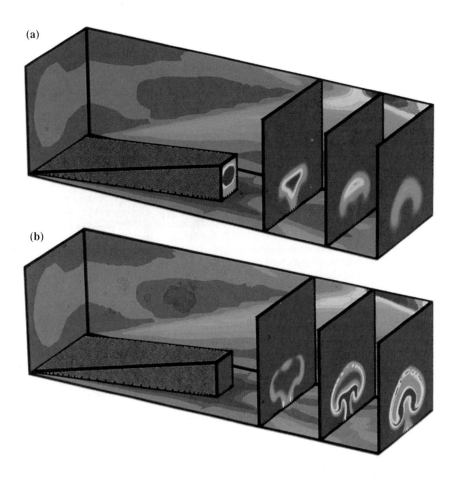

Plate 4 For $M_0 = 13.5$, mass fraction contours at three downstream crossflow planes and pressure contours on inflow plane, symmetry plane, and bottom burner wall, with swept ramp.[6.48] Colors red and violet represent the extreme values, with the former being high, and green represents approximately the middle of these extremes; (a) hydrogen mass fraction contours for a turbulent, cold flow simulation with air inflow at the burner entrance; (b) water mass fraction contours for a turbulent, reacting flow simulation with oxygen inflow at the burner entrance. A seven-reaction and seven-species H_2/O_2 chemistry model is used. (From Chapter 6.)

(a)

(b)

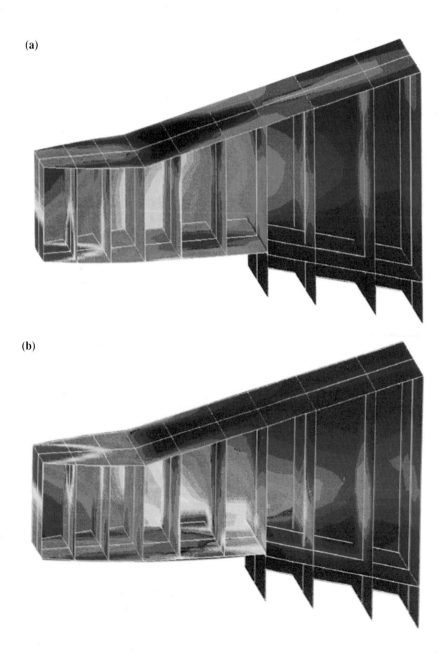

Plate 5   Pressure contours for $D_p=^\pm 0.31$ (Ref. 7.6): (a) $D_p=-0.31$; (b) $D_p=0.31$. Note that only a significant part of the expansion surface is shown. (From Chapter 7.)